Advanced Weaving Technology

Yordan Kyosev · Francois Boussu
Editors

Advanced Weaving Technology

 Springer

Editors
Yordan Kyosev
ITM
Technische Universität Dresden
Dresden, Germany

Francois Boussu
ENSAIT, ULR 2461 - GEMTEX - Génie et
Matériaux Textiles
University Lille
Lille, France

ISBN 978-3-030-91517-9   ISBN 978-3-030-91515-5   (eBook)
https://doi.org/10.1007/978-3-030-91515-5

This Springer imprint is published by the registered company Springer Nature Switzerland AG
The registered company address is: Gewerbestrasse 11, 6330 Cham, Switzerland

Editors' Preface

The weaving is a classical topic in the area of the textile technology, where several people thinks, "everything is done and known". Latest then, after someone starts to teach weaving, starts to understand how many basic background books are available and how less actual information for the current problems in the technology is written.

Exchanging regularly on such topics during meetings and conference, both editors, being as well university level teachers in weaving technology in Ensait, Roubaix and Narrow Weaving technology in Mönchengladbach, decided, that it is time such gap to become filled.

The book project started officially in 2018, during the 8th World Conference on 3D Fabrics and their Applications, after that several experts were invited to contribute.

The goal was to cover all topics in the weaving starting form weaving preparation with its problems, mechanics and limits of the weaving processes, pattern design, software, applications.

The latest developments and problems with their solutions are actually often not publically available, as these are keys for the success of the industrial companies.

Thats the reason why, several of the authors and topics which we liked to include remains "open". Some changes in the time plan were there too... After 13 years teaching several classes in narrow weaving, advanced narrow weaving and CAD for woven labels in Mönchengladbach, Yordan Kyosev received University Professorship position at TU Dresden and moved there. Changing living place and working topic required a lot of time, so from the initially planed five chapters in narrow weaving, the only one remains in the book.

COVID changed our teaching and daily live too, and presses us to concentrate on the creating new online materials.

Finally, we succeed to collect a set of contributions in this book, being sure, that it will help to our and your students to understand better the actual methods and problems in the weaving and become better specialists.

Some chapters have added exercises section to train you on the presented notions. Do not hesitate to give us your feedback in order to improve our evaluation of weaving compulsory notions.

We hope that you will enjoy the book and the weaving.

Roubaix, France   Francois Boussu
Dresden, Germany   Yordan Kyosev
October 2021

Contents

Basics of Weaving: Product and Process Parameters

Weaving Preparation

Jean-Yves Drean and Mathieu Decrette

Abstract This chapter will focus on weaving preparation which consists of winding, warping, sizing, drawing-in and/or tying in. This operation is essential because the idea of placing a creel with the exact number of warp yarn packages to be woven side by side at the back of the weaving machine is not realistic both for technical and economic reasons. This is why, the weaving process requires preparing weaver's beams with the exact number of warp ends that will be placed at the back of the weaving machine to produce a given fabric. Before obtaining the weaver's beam, the yarns coming from the spinning process require preparatory steps beginning by clearing and winding dense and uniform large size package, then by warping the yarns forming part or the whole of the final warp, furthermore by applying a non-permanent protective coating to increase warp ends resistance. The last operation consists in drawing-in when starting a new fabric style or tying-in in case of same fabric. Through the chapter, each step is precisely described and discussed and practical calculations are proposed.

Keywords Winding · Warping · Sizing · Drawing-in · Tying-in

1 Introduction

A fabric usually consists in the crossing of several thousand yarns (ends) in the warp direction with several thousand yarns (picks) in the weft direction [1]. A first idea to realize the warp would be to place a creel containing the exact number of warp yarn packages side by side at the back of the weaving machine. This idea is not realistic both for technical and economic reasons. This is why, the weaving process requires preparing weaver's beams with the exact number of warp ends that will be placed at the back of the weaving machine to produce a fabric with the given structure and specification. Before obtaining the weaver's beam, the yarns coming from the

J.-Y. Drean (✉) · M. Decrette
Université de Haute Alsace, Laboratoire de Physique et Mécanique Textiles UR4365, Ecole Nationale Supérieure d'Ingénieurs Sud Alsace, F-68200 Mulhouse, France
e-mail: jean-yves.drean@uha.fr

© Springer Nature Switzerland AG 2022
Y. Kyosev and F. Boussu (eds.), *Advanced Weaving Technology*,
https://doi.org/10.1007/978-3-030-91515-5_1

3

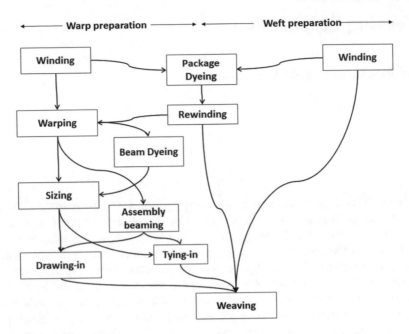

Fig. 1 Weaving preparation scheme

spinning process require preparatory steps to be transferred on the beam namely the process shown in Fig. 1.

At the very beginning, the spinner's packages need to be cleared to remove the main defects, and wound on a reasonably large package in a correct format compatible with the following operation. Then, these packages will be placed on a creel and several warper's beams will be produced during the warping operation. Then, the sizing operation will take place. During this operation, the warper's beams are placed at the back of the sizing machine and a single warp is drawn and impregnated with an adequate size to reinforce the yarns for withstanding the mechanical stresses and abrasion during the weaving process. The last operation of preparation consists in drawing-in the warp yarn through the heddle eyes and through the reed dents.

As mentioned in Fig. 1, variants of the preparation are put in place in the case of package dyeing (soft packages) and wrapper's beam dyeing for yarn dyed fabric weaving. The warp properties expected after the weaving preparation are summarised as follows:

- As strong and elastic as possible
- Uniform
- Smooth and hairiness-free
- Knot-free
- Thick, thin, slub places-free
- Withstanding abrasion of moving weaving machine parts
- Withstanding fatigue stresses due to weaving process.

2 Winding

Weaving is supplied by spinning. But, the spinning manufacturing process introduces defects and unevenness which is a natural stochastic phenomenon. These defects are harmful for the weaving process because they induce breakages and efficiency decrease. This is why, these defects as mentioned in Fig. 2, have to be removed and replace by knots or splices.

After clearing, dense and uniform large size package are wound from small spun cops or from open end packages. The quality of these large packages must have the adequate characteristics (shape, dimensions, hardness, density, yarn length) adapted to the subsequent operations i.e.: warping, sizing, tying-in, drawing-in and weaving. The functions of the winding operation are as follows:

- Clearing yarn defects: slubs, thin and thick places, neps, … to improve the aspect and facilitate the use in weaving
- Making larger wound packages,
- Winding soft packages for dyeing
- Facilitating transport

Figure 3 shows the classical winding process.

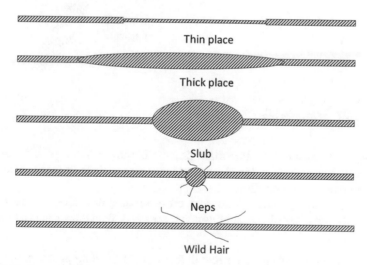

Fig. 2 Typical yarn defects

Fig. 3 Winding process
schematic

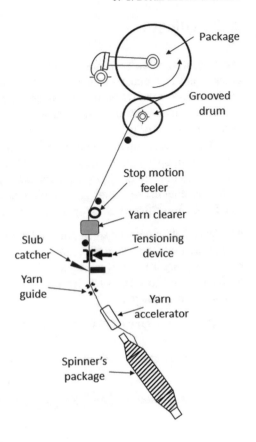

2.1 Types of Packages Winding

Winding is one of the most important process before weaving with an essential function which consists in making larger packages especially from small cops coming from ring-spinning process [20].

Two main winding processes are used, cross winding or parallel winding, which produce different types of packages as cone, cheese or flange packages as shown in Fig. 4.

- Parallel winding—Flange Package: in this type of winding package, the yarn is wound parallel on the package containing flanges on both sides of the package. The winding pitch is very close to the yarn diameter. The flanges provide a very good protection for the yarn and the winding density is very high but unwinding is done by side withdrawal (Fig. 5). In this process, the package rotates, the yarn does not rotate and as a result the yarn twist does not change, which is considered as an advantage. But the rotation of the package requires a specific equipment and due to the inertia, upon start-up high tension are developed and at high speed variations of

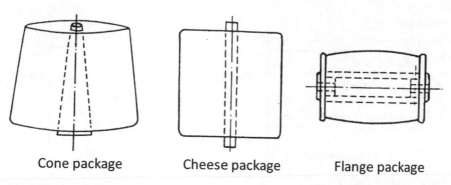

Cone package   Cheese package   Flange package

Fig. 4 Type of package

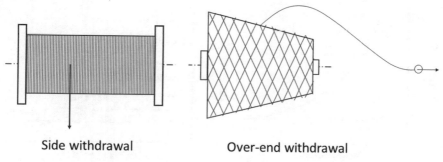

Side withdrawal   Over-end withdrawal

Fig. 5 Package withdrawal

tension can occur, which is really a disadvantage. This type of winding is usually used in case of brittle yarn as glass or carbon fibres.

- Cross winding—Three types of cross winding processes are commonly used, namely: random or open-winding, precision winding, Digicone® winding. Whatever the type of winding, a traversing mechanism is used to distribute the yarn axially along the package.

The winding pitch or the α helix angle (half of crossing angle) does not change during the package building in cross or random winding (Fig. 6). But the winding W_r ratio which is the ratio between the package revolution per minute and the double traverse changes continuously during the full winding process as shown in Fig. 7. As the double traverse remains constant, the increase of the package diameter has to be compensated with a constant decrease of the package rotation per double traverse, so that the winding ratio W_r can regularly decrease. But, when W_r becomes an integer, the yarn is wound in the same position as the yarn wound in the previous traverse. This results in "ribbons" or "patterns" formation. These zones have a high winding density and a poor unwinding behaviour which will decrease the efficiency of the subsequent processes. The prevention of the ribbon formation is done by different methods as the modulation of the yarn guide frequency, generating slippage between

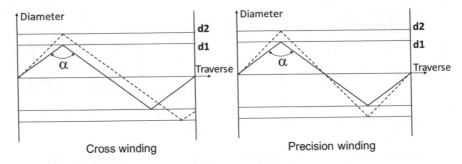

Fig. 6 Types of cross winding

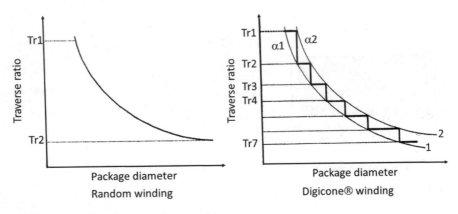

Fig. 7 Package diameter versus Traverse ratio

the grooved drum and the package or lifting the package from the driving drum at periodic intervals [21]. The density of the wound package is poor, in the order of 0.3 g/cm^3 for a classical cotton ring spun yarn [7].

As shown in Fig. 8, two types of devices can be used to produce random wound packages, plain cylinder and yarn guide or grooved drum. In both devices, the package

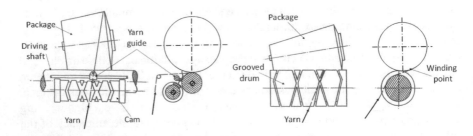

Fig. 8 Random winding

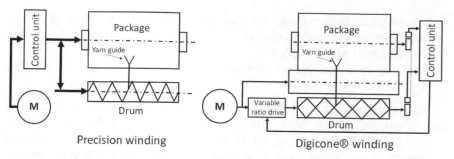

Fig. 9 Precision and Digicone® winding

is friction-driven by the plain cylinder or by the drum [31]. In the first case, the yarn is distributed along the length of the package with the help of a yarn guide controlled by a gear box, whereas in the latter case, the yarn is guided along the package by the groove of the driving drum running at a constant speed.

SSM's Digicone®2 [8] is a precision winding process where the W_r winding ratio incrementally varies as the package diameter increases, in order to maintain an almost constant crossing angle. Figure 9 shows the scheme of Digocone® device. The motor drives the drum at a constant speed inducing a constant winding speed. The variable ratio drive connected to the motor varies the velocity of the traverse cam. At the beginning of a winding package, the traverse ratio is high at Tr1 value which corresponds to the α1 winding angle. As the winding progresses, the traverse ratio is still maintained at Tr1 value until the winding angle reaches α2 value. At this point which is detected with the help of sensors placed on the two axes, the coupling ratio is changed when setting the winding angle at α1 value inducing the Tr2 traverse ratio. The winding process progresses and the above described procedure is repeated until the required package diameter is reached. This results in uniform and very well built packages. Digicone®2 precision winding is a ribbon-free winding process.

2.2 Clearing

A defect (like a thick place) not eliminates by winding process costs:

- 140 times more when it causes breakages at warping,
- 680 times more when it causes breakages at sizing,
- 56 times more when it causes breakages during weaving.

The breakages encountered during the warping operation show that for a 33 tex open end yarn:

- 55% of breakages originate from spinning (foreign yarns, foreign matter, thin places, thick places, piecing, lake of twist),

– 20% originate from package winding: pattern, tension, entanglements, shape defects, bad splices, bad winding after splicing, waste winding,
– 5% come from warping operation,
– 10% come from poor transportation conditions,
– 10% from unknown origin.

For ring-spun yarns, the first cause of breakage, during warping process, is the lack of resistance of the splices. The second cause is the stopping and restarting after stopping during a breakage on the winding machine (succession of forward and reverse movement when searching for the ends to be pieced, lack of tension and malfunction of the anti-pattern (ribbon) device at the very beginning of the winding process.

As above mentioned and as mentioned in Fig. 2, the defects like slubs, thin and thick places, neps… have to be removed and replaced by knots or splices. Before removing, these faults are electronically or mechanically detected. In the latter case, the yarn passes the slubs through a small slit and long thick places are cut and replaced by a knot or a splice.

Electronic fault detection is based on two technologies: optical and capacitive sensors.

In the optical yarn thickness measurement systems, the sensor consists of an infrared or visible light source and photodiodes. The yarn is passed through the emitted light, a part of the light is absorbed and a part is reflected by the yarn (Fig. 10). The amount of absorbed or reflected light is measured by photodiodes and converted to a signal proportional to the apparent yarn diameter [27]. New powerful sensors with multi-coloured light sources detect and differentiate coloured foreign fibres, vegetable matter and polypropylene content.

In the capacitive yarn thickness measurement systems, the sensor consists in two parallel metal plates (condenser electrodes) powered by an electrical alternative voltage (Fig. 10). The yarn is passed through the electrical field inducing a capacity change of the measuring condenser. This capacitance change is converted to a signal proportional to the mass of yarn and to the dielectric constant of the fibre. For some types of sensors, the material characteristic is automatically corrected by calibration

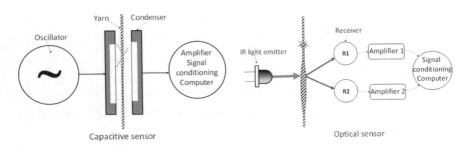

Fig. 10 Electronic fault detection

on the average dielectric value of the first yarn length controlled. Thus, some non-measured parameters are taken into account, such as the actual moisture content, the difference between the actual and nominal content of the fibre blends.

The process of piecing results from yarn breaks and yarn defect removal. An ideal piecing would be one which can withstand the subsequent processes without interruption and which does not lead to any deterioration of the quality of the finish product [18]. At the very beginning, the weaver's knot and the fisherman's knot were very popular due to their strength and the not-slip of the ends. But, considering the size of the knots as mentioned in Table 1, knots are obstructive and frequently cause breaks during the weaving process and have a detrimental effect on the finish fabric. As a result, knotless joining methods have been developed. The Table 1 summarises all the knotless joining methods used to piece two yarns ends.

Splicing is now commonly used because it produces strong enough piecing without affecting the weaving process and affecting the final aspect of the fabric. In this method, the two ends of the yarn are untwisted and then intermingled by electrostatic, mechanical or pneumatic means [18]. In the latter, which is currently the most used, yarn ends are inserted into the groove of the splicing chamber (Fig. 11(1)) and the yarn ends are clamped and cut (Fig. 11(2)). The air nozzles suck the yarn ends to be untwisted (Fig. 11(3)). The pressing block extracts and positions the two untwisted ends in the groove and the splicing chamber is sealed in the front side by means of a cover. A strong vortex of compressed air is blown which entangles and intermingles the fibres, joining the two ends together (Fig. 11(4)). The time and air pressure needed in air splicing are determined by the fibre type and the yarn characteristics. Air spliced yarn produces a joint that in most cases will meet all requirements of subsequent weaving preparation processes, both in terms of strength and appearance. Moreover, this type of splicing can also be applied to a large range of fibre and yarn types without requiring a lot of settings or adjustments [15, 16] (Fig. 11).

As above mentioned, the process of piecing results from the yarn breaks and yarn defect removal, so that the break rate has to be taken into consideration. Breakage causes one stoppage resulting in a loss of production, one defect (knot or splice) inducing a loss of strength. The rate of breakage or clearing rate depends on the quality of the yarn before winding and on the required quality of the yarn after winding.

The quality of the yarn before winding could be based on the USTER® STATISTICS which give for each material and each yarn count the global average quality for unevenness, tenacity, elongation at break, thin places, thick places, neps, ...

The quality after winding depends on the later manufacturing processes; e.g. in knitting and in tufting, a knot is more harmful than a thick place and depends on the intended subsequent uses. The severity of a fault (defect) depends of its size and its length.

For example, let us consider a fabric structure composed of 50 tex wool warp yarns and 50 tex weft yarns, with a yarn density 24 ends and 20 picks per cm, which has the following defects:

Table 1 Various yarn piecing methods

Types	Dimensions		Tenacity	Other characteristics
	Thickness	Length		
Weaver's knot	2–3 × yarn diameter	4–8 mm	Good	Average fatigue properties, reed obstructions, visible on finished fabric
Fisherman's knot	3–4 × yarn diameter	4–8 mm	Very good	Important reed obstructions, visible on finished fabrics, suitable for all types of yarns and materials except very stiff fibres (glass, carbon)
Wrapping	3 × yarn diameter	10 mm	Good	Complex and slow technology, local stiffness, reed obstructions
Gluing	2 × yarn diameter	10 mm	Good	Slow technology, local stiffness, local dyeing problem, well adapted for continuous filament yarns
Welding	1 × yarn diameter	Small point	Good	Only suitable for thermoplastic material, well adapted to monofilament yarns, local dyeing problem
Mechanical splicing	1–1.5 × yarn diameter	10–20 mm	Up to 80% yarn strength	Complex technology, not adapted to plied yarns, significant maintenance required
Electrostatic splicing	1–1.2 × yarn diameter	10–20 mm	Up to 80% Yarn strength	Required mechanical opening and separation of fibres, not adapted to plied yarns, sensitive to room climatic conditions, slow technology
Pneumatic splicing	1–1.2 × yarn diameter	10–20 mm	Up to 80% yarn strength	Well adapted to a wide range of fibre and yarn types, fast technology, compressed air supply required

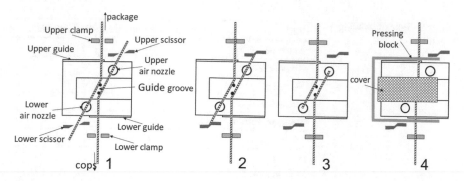

Fig. 11 Pneumatic splicing

- 100 thin places,
- 40 thick places and
- 10 neps detected for 1000 m of yarn, resulting in 660 faults/m².

Therefore, it is important before winding to determine:

- the rate of clearing,
- the dimensions and number of defects to be eliminated,
- the dimensions and number of defects which are to be left in the yarn.

This choice results in the yarn clearing limits. But, the optimisation of this yarn clearing limits needs a powerful analytical tool. The classification of yarn defects in classes developed by USTER® provides a good solution to reach the right yarn clearing limits. The classification of defects into classes will give the means for spinners to optimize their yarn clearing and the users of yarn to specify the required quality levels. The last development of the USTER® CLASSIMAT 5 extends the existing classification table to include a total of 45 classes (Table 2). The USTER®

Table 2 USTER® CLASSIMAT fault classes

Fault size %									
+400	A4	B4	C4	D4					
+250	A3	B3	C3	D3		E			
+150	A2	B2	C2	D2					
+100	A1	B1	C1	D1					
+75	A0	B0	C0	D0					
+45			CP1	DP1		F		G	
+30				DP2	FP21	FP22	GP21	GP22	
+20									
+10	N		S				L		
0	1	2	4	8	16	32	64		Length cm
-10									
-20					T				
-30				TD0	HD1	HD2	ID1	ID2	
-45		TB1	TC1	TD1	H1		I1		
		TB2	TC2	TD2	H2		I2		

Table 3 Type of winding machine

Task	Manual	Automatic	Full automatic
Machine feeding	O	O	M
Spindle feeding	O	M	M
Winding	M	M	M
Breakage repair	O	M	M
Package doffing	O	O	M
Monitoring	O	O	O

O = Operator
M = Machine

CLASSIMAT [29] can be used as a sophisticated classifying system which can help the winder to determine and set his own yarn quality standards [4, 17] (Table 3).

Manual winding machine: all operations, except winding, are done manually.

Automatic winding machine: winds, replaces empty cops with full cops, pieces cops having had a break.

Full automatic winding machine: feeds automatically cops, replaces empty cops with full cops, winds, pieces cops having had a break, doffs automatically full packages.

- **Manual winding machine**

Calculation of the number of spindle per operator:

Let's reduce t_1, t_2, t_3 t_4 times to the same production unit (kg, cops, package), respectively necessary to:

- replace of the empty cops
- repair breaks
- doff full packages
- wind

The corresponding operator time is as follows:

$$T_0' = t_1 + t_2 + t_3$$

This time is corrected, with an increase k, to take the monitoring into account, the supervision, the travels within the workshop, the operator performance (fatigue and simultaneous stoppages) and any other work. Therefore:

$$T = k \times T_{oo}'$$

The number of spindles to be entrusted to an operator is then:

$$N = \frac{T_0 + T_m}{T_0}$$

(with T_m: time of the machine)

Example: Winding a batch of 500 packages. The weight of each package is $P = 2000$ g with a yarn count $T_t = 25$ tex

- cops net weight, $p = 150$ g
- Winding speed, $V = 1000$ m/min
- Breaks rate, $c = 13$ c/kg

Allocated times:

- Package doffing, 12 s
- Cops feeding, 10 s
- Breakage repair, 8 s
- Correction coefficient, 1.2

Let us apply these times to a cops:

$$t_1 = 1 \times 10 = 10 \, s$$

$$t_2 = \frac{c \times 8 \times p}{1000} = \frac{13.8 \times 150}{1000} = 15.6 \, s$$

$$t_3 = \frac{12 \times p}{P} = \frac{12 \times 150}{2000} = 0.9 \, s$$

$$t_4 = \frac{p \times 10^3 \times 60}{T_t \times V} = 360 \, s$$

Hence, the number of spindles to be entrusted to an operator is then:

$$T_0 = k(l_1 + l_1 + l_1) = 1.2(10 + 15.6 + 0.9) = 31.8s$$

$$T_m = t_4 = 360s$$

$$N = \frac{T_0 + T_m}{T_0} = \frac{391.8}{31.8} = 12,32 \approx 13 \, spindles$$

Production calculation:

Time required to wind a package:

$$\frac{(T_m + T_0) \times P}{p} = \frac{391.8 \times 2000}{150} = 5224 \, s = 87 \, min$$

And so the operator can wind… 13 packages in 87 min. If he is alone to make this batch, he will need:

$$\frac{500}{13} = 38.5 = 39 \, doffings$$

namely $39 \times 87 = 3393$ min $= 56.55$ h.
The following conclusions can be deduced:

– Production per spindle, P_{sp}:

$$\frac{500 \times 2000}{13 \times 56.55 \times 1000} = 1.360 \text{ kg/h}$$

– Production per operator, P_{op}:

$$\frac{500 \times 2000}{56.55 \times 1000} = 17.56 \text{ kg/h}$$

– Efficiency:

$$\frac{P_{sp}}{V \times 60 \times T_t \times 10^{-3}} = \frac{1360}{1000 \times 60 \times 25 \times 10^{-3}} = 0.907 \approx 90.1\%$$

• **Automatic winding machine**

The following data are given for the automatic winding machine.
The machine:

– replaces an empty cops or repairs a breakage in 5 s
– winds at a velocity V $= 1000$ m/min
– the piecer makes a round trip behind 10 spindles (without intervention) in 15 s

The operator:

– can feed 13 up to 18 cops per minute: 15 cops will be chosen
– can replace a package in 18 s
– is involved in 1% of stops at the rate of 12 s
– must have 14% of normal working time for rest and auxiliary works

Winding machine with one piecer per spindle

– Package doffing:

$$\frac{p \times 18}{P} = \frac{150 \times 18}{2000} = 1.35 \text{ s}$$

– Cops feeding: 5 s
– Breakage repair:

$$\frac{5 \times c \times p}{1000} = \frac{5 \times 13 \times 150}{1000} = 9.75 \text{ s}$$

– Down time: $1.35 + 5 + 9.75 = 16.10$ s.

  Winding:

$$\frac{p}{T_1 \times 10^{-3}} \times \frac{60}{1000} = 360 \text{ s}$$

Time to produce p grams: $360 + 16.1 = 376.1 \approx 376$ s.
The following conclusions can be deduced:

– Production per spindle, P_{sp}:

$$\frac{0.150 \times 60 \times 60}{376} = 1.433 \text{ kg/h}$$

– Efficiency:

$$\frac{P_{sp}}{V \times 60 \times T_t \times 10^{-3}} = \frac{1433}{1000 \times 60 \times 25 \times 10^{-3}} 0.955 = 95.5\%$$

Winding machine with one piecer for 10 spindles

If there is no intervention on the other 9 spindles, the piecer is only present behind the concerned spindle after an average waiting time of 7.5 s. It can be assumed that during the round trip of the piecer, the 9 pins nearby, work in the same conditions as the concerned spindle.

So, in 376 there will be:

$9 \times 5 = 45$s of cops feeding
$\frac{9 \times 13 \times 150}{1000 \times 5} = 87.8$s of break repair
Namely $45 + 87.8 = 132.8$ s of piecer stop

As a first approximation, we can therefore deduce that for one piecer passage, in addition to travel time (7.5 s), there will be:
$\frac{7.5 \times 132.8}{376} = 2.65$ s of intervention on the other spindles; the average real waiting time will be therefore:

$$7.5 + 2.65 = 10.15 \text{ s}$$

Therefore, during the cops unwinding time:

– 1 cops feeding + waiting time $= 5 + 10.15 = 15.15$ s.
– $\frac{c \times p}{1000}$ breakage repair with waiting time $= \frac{13 \times 150}{1000} \times (5 + 10.15) = 29.54$ s.
– $\frac{P}{p}$ package doffing $= \frac{150}{2000} \times 18 = 1.35$ s.
– Winding $= 360$ s.
– Winding time to produce p grams $= 15.15 + 29.54 + 1.35 + 360 \approx 406$ s.

  The following conclusions can be deduced:

– Production per spindle, P_{sp}:

$$\frac{0.150 \times 60 \times 60}{406} = 1.330 \, \text{kg/h}$$

– Efficiency:

$$\frac{P_{sp}}{V \times 60 \times T_t \times 10^{-3}} = \frac{1330}{1000 \times 60 \times 25 \times 10^{-3}} = 0.887 = 88.7\%$$

Number of spindle per operator

Based on one cops, the operator:

– feeds one cops, or $\frac{60}{15} = 4 \, \text{s}$.
– replaces $\frac{P}{p}$ package or $\frac{150}{2000 \times 18} = 1.35 \, \text{s}$.
– is involved in 1% of stops at the rate of 12 s, or

$$\left(1 + c \times \frac{P}{1000}\right) \times 0.01 \times 12 = \left(1 + 13 \times \frac{150}{1000}\right) \times 0.01 \times 12 \approx 0.35 \, \text{s}$$

Or a total of: $4 + 1.35 + 0.35 = 5.70 \, \text{s}$.

– must have 14% of normal working time for rest and auxiliary works

$$5.70 \times 0.14 = 0.80 \, \text{s}$$

Hence the time $T_0 = 5.70 + 0.80 = 6.50 \, \text{s}$.

Depending on the type of package, during the 376 or 406 s winding of the cops, the operator intervenes for 10.14 s, hence the number of spindles:

$N = \frac{376}{6.5} \approx 58 \, spindles$ (1 piecer/1 spindle)

$N = \frac{406}{6.5} \approx 62 \, spindles$ (1 piecer/10 spindles)

Hence, the production per hour per operator is:

$1.433 \times 58 = 83.1 \, \text{kg/h}$ (1 piecer/1 spindle)

$1.330 \times 62 = 82.5 \, \text{kg/h}$ (1 piecer/10 spindles)

• **Production comparison**

Based on Table 4, the following remarks can be highlighted:

– the hourly production per spindle is approximately identical, provided however that the number of spindles/operator is well calculated.
– the number of spindles/operator is quadrupled on the automatic winding machine, the hourly production per operator too.
– the number of spindle per piecer has no influence on the operator production.
– the best performance is given by the winding machine with 1 piecer per spindle, it is the ideal case corresponding to one operator/spindle.

Table 4 Winding machine production

	Winding machine	Automatic winding machine	
	Manual	1 piecer/spindle	1 piecer/10 spindles
Spindles/operator	13	58	62
Prod./hour/spindle	1.360 kg	1.4330 kg	1.330 kg
Prod./hour/operator	17.68 kg	83.1 kg	82.5 kg
Efficiency	90.5%	95.5%	88.5%

– the comparison of the two automatic winding machines shows that the efficiency of a machine is not the most important criterion.
– As regards full automatic winding machine, the operator does not intervene very much. The same principle of calculation gives about 1000 spindles/operator which is totally absurd. It is necessary to take simultaneous stops and displacements into account and to give a greater part to monitoring.

3 Warping

Warping consists in winding the yarns forming part or the whole of the final warp with the same tension, parallel to each other.

At the very beginning, it seems preferable to warp the entire warp sheet in as a single operation. We could then imagine feeding the sizing machine from a creel with as many packages as yarns in the warp sheet. Then, there would be no need for warping. But as to production, the time saved would not be significant and as to the product quality, this procedure is to be rejected. Usually, it is admitted that as far as the regularity of the yarn tension is concerned, a creel supplied with 800 to 1000 packages constitutes the technical limit. Considering that he warps are generally composed of several thousands of yarns, the warping conditions related to the tension would be no longer achieved. Furthermore, it is impossible to repair the breaks occurring during the sizing process, so that the whole breakages repaired on the warping machine would turn into a missing yarn in the final warp beam. It is therefore necessary to split the warp to fulfil the above mentioned conditions.

Thus, the warping by parts of the weaver's beam is done on direct warping or beam warping while warping of whole weaver's beam is done on indirect warping or sectional warping. The choice of the type of warping process depends on many factors that will be discussed separately. Contrary to what has been often said, the need for having to size the warp down later is not the main determining factor.

Whatever the warping process, the warping equipment consists of as shown Fig. 12:

– The headstock which is the driver
– The creel which is the provision of packages.

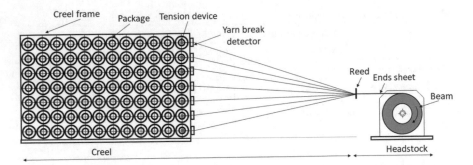

Fig. 12 Warping principle

3.1 The Creel

The creel is the set of holders for the packages supplying the warping machine.

The creel which has a plurality of yarn package holders, is equipped with devices for guiding the yarn along the creel, tensioners for maintaining the regularity of the yarn tension and auto stop sensors ensuring the control of the yarns by detecting the yarn breakages.

The creel capacity is the maximum number of packages it can hold. Its power is the maximum number of yarns that can be simultaneously drawn from it. Capacity and power are not always equal. The capacity depends on the dimensions of the packages that the creel must hold. The spacing of the holder (from 210 up to 420 mm) varies with the destination of the creel (silk, thin filament yarns, cotton, wool…). For a staple fibre creel, the spacing is generally of 240 mm, which corresponds to a maximum package diameter of around 210 mm.

The height of the creel is such that the operator can still access to the upper tiers of packages at least 190 to 210 cm off the floor. It is generally accepted that a creel must not exceed 12 to 15 m in depth. Beyond that value, the tension differences between front and rear packages become excessive. Similarly, the operator efficiency drops considerably due to the travel time. Roughly, we can therefore estimate the N maximum package capacity for each side of the creel with the following formula:

$$N = \frac{h.d}{s^2}$$

With:
   h = height of the creel
   d = depth of the creel
   s = spacing, distance between holders
   Thus, for s = 24 cm, h = 200 cm and d = 1200 cm, N $\cong$ 400 for a total capacity for the creel (two sides) around 800 packages.

Table 5 Example of creel dimensions

Truck number	Capacity e = 240 mm	Depth for 8 tiers (mm)	Capacity e = 270 mm	Depth for 7 tiers (mm)
3	288	5630	336	7790
4	384	7070	420	9410
5	480	8510	504	11030
6	576	9950	588	12650
7	672	11390	672	14270
8	768	12830	–	–

For documentary purposes, Table 5 shows the data extracted from a manufacturer's catalogue.

Thus, with a distance of 330 mm, between two package holders, the creel capacities range from 288 packages with a depth of 8100 mm up to 578 packages with a depth of 16,200 mm. The packages are split in 7 tiers.

Before examining the different types of creel, it is interesting to have an idea of the duration of use and the running time of the warping machine.

The running time for a direct warping machine are as follows:

T_C: Creel supplying time

- Around 20 min for 100 packages with two operators in case of piecing
- Around 60 min for 100 packages with two operators in case of a new batch

T_M: The warping machine running time which depends on the velocity of the machine and on the size of the batch.

T_R: The warper's beam replacing time (2 up to 4 min per beam).

T_{BR}: The time to repair a yarn breakage during a warping operation (2 up to 4 min per breakage).

The total time will be as follows:

$$T_T = T_C + (T_M + T_R + T_{BR})$$

Consider the warping of 600 ends, end count T_{tex} = 40tex on 4 warper's beams from packages of P = 2200 g at a velocity V = 900 m/min.

$$T_c = \frac{600 \times 20}{100} = 120\,\text{min}$$

$$T_M = \frac{L}{V} = \frac{P \times 10^3}{V \times T_{tex}} = \frac{2200 \times 10^3}{900 \times 40} = 61.2\,\text{min}$$

$$T_R = 4 \times 4 = 16\,\text{min}$$

$$T_{BR} = P \times C \times 4 = 2.2 \times 600 \times 0.01 \times 4 = 52.8\,\text{min}$$

where C = 1 breakage for 100 kg warp
   Therefore, total time is:

$$TT = TC + (TM + TR + TBR) = 120 + (61.2 + 16 + 52.8)$$
$$= 120 + 130 = 250 \text{ min}$$

This calculation shows that the T_C creel supplying time (120 min) is very high compared to the warping machine running time (130 min).

The creel successive upgrades have always had for purpose to reduce the downtime during the process.

Depending on the type of beams to be warped, different types of creel are available like creel, carriage creel, magazine creel, swivel frame creel, ...

- The standard creel: It is the simplest and least convenient creel. The package holders are fixed on both sides of a fixed frame. After the warping operation, the operators have to break each yarn close to the tensioner, then remove all the packages of the current batch, before putting all the packages of the new batch on the holders and finally knot the yarns of the new batch with the yarns still engaged in the tensioner. The loss of time is therefore maximal.
- The carriage or truck creel: It's the most common one. The fixed frame is replaced by a series of carriages as shown in Fig. 13. During the warping operation, reserve carriages enable carriage supplying with the packages of the following batch. At the end of the warping operation, after having "broken" the yarns, replacing the carriage and knotting is just required. While one set of carriages is used during the creel and warping operations, it is possible to prepare and supply another set of

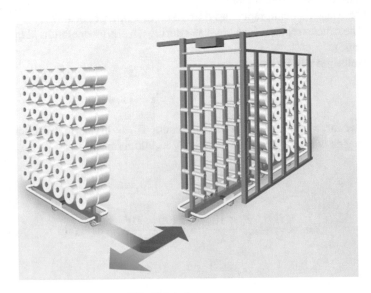

Fig. 13 Carriage creel (Courtesy of Karl Mayer)

Fig. 14 Magazine creel
(Courtesy of Rius)

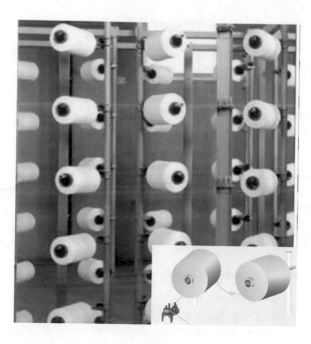

carriages at any given location in the mill. Thus, the suppling time will be reduced by more than 50% and the machine downtime is reduced too.

- Magazine creel:
  The 2-holder axes converge to the same yarn output eyelet. The two matching packages are tied end-to-end, the tail end of the yarn form one package with the tip of the yarn of the other one, so that the package is made non-stop (Fig. 14). This is a truly continuous warping operation without any loss of time in the creel supplying since, knotting included, it is done the warping machine running. Magazine loading prevents stops caused by packages running empty.

  However, this system has three serious disadvantages: for a given depth of creel, the power is only half its capacity, the knots are distributed over a too large length of warp sheet and cannot be eliminated, and during the transfer from one package to the other, transfer failures can occur due to a sudden change of tension.

- Swivel-frame creel:
  It consists of 2 parallel package holders (Fig. 15). The holders on external side are in working position while the holder in inner side are in a reserve position. During the warping operation, an operator supplies the package holders of the inner side. After the warping operation, the yarns are broken and the faces are turned around (peg or drive chain). Two types of creel are available: yarn draw-off outside which is the standard setting or yarn draw-off inside. In the latter case, the yarn path is more direct and is mainly appreciated in sectional warping where the end sheet is not as wide as direct warping end sheet. Supplying is thus done

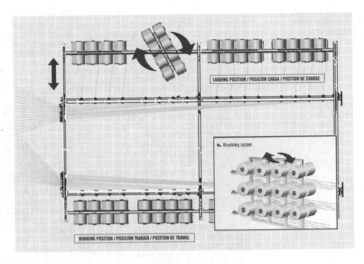

Fig. 15 Swivel-frame creel (Courtesy of Rius)

from outside. This creel is suitable for confine space; the floor space requirement is smaller than for the carriage creel.

- V-creel:

    The principle is the same but the 2 package frames form a V-shape opening towards the back instead of being parallel (Fig. 16). This has the advantage of drawing-off easily the yarn from the package and giving a more rectilinear

Fig. 16 V-creel (Courtesy of
Karl Mayer)

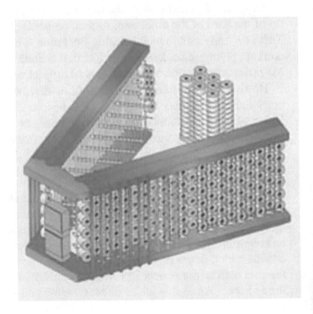

path to the yarns so that they no longer need to be supported which allows to reduce the tension, therefore to increase the running speed. These creels are rather recommended for the fast direct warping.

- Automatic creel:
  This is a carriage creel which is pieced automatically by a mobile device equipped with as many knotter heads as there are horizontal tiers of holders. The complete creel supplying is ensured by a movable grid and a trolley that leads the ends to the reed of the warping machine. The downtime for a complete supplying is about 1/20th of the downtime of an ordinary carriage creel.

As above mentioned at the beginning of this chapter, it is absolutely essential to have a uniform tension for all the yarns of the same batch [10]. In actual fact on the sizing machine, the end sheet is pulled and It is obvious that it is the most stretched yarns that distribute the tensile strength. Then, a disproportionate elongation of those yarns appears, whereas non-stretched yarns can conversely undergo shrinkage in particular during size impregnation and drying.

The tension which has to be taken into account is the dynamic tension at the entrance of the warping machine. Usually, we are considering that the tension in cN must be equal to the yarn count in tex. This formula may be applied to all kinds of material except for elastic yarns. In this case, the tension value is taken at the beginning of the stress–strain curve plateau.

The tension T at the entrance of the warping machine is the result of the tension t given by the tensioning system, the tension t_p generated by the weight p of the free yarn length running along the creel and the tension t_f due to the braking force f of this yarn induced by the various supporting devices (ceramic yarn guides) and by the stop motion devices. Therefore:

$$T = t + t_p + t_f$$

However, since the free length of the yarn is a function of the position of the package on the creel, it results that t_p and t_f vary from one yarn to another one, thus having the same tension T, it is necessary to give different tension t according to the position of package on the creel.

To meet the required above mentioned conditions, variable tensioning tensioner devices with progressive adjustment are used, based on variable tension rails, adjusted once according to the position of the tensioner. The tension is modified by the variable tension rails located at the delivery of the creel. The adjustment is thus centralized and very fast.

The tensioners currently used are of different kinds:

- Wrapping disc tensioner with disc weight or spring. The tensioner device generates the required yarn tensile force based on the combination of wrapping and disc weight (Figs. 17 and 18).
- Rotating disc: two positive disc drives placed on either side of the yarn are pressed against each other by a spring and are rotated by the yarn as shown in Fig. 19. The

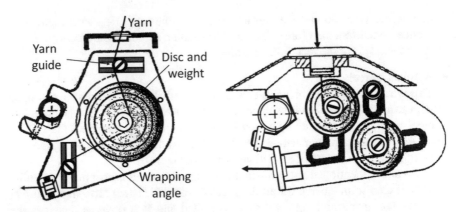

Fig. 17 Disc tensioners

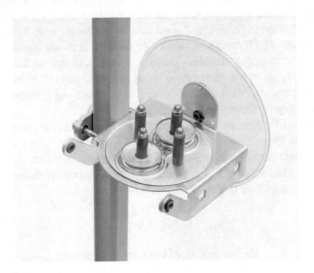

Fig. 18 Tensioner U2V (Courtesy of Karl Mayer)

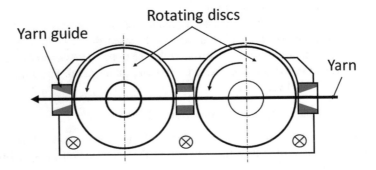

Fig. 19 Rotating disc tensioner

Fig. 20 Multitens® tensioner (Courtesy of Karl Mayer)

advantages are that they smoothen the tension shocks coming from the upstream and they are self-cleaning preventing dust deposits inside the tensioner.
– Electromagnetic yarn tensioner: The tension is given with the help of an electromagnetic device instead of a spring.
– Motor-driven tensioner: This tensioner is based on a sensor/motor device (Fig. 20). This motor-driven single-end yarn tension an end break detector as well as a stop motion for tight yarns and loose yarns.

Taking the warping speed (up to 1000 m/min) into account, the inertia of the beam, as well as its diameter (up to 1000 mm) and the irreversibility of the warping process after detecting a yarn breakage, the beam must stop in less than 2 to 3 turns. In fact, unwinding the beam would cause irreversible damages such as yarn entanglements. Thus, the time required for stopping down (signal transmission and braking) must not exceed 0.1 s. Hence, the creel is equipped with a stop-motion, based on electric, electronic or triboelectric technologies, located very close to the creel. The stop motion electrically links each yarn to the warping machine braking system, so that, in case of a yarn breaks, the warping machine will stop. Moreover, are positioned above the creel to prevent dust and fibre accumulation when processing spun yarns.

3.2   The Warping Machine

3.2.1   Direct Warping

This involves warping sheets of 300 to 800 ends (depending on the creel capacity) wound parallel to each other with the same tension on the same warper's beam as shown in Fig. 21.

The sheet of the same batch will be overlapped during the subsequent operation like assembly beaming or sizing.

Due to the construction of the warping machine, the beams, i.e. the end sheets, have the same width. The final weaver's sheet will be brought to the required width (reed width +2 up to 3 cm) on the assembly beaming machine or on the sizing machine.

Since the tension of the ends must be constant, the warping of the sheet on the warper's beam must be done at a constant circumferential speed. The ends from the different tiers of the creel are gathered in a single sheet by a condenser composed of 2 or 3 rods. The end sheet is then drawn in an adjustable V or zig-zag reed with 1end/dent reeding ratio. From the reed, the end sheet is rotating around the beam so as to drive it around the measuring device (Fig. 22). This friction drive explains that a minimum amount of ends per warper's beam is necessary to avoid any slippage between the end sheet and the measuring device, this slippage will induce measurement inaccuracies at start-up and especially at stop. For making the slippage less important, the cylinder of the measuring device (meter counter) is as light as possible while being of a diameter favourable to a large surface of contact with the end sheet. The axis of the warper's beam is equipped with a swivelling device for the wrapper's beam loading and unloading operations (doffing device). Taking the need of a very quick stop in case of end breakage into account, a powerful braking is required. Thus, the headstock will be equipped with an electro-magnetic braking system on both sides and the brakes are activated through the emergency stop, yarn missing, meter pre-selection or security stop.

The density and the hardness of the warper's beam is given by the yarn tension and by the packing of the pressure roller. The increasing yarn batch on the warper's beam shifts the pressure roll backwards against the resistance set for contact pressure. This

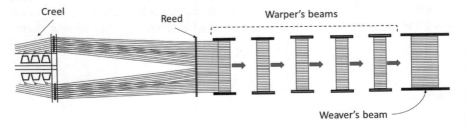

Fig. 21 Direct warping principle

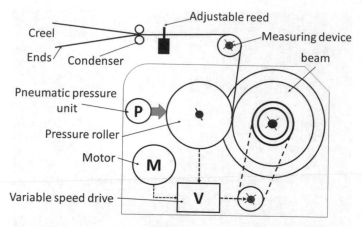

Fig. 22 Direct warping principle—Headstock

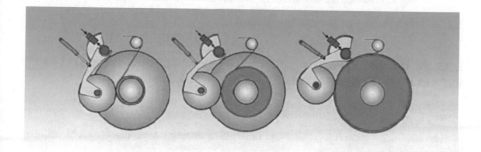

Fig. 23 Indirect pressure with active kick-back (Courtesy of Karl Mayer)

indirectly acting pressure ensures to achieve completely cylindrical beams. Since the press roll is pneumatically driven, it will swing out of the way immediately on braking (Fig. 23). Moreover, the warping process must be done at a constant adjustable linear velocity of the end sheet, and so the motor will be driven typically with the help of an electronic frequency inverter and PLC (Programmable Logic Controller). The warping speed is up to 1000 m/min.

Figure 24 shows a direct warping system.

3.2.2 Sectional Warping or Indirect Warping

Contrary to the direct warping system, in sectional or indirect warping, warping and weaver's beam preparation take place on the same machine but in two successive steps. The warp is prepared in sections with more or less the same number of ends, and these sections are warped one by one by juxtaposing them on the same drum.

Fig. 24 Direct warping system (Courtesy of Rius)

After the last section has been warped, the entire warp sheet is rewound on the weaver's beam. Thus the sectional process is composed of:

– The creel suplying
– The warping operation
– The beaming operation (which can take place during supplying the creel with the next batch).

Figure 25 summarizes the sectional warping principle.

Sectional warping is particularly well adapted for preparing the wearver's beam with multi-coloured patterns in case of yarn dyed fabrics [13].

The lease reed: one of the main advantage of the sectional warping is the warping lease. The warping reed, as shown in Fig. 26, allows such an operation and allows the warp sheet to be split in several sheets for drawing-in the separating strings during one-and-one leasing in order to facilitate the end separation during weaving and/or sizing operation. These separating strings are later replaced by split rods during the sizing operation if the latter is necessary. The leasing is also necessary for maintaining the correct order of the warp ends as required by the multi-coloured pattern in case of yarn dyed fabrics.

The ends drawn from the creel individually go through the dent of the reed. The dents are welded in groups of two according to a specific design according to the number of sheets envisaged. The lease reed design shown in Fig. 24 allows up to 8 separations. The drawing-in of the separating strings for the weaving and/or sizing operation can be fully automatic achieved.

The adjustable reed: the role of this reed, located just after the lease reed, is to maintain the whole ends of the section at the required width during the warping operation (Fig. 27). Then, the ends of the section are fed under the feeler roller.

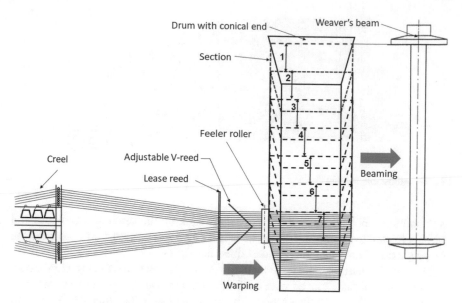

Fig. 25 Sectional warping principle

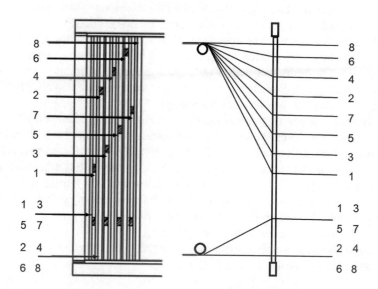

Fig. 26 Lease reed principle

The feeler roller: The height of the section warped on the drum depends on the yarn tension. The tension given by the tensioners on the creel, as even as it is, cannot provide an exact uniform height of each section on the drum. The exact height is achieved with the help of the feeler roller which is pressed firmly against the drum as shown in Fig. 25. Moreover, the feeler roller allows the section height measurement

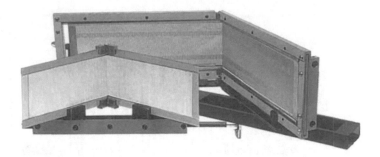

Fig. 27 Adjustable V-Reed

as it building up. The feeler roller is pushed back by the section in progress and the data related to the section height measured by a sensor feed a computer for the warping parameters calculation as described here after.

The warping carriage: the lease reed, the adjustable reed and the feeler roller are mounted on a carriage moved with a traverse motion and kept at constant distance between the carriage and the section surface on the drum. The traverse velocity of this carriage will determine the slope of the end sections on the drum.

The drum: warping drum is round-shaped with a cone at one side (fix or adjustable) which has a small angle preventing the yarns of the first section from slippage. The drum is generally made up of composites such as synthetic resin-bonded glass fibres or in a heavy duty steel perfectly cylindrical and balanced that withstand extreme pressures [1]. The drum is equipped with quick response hydraulic disc brakes at both sides of the drum, as well as brakes activated through the emergency stop, yarn missing or reaching the pre-selected warp length.

The beaming: when the warping operation has been completed and all sections wound on the drum, the end sheet will be unwound and transferred onto the weaver's beam with flanges. This operation is called beaming. The beaming process is totally synchronized with the warping drum and the braking pressure is adjustable for an accurate tensioning control during the beaming process. This tension will give the required hardness of the weaver's beam. The beaming speed ranges from 200 m/min up to 500 m/min depending on the warp characteristics.

Figure 28 shows a sectional warping system.

Before starting the warping operation, from the basic information, the warping parameters have to be calculated as follows:

– The number of ends per section
– The section width
– The conicity in case of variable cone drum
– The section height
– The transverse carriage velocity
– The number of drum revolutions.

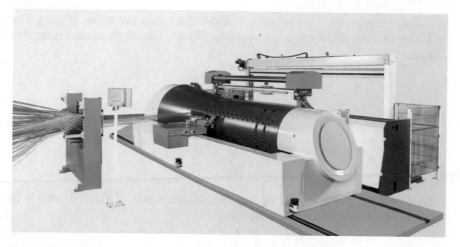

Fig. 28 Sectional warping system Prowarp® (Courtesy of Karl Mayer)

The number of ends per section: whatever the number of ends in the warp, the number of ends per section can be of any value. However, this number has to be equal or lower than the creel capacity, and has to be even to keep the lease warp (except for the last section also known as cut-able section).

The section width: the width of each section is to be accurately determined.

If: n_1 = number of yarn of the first section.

w_1 = the width of the first section.

N = the number of warp ends.

W = the width beteween the weaver's beam flanges.

Then the width of the first section is:

$$w_1 = W \times \frac{n_1}{N}$$

The width of the next sections is calculated relative to the first one:

$$w_i = w_1 \times \frac{n_i}{n_1}$$

The section height: the height of the section depends on the number of coils, therefore on the length of the warp and on the average thickness of one coil. This thickness **e** varies with the material, the yarn count and number of ends per cm. Determining this value **e** with a mathematical approach is very difficult and uncertain. However, a preliminary test must first be carried out to measure the **e** thickness This thickness can be automatically and directly measured on the drum of the warpng machine, while the first coils of the first section are processed, or before warping with the help of a specific pulley with a circular groove. In this case, the measurement will consist in mesuaring the heigth **h** of a predetermined revolution of the pulley i.e. 100 revolutions in the warping process conditions. The height **e** is determined as follows:

$$e = \frac{h}{100}$$

Assuming that **n** revolutions are necessary to warp a section of **H** height, then:

$$e = \frac{h}{100} = \frac{H}{n}$$

Unfortunately, **n** is unknown, but each end is wound in the form of a spiral to be described in the form of the following equation:

$$\rho = R + k\theta$$

R is the radius of the drum.
After one revolution, the spiral equation becomes:

$$\rho = R + e = R + k2\pi$$

Hence:

$$\rho = R + \frac{e}{2\pi}\theta$$

The length **L** of the spiral can be deduced as follows:

$$L = \int_0^{2\pi n} \rho.d\theta = \int_0^{2\pi n} R.d\theta + \int_0^{2\pi n} \frac{e}{2\pi}.\theta.d\theta = 2\pi nR + e\pi n^2$$

Since: $n = 100.\frac{H}{h}$ and $e = \frac{h}{100}$.
Then:

$$L = 2\pi.100\frac{H}{h}.R + \pi.100.\frac{H^2}{h}$$

Hence: $100\pi.H^2 + 200\pi.R.H - L.h = 0$.

and

$$H = -R + \sqrt{R^2 + \frac{L.h}{100\pi}}$$

The conicity: S since the different sections do not have lateral supports (flanges), they must be wound on the drum in successive layers laterally displaced so that the angle thus formed eliminates the risk of sloughing off at the end of the drum. Each section will be based on the conical shape of the previous section, with the exception of the first section that will be supported by the conical end of the drum. Adjustable cone angles are possible, but fixed cone are usually preferred to prevent pressure marks. The length of the weaver's beam is proportional to the length of each section and proportional to the section height H as above describes d. The H value is also proportional to the cone angle α. The higher α, the higher H, and the longer the weaver's beam but with a lower section stability. The α angle depends on the material. It is usually acknowledged that:

- For low friction coefficient yarns (silk, synthetics) $\alpha \leq 10°$
- For average friction coefficient yarns (cotton) $\alpha \leq 16°$
- For high friction coefficient yarns and hairy yarns (wool) $\alpha \leq 24°$

The α angle will be chosen according to the useful l_c length of the cone. In case of an adjustable cone, the calculation of α will be as follows:

$$\sin \alpha = \frac{H}{l_c}$$

The transverse carriage velocity: this velocity δ is expressed in translation distance (mm) for one drum revolution. This velocity is expressed as follows:

$$\delta = \frac{e}{\tan \alpha} = \frac{h}{100.\tan \alpha}$$

The number of drum revolutions: knowing the H heigth, the corresponding n number of drum revolution is calculated as follows:

$$n = 100.\frac{H}{h}$$

Nowadays, all above mentioned calculations are performed by the sectional warping system computer. Based on the warping data, the computer automatically calculates, the number and the width of sections, the transverse carriage velocity, the automatic stop for leasing, ... The computer sets the running parameters, velocity, velocity regulation, beaming traverse motion, ... and monitors the production data during warping [9].

3.2.3 Sample Warping

Warp sampling machines are intended for manufacturing samples and warps production either of short or medium length, especially the warps utilizing coloured yarns or yarns of different counts and materials. The warping process which is derived from

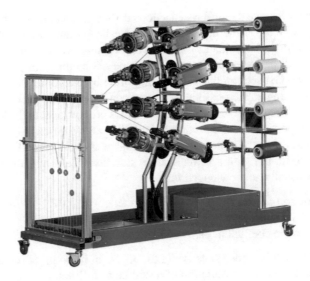

Fig. 29 Stationary creel (Courtesy of CCI)

the sectional one, consists in winding up a limited end length and placing them on the warping width with several sections of different colours to get the colour variants of the fabric. In practice, a package per colour is sufficient to obtain any required warping sequence [26]. Like conventional warping machines, a sample warping machine is also divided into two parts namely the creel and the headstock. Two types of creels can be used for sample warping: Stationary or fixed creel (Fig. 29) or rotatory creel (Fig. 30).

The fixed creel is used to perform pattern warping with a single end for complex pattern warping, the rotatory creel is used to achieve simple pattern warping with two or more ends.

The headstock consists of a warping drum, an automatic leasing device, a yarn selector, a push yarn device, a loose yarn-preventing device, a stopping device for yarn breakage and a beaming device. The warping drum is used to wind the patterned warp yarns. The yarn is wound directly on a drum by means of a yarn winding device. The yarn winding device is mounted on a circumferential surface of the warping drum so as to be circumferentially rotatable. The yarn is thus wound around the drum (Fig. 31). When the warping operation is completed, the warp will be wound on the beam.

3.2.4 Comparison Between Direct Warping and Sectional Warping

Constant factors: taking the two warping processes into account, some factors are held constant.

Fig. 30 Rotatory circular creel (Courtesy of Karl Mayer)

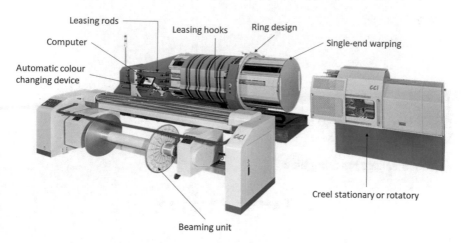

Fig. 31 Single-end warping machine Lutan® (Courtesy of CCI)

- The creels are the same, but the investment in a sectional warping machine is more than twice the amount of the investment in a direct warping machine.
- The floor space requirement for installing a sectional warping machine is approximately twice as much as a direct warping of the same capacity.
- The succession of fairly complex operations and the settings of the sectional warping machine require more qualified operators than that of the direct warping machine.

– The one-and-one lease which gives a perfect arrangement of all warp ends according to a given pattern is one of the main advantage of the sectional warping especially in case of a multicoloured pattern.

Production calculation—Case of direct warping without sizing: The warper's beams assembly is done on the sizing machine without sizing.

– Creel supplying time

$$T_{csd} = k \times \left(\frac{t_1}{100} \times \frac{N}{n} + t_2 \right)$$

With

K   number of creel supplying for a warp batch of L length.
t_1 elementary time for supplying 100 packages.
N   total number of ends.
n   number of warper's beams.
t_2 warping process preparation time.

– Warping time:

$$T_{dw} = \frac{L \times n}{V_{dw}}$$

With

L   length of the warp batch.
V_{dw} warping velocity.

– Yarn breakage repairing time:

$$T_{bd} = t_3 \times B \times \frac{W}{100} \times L$$

With

t_3 yarn breakage elementary repairing time.
B   breakage number per kg of warp.
W   Weight of 100 m of warp.

– Warper's beam doffing time:

$$T_{db} = t_4 \times n$$

With

t_4 warper's beam doffing elementary time.

– Warper's beam installation time on sizing machine:

$$T_{si} = t_5 + n \times t_6$$

With

t_5 sizing machine creel preparation time
t_6 warper's beam preparation elementary time

– Warper's beam assembly time

$$T_{ba} = \frac{L}{V_a \times R_a}$$

With

V_a assembly velocity
R_a assembly efficiency

– Weaver's beam doffing time:

$$T_{dt} = t_7 \times \frac{L}{l}$$

With

t_7 elementary time
l Warp length par weaver's beam

– Direct warping total time:

$$T_{TD} = T_{csd} + T_{dw} + T_{bd} + T_{db} + T_{si} + T_{ba} + T_{dt}$$

$$T_{TD} = L\left(\frac{n}{V_{dw}} + \frac{t_3 \times B \times W}{100} + \frac{1}{V_a \times R_a} + \frac{t_7}{l}\right) + T_{csd} + T_{db} + T_{si}$$

Production calculation—Case of sectional warping without sizing:

– Creel supplying time

$$T_{css} = k' \times \left(\frac{t_1}{100} \times \frac{N}{n'} + t_8\right)$$

With

k' number of warping for a warp batch of L length.
t_1 elementary time for supplying 100 packages.
N total number of ends.

n'   number of sections.

t_8 warping process preparation time.

- Warping time:

$$T_{sw} = k' \times \frac{L \times n'}{V_{sw}}$$

With

L   length of the warp batch

V_{sw} warping velocity

- Yarn breakage repairing time:

$$T_{bs} = t_3 \times B \times \frac{W}{100} \times L$$

   With

t_9 yarn breakage elementary repairing time

B   breakage number per kg of warp

W   Weight of 100 m of warp

- Section change time:

$$T_{sc} = k' \times t_{10} \times n'$$

With

t_{10} section change elementary time

- Beaming time:

$$T_{sb} = k' \times t_{11} + \frac{L}{V_b}$$

With

t_{11} beaming preparation time

V_b beaming velocity

- Sectional warping total time:

$$T_{Ts} = T_{css} + T_{sw} + T_{bs} + T_{sc} + T_{sb}$$

$$T_{ts} = L \times \left(\frac{k' \times n'}{V_{sw}} + \frac{1}{V_{sw}} + \frac{1}{V_b} + t_9 \times \frac{B \times W}{100} \right)$$
$$+ k' \times \left(t_8 + \frac{t_1}{100} \times \frac{N}{n'} + n' \times t_{10} + t_{11} \right)$$

Production calculation—Case of direct warping with sizing:

- Warping: the calculation is the same as: **p**roduction calculation—Case of direct warping without sizing
- Warper's beam installation time on sizing machine:

$$T_{si} = t_5 + n \times t_6$$

With

t_5 sizing machine creel preparation time
t_6 warper's beam preparation elementary time

- Sizing time:

$$T_{sm} = \frac{L}{V_{sm} \times R_{sm}}$$

With

V_{sm} sizing machine velocity
R_{sm} sizing machine efficiency

- Weaver's beam doffing time:

$$T_{dt} = t_7 \times \frac{L}{1}$$

With

t_7 elementary time
1   Warp length par weaver's beam

- Direct warping total time with sizing:

$$T_{td} = L \times \left(\frac{n}{V_{dw}} + \frac{t_3 \times B \times W}{100} + \frac{1}{V_{sm} \times R_{sm}} + \frac{t_7}{1} \right) + T_{cs} + T_{db} + T_{si}$$

Production calculation—Case of sectional warping with sizing:

- Warping: the calculation is the same as: **p**roduction calculation - Case of sectional warping without sizing
- Weawer's beam installation time on sizing machine:

$$T_{sc} = k' \times (t_5 + .t_6)$$

With

t_5 sizing machine creel preparation time
t_6 weaver's beam preparation elementary time

– Sizing time:

$$T_{sm} = \frac{L}{V_{sm} \times R_{sm}}$$

With

V_{sm} sizing machine velocity
R_{sm} sizing machine efficiency

– Weaver's beam doffing time:

$$T_{sd} = k' \times t_7$$

With

t_7 elementary time

– Sectional warping total time with sizing:

$$T_{TS} = L \times \left(\frac{k' \times n'}{V_{sw}} + \frac{1}{V_b} + \frac{t_9 \times B \times W}{100} + \frac{1}{V_{sm} \times R_{sm}} \right)$$
$$+ k' \times \left(\frac{N \times t_1}{100 \times n'} + t_5 + t_6 + t_7 + t_8 + n' \times t_{10} + t_{11} \right)$$

3.2.5 Exercise

- We were provided with the following fabric structure:
    22 ends/cm / 18 picks/cm in Tex 2×25 / 50.
    The maximum creel capacity is up to 700 packages.
    The reed width is 180 cm.
    You will be asked to calculate the total number of ends warped on the weaver's beam, the number of warper's beam, the number of ends on the warper's beam, the width of the warper's beam (distance between flanges) and the warper's beam end count/cm.

Table 6 Warping without sizing data

Direct warping data	Assembly data	Sectional warping data
k = 1	t_5 = 15 min	k' such as the maximum warp length = 1400 m
t_1 = 20 min	t_6 = 6 min	
n = 6 warper's beam	Va = 120 m/min	n' = 6
t_2 = 10 min	Ra = 0.95	V_{sw} = 800 m/min
V_{dw} = 1000 m/min	t_7 = 2 min	t_8 = 15 min
t_3 = 3 min	l = 1200 m	t_9 = 6 min
B = 0.5 breakage/100 kg		t_{10} = 6 min
W = 17.820 kg/100 warp m		t_{11} = 15 min

- We were provided with the following fabric structure:
  36 ends/cm / 28 picks/cm in Tex 25 / 31.
  Warp repeat: 36 ends colour A, 20 B, 64 C, 36 A, 24 D.
  The maximum creel capacity is up to 600 packages.
  The reed width is 180 cm.
  The length of the weaver's beam = 1200 m.
  You will be asked to calculate the parameters necessary to carried out the sectional warping of this warp. The drum diameter **R** = 500 mm, the useful length of the cone l_c = 500 mm and the heigth of 100 revolutions **h** = 22 mm.

- Based on the equation here above mentioned and the hereafter data (Table 6), calculate the direct and sectional warping total time without sizing and plot the curve total time (min) versus weaver's beam length up to 10,000 m. Draw the appropriate conclusions on which kind of warping process to be used.

  Fabric structure: 22 ends/cm / 18 picks/cm in Tex 2 × 25 / 50.
  Overall width = 150 cm.
  Number of ends N = 3300.

- Based on the equation here above mentioned and the hereafter data (Table 7), calculate the direct and sectional warping total time with sizing and plot the curve total time (min) vs. weaver's beam length up to 10,000 m. Knowing that a sizing elementary time unit is more expensive than a warping elementary tine unit, draw the appropriate conclusions on the different processes.

  Fabric structure: 36 ends/cm / 28 picks/cm in Tex 25 / 31.
  Overall width = 150 cm.
  Number of ends N = 5400.

Table 7 Warping with sizing data

Direct warping data	Sizing data	Sectional warping data
k = 1	t_5 = 30 min	k' such as the maximum warp length = 1800 m
t_1 = 20 min	t_6 = 6 min	
n = 9 warper's beam	Vsm = 80 m/min	n' = 9
t_2 = 10 min	Rsm = 0.90	V_{sw} = 800 m/min
V_{dw} = 1000 m/min	t_7 = 2 min	V_b = 200 m/min
t_3 = 3 min	l = 1600 m	t_8 = 15 min
B = 1.5 breakage/100 kg		t_9 = 6 min
W = 14.310 kg/ 100 warp m		t_{10} = 6 min
		t_{11} = 15 min

4 Sizing

4.1 Introduction

As the weaver's proverb goes, "you weave what you size".

Sizing: "Non-permanent protective coating applied before weaving to warp ends to increase their resistance to weaving stresses".

During the weaving process, the warp yarns are subjected to high tension combined with a strong abrasive action. A warp yarn, while passing from the weaver's beam to the cloth fell, is submitted to a complex mechanical action which is a combination of cyclic extensions, abrasion and bending. The cyclic extensions are due to shed opening and closing as well as bending while abrasion is due to the yarn friction against the back rest roller, lease rods, drop wires, heddle eyes, reed wires and picking device as shown in Figs. 32 and 33.

The sizing operation will consist in encapsulating the yarn with a thin and smooth but robust size film to withstand the weaving process stresses. Thus, the main purposes of sizing will be as follows:

- increasing the strength of the yarn while maintaining the yarn extensibility and cyclic extension behaviour,
- improving the working capacity of the yarn,
- improving the abrasion resistance against other yarns and weaving machine elements,
- reducing the yarn hairiness by laying in the protruding fibres on the body of the yarn,
- reducing the fluff and dust during the weaving process,
- easy elimination during finishing operations.

These purposes are achieved by the penetration of the size into the yarn structure. A sized yarn can be considered as a composite material with size as a matrix and fibres as a reinforcement and the size tends to bind the fibres together. The size

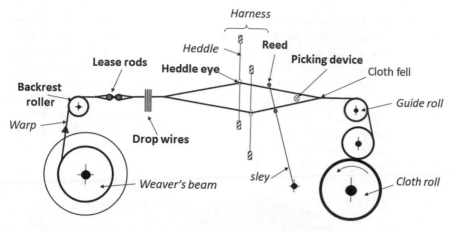

Fig. 32 Weaving machine schematic

Fig. 33 Weaving machine main abrasion points

penetration has to remain under control, and an excessive penetration will increase the bind between the fibres and will increase the stiffness of the yarn and whereby decreasing the working capacity of the yarn while a weak penetration will give poor sized yarn properties. Figure 34 shows the theoretical cross section of a sized yarn.

Figure 35 shows the cross section of a 60 tex cotton ring-spun yarn sized with 12% add-on of PVA. As shown Fig. 36 [28], the zones 1 to 3 highlight the gradient of size penetration in the yarn, the film formation by the encapsulation of fibres in zone1, a good penetration in zone 2 and a weak penetration in zone 3. The proportion of fibres, size and porosity represent respectively 60%, 20%, 20% in zone 1, 50%, 10%, 40% in zone 2 and 20%, 5%, 75% in zone 3.

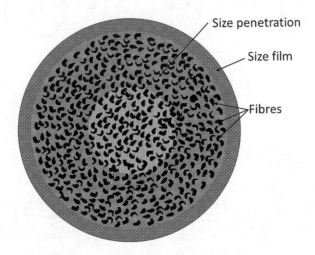

Fig. 34 Theoretical cross section of a sized yarn

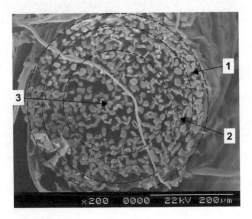

Fig. 35 Cross section of a 60 tex sized cotton yarn

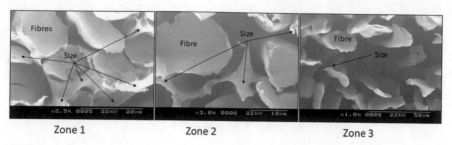

| Zone 1 | Zone 2 | Zone 3 |

Fig. 36 Size penetration

Fig. 37 Sizing-weaving curve

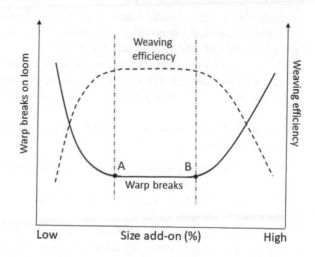

The size penetration depends on the type of size mix, on the pressure on squeeze rolls and on the size add-on. The influence of size add-on on the weaving efficiency is shown on Fig. 37 [24]. At the beginning, the warp breaks decrease with the increase in size add-on. The size creates a protective film around the yarn which increases the yarn strength and the abrasion resistance. Then the optimum add-on is reached between points A and B. Beyond point B, the increasing add-on creates a very thick film and deteriorates the mechanical properties of the yarn by increasing the stiffness and decreasing the elongation (extensibility).

The ultimate goal of sizing is to reduce the warp end breaks during the weaving operation. The efficiency of sizing will result in less warp breaks and consequently an increase of the fabric quality and a gain of productivity.

The break ratio is usually low, even for a warp with poor sizing, from 1 up to 2 breaks per 10^4 picks. This ratio follows a Poisson law. Due to the Poisson law, the evaluation of a statistically significant difference for an improvement of 10% of the warps breaks due to a new sizing process, needs the observation of a minimum of 400 warps breaks.

The corresponding production related to a **b** break ratio per 10^4 picks is consequently as follows:

$$Pr = \frac{400.10^4}{b} \text{ picks}$$

For a weaving machine running at **P** picks/min with an **E** efficiency, the **Pt** production time necessary to meet the Poisson law requirements is:

$$Pt = \frac{400.10^4}{60.P.E.b} \text{ h}$$

If the weaving machine is running at 800 picks/min with an efficiency of 95% and 1.5 breaks per 10^4 picks, a pick count of 28 picks/cm, the Pt production time is around 59 h and the produced fabric length around 1000 m. This observation needs a to be monitored for a quite long time, which can be successfully assisted with the help of a computer-aided-manufacturing system.

On the other hand, when taking assessment of a sizing process, account must be taken not only of the increase in term of weaving productivity but also of variations in the general cost, prices of the product and the additives, energy, labour, desizing, etc....

The need of sizing or the need of modifying the sizing process depends on many factors as follows:

- The type of weaving machine, the type of filling insertion system, warp tension, ...
- The type of textile material, mechanical behaviour, affinity for size, risk of electrostatic charges, ...
- The type of yarns: continuous filament or spun yarn,
- The thread count: the weaving difficulties, thus the need of sizing increases, the thinner the warp yarn count, the higher the warp setting and the lower the warp twist, the higher the yarn hairiness and so the lower the pick count, the higher the need for sizing.

But the characteristics of a good sizing process and a good sized yarn will depend on the user's point of view. This point of view is not the same as far as sizers, weavers, finisher and financial manager are concerned.

From the point of view of the sizer, a good sizing process involves a good machine running regularity, a small yarn sticking, and thus providing a weak hairiness at the output of the sizing machine, no fluff and dust and no size deposit on the drying devices.

From the weaver's point of view, an increase of the yarn mechanical properties, and cyclic extension behaviour, an improvement of the working capacity of the yarn, as well as the abrasion resistance, a reduction of the yarn hairiness, a size add-on as low as possible, a moisture content according to the material and no generation of electrostatic charges will have to be considered.

From the finisher's point of view, the sizing process must not cause any defects during the dyeing and finishing processes, and the size and chemicals must be eco-friendly.

From the financial manager's point of view, the overall cost balance, size, chemicals, labour, energy, desizing process must be as low as possible.

4.2 Sizing Materials

Sizing is a non-permanent protective coating applied before weaving to warp ends, it must be eliminated by desizing as easily as possible after the weaving process. Thus, the sizing material will form a film around the yarn without any chemical bonding. Moreover, the sizing material needs good adhesive properties as well as elastic and plastic cohesion compatible with the mechanical properties of the yarn. Consequently, the sizing material must have good film-forming properties and the film must have weak chemical bonds with the yarn being sized like hydrogen and Van der Waals bonds. Hence, there are a number of needed properties which a warp size should possess. Table 8 indicates a non-exhaustive list of these desirable properties.

The sizing materials, the film forming polymers can be natural or synthetic material. The natural materials include starch and modified starch, cellulose derivatives while synthetics sizing materials include polyvinyl alcohol, acrylate, acrylamide and polyester resins.

Starch: it is the oldest sizing material used in sizing and it is still widely used either alone or combined with other sizing agents due to its low cost and availability. It is one of the most abundant and renewable agricultural resource. The origin of the starch depends on the geographical location, thus, in Europe, starch is derived from potatoes, in the United States from yellow dent corn, rice is used in Asia while long roots like tapioca and manioc are used in Africa. Each starch type has its own characteristics and specifically its own viscosity due to different molecular weight [3, 30]. The propensity of raw starches (pearl or unmodified) to produce high viscosity sizes entails a high stiffness film, difficulties in water solubility and film forming as well as poor adhesion on polyester. The starch film cracks and splinters under

Table 8 Properties of a good sizing material

Improve weaving efficiency	Water soluble or water dispersible
Good film former	Controllable viscosity
Good specific adhesion	No skimming tendency
Good frictional properties	Compatible with other ingredients
Good film flexibility	Rapid drying
Good abrasion resistance	No size redeposition
Low electrostatic charge propensity	Eco-friendly non-polluting
Easily prepared	Reasonable cost

bending stresses. Reducing viscosity by one of the following processes will improve the sizing properties of starch:

– pearl (unmodified)
– acid modification
– oxidized (chemical modification)
– hydroxy ethylated (chemical modification)
– carboxy methylated (chemical modification)
– Conversion to gums.

Modified starches: to obtain acid modified starches, the molecular weight of raw starch (pearl) chains will be reduced by an acid as hydrochloric acid. This modification reduces substantially the viscosity of the starch and makes sure that the size has a high solid content and low viscosity size solution that is best adapted to the sizing process. In the same way, oxidized starch can be obtained by treating the raw starch with an oxidative agent like sodium hypochlorite. Oxidized starches have more or less the same properties as acid modified starches. Starch esters (acetates) and starch ethers have been developed [6] by using several chemical routes. The acetate derivatives can be considered as a partially hydrolyzed polyvinyl alcohol and give size with a good viscosity stability, lower pasting temperature, strong flexible films and a good adhesion on hydrophobic materials such as polyester. Starch ethers like methyl, ethyl, carboxymethyl, hydroxyethyl, hydroxypropyl starches have been developed. Low DS (degree of substitution) starch ethers are soluble in hot water and consequently easily desized [11]

Carboxyl methyl cellulose (CMC): The synthesis of CMC is obtained from purified alkali cellulose treated with monochloroacetic acid under controlled reaction conditions knowing that the starting materials are either wood pulp or cotton waste like cotton linters. By replacing one or more of cellulose hydroxyls with carboxymethyl groups by etherification, water soluble CMC is obtained. The degree of substitution, thus, ranges from 0.7 up to 1.2. The DP of the cellulose source and the DS determine the solubility or the insolubility of the solution. A size grade of CMC with a DS from 0.65 up to 0.85 is the best adapted to warp sizing and it is an excellent film-former with a moderate film strength and abrasion resistance. Due to its high water solubility, CMC size can be easily desized.

Polyvinyl alcohol (PVA): PVA is one of the most versatile and one of the mostly used warp sizing agent that is more specific for polyester and polyester/cotton blends. Contrary to starches which need enzymatic treatment for desizing, PVA can easily be desized in hot water which provides a competitive advantage. PVA is a very good film former, PVA coating compared to starch coating is stronger, more flexible and more abrasion resistant. PVA is manufactured by the hydrolysis of vinyl acetate under specific conditions meant to remove varying amounts of acetate radical and replace them with hydroxyl groups [12]. By controlling the PVA DP and the hydrolysis degree of acetate groups, the viscosity can be controlled and a large tailored PVA size family can be manufactured to meet the requirements of specific sizing needs.

Acrylics: Polyacrylic acid and its derivatives (polyacrylates and polyacrylamides) are used solely as film-former or as a binder to improve the adhesion between a primary film-former and the yarn. Acrylics produce low viscosity solutions which give thermoplastic film having a good strength, a good elongation and a low stiffness. These sizes can be made water soluble or insoluble and they are exceedingly compatible with other sizing materials. As a consequence, acrylics are used to improve the performance of size blends. These acrylic based sizes are commonly used to size hydrophobic fibres and their blends such as polyamides, acrylics, polyester, etc.

Binders: Binders can be used as real film-formers like acrylics but, generally, they are added to a primary film-former to improve the weaveability of the sized yarn. The binders promote the adhesion of the primary film-former like starch, PVA, CMC, acrylics to the fibre surface while reducing the inter sized yarn cohesion. among the advantages provided by binders using, the following can be cited:

– Easier size application
– Sizing application improvement
– Hairiness reducing
– Easier splitting at lease rods
– Weaving efficiency improvement
– Size removal and desizeability improvements
– Etc, …

Acrylics, polyester resins, vinyl acetate resins can be used as binders. Acrylics are well adapted to be blended with starch to promote adhesion on hydrophobic fibres or hydrophobic/natural blended yarns. Polyester resins (PET), as expected, is an excellent binder for polyester fibres for both filament and spun yarns. Vinyl acetate resins are used as sizing agent for polyester and filaments and as binders for glass, polyester and bended spun yarns.

Other additives: additives are products that may be added to size mix to modify their properties to meet specific weaving requirements. These additives include the following:

– Humectants such as urea, glucose and glycerin, are used to retain some moisture in the size product specifically for starch based sizes. Moisture makes the size film more flexible and less brittle preventing shedding and dust production.
– Antiseptics and fungicides as metallic salts or phenol and phenol derivatives are used to prevent mildew if the warp is to be stored for a long time
– Antistatic agents like cationic agents or humectants are used to prevent electrostatic charges accumulation in case of synthetic warp sizing.
– Softeners and lubricants like soaps and waxes are used to improve the yarn abrasion resistance and increase the elastic modulus if possible. They also prevent size cracking during weaving.

Various other additives may be included in the size as anti-foam agents, binders, penetrating agents, viscosity modifiers, weighing agents, anti-skin, etc. to improve the size properties. But, such additives should not disturb or interfere with the desizing and finishing operations in any way.

Choosing the right size mix, taking all the ingredients available into account, is quite complex but the key is to make the size mix as simple as possible. This choice

Fig. 38 Yarn structures

should be considered in relation with the yarn parameters, the type of sizing machine, the type of desizing process available and the environmental policy. Some factors can be cited as follows:

- The yarn material (cotton, wool, flax, synthetics, blend)
- The yarn structure (ring-spun, open-end, air-jet, vortex-yarn, hairiness, stiffness)
- The type and the velocity of weaving machine (rapier, projectile, air-jet, water-jet)
- The sizing machine design and yarn occupation in the size box and in the drying zone
- The % add-on (and % solids) required and the water demand
- The type of desizing process
- The environmental policy (restrictions).

Figure 38 shows the structure of three kinds of yarns, ring-spun yarn, open-end yarn and air-jet yarn [5]. It can be seen that even the count is the same, the apparent diameter and the bulk density (voluminosity) of the yarn is not the same [19] and the sizing requirement is different. All other things being equal, compared to a ring-spun yarn, an open-end yarn has a diameter (a bulk density) 10 up to 15% larger inducing a higher surface porosity [24], thus a higher size mix viscosity to prevent excessive size penetration in the yarn structure. The same reasoning would apply to the twist which is also a yarn structure parameter. The higher the yarn twist, the higher the yarn tightness and the lower the yarn porosity, the lower viscous size solution.

Table 9 gives typical sizing examples of a viscose warp sheet and an organic cotton warp sheet. (Didier, C. and Galvan, L., Emanuel Lang Weaving SAS, Personal communication, June 4, 2020).

4.3 Sizing Machine

The sizing machine must allow successively withdrawing and overlapping the primaries warp sheets to form the final warp sheet, the impregnation of the warp sheet by immersion in a size mix, the drying, the separation of sized yarns after the dryer, the final warp width setting and the winding onto the weaver's beam at constant tension and high density.

Table 9 Example of sizing parameters

Material	Viscose (grey)	Organic cotton (yarn dyed fabric)
Yarn count	20 Tex (Nm50/1)	17 Tex (Nm 60/1)
Number of warp ends	7308 ends	6136 ends
Width in the size box	168 cm	168 cm
Warp setting in the size box	43.5 ends/cm	36.5 ends/cm
Type of size mix	75% PVA + 25% CMS	CMS + low viscosity PVA resin
Wetting agent	0.5% (0.5 l/100 l size mix)	0.5% (0,5 l/100 l size mix)
Fatty agent-Lubricant	0%	0.5% (0,5 l/100 l size mix)
Size add-on	7%	8%
Residual moisture	11%	10%
Sizing velocity	70 m/min	80 m/min

Hence, the main components of a sizing machine to size (slash) spun warp yarns are shown in Fig. 39. These components may be described as follows:

- The creel housing warper's beams—unwinding zone
- The size boxes—sizing zone
- The drying cylinders—drying zone
- The lease rods—splitting zone
- The head stock—weaver's beam preparation zone

The warper's beams placed on the creel are withdrawn and are conducted to the size box with the help of a feed roll. Then the yarns are impregnated with the size mix and the excess of size is squeezed out by passing the yarn through a pair of rolls.

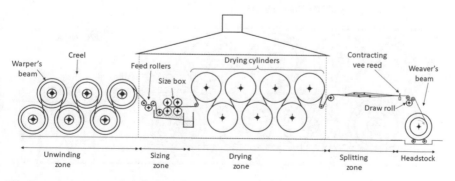

Fig. 39 Schematic of sizing machine

The warp sheet passes next through the drying cylinders to be dried at the required moisture. The dried warp sheet is then passed through a series of bust rods to separate the yarns. And finally, the separated yarns are passed through a guide reed and are wound onto a weaver's beam.

The creel: Basically, the creel is a well-built and robust framework capable of carrying heavy warper's beam usually up to 32. The primary function of the creel is to smoothen and keep unwinding the warp yarn sheet on a predetermined constant tension with an exact overlapping and without entangling to adjacent warp sheets being unwound. Thus, the lateral position of each warper's beam has to be accurately set to ensure the perfect overlay of the warp sheets and to avoid the warper's sheet to touch the beam flanges. The constant tension of the sheets regardless of the relative position of the rolls and their diameters is controlled both by a specific sheet path and a specific braking device. The warp yarns pass over one beam, under the next beam, again over the next beam and so on. This is called over/under creel as shown in Fig. 40.

Another possibility to maintain a constant tension in case of light weight fabric is to use an equitension creel as shown in Fig. 41. In this type of creel, the yarn sheet of each warper's beam is withdrawn individually, passed over a guide roll, joins the

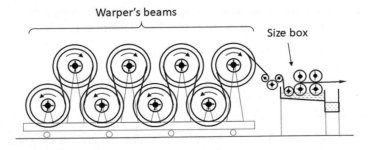

Fig. 40 Schematic of over/under creel

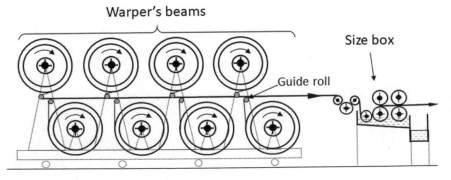

Fig. 41 Schematic of equitension creel

yarns coming from the other beams of the top or bottom tiers and then passes forward directly to the size box.

The creel is equipped with a braking system. The braking system has two functions: it allows the control of the yarn tension between the creel and the size box during the normal constant velocity sizing operation taking the beam diameter decrease into consideration. Moreover, the braking device prevents the over-running of the beams, due to beam inertia especially when a break occurs or while doffing the weaver's beam which induces a strong reducing of the sizing machine speed. The braking device is usually a combination of band brake and pneumatic cylinder. The yarn sheet tension is measured in front of the size box for feedback control. Adding the non-contact diameter detector, the tension can be uniformly controlled even during acceleration and deceleration as well as from the start to finish of warper's beam. This computerized system assumes a constant unwinding predetermined tension of the warp sheet by precisely controlling the braking power of pneumatic cylinder and band brake.

In view of the downtime during the warper's beam loading on the creel, the sizing machine is equipped with movable creels as shown in Fig. 42. These movable creels, instead of stationary creels can be side to side or rotatory. The advantage of such movable creels is that, while sizing is in progress from one set of beams, the loading of another new one set can be done on another stand-by creel. Thus, the next set of beams on the sizing machine can be started with much less downtime that considerably increasing the sizing machine efficiency.

Moreover, a specific package stand is often associated with the creel to be able to add the selvedge yarns during the sizing operation and thus avoiding to warp them.

The feed rolls: this device generally integrated into the size box, as shown in Fig. 39, withdraws the warp sheet under a uniform controlled tension. A tensioning

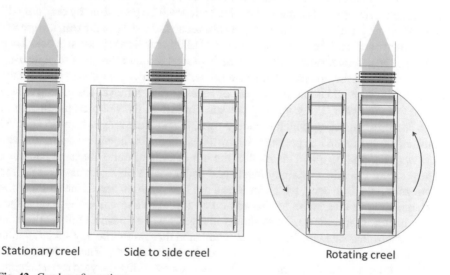

Stationary creel   Side to side creel   Rotating creel

Fig. 42 Creel configurations

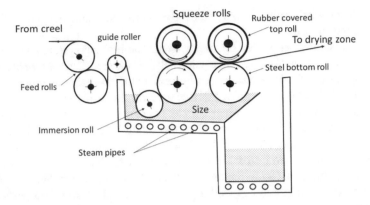

Fig. 43 Schematic of size box

compensation device composed of a dancer roll, could sometimes take place between the creel and the feed rolls to compensate the inertia of the warper's beams during the acceleration and deceleration of the sizing machine.

The size box: The basic function of a size box is to impregnate the warp sheet in the size solution at a given application temperature and then to squeeze out the excess of size solution before the yarn sheet is moved to the drying zone. Most sizing machines are equipped with a single sizing box. The typical sizing box as shown in Fig. 43 is composed of an immersion roll, two pairs of squeezing rolls and a size bath heated with the help of steam pipes.

The immersion roll and the squeeze rolls are movable and can be raised and lowered at the very beginning of the sizing process. The size impregnation starts at the nipping point between the immersion roll and the stainless steel roll of the first pair of squeezing rolls. The amount of size which will be picked-up by the yarn will depend on the position of the immersion roll, because the deeper the roll, the greater the amount of size on the yarn, and on the level of the size liquor in the size box. Then the yarn sheet impregnated with wet size is moved through the pairs of squeezing rolls. The role of the first pair of squeeze rolls is to push deeper the size between the fibres of the yarn while the second pair of squeeze rolls removes the excess of size liquid from the yarn. The squeezing pressure influences the wet pick-up. In case of high pressure squeezing, which is widely-used, the pressure ranges from 45 up to 60 daN/cm along the nip line pressure, which allows high size liquid concentration and low wet pick-up [14, 24]. This implies the use of rolls of quite the same hardness. The pressure is applied by air cylinders and under such a high pressure that the top squeeze roll tends to bend, which induces a nip size variation across the width of the roll, resulting in a variation in size add-on on the warp sheet from both selvedges to the centre. A specific design as shown in Fig. 44 will achieve a uniform pressure with a controlled size nip along the roll width and there won't be any space at all in the centre when both ends of the roll are under high pressure. The pressure in the cylinder is automatically controlled and monitored by taking the yarn sheet velocity into account, thus keeping the wet pick-up constant.

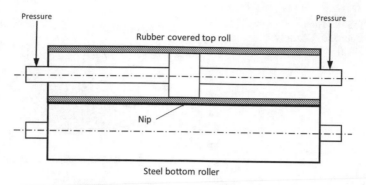

Fig. 44 Schematic of a pair of squeeze rolls

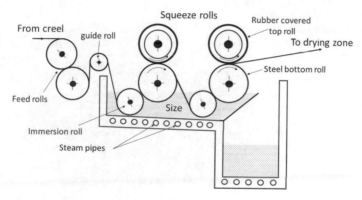

Fig. 45 Schematic of a double squeeze size box

The double squeeze size box as shown in Fig. 45 is equipped with one set of immersion roll and squeeze rolls followed by another set of immersion roll and squeeze rolls. The twin immersion rolls and two pairs of squeeze rolls increase the penetration of size liquor in the yarn whereby a uniform coating of the yarn will be obtained. The double squeeze size box is commonly used for sizing heavy warp spun yarns but could be also used for lighter warp sheet. In fact, for "light" sizing, instead of reducing the size concentration, the warp sheet should not be immersed, thus allowing the size penetration in the yarn within the nip of the squeeze rolls.

Other size box configurations are used with the aim of improving the warp sheet tension control as well as the size concentration and viscosity evenness, reducing that way the size loss and decreasing the quantity of water to be evaporated. Figure 46 shows a size box design by Tsudakoma® and Fig. 47 a size box design by Karl Mayer®

More recently, a new sizing box design has been proposed by Karl Mayer which could be considered as a breakthrough innovation in the sizing technology. The VSB® sizing box shown in Fig. 48 differs from the classical size box which is based on the immersion bath technique that there is no immersion of the yarn sheet. It is a

Fig. 46 TTS10S High
pressure squeeze® (Courtesy
of Tsudakoma)

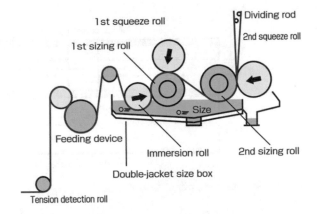

Fig. 47 CSB type size box®
(Courtesy of Karl Mayer)

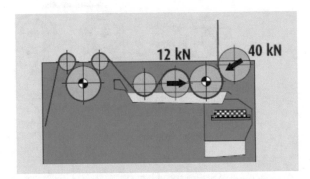

Fig. 48 VSB® type size box
(Courtesy of Karl Mayer)

free-immersion process. It consists of three highly turbulent and uniform application
zones with a spray bar system and final application with squeeze rolls.

Compared to the immersion bath technique, the application of the size with the
help of spray nozzle under pressure results in a uniform film coating with less sizing

additives. This technique guarantees uniform size concentration, temperature and viscosity due to the high size circulation rate. Moreover, the special VSB® design enables the warp sheet to be operated vertically during delivery, which provides advantages in terms of ergonomic as well as sufficient space and for easy access to the machine.

The drying zone: The role of dryer is to remove the water contained in the size impregnated in the warp sheet. The drying process must be done:

- at a lower cost, that is, with the best thermal efficiency,
- without overheating the warp sheet, i.e., without going below the moisture regain of the sized material; so as to avoid any risk of yellowing or plastic deformation in the case of synthetic materials. The drying should however be sufficient not to create any damage on the warp beam (sticking, mildew, etc....). This imperative determines the running speed of the sizing machine.
- homogeneously, both at the level of the individual yarn and at the level of the yarn sheet with the same conditions of drying at the selvedges as well as in the centre of the warp sheet.

The different drying modes are the following:

- conduction drying: cylinder, can dryer
- convection drying: hot air
- radiation drying: Infrared radiation (IR), microwave radiation
- mixed drying: combination of different drying modes

Conduction dryer: The cylinder dryers are the most commonly used as they are the most energy efficient. The warp sheet to be dried passes successively on 5 up to 11 (Fig. 39) drying cylinders (cans) heated by overheated steam, up to about 140 °C (3.5 bar). The temperature of each cylinder is individually adjustable so that it can be dried gradually if necessary. The first 3 cylinders are usually coated with nonstick coating e.g. Teflon® to prevent the size and yarn from sticking while the warp is partially dried. The steam from the water evaporation in the warp sheet is collected by a hood to be evacuated. Depending on the characteristics of the warp sheet, and depending actually on the amount of water to be evaporating and the number of cylinders, this allows sizing speeds up to 150–180 m/min.

Convection dryer: in the past, the drying fluid used to be hot air, which required a long warp sheet layer (at least 80 m) in the dryer and created an overheating risk or condensation depending on the flow direction of the fluid in relation to the warp sheet. Currently, convection dryers operate with a drying fluid composed of superheated air and steam (140 °C up to 160 °C) which limits the warp sheet temperature to about 80 °C. The steam of the drying fluid is supplied by the evaporation of the warp sheet to be dried. The dryer is composed of 2 or 3 chambers in series, each containing a few meters of warp sheet. The drying fluid is reused after removing part of it and reheating the rest of it. The advantage of the hot air convection drying system is that the whole surface of the yarn is subjected to a uniform temperature, which is not the case of a conduction drying system where only a part of the sheet is in contact with a hot cylinder. But, the major disadvantage of this convection drying system is that it requires a strong mixing of the fluid close to the warp sheet, whereby yarn entanglements can be created.

Radiation dryer: Other forms of drying methods, not widely used in practice for sizing, are based on infrared drying and microwave drying systems. The infrared radiations can be generated by a series of infrared lamps with a reflector or by a refractory material heated to incandescence by gas burners. The warp sheet to be dried passes through a number of infrared-radiating emitters. The microwave drying requires a specific waveguide design to avoid overheated or underheated areas on the warp sheet. Based on this improvement, microwave drying could be a good opportunity to dry yarn sheet. But, the cost of energy, the cost of installation as well as the operation proper is still prohibitive for drying flat textiles like yarn sheets. Drying by IR radiation is sometimes used in pre-drying chambers for untwisted filament yarns.

By comparing the two drying techniques, conduction and convection drying, the following conclusions can be drawn:

– from a thermal efficiency point of view, conduction is better,
– regarding the flexibility of use (running conditions), a convection chamber can be cooled instantaneously, by blowing cold air when there is a considerable, risk of overheating and yellowing when a cylinder is stopped or even slowed down,
– from the point of view of the drying quality, the yarn sheet is better supported on the cylinder, without any subsequent risk of entanglement.
– The drying in the convection chamber is more regular because it is done on the entire surface of the yarn sheet and it is more progressive resulting in a better size adhesion on the yarns.

The splitting zone: Figs. 39 and 49 show schematics of the splitting zone.

The first role of the splitting zone is to separate the individual yarns of the warp sheet that are more or less stuck together during the drying of the size film in the drying section. This is done by means of a series of lease rods with one large diameter bust rod positioned at the input of the warp sheet. These rods are flattened at both ends to be clamped firmly in the brackets. As shown in Fig. 49, the number of lease rods used is determined by the number of warper's beams being used in the creel. The second role of the splitting zone is to allow, with the help of a contracting vee reed (Fig. 39), positioned close to the headstock, the setting of the warp sheet to

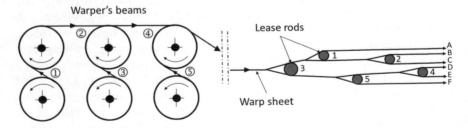

Fig. 49 Schematic of the splitting zone

Fig. 50 Head stock side view

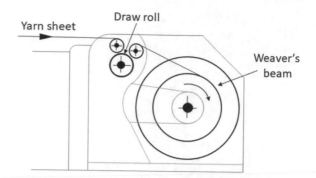

Yarn sheet · Draw roll · Weaver's beam

the final width. Thus, the residual moisture content is controlled at the outlet of the drying zone with the help of sensors that measure the moisture at the selvedges and centre of the warp sheet. These sensors allow the automatic regulation of the running speed.

The head stock: the head stock is a take-up unit supporting the weaver's beam. Figure 50 shows the side view of a typical head stock. The role of the head stock is to play an important role, with the splitting zone, in the warp sheet final width setting. For this purpose, the head stock is equipped with a contracting vee reed to adjust the warp sheet yarn density and to align precisely the warp sheet with the flanges of the beam as well as a delivery device including a positively driven draw roll. The warp sheet is well wrapped around the draw roll to be draw without slippage. The delivery device moves the warp sheet at a constant velocity and the velocity of the weaver's beam is continuously adjusted by taking the winding tension into account. The sizing machine control panel is attached to the head stock.

The tension and stretch: The warp sheet passes through a succession of dry and wet conditions between the unwinding zone and the weaver's beam winding zone (headstock). The mechanical behaviour of the warp sheet in each of these phases, dry or wet, depends essentially on yarn material. Some textile materials are slightly reinforced in the presence of water while others will lose a part of their mechanical properties.

In any case, efforts should be made to:

- maintain or even improve all the mechanical properties of the warp sheet,
- obtain the optimal sizing by allowing the precise quantity of size liquid to penetrate regularly the individual yarns and form a protective film
- dry under the best conditions,
- wind the weaver's beam regularly,
- not to cause breakage.

For this purpose, the sizing machine is split into several tension zones as shown in Figs. 39 and 51, the undwinding zone, the sizing zone, the drying zone, the splitting zone and the beaming zone (headstock). The yarn tension must be most accurately controlled in those zones. Figure 51 shows the diagram of a digital tension control based on a multi-motor system. In this configuration, individually driven motors are

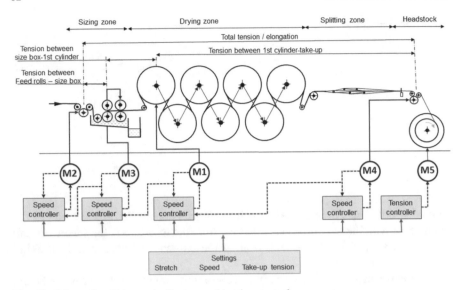

Fig. 51 Schematic of kinematic diagram and tension control

located in each sizing, drying and headstock (beaming) zone. Based on the sizing machine data controller, the dryer is driven by an M1 motor which determines the average velocity of the sizing machine. From the M1 motor controller, the speed set points are given to M2 and M3 motors and the tension on the warp sheet between the feed rollers and the first pair of squeeze rolls is given by the speed differential between M2 and M3, the tension on the impregnated warp sheet between the size box and the first drying cylinder is given by the speed differential between M1 and M2, the tension on the warp sheet being dried between the dryer and the draw roll of the headstock is given by the speed differential between M1 and M4 while the beaming tension and the velocity of the head stock as a function of the diameter of the weaver's beam is given by the M5 motor speed [2].

The instrumentation and controls: there are a wide variety of controls on the sizing machines whose essential functions are to provide the optimal quality of the warp in the best running conditions and at a cost as low as possible. The controls operate on the basis of information given by specific sensors placed on the sizing machine. These sensors should be installed and located in the sizing machine as shown Fig. 52.

– Tension controls: The sensors are placed in the direction of the yarn sheet path from the unwinding zone to the headstock to monitor the tension and regulate the running of the sizing machine. The sensors in the unwinding zone monitor the creel brake for preventing any warper's beam over-running and maintaining the warp sheet tension constant whatever the running speed.
– The level and the temperature of the size liquid in the size box are controlled to guarantee the viscosity and size add-on as well as the stretch level. An excessive tension on the warp sheet will induce an excessive elongation of the yarn resulting

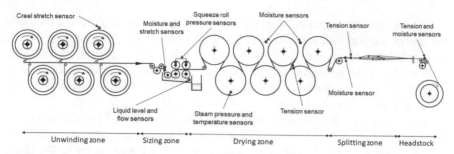

Fig. 52 Sensors implementation on a sizing machine

in an increase in warp breaks during the weaving process and together with an efficient decrease.

– The steam pressure and temperature are controlled to maintain the accurate parameters for an optimal drying and to monitor the velocity of the sizing machine in case of slow down.

– The squeeze roll pressure is controlled and monitored with respect to the sizing machine running conditions (operating speed).

– The size add-on and the warp sheet moisture, wet-up pick up are on line controlled. Based on those measurements, the size add-on can be automatically corrected by adjusting the squeeze roll pressure and thus maintaining the size add-on practically constant during the sizing operation.

Implementation and sizing machine parameters effect: at the end of the sizing process, the weaver's beam must have a suitable residual moisture adapted to the yarn material and to the climatic conditions of the workshop in which the beam will be stored before weaving. Indeed, as drying operation is a very expensive operation, a warp sheet over dried will result in regaining workshop atmosphere moisture which is counter-productive. Furthermore, the rate of evaporation (drying rate) determines the running speed of the sizing machine. It is therefore imperative to control the residual moisture accurately. This control, as shown Fig. 52 is done at the exit of the drying zone.

Otherwise, the whole textile process imperative rule applies to sizing, and the tension must be as low as possible in order to preserve the mechanical properties of the yarn.

Creel zone: the tension is given by the inertia of the rolls to be driven which is variable according to their winding diameter, and by the braking force exerted on those rolls. The brakes are also used to prevent the rolls from over-running during decelerations due to slowdowns. Moreover, what matters more than the intrinsic value of tension and what is the most important lies in drawing the warp sheet from each warper's beam under the same tension. The equal distribution of the yarns on the rollers allows an identical warper's beam diameter as well as an identical mass and contributes to even tension distribution.

The quality of the beams has to be perfect. The flanges and edges should be smooth and cracks and burrs alike should be free to prevent warp ends from clinging during the unwinding process. This may cause warp ends abrasion and worst, yarn breaks. If the quality of the beam is good and the tension accurately controlled, the density of the warping beam should be as uniform as possible without any cross-ends nor buried or embossed ends at the edges close to the flanges.

Sizing zone: in order to facilitate the uniform penetration of the size thorough the yarn and to avoid an over-stretching of the wet ends, the tension must be as low as possible, although sufficient enough to allow the immersion of the sheet and to prevent the formation of ropes at the first pair of squeeze rolls. The sizing box is a very complex device and many parameters have to be taken into account such as the size box level, size solution parameters (viscosity, temperature), squeeze parameters (squeezing pressure, squeeze roll hardness and diameter), yarn count and warp density in the size box... The contact time between the size solution and the yarn is determined by the size box level in the size box. The lower the size level, the shorter the contact time and vice-versa. A shorter contact time results in a weak size penetration into the yarn and an inadequate film-forming. Moreover, a low size level in the size box influences the size add-on, the size coating the film characteristics, the yarn hairiness... It should be remembered that the speed being regulated according to the residual moisture of the warp sheet at the exit of the dryer, the contact time varies.

The size viscosity depends on the concentration of the size and on the application temperature in the size box. Moreover, the temperature influences not only the quality of squeezing and the surface tension, but also the evaporation and as a way of consequence the size concentration as well as the elimination of air trapped in the yarns. It is therefore important for the temperature to be kept constant. The size add-on decreases, the higher the temperature, the higher the size concentration.

The dimension of the nip influences the level of size add-on of the yarn. The nip depends on the hardness of the rubber-covered squeeze roll and on the squeezing pressure. Harder rolls lead to a sharper nip and lower pressure than softer rolls, all other things being that way equal by squeezing more size from the yarn, resulting in a lower pick-up while, all other things being equal, soft rolls make a flatter nip resulting in a higher pick-up. For a given characteristics warp sheet (material, count, density), the squeezing pressure determines the wet pick-up thus the size add-on:

$$\text{Size ad on} = \text{wet pick} - \text{up} \times \text{size concentration}$$

$$\text{Wet pick} - \text{up} = \frac{\text{weight of sizing solution (kg)}}{\text{weigth of unsized yarn (kg)}}$$

Or:

$$\text{Wet pick} - \text{up} = \frac{\text{size add on}(\%)}{\text{solids concentration in sizing solution}(\%)}$$

For a given size add-on, it is therefore advisable to increase the squeezing pressure and the size concentration, consequently the wet pick-up increase leads to decreasing the quantity of water to be evaporated, saving energy at drying and increasing the sizing machine running speed. This is the interest of high-pressure squeezing, which, moreover, leads to an improvement of the quality of the sized yarn, a better penetration into the yarn and less hairiness, …

The density of the yarn warp sheet in the size box determines the quality of the yarn encapsulation (Film forming). In fact, an adequate space between the yarn in the size box ensures a good size penetration and a good yarn encapsulation but if the yarns are too close, the protruding fibres of two adjacent yarns become entangled and cemented together. The latter has a particularly detrimental effect during splitting after drying, increases the energy consumption, the yarn hairiness and the yarn breakage. It is accordingly necessary to calculate the distribution of the yarns in the size box which could be expressed as the percent yarn occupation or warp sheet density (WSD) as follows:

$$WSD = \frac{\text{Width occupied by the yarns}}{\text{Total yarn sheetwidth}} \times 100(\%)$$

Or: $WSD = 100 \times N \times d$ with N: Warp setting, number of yarns/cm
d: Yarn diameter (cm)

Assuming that the yarn is cylindrical, even and homogeneous, the yarn diameter can be calculated as follows:

$$d = 0.0036 \times \sqrt{\frac{C(Tex)}{\rho \text{yarn}}}$$

With C   yarn count in Tex
ρ yarn density in g/cm^3

Assuming that the yarn density is approximatively 60% of the fibre density, WSD can be deduced as follows:

$$WSD = \frac{100 \times N \times 0.0036}{\sqrt{0.6}} \times \sqrt{\frac{C(Tex)}{\rho \text{ fibre}}}$$

Hence

$$WSD = 0.45 \times C \times \sqrt{\frac{C(Tex)}{\rho \text{ fibre}}}$$

Taking the quality of the sizing into account, the industrial experience shows that the optimal drying conditions correspond to a WSD warp sheet density of 25–30%.

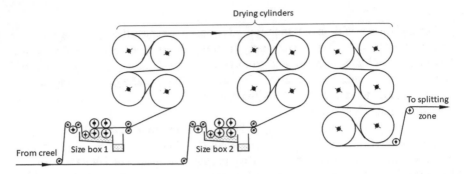

Fig. 53 Schematic of a two-size box sizing machine

However, a classical sizing machine equipped with one size box allows sizing in quite good conditions up to 80% (or even 100%) without any wet separation before drying as shown in Fig. 53. If WSD is higher than 80%, using 2 size box sizing machine proves to be necessary (Fig. 54).

Example: The characteristics of the fabric being woven are as follows:

– Overall width: 180 cm
– Thread count: 27 ends/cm—18 picks/cm
– Warp count: 83 tex (Nm 12)—weft count: 83 tex (Nm 12)
– Warping width: 198 cm
– ρ fibre: 1,5 g/cm^3 (cotton)

Calculation of warp setting in the size box:

$$27 \times \frac{180}{198} = 24.5 \text{ends/cm}$$

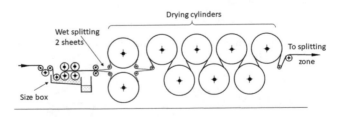

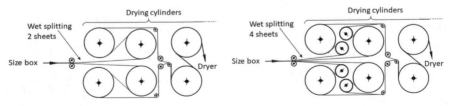

Fig. 54 Schematic of wet splitting

WSD calculation:

$$\text{WSD} = 0.45 \times 24.5 \times \sqrt{\frac{83}{1.5}} \cong 82\%$$

Conclusion: The warp sheet could be sized on a single size box sizing machine.

Drying zone: drying is the most energy-intensive operation of the sizing machine with 75 to 80% of the total energy used during the sizing process. As above described, the conduction method is the most commonly used and the most efficient. The amount of water to be evaporated from the wet yarn sheet depends on the size add-on levels and the solid content of the size. The dryer configuration depends on the number of size boxes used, the number of yarn in the warp sheet, warp sheet density (WSD), the size add-on level, the solid content of the size mix, and the number of available drying cylinders. Depending on these parameters, the yarn sheet can be split in two or four sheets as shown in Fig. 54, each sheet is dried on a separate set of cylinders and combined again for the final drying.

The drying cylinders are commonly non-stick coated i.e. Teflon® coating. The temperature of the drying cylinder increases from the first up to the last cylinder. The temperature of the first one is set at the lowest possible temperature to avoid the sticking of the warp sheet on the wall of the cylinder and prevent hairiness increase and then the temperature is gradually increased to ensure the adequate drying. The last drying cylinder is usually set at a relatively cooler temperature to prevent false moisture regain and to control the tension variations. The tension must be set so as to be clamp sufficiently and properly flatten all yarns against the cylinder walls. A low tension level should induce a good regeneration of the elasticity of the yarns.

As an example, the drying energy consumption can be calculated based on the following example (Table 10):

The steam boiler energy consumption is: $\frac{0.537}{0.35} = 1.53$ th/kg of water to be evaporated.

Hourly production (kg/h):

Table 10 Sizing characteristics

Warp characteristics	
Warp weight	156 g/m^2
Overall width	180 cm
Moisture regain	8.5%
Sizing characteristics	
Wet pick-up	140%
Size add-on	11%
Sizing speed	80 m/min
Sizing machine efficiency	80%
Latent heat evaporation (water)	0.537 th/kg of water
Steam boiler efficiency	35%

$$P = \text{warp weight}\left(\frac{kg}{m^2}\right) \times \text{overall width (m)} \times \text{sizing speed}\left(\frac{m}{min}\right)$$
$$\times 60 \times \text{sizing machine efficiency (\%)}$$

$$P = 0.156 \times 1.8 \times 80 \times 60 \times 0.8 = 1078 \text{kgofwarp/h}$$

Quantity of water to be evaporated:

$$Q = \text{Hourlyproduction} \times \left(\frac{\text{wetpickup} - \text{sizeaddon} - \text{moistureregain}}{100}\right)$$

$$Q_{140} = 1078 \times \left(\frac{140 - 11 - 8.5}{100}\right) = 1299 \text{kg/h}$$

Required energy for drying:

$$E_{140} = 1299 \times 1.53 = 1987 \text{th/h}$$

Knowing that 1 th $= 1.16$ kWh, the required energy consumption, is as follows:

$$E_{140} = 1987 \times 1.16 = 2305 \text{kWh}$$

As above calculated, the drying operation is energy-intensive and one way to achieve significant cuts in energy consumption is to decrease the wet pick-up. All things being equal, lowering the wet pick-up at 80% instead of 140%, the energy consumption will be as follows:

$$Q_{80} = 1078 \times \left(\frac{80 - 11 - 8.5}{100}\right) = 652 \text{kg/h}$$

$$E_{80} = 652 \times 1.53 = 998 th/h$$

$$E_{80} = 1056 \times 1.16 = 1158 \text{kWh}$$

Thus, an energy saving of 50% will be obtained.

Figure 55 shows the energy consumption versus wet pick-up based on the Table 10 characteristics.

Splitting zone and head stock: the dried warp sheet coming out of the dryer has to be separated in individual ends and then wound on the weaver's beam onto a uniform controlled tension. This is done by means of a series of lease rods with one large diameter bust rod positioned at the input of the warp sheet (Fig. 49). These rods are well polished to prevent the damage of the warp sheet that will increase the hairiness of each individual ends. Then, the individual ends pass through the front reed. The

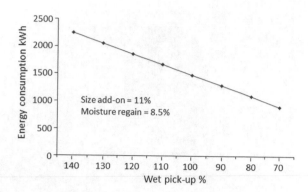

Fig. 55 Energy consumption versus wet pick-up

density of the reed is adjustable (slant or contracting vee reed) to ensure a uniform distribution and the final width of the warp sheet.

The winding of the warp sheet on the weaver's beam is effected at a constant velocity onto a uniform and controlled tension.

Yarn stretch: The most critical parameter in sizing is the yarn tension. The yarns move through a long path from the creel to the head stock, from dry state to wet state and dry state, and the tension applied on the yarn in the process will tend to elongate it. If this elongation is not controlled, the deformation applied that way on the yarn will be permanently set inducing poor yarn mechanical properties and yarn breakages. As an illustration Table 11 gives the tension to be applied on an average yarn tenacity Ty (cN/tex). The tension could be electronically controlled and monitored as mentioned in Fig. 51.

Figure 56 shows a two size box sizing machine.

Prewetting of spun yarns: The concept of prewetting consists in padding the spun yarns warp sheet in a boiling water bath before impregnating them with the size solution as shown in Fig. 57. Thus, the prewetting operation can be considered as a washing process which improves the specific adhesive force of the size on the surface of the yarns by dissolving waxes, dust and other materials deposited on the fibre.

As an illustration, Fig. 9 shows the size spreading on a dry and wet yarn [23]. Figures 9a, b show the photographs of a 29 tex 100% cotton open-end spun yarn on which two drops of CMC size solution (concentration, 60 g/l; temperature, 60 °C;

Table 11 Tension set-point

Sizing machine zone	Tension cN/tex	
	Minimum	Maximum
Unwiding zone – Creel	0.025 Ty	0.050 Ty
Sizing zone	0.010 Ty	0.025 Ty
Drying zone	0.010 Ty	0.030 Ty
Splitting zone	0.060 Ty	0.100 Ty
Beaming zone—Head stock	0.040 Ty	0.100 Ty

Fig. 56 ISOSIZE® two size box sizing machine (courtesy of Karl Mayer)

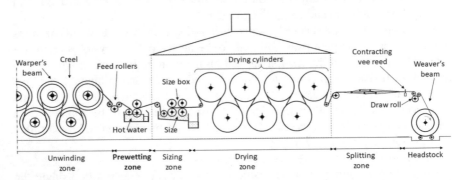

Fig. 57 Schematic of the prewetting concept on a single size box machine

volume, 0.028 ml) have been deposited. These figures illustrate the faster spreading in the case of the wet yarn (b) compared to dry yarn (a).This means that the existing water in the yarn improves its wetting by due to the size solution. Therefore, the size bridges a prewetted yarn in a better way, as shown in Figs. 58, 59 and 60.

Furthermore, the examination of the cross-sections of the two yarns (Figs. 61 and 62) show that these bridges between fibres have made it possible to obtain better

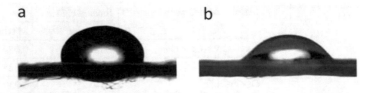

Fig. 58 Spreading of a drop of a CMC size solution on a dry (**a**) and a wet cotton yarn

Fig. 59 Scanning electron micrograph of a sized yarn longitudinal view without prewetting

Fig. 60 Scanning electron micrograph of a sized yarn longitudinal view with prewetting

Fig. 61 Scanning electron micrograph of a sized yarn cross section without prewetting

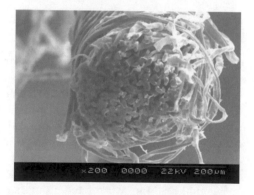

sized yarns. In fact, the size is better distributed on the surface of the sized yarn with pre-wetting [22]. The organization of the fibres within the yarn are quite different, greatly compacted in the core and outside without prewetting, tight on the surface and crown shaped with prewetting. This difference of structure could be attributed to the prewetting operation with the hot water which "washed" and "cleaned" waxes off the fibre surface. This induces an increase in the surface energy and explained the improvement of the physical wetting. In addition, the size solution in contact with the

Fig. 62 Scanning electron
micrograph of a sized yarn
cross section with prewetting

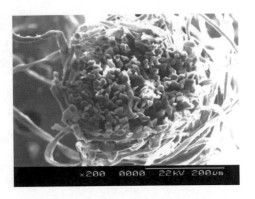

pre-wetting water would have a reduced viscosity and the prewetting water prevents an excessive penetration of the size into the yarn. This observation explains why the bridges between fibres were larger and better distributed. These results could explain the increase of Young's modulus and the elastic recovery compared with the yarn conventionally sized [23].

Thus, sizing with prewetting presents competitive advantages both in terms of weaving efficiency, since breaks are lower, as well as for desizing, since effluent loads and pollution decrease.

5 Drawing-In and Tying-In

5.1 Drawing-In

After the sizing operation, the sized warp beam is prepared to be placed on the weaving machine. As shown Fig. 63, the drawing-in operation consists in passing a yarn from a new warp through the weaving elements namely drop wires, heddles (drafting) and reed (denting) when starting a new fabric style. The drawing-in operation is mostly performed manually but in large scale textile industries, automatic drawing-in machines are used to improve the productivity. When the same fabric design is repeated regularly, the warp tying process is applied to change a weaver's beam. This process can be done manually or with the help of knotting machine.

Manual drafting: when the drafting is performed manually, two operators are required. The "reacher" who selects the ends and presents them for drawing. The "drawer" who draws the ends through the heddle eyelet by using a drawing hook. The required number of heddle frames are prepared with the right number of heddles corresponding to the total number of yarns on the weaver's beam. Then, these frames are hung on the stand as shown in Fig. 64 and the weaver's beam is put just behind the stand. The reacher prepares the yarns with the help of dressing brush and thus selects the yarn according to the lease and presents it in front of the heddle eyelet

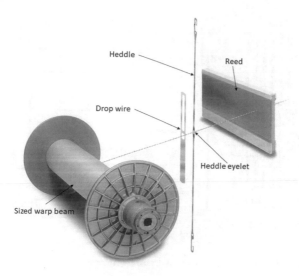

Fig. 63 Schematic of drawing-in (Courtesy of Stäubli)

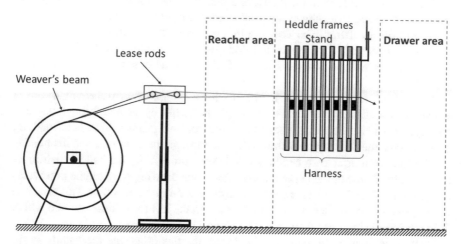

Fig. 64 Schematic of manual drafting

selected by the drawer. The drawer inserts the drawing hook into the heddle eyelet and the reacher puts the selected yarn in the hook. By pulling the hook, the yarn is drawn through the heddle eyelet. This action is repeated till the full warp sheet drafting has been completed. When the drafting is completed, the yarns are passed through the reed dents.

In case of closed drop wires, the yarns are drawn through the eye of the drop wires before passing through the heddle eyelet while, they are pinned during the warp changing, in case of opened drop wires.

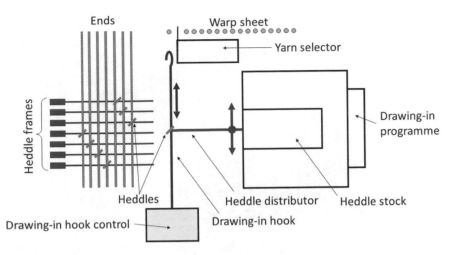

Fig. 65 Schematic of an automatic drawing-in system principle

This method, performed manually, is a lengthy time consuming process, and so automatic drawing-in machines are used to improve productivity.

Automatic drafting: The drawing-in process is fully automated based on a robotized system which manages the ends drawing-in through closed drop wires and heddles as well as denting. Figure 65 shows an automatic drawing-in system principle.

The ends are taken one by one in one previously prepared warp sheet by means of a selector. The drawing hook passes through the heddle eyelet and through the drop wire eyelet. Then, the drawing hook moves the front and catches the selected end. By moving back, the drawing hook draws the end through the drop wire and the heddle. The heddle, coming from a heddle stock is then brought by the heddle distributor on one of the frame of the harness according to the drawing programme previously stored in a file. The denting operation can be done at the simultaneously. This principle implies that one of the lateral support of each frame has to be disassembled. Other drawing-in principles can operate with classical heddle frames.

Automatic drawing-in increases the speed, the flexibility and the quality in the weaving preparation compared to the manual drawing-in. A drawing-in machine can operate with up to 28 weaving frames, 6 up to 8 drop wire distributions, a width up to 400 cm and a rate up to 50,000 ends per 8 h is possible.

Two types of machine design are available: the drawing-in system with stationary drawing-in machine and the mobile drawing-in trucks. The mobile machine, docked onto the stationary drawing-in station, draws yarns into the reed, heddles and drop wires. This design proven for decades is illustrated in Fig. 66. And newly developed system with mobile a drawing-in machine and stationary drawing-in stations are shown in Fig. 67.

This machine (Fig. 67) consists of a stationary drawing-in machine and typically 2 mobile drawing-in trucks on which the warp sheets to be drawn in are stretched. It

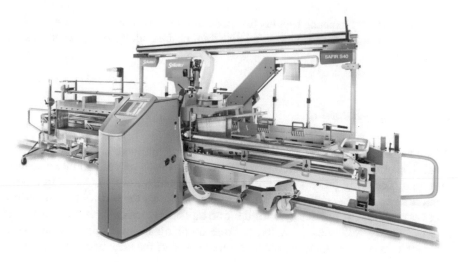

Fig. 66 SAFIR S40® automatic drawing-in machine (Courtesy of Stäubli)

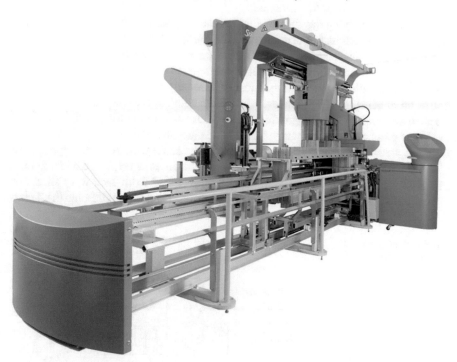

Fig. 67 SAFIR S80® automatic drawing-in machine (Courtesy of Stäubli)

offers significantly expanded application possibilities, handling up to 2 warp beams
with a total of up to 4 ends layers and distributing the heddles on up to 28 heddle
frames.

5.2   Tying—Knotting

When the fabric design is repeated on the same weaving machine after a weaver's
beam depletion, there is no need to perform a drawing-in process on that weaving
machine. The tying process is used to replace the old weaver's beam with a new
weaver's beam. Tying or knotting is the process of joining the ends of the exhausted
weaver's beam, cut previously with the corresponding ends of the new weaver's
beam. Usually, this tying operation is done on or off the weaving machine with
the help of a portable knotting robot. Figure 68 shows the schematic of a knotting
machine.

The tying operation assumes that the drawing-in, the denting, the number of ends
as well as the yarn count are the same.

The success of tying depends mainly on the preparation of the warp sheet. The
old warp sheet is cut and folded down on the harness. The empty weaver's beam is
removed and the new beam is set up. The warp sheet coming from the new weaver's
beam is prepared first. This warp sheet is detangled, brushed, drawn and combed to
improve the parallelism of the warp ends, without any crossing, and an end distribution as homogeneous as possible over the whole width of the warp sheet. These ends
are then clamped by two longitudinal clamps. A separating string is tightened and
the lower parts of the upper warp sheet clamps are placed. The warp sheet coming
from the weaving machine is then prepared in the same way. After the upper parts
of the clamps have been installed, the knotter head can now be placed on its guides.
The lower and upper selectors (Fig. 68) separate an end from each warp sheet and
give it to a clamp-scissor that cuts the back ends of the ends and presents the 2 ends

Fig. 68 Schematic of a
knotting machine

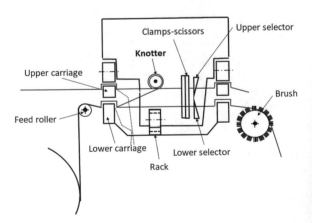

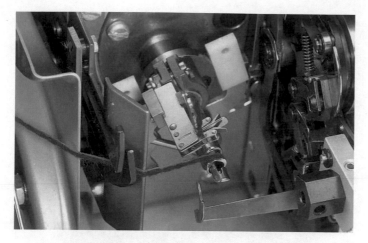

Fig. 69 Detail of a knotter (Courtesy of Stäubli)

clamped together to the knotter (Fig. 69). The ends are wrapped around the body of the knot. After being released from the clamp, the ends of the yarns are pulled by the knotter needle through the ends loop, thus forming the knot. The knotter moves one end and the cycle can start again.

Classical tying machines are able to tie any staple fibre or filament yarn in the range from 0.8 up to 500 tex (Nm 1250 – 2), making single or double knots as shown in Fig. 70 and requiring no elaborate resetting.

Specific warp tying machines are dedicated to technical textiles as shown in Fig. 71. These machines tie polypropylene ribbons with up to 5 mm width, aramid yarns, monofilaments up to a diameter of about 0.50 mm and coarse multifilaments as well as yarn counts ranging from 20 up to 2000 tex (Nm 50 – 0.50).

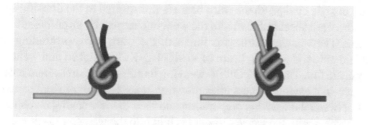

Fig. 70 Single knot and double knot (Courtesy of Stäubli)

Fig. 71 MAGMA warp tying machine for technical textiles (Courtesy of Stäubli)

5.3 Process Sequence Between the Warp Beam Storage and the Weaving Mill

The drawing-in or tying operations are one of the operations in the process flow from the weaver's beam storage to the weaving machine. Based on Stäubli Weaving Preparation Systems (WPS), Fig. 72 shows the process sequence between the warp beam storage and the weaving department.

In case of style change Ⓐ, the warp beam is transported to the drawing-in department and the warp sheet is drawn into the weaving harness by an automatic drawing-in machine. The weaving harness is finished, the warp yarns protruding from the reed are laid onto the welding beam of WARPLINK®, stretched and welded onto a sheet of plastic foil. The UNI-PORT® weaving harness transport system receives the entire drawn-in weaving harness after drawing-in for temporary storage and secure transport between the weaving preparation and then to the weaving machine. Newly drawn-in harness with the weaver's beam is laid into the weaving machine; weaving will start.

In case of warp changes Ⓑ, a new warp is transported from the warp beam storage to the weaving machine. The old warp beam is removed from the weaving machine and the new warp sheet is laid in. The old and new warp sheets are tied together by a warp tying machine and then the weaving process will restart.

In case of preliminary process requirements Ⓒ, the warp sheet is prepared to avoid any problems in automatic drawing in or tying operation. The yarn elasticity can be

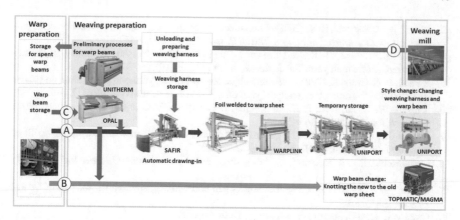

Fig. 72 Process sequence between warp beam storage and weaving department (courtesy of Stäubli)

thermally fixed by UNI-THERM®. The thermal process fixes warps with elastic or high-twist yarns. The yarns treated in this way can easily be automatically drawn-in or tied like any normal warp yarns. A 1:1 lease is inserted in warp sheets of fancy or single colour ends by an OPAL® leasing machine.

Return transport ⓓ, woven-out weaving harnesses and spent warp beams are transported back to the weaving preparation department and the warp beam storage for cleaning, prepping, storage and reuse.

References

1. Adanur, S. (2001). *Handbook of weaving*. CRC Press.
2. Ashok Kumar, L., & Senthilkumar, M. (2018). *Automation in textile machinery: Instrumentation and control system design principles*. CRC Press.
3. BeMiller, J., & Whistler, R. (2009). *Starch*. Academic Press.
4. Bihola, D. V., & Hiren, N. A. (2015). Classimat yarn faults. *International Journal of Advance Engineering and Research Development, 2*(2), 286–295.
5. Cotton Guide. (2020). *Cotton Guide Yarn Formation*. Retrieved from http://www.cottonguide.org/cotton-guide/cotton-value-addition-yarn-formation/.
6. Davidson, R. L. (1980). *Handbook of water soluble gums and resins*. McGraw-Hill.
7. Drean, J.-Y. (2019). *Weaving preparation [Lecture]*. ENSISA University of Mulhouse, Mulhouse, France.
8. Durur, G. (2000). *Cross Winding of Yarn Packages*, PhD thesis, School of Textiles Industries, The University of Leeds, UK
9. Eren, R., Suvari, F., & Celik, O. (2018). Mathematical analysis of motion control in sectional warping machines. *Textile Research Journal, 88*(2), 133–143.
10. Fernando, E, & Kuruppu, R. U. (2015). Tension variation in sectional warping, Part I: Mathematical modeling of yarn tension in a creel. *International Journal of Engineering and Advanced Technology, 4*(3), 158–163.
11. Goswami, B., Anandjiwala, R., & Hall, D. (2004). *Textile sizing*. Marcel Dekker.
12. Hallensleben, M., Fuss, R., & Mummy, F. (2015). Polyvinyl compounds Others. *Wiley on line library*. https://doi.org/10.1002/14356007.a21_743

13. Hari, P. K., & Beheraa, B. K. (2008). Computer controlled warp patterning on sectional warping machine. *Indian Journal of Fibre & Textile Research, 33*(9), 318–325.

14. Hari, P. K., Behera, B. K., & Dhawan, J. P. (1989). High pressure squeezing in sizing - performance of cotton yarn. *Textile Research Journal, 59*(10), 597.

15. Khaled, I., & Grutz, R. (2005). New technique for optimising yarn-end preparation on splicer, and a method for rating the quality of yarn-end. *AUTEX Research Journal, 5*(1), 1–9.

16. Kaushik, R. C. D., Hari, P. K., & Sharma, I. C. (1988). Mechanism of the Splice. *Textile Research Journal, 58*(5), 263–268.

17. Locher, H., & Ernst, H. (1971). Contrôle de qualité et surveillance de la propreté du fil en filature », Zellweger *Uster Uster News Bulletin*, n° 17, 15 p.

18. Luenenschloss, J., & Thierhoff, J. (1984). Yarn Forming. *International Textile Bulletin, 30*(4), 54.

19. Moreau, J. P. (1983). Comparison of sized open end and ring spun yarns. *Journal of Coated Fabrics, 13*(7), 12.

20. Oxtoby, E. (1987). *Spun yarn technology*. Butterworths.

21. Rebsamen, A. (1988). New possibilities of package building with microprocessors. *International Textile Bulletin, Issue, 3*, 18–32.

22. Sejri, N., Harzallah, O., Viallier, P., Ben Amar, S., & Ben Nasrallah, S. (2008). Influence of pre-wetting on the characteristics of a sized yarn. *Textile Research Journal, 78*(4), 326–335.

23. Sejri, N., Harzallah, O., Viallier, P., & Ben Nasrallah, S. (2010). Influence of wetting phenomenon on the characteristics of a sized yarn. *Textile Research Journal, 81*(3), 280–289.

24. Schutz, R. A. (1964). *Theoretical and practical aspects of sizing*. In J. B. Smith (Ed.), *Technology of Warp Sizing* (pp. 65–92). Columbine Press.

25. Schutz, R. A., Drean, J. Y., & Carrière, D. (1981), Characterization of yarns. *Textilveredlung, 16*, 9

26. Shaker, K. (2020). *Sample warping preparation [Lecture]*. National Textile University, Faisalabad, Pakistan.

27. Sparavigna, A., Brogliab, E., & Lugli, S. K. (2004). Beyond capacitive systems with optical measurements for yarn evenness evaluation. *Mechatronics, 14*, 1183–1196.

28. Turkawi, M. (2007). *Comportement mécanique des filés de coton encollés soumis aux sollicitations de traction à grande vitesse*, PhD thesis, Université de Haute Alsace, Mulhouse, France.

29. Uster. (2020). Retrieved from https://www.uster.com/fileadmin/user_upload/customer/customer/Instruments-use_folder_PRODUCTS/Broschueren/en_USTER_CLASSIMAT_5_web_brochure.pdf.

30. Whistler, R., BeMiller, J., & Paschall, E. (1984). *Starch: Chemistry and technology* (2nd ed.). Academic Press.

31. Xunxun, M., Shujia, l., Haiyan,Y., Shengze, W., & Yongxing, W. (2021). Identification of dynamic contact parameters between contact roller and filament package. *Textile Research Journal, 91*(21/22), 2430–2447.

Basic Notions to Design a Dobby Fabric

Francois Boussu

Abstract This chapter is dedicated to provide the basic knowledge to design a dobby fabric. A first part of this chapter will provide a brief classification of the different types of existing 2D and 3D fabric. Then, basic knowledge on the weave diagram theory and the different rules to model a fabric will be introduced. General definitions with examples of the different fundamental weave diagrams will be given. Some exercises will be proposed on the weave diagram construction with responses provided in the appendix. A second part of this chapter will be dedicated to the design of a dobby fabric with the necessary parameters as the peg plan, the drawing-in, the tie-up, treadling plan, warp and weft formulas. Some exercises will be proposed on the dobby fabric construction with responses provided in the appendix.

1 Introduction to 2D and 3D Fabrics

According to Laurence Camaro [1], there are different existing technologies allowing to produce textile architectures, whom properties depend only on the yarn orientation and their fibres which are they made. In general, textile surfaces correspond to constitutive yarns located in the fabric plan and 3D textile structures represent the location of yarns in the three main directions of the material. For each of these categories (Surface or 3D textiles), the yarns direction in the plane can be uni-axial, bi-axial or multi-axial.

Different 3D textiles can be obtained by weaving, braiding, knitting, non-woven processing and sometimes with a combination of these technologies. A general clustering of textile structures has been proposed by Nemoz [2] by using two main criterions as:

- The number of axis or yarns directions of the textile structure
- The number of handling directions of the yarns inside the textile structure

F. Boussu (✉)
ENSAIT, ULR 2461, GEMTEX–Laboratoire de Génie Et Matériaux Textiles, University Lille, 59000 Lille, France
e-mail: francois.boussu@ensait.fr

© Springer Nature Switzerland AG 2022
Y. Kyosev and F. Boussu (eds.), *Advanced Weaving Technology*,
https://doi.org/10.1007/978-3-030-91515-5_2

Thus, taking into account this clustering approach, a plain weave woven structure will be considered as a planar material (2D structure) with two different directions to handle the yarns inside the fabric plan (warp and weft yarns directions). The higher the number of yarns directions to handle, the more complex the machine is. As regard the woven textile structures, several existing and basic constructions are available to provide fabric thickness or additional shape in the fabric thickness direction. This list of fabrics, which is not exhaustive, is given as:

- 2D fabrics with diamond woven pattern
- 2D wrinkled woven fabrics
- 2D spiral woven fabrics
- 2D terry fabrics
- 2D velvet fabrics
- 2D double faces woven fabrics
- 2D tubular woven fabrics
- 2D double-wall fabrics
- 3D warp interlock fabrics
- 3D shape woven fabrics.

1.1 Fabrics with Diamond Woven Pattern

The weave diagrams with a diamond pattern as referenced in [3], allow to deform the woven structure in the perpendicular direction of the fabric, while woven flatly on the weaving loom with a constant tension on warp yarns. However, this deformation is due to the fabric relaxation store on the fabric roll by the differences of warp and weft shrinkages.

The obtained effect provided by the diamond woven pattern led to pyramidal shape after fabric relaxation, which means without any tensions on warp yarns [4] (Fig. 1).

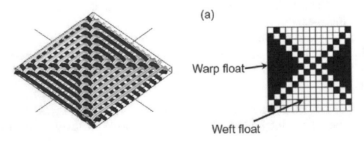

Fig. 1 (Left) 3D view of the woven fabric (middle) diamond pattern weave diagram (right) final aspect of the deformed 2D fabric

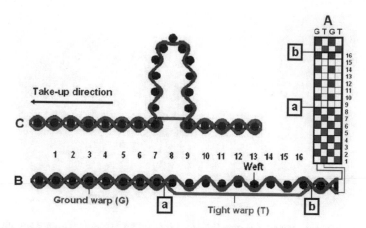

Fig. 2 **A** weave diagram of the wrinkled woven fabric—**B** difference of warp shrinkage values between ground and tight warp yarns—**C** wrinkles forming step [5]

1.2 2D Wrinkled Woven Fabrics

The 2D wrinkled woven fabrics allow to form wrinkles perpendicular to the fabric surface by combination of different weave diagrams for the odd and even warp yarns with very different warp shrinkages.

The Ground (G) warp yarns of the 2D wrinkled woven fabric have a higher warp shrinkage value than the Tight (T) warp yarns, as represented in Fig. 2A. This warp shrinkage difference allows to obtain a long float of tight warp yarns during the weft yarns insertion between locations "a" to "b", compared to the ground warp yarns, as represented in Fig. 2B. By combining a retreat of the ground warp yarns with a rear motion of their warp beam and the beat-up of the weaving reed on the fabric line, the woven wrinkle is perpendicularly formed to the fabric plan, as given in Fig. 2C.

1.3 2D Spiral Woven Fabrics

During the weaving process, the fabric roller has a conical shape to collect the fabric and tensions applied to warp yarns are manage independently. Then the obtained fabric has spiral shape as represented in Fig. 3.

1.4 2D Terry Fabrics

Terry cloth loops can be formed on one side or on two sides of the fabric. Two warp beams are necessary to produce terry fabrics. The first beam dedicated for the ground

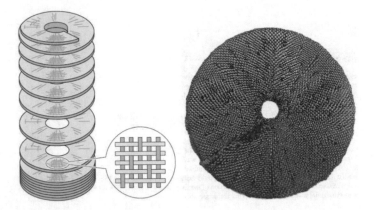

Fig. 3 (Left) 2D spiral weaving process [2]—(right) Resulted 2D spiral woven fabric [6]

warp yarns and the second beam to the loop warp yarns. A higher tension is applied to ground warp yarns and a looser one on the loop warp yarns.

The different phases of the one side loop forming of a terry fabric are described respectively in Fig. 4:

- Phase A: insertion of the first weft yarn followed by a first beat-up of the weaving reed to its intermediate position
- Phase B: insertion of the second weft yarn followed by a second beat-up of the weaving reed to its intermediate position
- Phase C: insertion of the third weft yarn
- Phase D: third beat-up of the weaving reed up to the intermediate position
- Phase E: forming of the terry cloth loops by beating-up until the fabric forming line

Fig. 4 One-side forming process of the terry cloth loops [5]

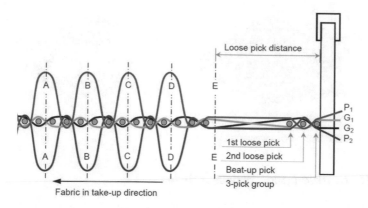

Fig. 5 Double-side forming process of the terry cloth loops [7]

The double side forming process of the terry cloth loops is similar to the one-side process, only by adding loop warp yarns on the two sides of the terry fabric, as represented in Fig. 5.

1.5 2D Velvet Fabrics

According to the weaving process of velvet fabrics by binding warp yarns, two parallel fabrics are woven simultaneously and maintained at a variable distance. The two fabrics are linked by pile warp yarns coming both from the two fabrics. Just after the weft insertion, a cutting device helps to separate the two fabric and the cut pile warp yarns contribute to the final aspect of the two 2D velvet fabrics, as represented in Fig. 6.

Fig. 6 2D velvet weaving process [2]

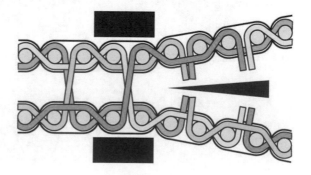

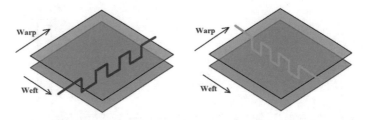

Fig. 7 2D double face fabrics weaving process

1.6 2D Double Faces Woven Fabrics

2D double faces woven fabrics can be produced by linking the two sides in the warp yarns direction (Fig. 7—left) or in the weft yarns direction (Fig. 7—right).

1.7 2D Tubular Woven Fabrics

2D tubular woven fabrics can be produced without edges on ribbon weaving loom or continuously on large width fabric weaving loom, as represented in Fig. 8.

For instance, the four main warp yarns evolutions to produce a 2D tubular woven structure based on a plain weave diagram are represented in Fig. 9.

Fig. 8 (Left) Single tube woven fabric—(right) multi-tubes woven fabric [6]

Fig. 9 Evolutions of warp yarns to produce 2D tubular woven fabric

Fig. 10 (Left) continuous [8] and (right) non-continuous [6] binding warp yarns insertion into the double-wall fabrics

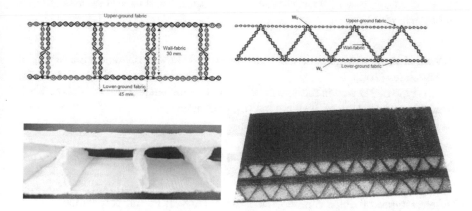

Fig. 11 (Left) perpendicular or (right) inclined woven structures between the two walls of the 2D double-wall fabrics [9]

1.8 2D Double-Wall Fabrics

Based on the 2D velvet weaving process, if the cutting device is removed and the distance between the two fabrics adjustable to a given height, then 2D double-wall fabrics can be produced, as represented in Fig. 10.

By combining the double-wall weaving process with the warp yarns sliding device of the 2D wrinkled woven fabrics process, double wall woven structure can produce perpendicular or inclined woven structure between the two walls, as represented in Fig. 11.

1.9 3D Warp Interlock Fabrics

A 3D warp interlock fabric is composed of several layers of 2D fabrics linked through the thickness by a binding warp yarn as represented in Fig. 12.

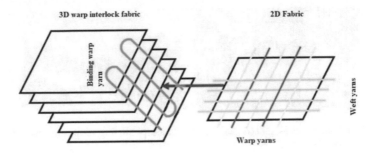

Fig. 12 Geometric view of 3D multi-layers woven structure linked through the thickness by a binding warp yarn

Several yarns arrangements can be achieved and allow to obtain a wide variety of 3D structures [10].

This type of 3D woven fabric will be detailed in the chapter of the book entitled: "Definition and design of 3D warp interlock fabric".

1.10 3D Shape Woven Fabrics

Different types of shape can be woven, as: T, H, X, I, Pi and perpendicular tubes shapes, on specific and adapted 3D weaving looms (Fig. 13). The thickness of these 3D shapes can vary to several millimetres to centimetres.

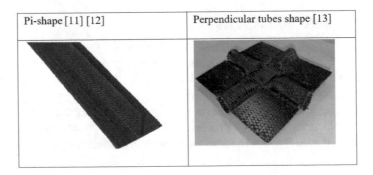

Pi-shape [11] [12]	Perpendicular tubes shape [13]

Fig. 13 Different shape types of 3D woven fabrics

1.11   Weave Diagram Theory

The weave diagram of 2D fabric corresponds to the elementary pattern of warp and weft yarns interlacement that can be replicate all along the fabric width and length. One of the major interest of the weave diagram theory consists in the introduction of geometrical modelling with simple rules. Thanks to this geometrical theory, every weavers of the world can communicate by exchanging simple diagrams [11–13].

   To understand the basic notions of the weave diagram theory, a step-by-step methodology is explained to model a fabric from an optical observation. Then, based on this theory fundamental weave diagrams as plain weave, twill weave and satin weave will be defined and their construction mode explained.

1.12   Observation of a Woven Structure

Three main directions can be used to observe a fabric and the different crosslinking points between warp and weft yarns, as represented in Fig. 14. The perpendicular direction to the fabric plan allows to observe the different warp and weft interlacements and the resulted weave diagram. However, direction views inside the fabric thickness may be helpful to locally follow the evolution of a yarn through the woven structure.

Fig. 14 Mode d'observation d'un tissu [14]

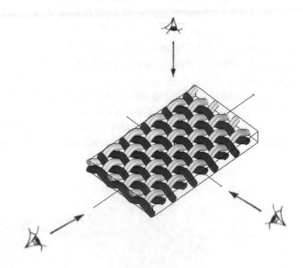

1.13   Geometrical Modelling of 2D Woven Fabrics

Basis of the weave theory diagram lies on three main rules:

1. The warp yarn evolution will be considered in priority than the weft yarn evolution
2. At a crosslinking point between a warp yarn and a weft yarn; two positions can be found as:

    a. When the warp yarn is above the weft yarn then it's consider as a UP position and the intersection between the warp and weft yarns of the geometrical model will be filled
    b. When the warp yarn is below the weft yarn then it's consider as a DOWN position and the intersection between the warp and weft yarns of the geometrical model will be empty.

3. The weave diagram corresponds then to the elementary woven pattern that can be repeated in both warp and weft directions.

To apply these essential rules, let consider the following step-by-step methodology to model the 2D fabrics.

First step: Based on the observation of the fabric and the numbering of warp and weft yarns, a table with 10 columns (warp yarns) and 6 rows (weft yarns) can be draw as represented in Fig. 15.

Second step: Isolate the first warp yarns, check each of this position with weft yarns and apply the basic rules of the weave diagram theory to find the resulted geometric model as represented in Fig. 16.

Based on the geometric modelling of the first warp yarn, by isolating the second warp, the same approach can be applied to obtain the resulted geometric model as given in Fig. 17.

Third step: by isolating each warp yarns, build the final geometric model of the observed fabric, as represented in Fig. 18.

Fourth step: Recognize the elementary woven pattern that can be repeated in all the fabric directions and isolate the resulted weave diagram as represented in Fig. 19.

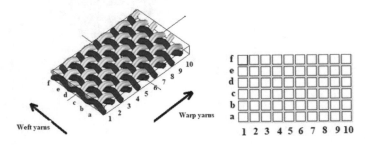

Fig. 15 (Left) Numbering of warp and weft yarns of the 2D woven fabric—(right) corresponding table of the observed fabric

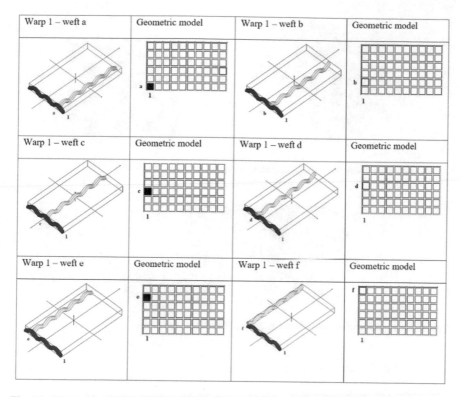

Fig. 16 Observation and modelling of the 1st warp yarn position with each weft yarns

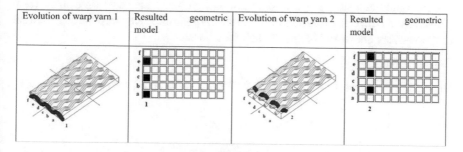

Fig. 17 Geometric modelling of the evolution of the 1st and 2nd warp yarns

1.14 Fundamental Weave Diagrams

Fundamental weave diagrams are the basis for all the other types of existing weave diagrams and are composed of: the plain weave, the twill weave and the satin weave.

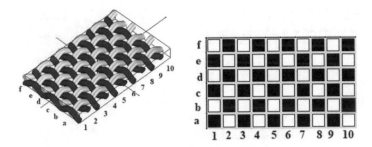

Fig. 18 Representation of the final geometric model of the observed fabric

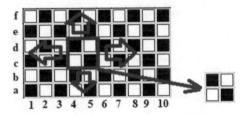

Fig. 19 Recognition of the final weave diagram of the observed fabric

1.14.1 Plain Weave Diagram

Description

It's an unique and simple weave diagram and probably the most used. Odd warp yarns have the inverse evolution of the even warp yarns. Each warp yarns evolution is alternatively above and below of each weft yarns.

Construction

Dimension of the weave diagram R = 2 (2 warp yarns and 2 weft yarns) (Fig. 20).

Fig. 20 Plain weave geometric model and the associated 3D view

1.14.2 Twill Weave Diagram

Description

The twill weave diagram is composed of float warp yarns or float weft yarns linked by a crosslinking point with a shift value of 1. The twill weave diagram is defined by its own: dimension R, (warp of weft) effect and (left or right) shift sense.

Construction

Let's consider the dimension R of the twill weave diagram.

The warp effect means that there will be (R-1) UP and 1 DOWN positions of the 1st warp yarn with all the weft yarns.

The weft effect means that there will be 1 UP and (R-1) DOWN positions of the 1st warp yarn with all the weft yarns.

Left shift corresponds to a displacement of 1 step of the crosslinking points of one warp yarn of the weave diagram inversely to the sense of the weft direction.

Right shift corresponds to a displacement of 1 step of the crosslinking points of one warp yarn of the weave diagram in the sense of the weft direction.

Twill 3

To better understand these construction rules of a twill weave diagram, let's consider the following definition: Twill 3 warp effect right shift.

Step 1. Twill 3 means 3 warp and 3 weft yarns for the weave diagram as R = 3.

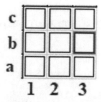

Step 2. Warp effect for warp yarn 1 means (R-1) 2 up warp yarns with weft yarns a and b and 1 down warp yarn with weft yarn c.

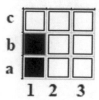

Step 3. The will weave diagram implies of shift value of 1 with a right direction which means in the sense of weft yarns insertion direction. Then, the crosslinking points of the warp yarn 1 are slided onto the warp yarn 2 with a shift value of 1.

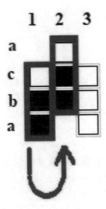

Step 4. The crosslinking point between warp yarn 1 and weft yarn a is integrated into the weave diagram.

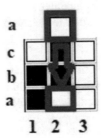

Step 5. The crosslinking points of the warp yarn 2 are slided onto the warp yarn 3 with a shift value of 1.

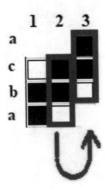

Fig. 21 Twill 3 weave geometric model and the associated 3D views

Twill 3	Warp effect	Weft effect
Left shift		
Right shift		

Step 6. The crosslinking point between warp yarn 3 and weft yarn a is integrated into the weave diagram.

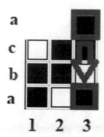

By applying these general rules to the twill 3 (R = 3), then 4 main different types of twill 3 can be considered as represented in Fig. 21.

1.14.3 Satin Weave Diagram

Description

The satin weave diagram helps to avoid any inverse interlacements of warp and weft yarns as for the plain weave diagram and reduce any straight diagonal effect as for the twill weave diagram. The shift values of the satin are determined by a specific rule, except for the satin 4 and satin 6 which have irregular shift values.

Construction

The satin weave diagram is defined by its own: dimension R, (warp of weft) effect and different shift values.

For the warp effect, the shift of the crosslinking point is done in the warp yarns direction.

For the weft effect, the shift of the crosslinking point is done in the weft yarns direction.

Regular shift values of a satin weave diagram of dimension R, can't be 1, R, R-1 and having no common multiplier.

For instance, the different affordable shift values of satin weave diagram are listed in Table 1.

Weave diagrams satin 4 and 6 have irregular shift values.

The order of shift values for satin 4 is: $+1; +2; -1; -2$.

The order of shift values for satin 6 is: $+2; +2; +3; -2; -2; -3$.

Satin 4

By applying these general rules to the satin 4 (R = 4), then 2 main different types of satin 4 can be considered as represented in Fig. 22.

Table 1 Regular shift values of satin weave diagrams

Satin	Shift values
4	irregular
5	1 2 3 4 5
6	irregular
7	1 2 3 4 5 6 7
8	1 2 3 4 5 6 7 8
9	1 2 3 4 5 6 7 8 9
10	1 2 3 4 5 6 7 8 9 10
11	1 2 3 4 5 , 6 7 8 9 10 11
12	1 2 3 4 5 6 7 8 9 10 11 12
13	1 2 3 4 5 6 7 8 9 10 11 12 13

Fig. 22 Satin 4 weave geometric model and the associated 3D views

Satin 4	Warp effect	Weft effect
	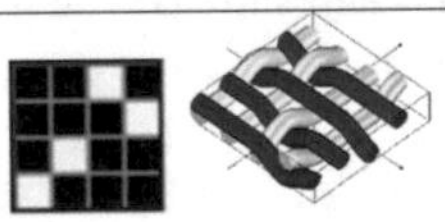	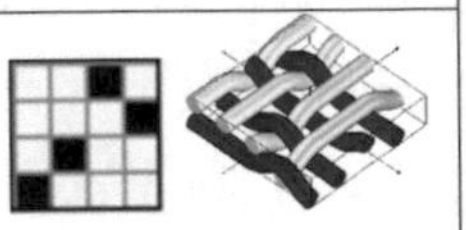

Fig. 23 Satin 5 weave geometric model and the associated 3D views

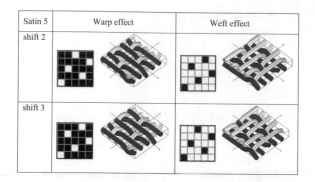

Fig. 24 Satin 6 weave geometric model and the associated 3D views

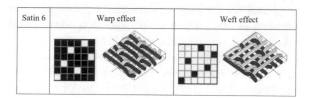

Satin 5

By applying these general rules to the satin 5 (R = 5), then 4 main different types of satin 5 can be considered with two effects (warp or weft) and two shift values (2 or 3) as represented in Fig. 23.

Satin 6

By applying these general rules to the satin 6 (R = 6), then 2 main different types of satin 6 can be considered as represented in Fig. 24.

2 Design of a Dobby Fabric

The main difference between a dobby and a Jacquard fabric lies on the selection warp yarns types. For the dobby fabric, the different warp yarns evolutions are restricted to the available number of shafts on the weaving loom. For the Jacquard fabric, the different warp yarns evolutions are restricted to the number of available hooks of the Jacquard machine attributed to the woven pattern.

Fig. 25 Warping formulae

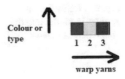

2.1 Description of Elementary Parts to Design a Dobby Fabric

To design a dobby fabric, several different necessary parts are needed. The design approach can be done into two different ways. For the first one, the necessary elementary items are: the warping formula, the drawing-in plan, the tie-up plan, the treadles plan and the weft formula. This approach allows to stick the weave diagram of the fabric into the tie-up plan, which helps to easy recover it. For the second one, the necessary elementary items are: the warping formula, the drawing-in plan, the peg plan and the weft formula. This approach allows to directly identify the peg plan of the dobby fabric which helps to drive the dobby weaving loom.

Let's consider the definition of all these necessary items to combine them and provide the final dobby woven pattern.

2.1.1 Warping Formulae

The warping formulae corresponds to the minimum combination of colour or type of warp yarns.

For instance, the warp formulae represented in Fig. 25 can be defined by the different colours of the warp yarns as: 1 blue 1 white 1 red;

2.1.2 Weft Formulae

The weft formulae corresponds to the minimum combination of colour or type of weft yarns.

For instance, the weft formulae represented in Fig. 26 can be defined by the different colours of the weft yarns as: 1 green 1 yellow.

Fig. 26 Weft formulae

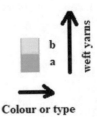

Fig. 27 Drawing-in plan

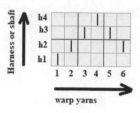

Fig. 28 Tie-up plan

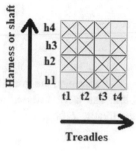

2.1.3 Drawing-in Plan

The drawing-in plane corresponds to the geometric model of the drawing-in operation which consist in inserting each warp yarn into the heedle eye of the heedle located in one of the dobby shaft (or harness).

For instance, the drawing-in plan represented in Fig. 27 is named: herringbone drawing-in plan. The warp yarn 1 is inserted into the heedle of the shaft 1, the warp yarn 2 is inserted into the heedle of the shaft 2, etc....

2.1.4 The Tie-up Plan

The tie-up plan corresponds to the link between the treadles and the shafts of the dobby weaving machine. Nowadays, treadles have been replaced on the loom by different types of dobby machines [15]. But, from the design point of view, it helps to directly see the weave diagram(s) used for the dobby woven fabric.

For instance, the tie-up plan represented in Fig. 28 corresponds to the Twill 4 warp effect right shift weave diagram.

2.1.5 The Treadling Plan

The treadling plan helps to select the different combinations of shafts defined in the tie-up plan according to the weft yarns.

Fig. 29 Treadling plan

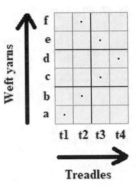

For instance, the treadling plan defined in Fig. 29 is named: herringbone treadling plan. The weft yarn a allows to select the treadle 1, the weft yarn b allows to select the treadle 2, etc.…

2.1.6 The Peg Plan

The peg plan links the weft yarn selection to the combination of selected shafts of the dobby weaving machine. Then, at each inserted weft yarn, the selection of shafts will be lifted or not, which corresponds to the main data driving the dobby weaving loom.

For instance, based on the peg plan represented in Fig. 30, when the weft yarn a is selected then the shafts h2, h3 and h4 are selected to be lifted and shaft h3 is not selected and not lifted to contribute to the shedding motion.

The peg plan can also been deduced from the combination of the treadling plan and the tie-up plan as represented in Fig. 31.

Fig. 30 Peg plan

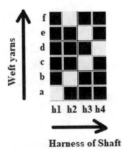

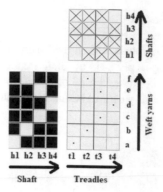

Fig. 31 Construction of peg plan based on treadling and tie-up plans

2.2 Design of Dobby Fabrics [16]

2.2.1 General Dobby Woven Pattern

To represent the general dobby woven pattern, two approaches can be applied:

- The first approach consists in using the following necessary items as: the warp formulae, the drawing-in plan, the tie-up plan, the treadling plan and the weft formulae, represented in Fig. 32 (left),
- The second approach consists in using the following necessary items as: the warp formulae, the drawing-in plan, the peg plan and the weft formulae, represented in Fig. 32 (right).

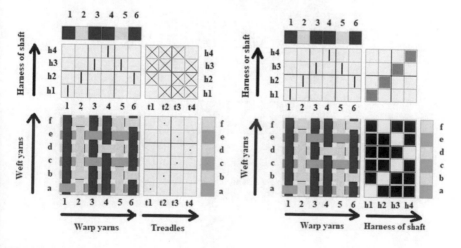

Fig. 32 (Left) dobby fabric with treadling and tie-up plans—(right) dobby fabric with peg plan

2.2.2 Methodology of Design

To understand how a dobby fabric is designed using the different items as: warp and weft formulas, drawing-in plan, tie-up plan, treadling plan and peg plan, let's consider a step-by-step methodology based on 4 steps.

Step 1. Selection of the weft yarn a to be inserted in the shed of the dobby woven fabric.

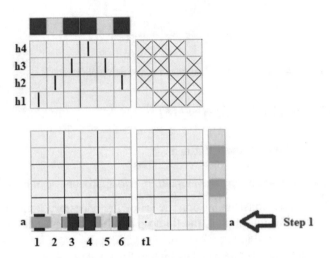

Step 2: Selection of treadle t1 according to the given treadling plan.

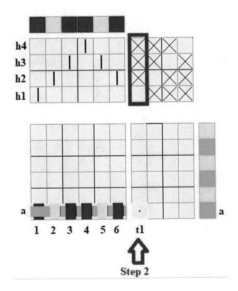

Step 3: Selection of shafts h2, h3 and h4 according to the given tie-up plan.

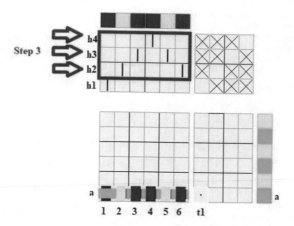

Step 4: Selection of warp yarns 2, 3, 4, 5, 6 to be lifted and warp yarn 1 to not be lifted according to the given drawing plan.

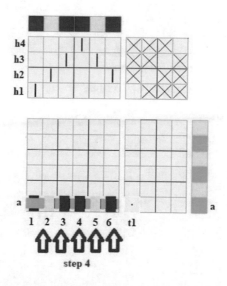

step 4

Based on this 4 steps methodology to design a dobby woven fabric, it could be possible to apply it for each of weft yarns and revealed the general dobby woven pattern as described in Fig. 33.

If we consider only the drawing-in, tie-up and treadling plans, the general woven pattern of the dobby fabric can be represented by the basic rules of the weave diagram theory as given in Fig. 34.

Design of woven pattern of dobby fabrics lies in combinations of N drawing-in plan × P tie-up plan × Q treadling plan, which can offer a total N × P × Q solutions. Design of dobby fabric lies in combinations of N drawing-in plan × P tie-up plan ×

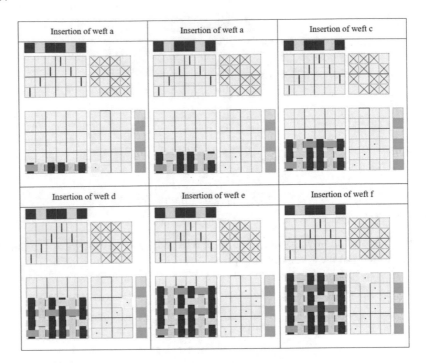

Fig. 33 Step-by-step design of the dobby woven fabric

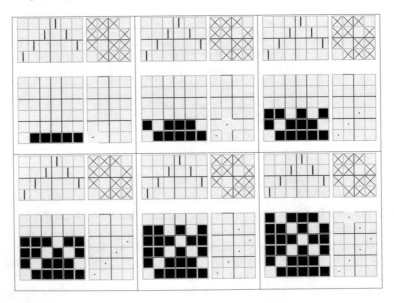

Fig. 34 Step-by-step design of the woven pattern of the dobby fabric

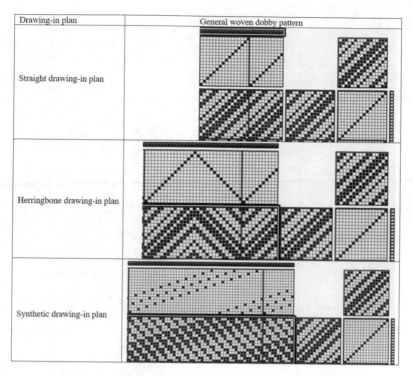

Fig. 35 Design of dobby woven pattern based on the same tie-up and treadling plans and variations of drawing-in plan

Q treadling plan × R warp formulae × S weft formulae, which can offer a total N × P × Q × R × S solutions.

2.2.3 Dobby Fabrics with Different Drawing-in Plans

Considering the same tie-up and treadling plans and variation of drawing-in plans, it can be possible to obtain different design of woven pattern or dobby fabric (considering basic warp and weft formulas) as represented in Fig. 35.

2.2.4 Dobby Fabrics with Different Treadling or Peg Plans

Considering the same tie-up and drawing-in plans and variation of treadling or peg plans, it can be possible to obtain different design of woven pattern or dobby fabric (considering basic warp and weft formulas) as represented in Fig. 36.

Treadling plan	General woven dobby pattern
Straight treadling plan	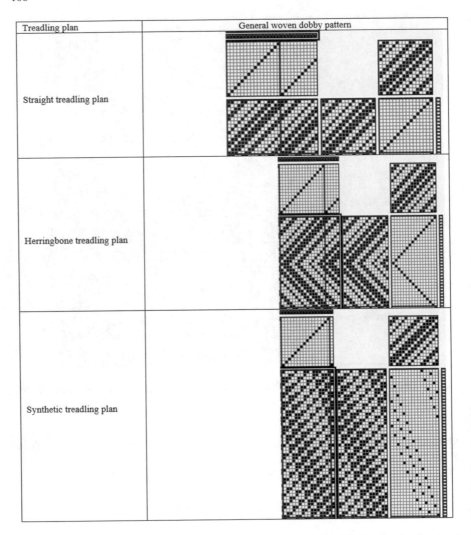
Herringbone treadling plan	
Synthetic treadling plan	

Fig. 36 Design of dobby woven pattern based on the same tie-up and drawing-in plans and variations of the treadling plan

2.2.5 Dobby Fabrics with Different Drawing-in and Peg Plans

By combining different drawing-in and treadling plans while keeping the same tie-up plan, different general dobby woven pattern with the resulted dobby fabric can be observed in Fig. 37. Indeed, these two dobby fabrics have the same initial weave diagram in the tie-up plan.

combination	General woven dobby pattern	Resulted dobby fabrics
Herringbon drawing-in and treadling plans		
Synthetic drawing-in and treadling plans		

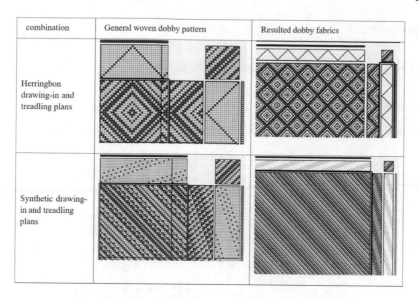

Fig. 37 Combinations of different drawing-in and treadling plans with the same tie-up plan

3 Conclusion

Different types of existing fabrics can lead to 3D shape in the thickness direction.

Then, an introduction to the weave diagram theory has provided the different basic and essential rules to get the geometric model of any fabric based on its observation. Three fundamentals weave diagram has been defined and explained.

Then, two different design methods have been introduced for the dobby woven fabric. The main parameters of the dobby fabric as: the warp and weft formulas, the drawing-in plan, the tie-up plan, the treadling plan and the peg plan, have been also defined and explained.

By combining the different parameters, several dobby fabrics have been designed.

Recommendations

All the geometric 3D view of woven patterns have been prepared using the software: Wisetex © [17].

All the general weave patterns of the dobby fabrics have been designed with the CAD weaving softwares: DBweave © [18] and Pointcarré © [19].

Appendices

Exercises on Weave Diagrams Construction

Problem 1

Based on the following fabric, find the elementary woven pattern and model it.

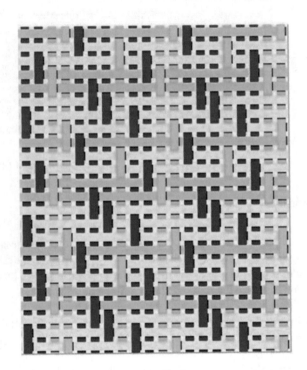

Problem 2

Find the different weave diagram of the Twill 5.

	Warp effect	Weft effect
Right shift		
Left shift		

Problem 3

Find the different weave diagram of the Satin 8.

	Warp effect	Weft effect
Shift value 1		
Shift value 2		

Exercises on Dobby Woven Fabric

Problem 4

Based on the different warp and weft formulas, drawing-in plan, tie-up plan and treadling plan, find the general woven fabric.

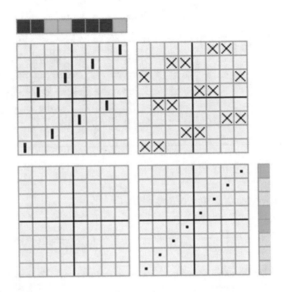

Problem 5

Based on the different warp and weft formulas, drawing-in plan and peg plan, find the general woven fabric.

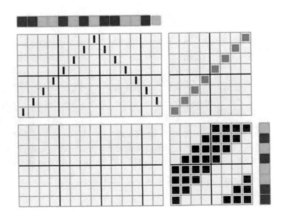

Solutions on Weave Diagrams Construction

Solution 1

The general woven fabric is:

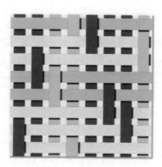

The general woven pattern is:

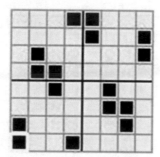

Solution 2

The different weave diagrams of the Twill 5 are:

	Warp effect	Weft effect
Right shift	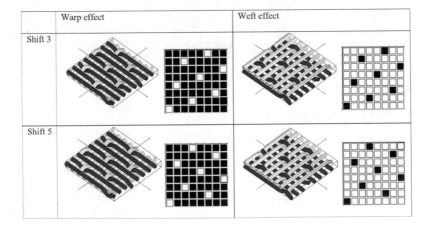	
Left shift		

Solution 3

Based on the rules of construction for a satin weave diagram, the regular shift values of a satin weave diagram of dimension R, can't be 1, R, R-1 and having no common multiplier.

The different regular shift values for satin $8 = 2^3$ (R = 8) are:

1	$2 = 2 \times 1$	3	$4 = 2 \times 2$	5	$6 = 2 \times 3$	7	8
1	Common multiplier	OK	Common multiplier	OK	Common multiplier	R-1	R

The different weave diagrams of the satin 8 are:

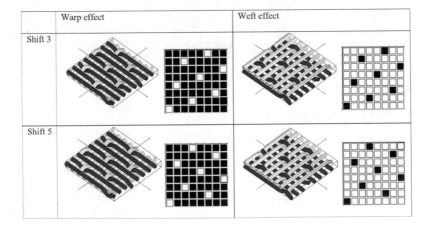

Solutions on Dobby Woven Fabric

Solution 4

The general dobby woven fabric is:

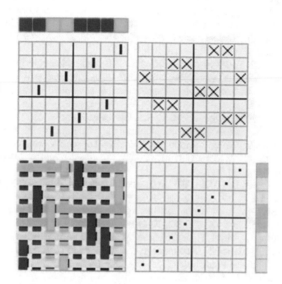

Solution 5

The general dobby woven fabric is:

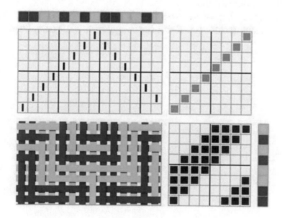

References

1. Camaro, L. (2006, Avril). Textiles à usage technique. *Techniques de l'ingénieur, 2,* (511), 1–8.
2. Nemoz, G. (2003, Octobre). Textures textiles tridimensionnelles. *Techniques de l'ingénieur,* 5(122), 1–16.
3. Delommez, A., & Popelier, E. (1951). Atlas d'armures Textiles, Roubaix: Duporge.
4. Xiao, X., Hua, T., Wang, J., Li, L., & Au, W. (2015). Transfer and mechanical behavior of three-dimensional honeycomb fabric. *Textile Research Journal, 85*(12), 1281–1292.
5. Kienbaum, M. (1996). *"Drehergewebe, Faltengewebe, Florgewebe und Jacquardgewebe", in Bindungstechnik der Gewebe.* Schiele & Schön.
6. Amirul Islam, M. (2015). 3D woven preforms for E-textiles and composites reinforcements. In Advances in 3D Textiles, X. Chen, (Ed.), (pp. 207–264). Cambridge: Woodhead Publishing.
7. Grosicki, Z. (1989). *Watson's advanced textile design, compound woven structures* (4th ed.). Butterworth & co. Ltd.
8. Girmes Company. (2003). Product-catalogues. Germany.
9. Badawi, S. (2007, November 06). Development of the weaving machine and 3D woven spacer fabric structures for lightweight composites materials. Dresden, Germany.
10. Tong, L., Mouritz, A., & Bannister, M. (2002). *3D Fibre reinforced composite materials* (p. 241). Elsevier Applied Science.
11. Schmidt, R., & Kalser, D. (2005). Woven preform for structural joints. TX, USA Patent US 6 874 543 B2, 5 April.
12. Schmidt, R., Bersuch, L., Benson, R., & Islam, A. (2004). Three dimensional weave architecture. USA Patent 6 712 099 B2, 30 March.
13. Boussu, F., Dufour, C., Veyet, F., & Lefebvre, M. (2015). Weaving processes for composites manufacture. In P. Boisse (Ed.), *Advances in composites manufacturing and processes - Part 1* (pp. 55–78). Woodhead Publishing.
14. Lord, P., & Mohamed, M. (1973). *Weaving, conversion of yarn to fabric* (1st ed.). Publishing, Merrow.
15. Trend-setting technology for the textile industry. Retrieved August 14, 2019 from https://www.staubli.com/en/textile/textile-machinery-solutions/.
16. Roussel, F. (1996). Variations d'armures par le rentrage,» Industrie Textile, n°1273, Février.
17. WISETEX SUITE. Retrieved August 14, 2019 from https://www.mtm.kuleuven.be/Onderzoek/Composites/software/wisetex.
18. DB-WEAVE. Retrieved August 14, 2019 from https://www.brunoldsoftware.ch/dbw.html.
19. Pointcarre CAD weaving software. Retrieved August 14, 2019 from https://www.pointcarre.com/.

Shedding Principles and Mechanisms

Mathieu Decrette and Jean-Yves Drean

Abstract This chapter will focus on shedding principles and mechanisms which are used for achieving shedding. This step is essential for the weaving process as it is necessary for weft insertion. Thanks to frame motion order and design resulting from it, shedding is also primordial for the woven fabric integrity. Shedding principle consisting in separating warp yarns into two layers, it leaves an empty space used for pick insertion. As a consequence, an opened shed may be considered as a geometric shape composed of the strained warp yarns. The parameters of shed geometry, their setting as well as their influence on yarn strain and stress will be discussed. Another part is concerning the several shedding mechanisms available: tappet, cam, dobby and Jacquard systems. In each case, a presentation of the mechanism principle is made on the base of original or basic examples; classification is made according to operation parameters, modern mechanisms composition and operation are precisely described and discussed and finally practical exercises are proposed all along the chapter.

Keywords Shed · Shedding mechanism · Cam · Dobby · Jacquard

1 Introduction

1.1 Shedding Principle

Shedding may be seen as one of the major steps during a weaving cycle. Each warp yarn being held in the eyelet of a heddle, the rising or lowering of groups of heddles (frame shedding) or single heddles (Jacquard shedding) separates warp yarns. Once

M. Decrette (✉)
Textile Physics and Mechanics Laboratory, ENSISA Textile Engineering School, Université de Haute Alsace, Mulhouse, France
e-mail: mathieu.decrette@uha.fr

J.-Y. Drean
Université de Haute Alsace, Laboratoire de Physique et Mécanique Textiles UR4365, Ecole Nationale Supérieure d'Ingénieurs Sud Alsace, F-68200 Mulhouse, France

© Springer Nature Switzerland AG 2022
Y. Kyosev and F. Boussu (eds.), *Advanced Weaving Technology*,
https://doi.org/10.1007/978-3-030-91515-5_3

weft insertion (filling) has been performed while the shed is opened, shedding will occur again while reed beat-up with a usually new configuration of warp yarns arrangement.

Shedding: constraints and issues.

Shedding step and its parameters settings are essential for weaving quality. As the main objective of shedding is enabling the fastest and easiest passage of the weft carrier, several conditions need to be met:

– prevent carrier slowing down,
– lead weft carrier,
– shed sufficiently kept opened (opening rest time + geometry) for the carrier passage,
– limitation of frames acceleration,
– Enable individual frame position and movement amplitude (shed geometry).

During shed opening, geometrical configuration tends to cause yarn strain, which is intrinsically linked to tension according to yarns mechanical properties. As a consequence, strain and tension are crucial parameters which will need to be adjusted according to woven yarns structural and material behaviour.

1.2 Introduction on Warp Tension [9]

It is well known that the several yarn structures available and textile materials can have very different responses to the solicitation caused by weaving. Solicitations on tension variation are very varied [10], this is why focus will only be made on warp geometry influence. Figure 1 is the graph of a yarn quasi-static tensile test for a constant rate of

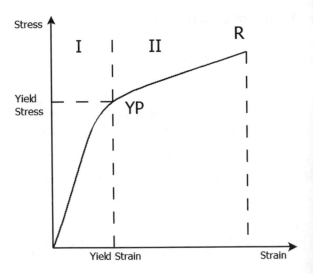

Fig. 1 Typical stress–strain curve of yarn main textile materials

elongation. On ordinates is placed the yarn stress response to solicitation according to its elongation in the abscissa. It may be observed in this example that a first part of the curve is quasi-linear: an initial material modulus can be found to describe material behaviour to minor solicitations. Then, as solicitations increase, the curve is reorienting to get to another behaviour. Most of common textile materials have such behaviour with a visco-elastic area (I) where elongation can be recovered and a visco-elasto-plastic behaviour area with unrecoverable strain. The interface between these two behaviours is called Yield Point (with corresponding Yield stress and Yield strain coordinates). Considering that materials may have different Yield coordinates and initial modulus, the woven material behaviour needs to be well-known in order to be sufficiently far from Yield strain and prevent yarn damages from permanent and repeated extensions.

This is the reason why, without reaching yarns breakages or permanent deformation, an excess of tension has to be avoided as it may already have several disadvantages:

on weaving process:

- increase of warp breakages,
- cause of warp defects,
- reduction of warp shrinkage and cause of an increase of weft shrinkage, with potential tearing near templets;

on fabric parameters:

- unbalanced fabric tension,
- fabric width and area density changes,
- visual defects,
- loss of fabric mechanical properties and fashionability.

Nevertheless, tension needs to be sufficient and well balanced in order:

- to separate warp yarns during shedding,
- to enable weft beating,
- to hold drop wires,
- to enable warp beam unwinding,
- to ensure the awaited pick density.

1.3   Shed Definition and Woven Fabric Construction

Warp is composed of parallel yarns (ends) usually winded on a beam at the back of the loom and will be winded on a fabric beam at the front of the loom after weaving process. Warp yarns are set horizontal thanks to the yarn carrier (back rest) and the breast beam (front rest). According to Fig. 2, any warp end passes through a drop wire for break detection, a heddle used for yarns raising or lowering and a reed dent for pick beat-up. Thanks to heddles motion, the shed is generated between warp stop motion (drop wires location) and cloth fell as the warp is divided into two layers

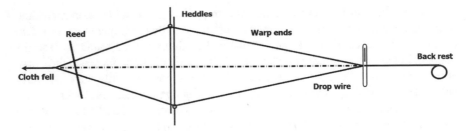

Fig. 2 Schematic illustration of the shed, and loom elements related to the shed

which lare eaving an empty space for carrier passage. In frame weaving, heddles are gathered in frames which raise some groups of ends while the rest is lowered.

In order to lock weft yarns after their insertion, the shed needs to be closed and opened in another configuration. As the alternation of frames raising or lowering generates a visual pattern on the surface of the fabric (case a) in Fig. 3), frames motion order needs to be set beforehand for obtaining the adequate and expected mechanical, hand and visual properties of the fabric.

This visual effect on the fabrics comes from the arrangement between warp and weft yarns. The pattern may be translated into a black or white squares table (case b) in Fig. 3). For flat standard fabrics, the colour of the square represents the single warp end (black square) or single weft end (white square) appearing on a precise weave surface location. As most of shedding technologies are meant for frame weaving with a limitation in pick repeat, a fabric design is generally composed of repeated pattern units. The smallest repeat unit which will be sufficient for the entire fabric description will be called the weave pattern (case c) in Fig. 3). As the fabric is represented as it is woven from weaver side, warp end number should be red from left to right and weft yarn order from bottom to top. As a result, in case c), first warp end will be hidden

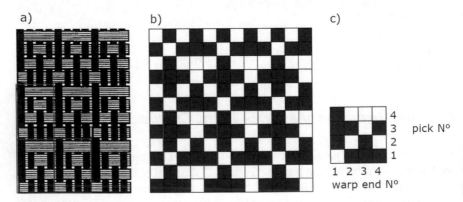

Fig. 3 Translation of a fabric into weave pattern: **a** Fabric, **b** Schematic representation of the fabric, **c** corresponding weave pattern

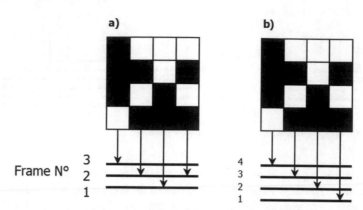

Fig. 4 Straight drawing-in for the Fig. 3 example pattern: **a** drawing-in draft (DID) for 3 frames, **b** DID for 4 frames

under first weft at the first pick and will be then over the three next weft yarns for pick 2, 3 and 4.

Once the weave pattern has been defined from the desired fabric, drawing-in, reed plan and chain draft parameters will need to be defined so that the fabric may be woven.

As each warp end should be drawn in a single heddle of a frame, the drawing-in draft (Fig. 4) is usually placed under the pattern and is composed of horizontal lines figuring frames and vertical arrows indicating in which frame is drawn in each warp yarn. Consequently, drawing-in repeat (number of vertical arrows) is at least the same as warp pattern repeat. Several pattern repeats may sometimes be drawn in only one drawing-in draft in order to spread tension over more frames and reduce the tension carried by each frame. On the other hand, the number of horizontal lines corresponds to the number of frames used, in the limit of the number of frames available. Draft may also be represented on the base of weave pattern principle with black and white squares.

Figure 4 shows two drawing-in drafts possibilities for the Fig. 3 example. Although warp repeat is four ends, only three frames may be enough as warp ends 2 and 4 have the same evolution. Nevertheless, case a) might cause warp defects due to tension dissimilarity between frames: as frame 2 would drive pair warp yarns, its tension would be twice higher than the two other frames. This is why a second straight draft with four frames in case b) may be chosen for a similar load raising. Such an example indicates that a careful draft should be set up. Figure 5 displays several basic and standard drawing-in draft possibilities. More complex drafts can derivate from one basic draft or a combination [1]. Straight draft (case a)) is set by drawing-in yarns in the order of frames. It is quite commonly used because it is very easy for yarn practical drawing-in, however it might cause high strain and tension difference between the first yarn and last yarn of the repeat. In case of weft effect fabric weaving, straight pattern can easily be used: potential visual warp defects can be insignificant as warp yarns are hidden by the majority weft yarns. Skip draft (case b)) is carried

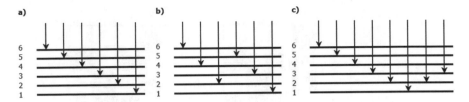

Fig. 5 Basic drawing-in drafts: **a** straight, **b** skip, **c** pointed

out by drawing-in yarns in one out of every two frames. Once odd frames have been drawn-in, even frames will be drawn-in. Pointed draft (case c)) is typically used for multiple pattern repeat drawing-in or weave patterns with multiple ends with same evolution. Besides and easy setting, this type of draft, when applicable, is ideal in terms of tension as tension difference can be constant between each warp end all over the fabric.

After frames drawing-in comes reed plan definition. Warp ends are successively drawn-in reed dents by group of two or more. Drawing in only a single yarn in a reed dent would be impossible as it would require a much larger reed than harness due to dents thickness. Moreover, reed dents should never be filled with a single yarn in order to prevent warp ribbing defects. As shown in Fig. 6, yarns grouped in a single dent is represented by linking the vertical arrows of corresponding yarns. the number of yarns in consecutive reed dents may be constant (or not) according to the pattern in order to increase its visual effect. Concrete parameters may also restrict the number of ends in a single dent such as yarn diameter or count, or width of the dent to ensure easy shedding.

The final step is the chain draft definition, which is the translation of the weave pattern into frames raising/lowering order for the successive picks according to drawing-in draft. As illustrated in the example in Fig. 7a, frames horizontal lines

Fig. 6 Drawing-in draft +
Reed plan for an example
pattern: constant reeding of 2
yarns/reed dent

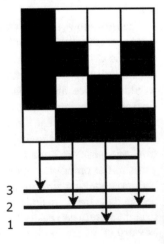

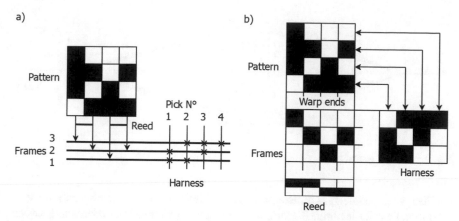

Fig. 7 Full definition of frames motion for pattern weaving: pattern, drawing-in draft, reed plan and chain draft, two methods of representation (**a**) and (**b**)

are prolonged and will cross vertical lines which represent picks. Marks are placed on intersections to figure frames raising whereas empty intersections are lowered frames. In the same way as drawing-in draft, chain draft representation may also be based on the weave pattern representation with black and white squares (Fig. 7b).

1.4 Shed Geometry

Shed geometry is an essential weaving parameter as it has a great influence on warp tension fluctuation during shed and thus, may have many repercussions on fabric quality (tension homogeneity, clean opening and untangling of yarns, yarns breakages, process fluidness). It may usually be set thanks to clamping loops position (Fig. 27) which link frames transmission components to shedding mechanisms lever. Basic shed geometries may be classified as "clear" / "unclear" [7] or "even" / "uneven" [3] but a more precise classification can be established:

– Clear shed (Fig. 8)

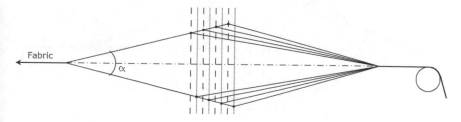

Fig. 8 Clear shed geometry

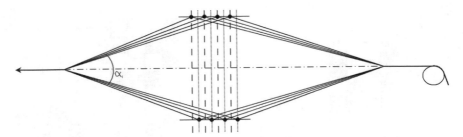

Fig. 9 Flat shed geometry

Clear shed geometry is quite easy to set and is the ideal shed configuration for an optimal carrier leading. A main disadvantage is yarn tension difference between first and last frames, especially for a high amount of frames. Even if the angle α is just sufficient for carrier, back frames amplitude will need to be very high in order to keep a constant angle α value with a risk of overstraining yarns drawn in back frames.

- Flat shed geometry (Fig. 9)
  Flat shed is easy to set as the opening amplitude is constant for all the frames; tension homogeneity is better than clear shed. Nevertheless, it is also called unclear shed as shed angle α is different for every frame, which implies a poorer carrier leading. The smallest shed angle α depends on the back frame (which has to be enough for carrier passage) whereas the front frames angles are very high, which may impact weaving quality and yarns wear. An optimal shed amplitude needs to be considered in order:

  - to ensure a sufficiently high angle from back frame for carrier crossing,
  - to ensure a sufficiently low angle from front frame for reasonable yarn strain and angle in heddles.

- Elliptic shed geometry (Fig. 10)
  Even if an elliptic geometry is the most difficult to set, it is the optimal solution for weaving quality as it guaranties a homogeneous yarn sstrain and tension

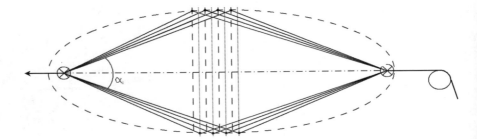

Fig. 10 Elliptic shed geometry

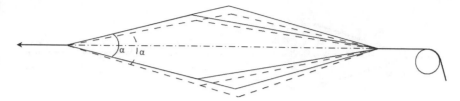

Fig. 11 Shifted clear shed geometry

between any frame. Although elliptic shed belongs to unclear shed family and may hamper carrier crossing, such a geometry should ensure a perfect fabric stability.

- Offset shed geometry
  Possibly compatible for any of the previous geometries (clear, flat, elliptic), an offset may also be set in such a way that the whole of the frames form two slightly shifted shed in the amplitude (shed layers are separated) Fig. 11 shows an example of shed offset. The main advantage is a better and easier yarns crossing during shedding, although carrier guiding may be drastically worsened especially for unclear sheds.

Any of these basic shed geometries have an influence on the previous conditions (carrier leading) and parameters (yarn strain, fabric quality, warp visual defects…). Consequently, geometry and ensuing tension or stress needs to be set carefully according to yarn mechanical properties. On the understanding that yarn length increase for an opening of shed, yarn stress is the result of yarn strain due to geometry change.

As depicted in Fig. 12, which shows a symmetrical opened shed geometry and for a given frame, the horizontal dotted line is the zero line. The zero line is corresponding to a closed shed and which length is $b = AB$ between cloth fell A and warp stop motion B. As a frame rises a yarn to C giving a shed half amplitude $h = CD$, yarn strain may be written as expressed in Eq. 1.

Equation 1 General expression of warp yarn strain.

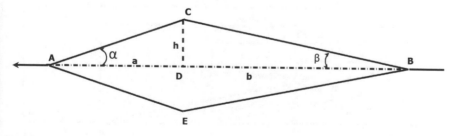

Fig. 12 Warp shed symmetrical geometry

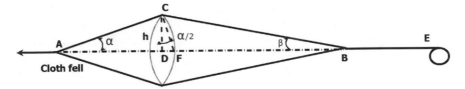

Fig. 13 Geometrical shed configuration for small α and β angles

$$\text{Strain} = AC + AD - AB = \sqrt{a^2 + h^2} + \sqrt{(b-a)^2 + h^2} - b \qquad (1)$$

Equation 1 is a general expression of shed half amplitude. An increase of shed amplitude leaves a larger space for carrier. However, in a given **a** and **b** configuration, the increase of amplitude **h** would lead to a significant rise of yarn strain and ensuing stress. The parameter **a** may also have a significant influence. Considering a constant shed angle, a smaller distance **a = AD** between cloth fell and the first frame allows a lower frame raising or lowering **h** (Fig. 13).

Equation 2 Relation between half shed amplitude **h** and shed angle α.

$$h = a \cdot \tan\alpha \qquad (2)$$

On the opposite and considering a constant shed amplitude **h**, reduced value of **a** and **b** will involve an increase of angles α and β and therefore a drastic increase of strain value as it will be shown on the following example in Fig. 15.

An approximated expression of strain may also be defined with the help of Fig. 13 for small α and β respectively front and rear shed angles.

An approximation of Strain may be defined as: $A = \delta_1 + \delta_1$,

With $\delta_1 = AC - AD = DF = h \cdot \tan\frac{\alpha}{2}$ and similarly $\delta_2 = h \cdot \tan\frac{\beta}{2}$

For small values of α and β: $\delta_1 = h \cdot \tan\frac{\alpha}{2} = \frac{1}{2} \cdot \tan\alpha$, and $\delta_2 = h \cdot \tan\frac{\beta}{2} = \frac{1}{2} \cdot \tan\beta$ which gives the final relation in Eq. 3:

Equation 3. Approximate expression of strain for small shed angles.

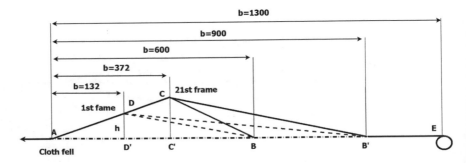

Fig. 14 Shed geometry example for a 20 frames loom

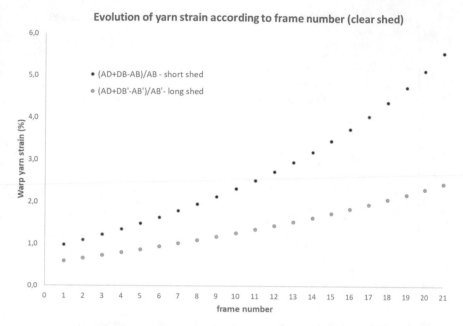

Fig. 15 Evolution of strain (%) according to frame number. first frame position from cloth fell = 132 mm, clear shed angle $\alpha = 14.9°$, shed short length = 600 mm (black curve), shed long length = 900 mm (grey curve)

$$A = \delta_1 + \delta_2 = \frac{h^2}{2} \cdot \frac{b}{a \cdot (b-a)} \tag{3}$$

Equation 3 shows that strain behave as a parabolic function of shed amplitude. It should be used theoretically for small shed angles but may be employed for common shed amplitude value.

Figure 15 is an example of a shed geometry of a 21 frames loom. Clear shed was chosen for this example as it may be the most commonly used and moreover strain variations are quite noticeable. In close shed position, 132 mm separate cloth fell from front frame and 372 mm for last frame with a 12 mm pitch. Total shed length may also be set from 900 mm (AB) to 900 mm (AB') between cloth fell and warp stop motion. Based on this geometry, graph of Figs. 16, 17 and 18 describe the evolution of warp yarn strain (%) according to changes in loom and shed geometrical configuration. It has to be noticed that this example is theoretical and is a result of calculations based on Eq. 1 applied in perfect conditions. This is why such an example should be comprehended as a simple illustration and that experimental reproduction of this example may not have as much significant trends.

Figure 16 shows the graph for the evolution of relative strain according to frame number, calculated thanks to Eq. 1, with several fixed parameters. Here $\alpha = 14.9°$ which consequence is a regular increase of shed half amplitude **h** frame by frame, the first frame is placed at 132 mm from cloth fell with a 12 mm pitch. The black

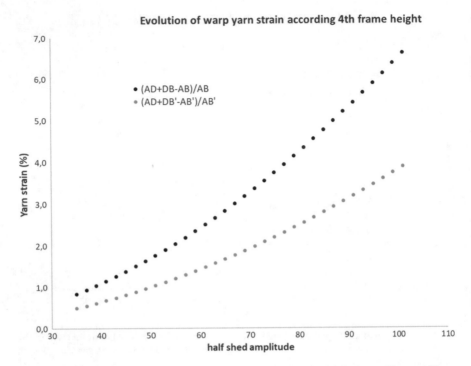

Fig. 16 Evolution of yarn strain according to the 4th frame height. 4th frame position $= 164$ mm from cloth fell, clear shed angle $\alpha = 14.9°$, shed short length $= 600$ mm (black curve), shed long length $= 900$ mm (grey curve)

curve shows the evolution for a 600 mm shed length and the grey for a 900 mm shed length. Both curves have a parabolic behaviour as tend to show Eq. 3. The 900 mm shed length case has a lower and more homogeneous value of yarn strain, as the increase of length tend to reduce the value of rear shed angle β. On the opposite a short shed may cause high strain and stress to yarns drawn in back frames.

The graph in Fig. 17 considers yarns relative strain of 4th frame according to shed amplitude for short shed length (black curve) and long shed length (grey curve). Despite the parabolic evolution of strain according to shed height, strain variation may be considered as close to linear for standard height values. The same trend as in Fig. 16 may observed for the comparison of shed length.

Figure 18 shows that strain reduces as shed length increases. Shed length increase as little influence on yarn strain for front frames as they are placed near cloth fell and it has no influence on front angle α. However, it may be noticed that whatever the length of shed, there is a significant gap (0.9% to 1.4% in this example) between front a rear frames. It should be taken account for drawing-in draft.

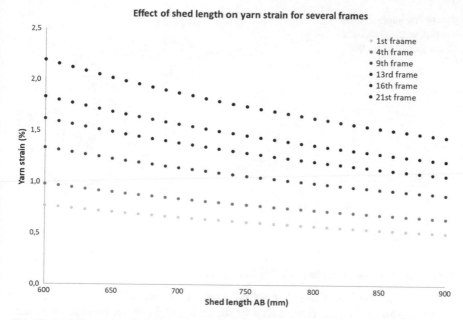

Fig. 17 Evolution of strain for several frames according to total shed length. Clear shed angle α = 14.9°, curve gradation from light grey to black are corresponding to frame position, respectively from front to back frames

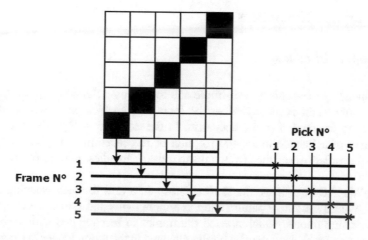

Fig. 18 Full combination of weave pattern, drawing-in-draft reed plan, and chain draft for a 5 picks and ends twill

Fig. 19 Example of tappet
weaving mechanism for a
height picks repeat plain.
[12]

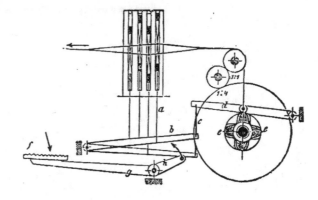

1.5   Exercise

A drawing-in draft example is visible in Fig. 19.

- Is this Drawing-in draft an ideal solution for such a pattern?
- What drawing-in draft could better fit for an optimal tension variation between yarns?

2   Shedding Mechanisms

2.1   Tappet Shedding

Tappet shedding mechanisms are simple and lower-cost solutions to perform shedding. A crank shaft is integrated to the loom and enables basic weave patterns according to the motion of the eccentric. In the example in Fig. 19 of the crank mechanism of a Lærson loom, the rotation of the eccentrics (e) gives a vertical motion of treadles (d) and then (b) thanks to rods (c). treadles (b) raise frames thanks to rods (a).

For plain and derivatives, the shaft rotates half a turn for each weaving cycles. Previously used for longer pattern such as simple twills and satins, the crank shaft was only rising frames, which needed the frames to be equipped with springs for lowering. As a crank shaft may be hardly changed in the loom, longer basic patterns are now woven thanks to external cam mechanisms which offer flexibility for pattern changes.

This type of mechanism is highly reliable and robust and can perform high-speed weaving, this is why positive tappet weaving is now used only for plain weave with some water and air jet modern looms such as Tsudakoma's ZW8100 or Toyota's LWT810 water jet looms.

3 Cam Motions

As well as crank mechanisms, cam mechanism both transfer movements to frame and contain weave pattern information. Cams are place on a shaft which gets its rotation from the main loom shaft. They are also simple and inexpensive solutions for high speed basic patterns weaving [6].

The cam shaft may be either part of the loom or a separate mechanism next to the loom. Such as crank mechanism, cam mechanisms integrated to the loom are fast, precise, powerful and highly reliable. Nevertheless, it can only be used for small basic patterns with a maximum of 5 cams and five picks patterns. The geometrical and mechanical reasons for repeat limitations have been discussed [8]. Owing to a low accessibility, cams may also be difficult or impossible to change. Such devices are still used for some water- or air-jet looms such as Toyota's JAT810 air-jet loom.

External cam mechanisms are widely used for modern looms as they have several advantages compared to internal mechanisms:

- good accessibility,
- easier cams changes,
- larger patterns (up to 10 lifting units and 6 picks report).

3.1 Cam Mechanisms Classification and Cams Design

Cam mechanisms may be classified according to its negative or positive movement principle.

Negative devices might be used for air-jet or wat-jet loom for basic pattern weaving. In such mechanisms, frames are raised or lowered thanks to cams, which imply the use of sets of springs for return movement. Even if they are less expensive than positive mechanisms, they are less and less used for modern loom due to the difficulty to proceed to any pattern change (springs tension release and loading).

Cam design is essential and may be classified according to four characteristics:

- Cam track for movement transmission

  A single cam track can be considered as typical of negative cam mechanisms. The cam track can push the follower, but a pressure has to be exerted on the follower (thanks to return springs for instance), to be sure that the contact between the cam track and the follower is maintained.

  In positive systems, cams may be either machined grooved, or composed of a pair of matching surfaces cams. In both cases cams raise and lowers thanks to the action of the cam or the pair of cams.

  In grooved cams mechanisms, such as the Hartmann system example in Fig. 20, the follower and corresponding arm (d) motion is generated by the guiding of the follower in the groove of the cam (g). The action of both surfaces of the groove enable frames (a) either raising or lowering (Fig. 21).

Fig. 20 Example of grooved
cam shedding mechanism.
[12]

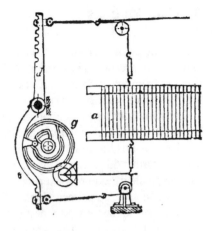

Fig. 21 Schematic
illustration of positive
mechanism with a pair of
matched cams

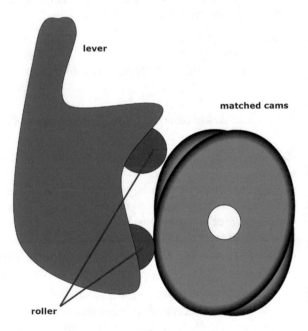

The illustration 21 represents a second type of positive cam mechanism. It is
composed of a pair of matched cams. A set of two rollers are kept in contact with
cams running surface; each roller is in contact with a dedicated cam. As a results,
the pair of cams rotation enables lever oscillation and frames movement.

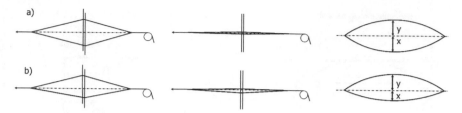

Fig. 22 Effect on shed geometry of: **a** symmetrical cams, **b** asymmetrical cams

- Symmetrical or asymmetrical cams

  Pair of matching tracks may either be asymmetrical (frames lowering and raising cinematic asymmetrical) or complementary to create a symmetrical shed geometry between lowered and raised yarns, as depicted in Fig. 22 where in case (a): $\frac{y}{x} = 1$ and in case (b): $\frac{y}{x} < 1$.

  For a given pattern, a symmetrical cam may be either used for a weft or warp effect (ex. $\frac{1}{4}$ or $\frac{4}{1}$) as it may be used on both sides thanks to their complementary surfaces.

  As an asymmetrical cam can only manage a single pattern, it has to be set on its correct side on the cam shaft. (ex. for $\frac{1}{4}$ or a $\frac{4}{1}$ pattern, two different cams will need to be purchased).

  The choice of a symmetric or asymmetric cam is generally ordered by the woven material. Symmetrical cams can be used for synthetic tows of filaments whereas asymmetrical cams would better fit staple yarns as strain solicitation are lessen.

- Cam profile

  Cam profile is initially and usually defined by the machinery constructor according to the loom and the product woven, in order to leave a maximum amount of time for carrier shed crossing. Some of the characteristics which need to be taken in account are:

  - the width of the loom
  - the carrier bulk
  - the type of material woven (staple yarns or filaments).

As a result, shed movement will be shorter for a higher symmetrical profile, causing a higher yarn movement acceleration, as shown in Fig. 23a and Table 1 which presents Stäubli symmetrical cams (called P*) characteristics. A higher proportion of shed opening is reached thanks to high profile cams (ex. 68.75% for a P3 cam) at 60° loom cycle. Such an acceleration can be helpful for staple yarns hairiness separation but may harm more sensitive yarns on the longer term.

Essential effects of Stäubli AL* asymmetrical cams is the displacement of cross shed height. A higher asymmetry of AL* cams tends to lower cross shed position in the lower part of the shed, as illustrated in Fig. 23b. Based on Fig. 22 geometry illustration, asymmetry is defined as: Asymmetry $= \frac{y}{x}$. A consequence of the

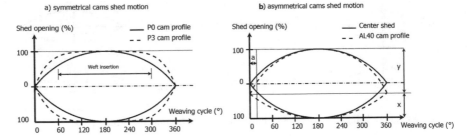

Fig. 23 Influence on shed motion of: **a** P* type symmetrical cams, **b** AL* type asymmetrical cams

Table 1 Effect of P* series Stäubli cams on shed opening proportion at 60° loom cycle

P* cam profile	Shed opening value (%) at 60° weaving cycle
P 0	50
P 1	60
P 2	65
P 3	68.75

lowered crossing position is the displacement of the centre shed crossing time, with a higher AL* profile giving a higher phase difference (angle a in Fig. 23b) for the 0% position, as listed in Table 2.

– Weft repeat
  Recent cams size and profile may be 2, 3, 4, 5 or 6 picks repeat. Figure 24 shows 3 examples of 4, 5 and 6 picks repeat symmetrical cams reversely used for warp or weft effect.
  More generally, as cam mechanisms are meant to be employed for basic and fast weaving, mechanisms are now limited to 6 picks repeat and 10 lifting units for heavy devices. More complex fabrics should be woven with more reliability thanks to dobby mechanisms.

Table 2 Effect of AL* series Stäubli cams on displacement of the cross shed position and the displacement of the shed centre yarn reaching. Cam ordering according to Asymmetry value

AL* cam profile	Asymmetry A/B	Angle a (°) at 0% shed	Displacement of cross shed (%)
P* (as reference)	1	0	0
AL 10	1.12	8	2.83
AL 20	1.41	17	5.80
AL 60	1.74	21	13.51
AL 40	1.98	30	15.44
AL 70	2.64	32	22.50
AL 50	2.73	40	23.20

Fig. 24 Examples of symmetrical cams for a: **a** 4 picks repeat pattern, **b** 5 picks repeat pattern and **c** 6 picks repeat pattern. (Courtesy of Stäubli)

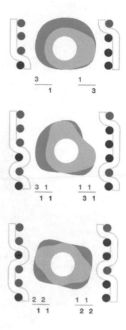

Cam mechanism type	Housing	Pick report	Loom
S1600	8 lifting units 12 mm pitch	2–6 picks	Rapier, airjet
S1700	10 lifting units 12 mm pith	2–6 picks	Rapier, airjet

Table 3 Stäubli cam mechanisms range

The only external mechanisms left on western markets are Stäbuli's positive cam mechanisms based on matched cam driving. Their general cam systems range is presented in Table 3.

3.2 External Cam Mechanisms Description

Figure 25 shows the parts constituting an external cam mechanism which will be precisely described owing to its widespread use.

The composition and operation of an external positive cam mechanism is as follows:

- the drive shaft brings movement to mechanism,
- the cam or lever shaft is fitted with cams and enables their rotation,
- cams rotation is moving levers and frames according to their profile,

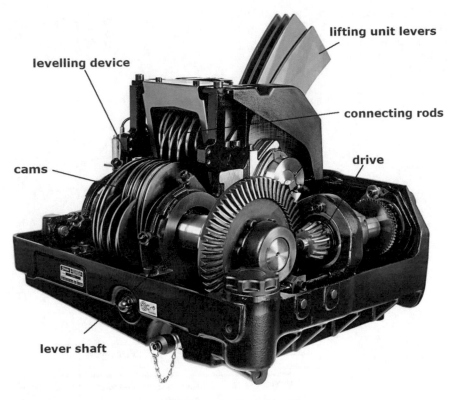

Fig. 25 178* cams mechanism main parts. (Courtesy of Stäubli)

- cams followers (or connecting rods) follow cams track and transmits movement to levers,
- levers of lifting units are linked to transmission components for frame lifting or lowering thanks to clamping loops (Fig. 27),
- the levelling device for shed closing (loom stop or cam shaft change) as shown in Fig. 26.

3.3 Cam Indexing

Setting a new weave pattern on the loom should begin with cam changes. After the levelling of frames has been made and the cam shaft removed from cam mechanism, a new shaft fitted with proper cams will be prepared thanks to an indexing plate (Fig. 28) and placed back into the cam mechanism.

According to Fig. 28, cams (3) are placed in the right order onto the cam shaft (1) and its spacer (used to enhance tightening and prevent cam slipping) with the cam weave reference facing the plate (4) or the user according to the indexing code. A

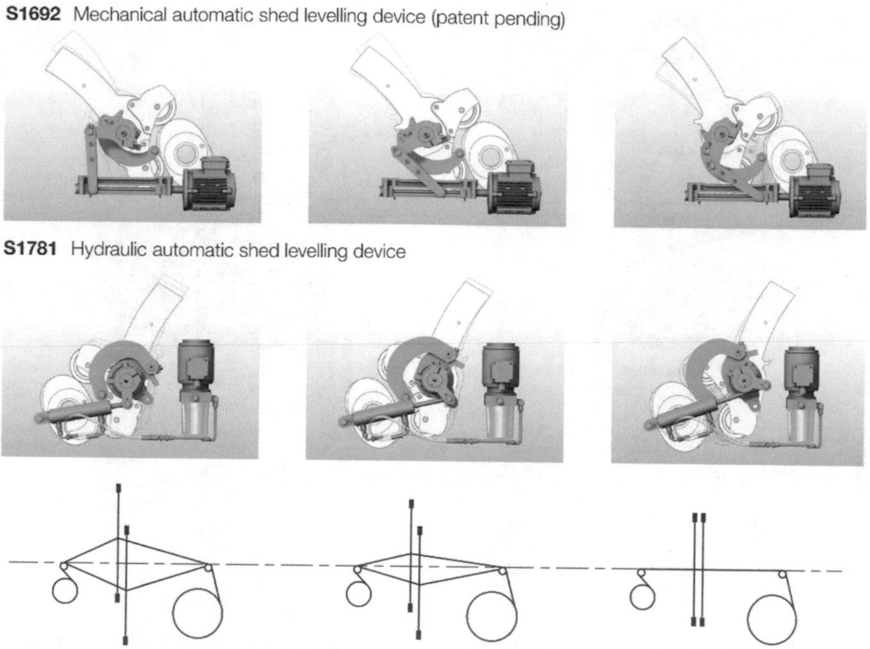

S1692 Mechanical automatic shed levelling device (patent pending)

S1781 Hydraulic automatic shed levelling device

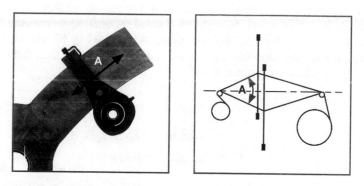

Fig. 26 Mechanical and Hydraulic levelling devices. (courtesy of Stäubli)

Fig. 27 Clamping loop position and effect on shed amplitude. (Courtesy of Stäubli)

wrong side positioning of cams would have a limited impact for symmetrical parts as the chosen pattern would be on the reverse side of the fabric. For the asymmetrical cams case, the side with reference needs necessarily to be positioned in the correct position. Then an index (5) is placed into the cam notch and positioned before (clockwise rotation) the rod (2). The amount of rods placed on the plate is equal to pick repeat. A stagger (or shed phase difference) may be set between cams in order to start shedding earlier or later between cams thanks to specific indexes. Stagger can be set from 5° to 30° with steps of 5°.

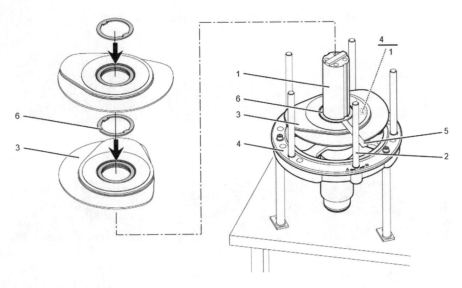

Fig. 28 Indexing plate. (Courtesy of Stäubli)

The indexing code, which is essential for cam side and index positioning, is defined by the setting of the mechanism. Available settings, chosen by loom constructors, are depicted in Fig. 29.

Cam mechanism can be either placed on the left or right of the loom, or on weaver or beam side, which influence direction of rotation (A or B) as well as the warp or weft effect (1 or 2), giving four different codes A1, A2, B1 or B2. As such a code depends of the loom constructor choice for setting up, the indexing code may be found on the cam motion's nameplate.

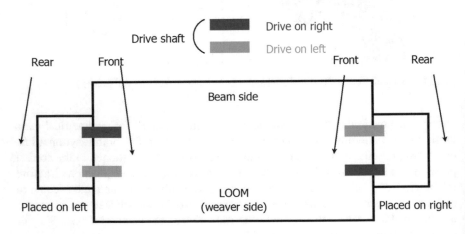

Fig. 29 Available establishments of a cam mechanism

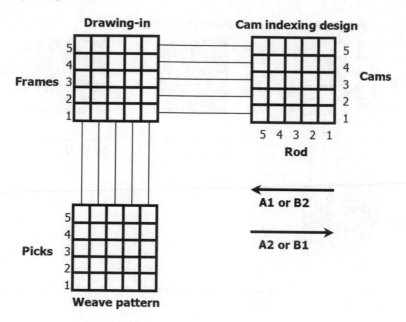

Fig. 30 Basic cam indexing design

Cam indexing can then be made thanks to the indexing design in Fig. 30, based on the same model as chain draft definition.

As illustrated in Fig. 30, after weave pattern and drawing-in have been transferred on the corresponding designs, the cam indexing design will be filled. The rod for index positioning will then be defined by researching and marking the first lift on cam profile thanks the direction A1/B2 or A2/B1.

Then the choice of cams is made thanks to its family (symmetrical or asymmetrical), pick repeat and the calculation of the ratio $R = \frac{\text{pick repeat}}{\text{number of movements}}$. The reference of the cam begins with the longest lift float.

In this example, five identical cams will be used, with a 5 pick repeat and $R = \frac{5}{2} = 2.5$, giving a $\frac{4}{1}$ cam reference.

Cam reference will face the indexing plate in case of a A1 or B1 code and face the user otherwise.

According to Fig. 31, order of rods for index positioning will be:

– 2, 4, 1, 3, 5 in case of a A1/B2 indexing code,
– 5, 2, 4, 1, 3 in case of a A2/B1 indexing code.

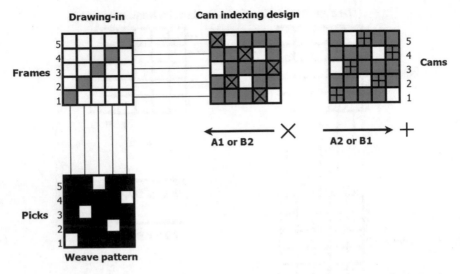

Fig. 31 Example of cam indexing with a 5 picks warp effect Satin and straight draft. Notice that in some cases, all the rods might not be used

Fig. 32 Exercise pattern

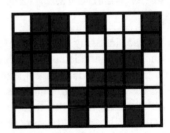

3.4 Exercise

Find the following elements: pick repeat, cam ratio R, cam identification, indexing order of rods for index positioning. Indexing code is B1, weaving pattern in shown in Fig. 32 and drawing-in draft in Fig. 33.

4 Dobbies

4.1 Principle

A dobby mechanism has the same objective as the previous shedding mechanisms, which is shed opening for weft insertion. For this reason, it needs to meet the specification described in the introduction part.

Fig. 33 Exercise drawing-in draft

The major difference and advantage of dobbies is the separation between power mechanism (frame movement) and weave design information (selection mechanism), which enables high speed weaving, fast pattern change and moreover quasi-infinite weave pattern length [6].

The characteristics for dobbies use are:

- a very good accessibility (over or next to the loom),
- easy and fast change of weave pattern (USB, network or wifi data transmission),
- medium or high speed weaving (depending on dobby type),
- more expensive than other frame shedding solutions,
- recent standard housing can be up to 24 lifting units with 12 mm pitch or 30 lifting units in 16 mm pitch.

As dobbies are originally later than Jacquard systems, they were initially placed over the frames part of the loom and later next to the loom. Its mechanical movement transmission and synchronization comes from the loom and pattern reading from punched card technology. A basic description of composition can be made with the help of Fig. 34:

- a weave pattern reading mechanism (1) composed of needles, which movement selects hooks (2) for frames lifting (or lowering) according to the red information (3),
- a set of hooks, associated to frames, which can be selected by knives (4),
- a main movement (5) mechanism (not described here) which enable translation movement of hooks,
- Transmission parts (6) which link hooks to frames (7) for their motion.

This original and general principle has been kept until today, with optimization and either complexification or simplification according to new dobbies designs and weaving market needs (weaving speed, reliability, evolution of technology, woven products diversification).

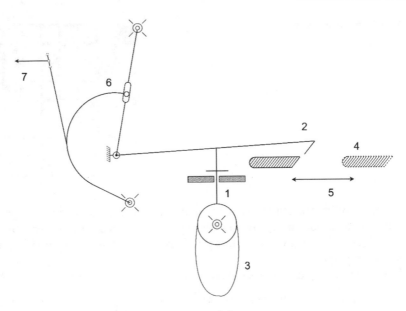

Fig. 34 Description of dobby basic movements and elements

4.2 Classification

Dobby mechanisms may be classified according to three characteristics:

- Single lift or double lift dobbies:
  This characteristic concerns information reading/selection part, as originally in single lift systems, one pick was red, whereas in double lift dobbies two reading hooks + needles and dedicated knives where dedicated for each frame, one hook for odd picks and the other for even picks. Thus, the two knives successively change frames positions (according to hooks selection) pick per pick. The consequence is that dobby speed becomes half of loom speed, excluding potential speed limitation of dobbies for high speed loom weaving. Owing to the higher performances of double lift dobbies, only the later are now commercialized.
- Close shed or opened shed dobbies:
  Close shed systems involve a movement back to the zero shed line for every pick even in case of warp floats (two or more consecutive high positions). As shows Fig. 35a, yarn raised or lowered in order to change from one open shed layer to another have a classical motion (R: raised or L: lowered). On the other hand, as depicted in Fig. 35a, warp yarns kept in up (WF: warp float) or down (HWF: hidden warp float, as on reverse side of fabric) position reach the zero shed line for each shedding before coming back to initial position. Figure 35b describes yarn's cinematic during 5 shedding cycles:

  - pick 0 to 1: top shed line to bottom shed line,
  - pick 1 to 2: down shed line to top shed line,

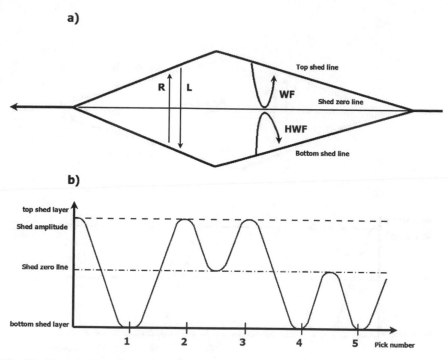

Fig. 35 Close shed **a** principle and **b** time diagram

- pick 2 to 3: top shed line, reaches shed zero line at 0° shedding time and gets back to top shed line,
- pick 3 to 4: top shed line to bottom shed line,
- pick 4 to 5: bottom shed line, reaches shed zero line at 0° shedding time and gets back to bottom shed line.

In open shed systems, a frame is maintained static in top or bottom position as long as needed for pattern weaving. As shown in Fig. 36a, yarn raised or lowered in order to change from one open shed layer to another have a classical motion (R: raised or L: lowered). Yarns are maintained on its top or bottom line as long as no change is needed. Figure 36b describes yarn's cinematic during 5 shedding cycles:

- pick 0 to 1: top shed line to bottom shed line,
- pick 1 to 2: down shed line to top shed line,
- pick 2 to 3: maintained on top shed line,
- pick 3 to 4: top shed line to bottom shed line,
- pick 4 to 5: maintained on bottom shed line.

Open shed dobbies are now the main used technology. Cloe shed mechanisms have been considered as obsolete, mainly due to wasted movement [8] and was considered to gradually disappear. Nevertheless, it is clear that close shed dobbies are still

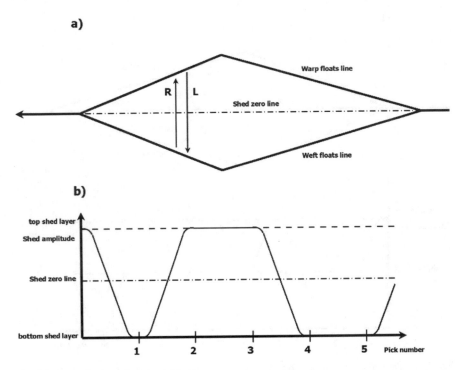

Fig. 36 Open shed **a** principle and **b** diagram

commercialized for specific woven product, like the Stäubli's 4080 dobby used for technical applications, especially forming fabrics. In addition, this parameter belongs to the UNIVAL technology parameters set, as it will be further discussed.
– Negative or positive dobbies:
   While negative dobbies only lift frames, these being lowered thanks to a set of springs, positive dobbies drive both lifting or lowering frames, giving a better control of frames motion. Only positive dobbies are now built, for obvious reasons of performances, and employees security, article changes and maintenance when removing springs.

4.3 Modern Dobbies Principle and Composition

Stäubli is now the leader in textile frame shedding machinery on the world market. Due to the increasing need on the weaving market for a reliable and precise heavy weaving or high speed weaving, the only left dobbies, since 2018 with the end of Stäubli's 2700 series of negative dobby, are rotary dobbies which are optimized positive double lift open shed dobbies. They may be placed over the loom for water jet weaving (Fig. 37 case a)) or next to the loom for other weft insertion technologies (Fig. 37 case b)).

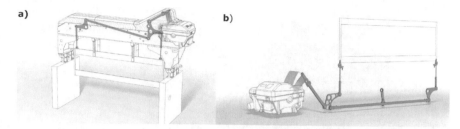

Fig. 37 Dobby setting: **a** over to the loom (waterjet loom), **b** next the loom. (Courtesy of Stäubli)

Table 4 Stäubli dobby series range

Dobby type	Housing	Loom/ application	Pick report
S3000	16/20 lifting units 12 mm pitch	Rapier, airjet, waterjet	>20 000
S3200	16–24 lifting units 12 mm pith	Rapier, airjet	>20 000
2688	Up to 28 units – 20 mm pitch Up to 56 units – 10 mm pitch	All types (technical fabrics)	
4080	Up to 24 units – 20 mm pitch Up to 30 units – 16 mm pitch	Technical fabrics looms	
UNIVAL 500	Up to 56 units – 10 mm pitch	Any technical or high density fabrics—rapier	

Stäubli dobbies general range is presented in Table 4. Thanks to this new dobby technology, an optimum performance has been reach thanks to:

– the elimination of any rectilinear alternative or unnecessary motions,
– the elimination of any continual movement except for the main shaft,
– the separation or elimination (thanks to electronic technology) of several motions occurring in the dobby (information reading, hook selection, main shaft motion transmission.

According to Figs. 38 and 39, the constitution and the motion of 3000 series dobbies are as follows: the main shaft (6), which carries the driver (2), is turning freely while weaving a float, with a stop every dobby 180° (= loom 360° with dobby speed being half of loom speed) thanks to a modulator placed at the entry of the main shaft which generates a precise working position for hooks engagement.

Frames and corresponding levers (9) are kept in position (here lever in upper position and frame in low position) thanks to the holding ramp (5) held by the upper arm (8). The upper arm (8) stays in position thanks to the non-selected position of the selector (10) which prevent the upper arm (8) from being disengaged with the ramp (5).

With no selection of the electromagnet (9), once the driver (2) as stopped and levelling bar (8) has come back, the lower control arm (8) will come back to its position, pushing on the casket (7), preventing locks (3) to engage the driver (2) and keeping the lever (9) in up position.

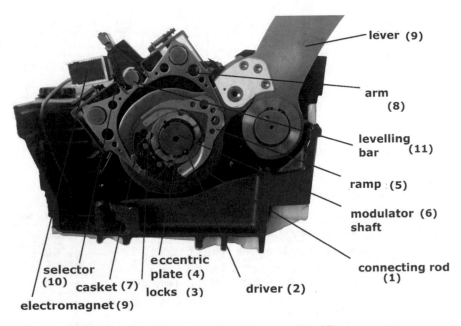

lever (9)

arm (8)

levelling bar (11)

ramp (5)

modulator (6) shaft

connecting rod (1)

selector (10)

casket (7)

electromagnet (9)

eccentric plate (4)

locks (3)

driver (2)

Fig. 38 S3000 series rotary dobbies composition. (Courtesy of Stäubli)

Fig. 39 Effect of electromagnet selection on S30** dobby motion. (Courtesy of Stäubli)

As depicted in Fig. 39, at no motion time due the modulator, once electromagnet (9) has attracted the selector (10), the pusher will push both control arms (8) enabling lever (9) position change to lower position. The upper arm (8) allowing the ramp (5) movement while the lower arm (8) releases pressure exerted on the casket (7) which releases the locks (3) and engaging the latter in the eccentric plate (4). At movement restart, the driver (2) drives the eccentric (4) thanks to the locks (3). The movement of the eccentric (4) converts the rotation motion into a translation which is transmitted to the lever (9) by the connecting rod (1).

After a 180° rotation of the eccentric, the ramp (5) will be engaged with the lower arm (8). The lever (9) will be held in lower position (warp float) as long as the electromagnet (9) is maintained active.

According to Figs. 40 and 41, the constitution and the motion of S3200 series dobbies is very similar to S3000 series. As well as previously, the selection of locks (3) mounted in the casket (7) is also maid thanks to control levers (8). The driver (3) carried by the main shaft (6) allows the eccentric (4) rotation and this way the movement of the connecting rod (1) and the lever (9). A major difference is the replacement of the holding function of the upper arm (8) by ergots (12). In order to weave warp or hidden warp floats, one of the two ergots (12) is locked by the pawl (5) so that the eccentric (4) may be kept in position.

As previously mentioned, a modulated motion enters dobby with a motion stop of main shaft and associate drive every 180° loom cycle. Although another modulator technology is available, the mechanism in Fig. 42 will be discussed. It is an illustration of a bronze pad mechanism modulator. The shaft (1) has a constant rotation coming

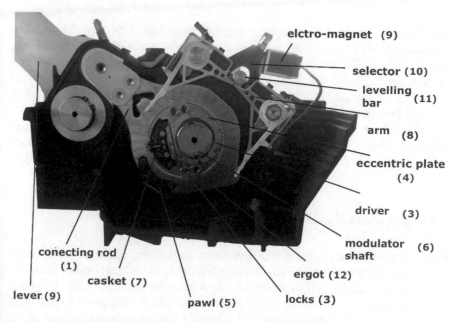

Fig. 40 S3200 series rotary dobbies composition. (Courtesy of Stäubli)

Fig. 41 Effect of
electromagnet selection on
S32** dobby motion.
(Courtesy of Stäubli)

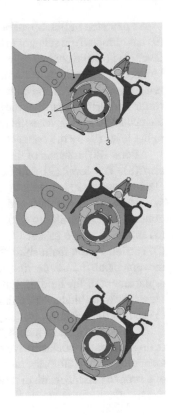

from the loom. It drives two opposite systems composed of the part (2) linked to
a bronze pad (6) which is free to run in the slide (5). The slide (5) is linked to a
pair of followers (3) which are in contact with the track of a pair of matched cams
(4) (one per cam) and linked (7) to the modulator shaft. As the main shaft drives
(2) and (6) rotation, (6) causes (5) and (3) rotation around the central axle: rotation
is transmitted to the modulator shaft and the dobby drivers. After the edge of front
cam (4) has been reached, the followers (3) and slide (5) set will topple, eliminating
rotation motion of (7): modulator shaft is stopped during topple (happening every
half rotation: 180°), which enables locks engaging in the driver at the dobby stage.

4.4 Motorized Dobbies

It should also be mentioned a new generation of dobbies, based on a motorized
technology like Toyota's E-shed or Stäubli's UNIVAL 500, which some of the specificities are presented in Table 4. As each frame is driven by a single actuator, E-shed or
UNIVAL technology offers a very wide setting range for shedding parameters and has
a performant speed and load capacities. Such performances are particularly adequate

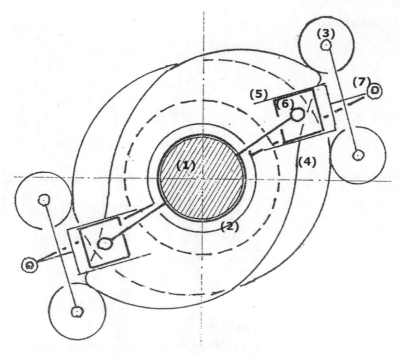

Fig. 42 Motion modulator principle

for high density fabrics (carpet weaving) or technical fabrics (multilayer composites reinforcements) applications. UNIVAL specific parameters will be discussed further in the UNIVAL 100 part.

5 Jacquard

5.1 A History of Jacquard [2, 5]

Jacquard shedding mechanism may be considered as fully automatized versions of original hand looms. After vertical warp have been placed horizontally, they were drawn in heddles, moved with hand thanks to weights or strings, then returned with tensioning springs (Fig. 43). These looms needed two workers, as the shedding part was fully separate from weft insertion process. Such a weaving process was quite long and moreover the risk of errors was high [2, 5].

The next inventions focused on process automation and the optimization of pattern information reading transmission:

Fig. 43 Hand selection
loom [5]

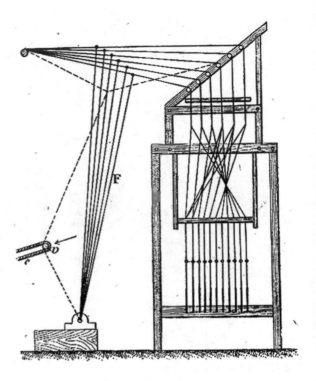

- the draw boy was replaced by paper (P) red by a needle (a) for selection in Bouchou's loom (Fig. 44). As the drilled plate (D) pushes the punched paper on the needle (a), the latter will stay still and the knot (r) will remain under the lever (E). Once the weaver has pressed the treadle (PD) pulling the lever (E), the string (c) will be pulled and corresponding heddle or frame will be raised.
- Bouchou's loom was improved with Falcon's loom (Fig. 45) were paper was replaced by more resistant punched cards (C), knotted strings replaced by metallic hooks (M) and where the treadle (PD) drives a knives box (M) to pull hooks.
- Reynier's loom (Fig. 46) pattern information is red on a drum. With Vaucanson placing this drum over the loom with a hooks and needle mechanism, all elements are gathered for future Jacquard's loom.

Jacquard invented his mechanism in 1801 by combining Bouchou's needles, Falcon's punched cards and Vaucanson's drum positioned over the loom. Even if first Jacquard looms (which example is shown in Fig. 47) were still manually operated, it was fully automated in 1860 thanks to mechanization.

A breakthrough was achieved with Verdol's Jacquard mechanism in 1884 where pattern reading and hooks actuating were separated. As Verdol mechanism shown in Fig. 48, needles (7) push hooks (Cr) according to punched card selection. A second vertical motion craven by knives (3) rise the hooks (Cr) which have not been pushed. The motion of hooks enables the rising of strings (5) and hooks (6) where are tied Jacquard heddles. The separation of reading and strings motion was achieved thanks

Fig. 44 Bouchou's loom [5]

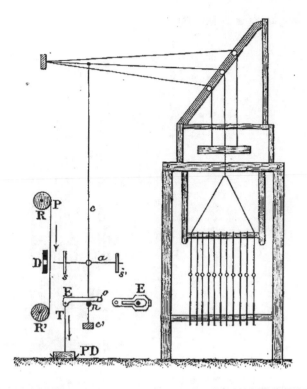

to a group of aglets (e) which replace Vaucanson cylinder and push selected needle (7).

5.2 Modern Electronic Jacquard [1, 11]

Another fundamental evolution was the invention of electromechanical selection of hooks by electromagnets. First electronic Jacquard were commercialised in 1983 for Bonas and in 1987 for Stäubli. Hooks electromechanical selection will be further described.

The two Jacquard main manufacturers on the western market are Bonas (Vandewiele) and Stäubli. The general set up and principle of a modern Jacquard establishment are explained thanks to Fig. 49. The Jacquard mechanism is placed on a superstructure precisely over the loom. The main motion is mechanically trans-ferred from loom to Jacquard thanks to a cardan shaft (called drive shaft in Fig. 49). The cardan shaft needs to be set in order to ensure synchronisation between the loom (pick insertion and beat-up) and the Jacquard (shedding). Cardan shafts tend to be replaced in some cases (Stäubli's ISD: Independent Stäubli Drive, Jacquard-equipped Dornier's loom and H3D Bonas Jacquard equipped with independent servo motor

Fig. 45   Flacon's loom [5]

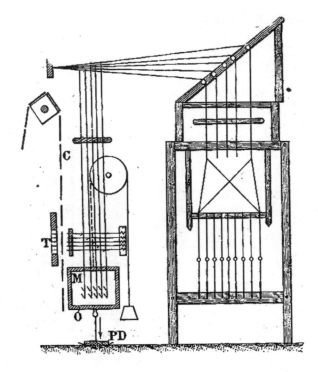

Fig. 46   Reynier's loom [5]

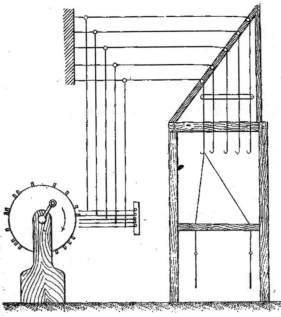

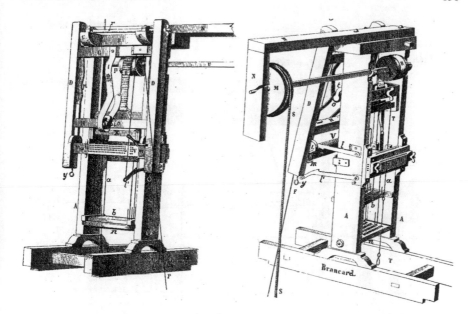

Fig. 47 First Jacquard mechanisms from Lyon [4]

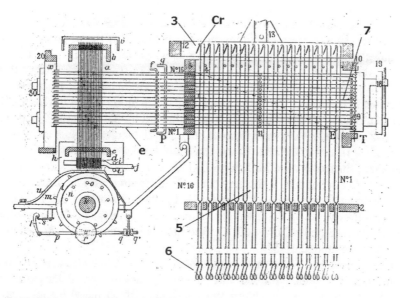

Fig. 48 Verdol Jacquard mechanism principle

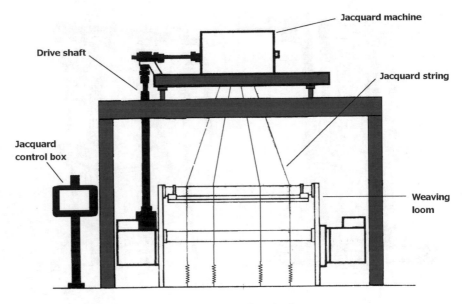

Fig. 49 General set of loom equipped with Jacquard mechanism

drive) by motors synchronised on loom rotation thanks to a resolver. Although cardan mechanism ensures an easy synchronisation, an independent motor brings precision and allows more load bearing for shedding as no power is taken from the loom. Rotation motion will be then transformed into an alternative and vertical motion in order to drive knives. According to Fig. 50 where the Stäubli SX Jacquard series mechanism is illustrated, the shaft (1) which gets its motion from the loom drives two sets of double eccentric parts (2). The horizontal motion of parts (3) generated by the rotation of (2) is converted into an alternatively vertical raising and lowering motion of (5) thanks to the axle (4) and transferred to Jacquard knives (6).

Another principle is used for Stäubli LXL Jacquard series mechanism shown in Fig. 51 where a set of complementary cams (2) are mounted on the drive shaft (1). Motion is transmitted to a coaxial system (4) thanks to Followers (3). LX Coaxial principle is the use of two shafts with the same rotation axis thanks to the rotation of a shaft inside a second hollow shaft. Coaxial motion is transmitted to knives thanks to the alterative motion of (5).

Selected hooks are raised thanks to modules (Fig. 52). A Stäubli Jacquard module is composed of 8 units, with one unit controlling one or several Jacquard cords (according to the number of pattern repeat). Double lift Jacquard modules, which principle is described with the help of Fig. 52, are composed of electromagnets (h) which convert pattern information into selection or non-selection of retaining hooks (d) and (e); knives (f) and (g) which are continuously exerting back and forth vertical movements for hooks (c) and (b) displacement; double rollers (a) move Jacquard strings and corresponding yarns to high or low shed position.

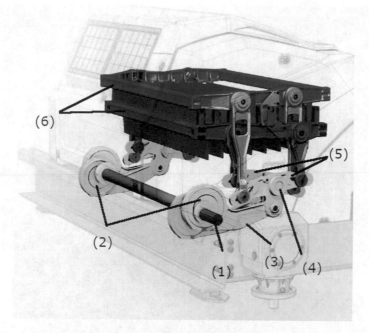

Fig. 50 Stäubli SX Jacquard mechanism. (Courtesy of Stäubli)

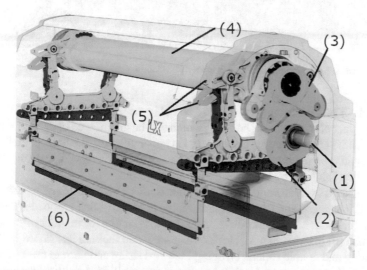

Fig. 51 Stäubli LX Jacquard mechanism. (Courtesy of Stäubli)

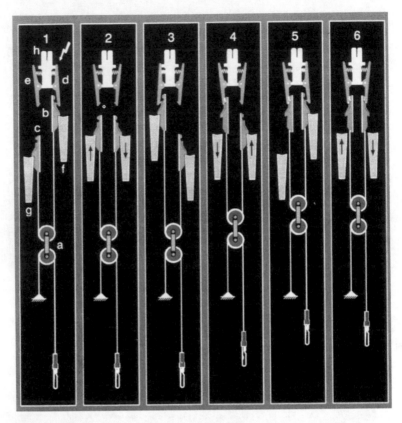

Fig. 52 Basic Stäubli module motion. (Courtesy of Stäubli)

A float in low position (case 1) may be achieved thanks to the selection by the electromagnet of the retaining hook (d) which will be maintained opened. As a result, Once the knife (f) has brought the hook (b) in upper position, the hook (b) will not be engaged and retained by the retaining hook (d) and the rising of the string attached to (f) will be balanced by the lowering of the knife (g).

For a change from down to up yarn position (cases 2 to 6), knife (g) is raising hook (c) until it is engaged and locked in the retaining hook (e) as there is no electromagnet selection (cases 2 and 3). As knife (g) goes down, (f) raises (b) in order to engage and lock it with (d) (cases 4 and 5). Hooks movement and locking has raised both sides of the string and double rollers, bringing Jacquard string and yarn to up position. An absence of electromagnet selection on next picks would lead to an up position float (case 6), whereas a selection would disengage hooks from retaining hooks and bring warp yarn on low position.

Two types of Stäubli modules with different capacity are available according to the shed amplitude, the speed of the process and the load applied by yarns. The choice of the appropriate module is defined with the calculation of the product of

Table 5 Stäubli Jacquard modules range

Module type	Jacquard type	Maximum A.L.S
MX-A	SX/LX	30,000
	LXL	25,000
MX-B	SX/LX	65,000
	LXL	45,000

these three parameters (which will be called A.L.S for Amplitude.Load.Speed) and modules A.L.S. range in Table 5.

As depicted in Fig. 53, Jacquard cords are arranged from Jacquard configuration to harness configuration thanks to respectively a perforated board (Fig. 54) placed right after modules and a comber board (Fig. 55) which sets harness density. Each cord is equipped with a spring which helps heddle lowering. Comber board configuration depends on the woven fabric family; Fig. 55 shows comber boards examples for double Jacquard pattern repeat, terry and carpet fabric. Drilling patterns depend on several parameters such as:

– applications: harness may need to be separated according to yarns evolution. A typical case is terry applications with ground warp having a different evolution from yarns dedicated to the visual aspect (here protruding loops),
– warp density,
– drawing-in draft (number of cords in a column, number of columns),

Fig. 53 Path of a Jacquard string

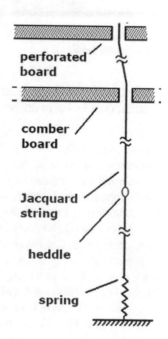

Fig. 54 Perforated guiding
board under modules.
(Courtesy of Stäubli)

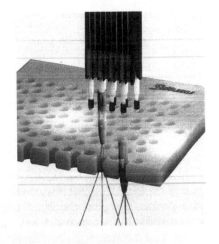

Fig. 55 Several comber board drilling frames: pattern repeat, terry and carpet applications.
(Courtesy of Stäubli)

– woven material (high count and technical yarns need larger Jacquard cords and
  larger comber board drilling).

The main advantage of Jacquard is the possibility to drive each yarn individually
compared to frame weaving. Jacquard woven fabric width is as large as the total
number of heddles or number of heddles in a repeat. Therefore, Jacquard weaving
is a relevant solution for large and complex patterns in fine furnishing or clothing
textile applications. Jacquard weaving is also an interesting process for complex and
3D fabrics thanks to its infinite yarn position possibilities. For conventional Jacquard
mechanisms, shed geometry is set and fixed and harness set up cannot be changed
unless harness is fully removed and replaced.

Harness geometry is essential for a long life use of Jacquard cords: a minimum
angle of $\alpha = 60°$ is to be respected for any Jacquard cord between the cord and the
comber board. This rule may be particularly difficult to observe in multiple pattern
repeats for cords going from modules of one side to yarns of the opposite side, as it
may be seen in Fig. 56. Cords angle generally depends on:

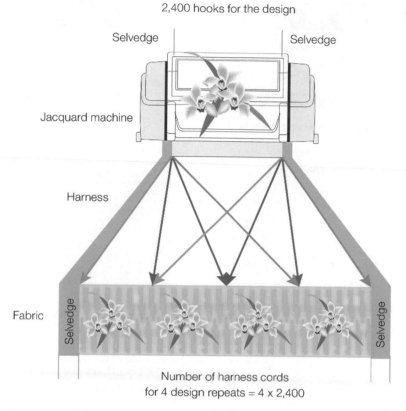

Fig. 56 Example of harness geometry set up for 4 design repeats. (Courtesy of Stäubli)

- the number of pattern repeat, i.e. the number of cords controlled by a single module unit. The ideal solution would be a setting of a cord per module unit, allowing any pattern width, although the price of a Jacquard depends of the number of modules;
- the width of the loom, considering that larger is the loom, smaller is the minimum cord angle α,
- the height available in the weaving building. A large loom may be compensated by a higher Jacquard mechanism position (G height dimension between Jacquard and comber board, visible in Fig. 57).

In some particularly difficult weaving environment (small building height, large loom width, impossibility for individual cord height control), smaller cords angle might be tolerated in some conditions of low load and low speed weaving, which would not imply too drastic cord wear. Nevertheless, other more elegant solution should first be taken in account like harness optimisation. The example in Fig. 57 depicts such an optimisation which solves a cord angle issue. The angle may become too small on repeat order, a solution may consist in exchanging repeats, which tend to increase cord angle. Case (a) shows a standard patterning with an obvious harness

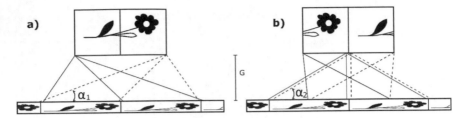

Fig. 57 Example of harness optimisation

Jacquard type	Format (number of hooks)
SX	1408 and 2688
LX	3072 to 6144
LXL	6144 to 14,336
LXXL	12,800 to 25,600
UNIVAL 100	512 to 15,360

Table 6 Stäubli Jacquard machine range

Jacquard type	Format (number of hooks)
Ji2	1920 to 2688
Ji5	3456 to 5760
SI	2304 to 31,104
H3D	Up to 16,128

Table 7 Bonas Jacquard range

issue as cord angle α_1 is too small. Exchanging pattern paths generates a significant increase for $\alpha_2 \geq 60° > \alpha_1$, enabling an optimal process.

Tables 6 and 7 display the products specificities of the two main Jacquard machinery constructors. As Today's market trend leads to higher number of hooks, Jacquard constructors also offer the possibility to combine machines in order to increase hooks capacities for high density/definition fabrics.

5.3 Motorized Jacquard

Stäubli's motorized Jacquard UNIVAL 100 was a breakthrough as it was first commercialized in 1999. The main principle was the replacement of mechanical hooks by individual "Jactuators"-called actuators, as shown in Fig. 58. An UNIVAL actuator (1) is composed of:

- a code wheel (2) in order to monitor actuator position,
- a connexion (3) with UNIVAL module instructions and feedback transmission,
- actuator shaft (4),

Fig. 58 UNIVAL 100
actuator. (Courtesy of
Stäubli)

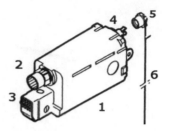

- a pulley (5) which is clipped on the shaft (4), and linked to a Jacquard cord (6)
- a Jacquard cord which is raised or lowered thanks to pulley (5) rotation.

Then, similarly to standard modern Jacquard, cord coiling on the pulley raises its corresponding warp yarn thanks to a heddle and eyelet while spring helps yarns lowering with pulley unwinding.

Similarly to mechanical modules, a UNIVAL module gathers and controls groups of eight actuator (Fig. 59). They have a modular configuration in a UNIVAL set up according to the number of cords needed. The individual motorized yarn control necessarily implies only one repeat harness.

In addition to the elimination of mechanical transmission from loom to Jacquard, this fully motorized technology offers new parameters for Jacquard shedding. Talking

Fig. 59 UNIVAL 100
Jacquard module. (Courtesy
of Stäubli)

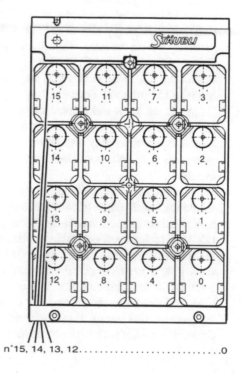

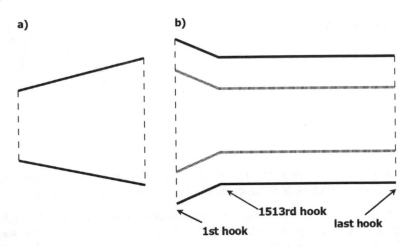

Fig. 60 Example of specific harness geometry made with the UNIVAL technology: **a** lateral view of shed amplitude at 1513rd hooks, **b** shed geometry view at weaver position

about new parameters might actually be overstated, as they may be seen as a transposition of other shedding mechanisms (cam and dobby) to Jacquard technology. Four main parameters can be set and can potentially be changed for each pick.

- Shed geometry parameter is obviously corresponding to clamping loop setting on level in frame weaving, adapted to single yarns. Contrary to mechanical Jacquard, UNIVAL technology allows any individual yarn position and any global shed geometry within the limit of shed maximum amplitude. Figure 60 is an example of a specific geometry. The (b) part shows visual aspect of opened shed from weaver point of view with a first amplitude set on first hooks and a smaller one from the 1513rd hook. Such a setting implies a linear decrease of amplitude from 1st to 1513rd hook and a constant geometry from 1513rd to last hook. The (a) part is a lateral view of the shed at the 1513rd hook. It is globally a clear shed. Moreover, an offset may also be set; originating from shifted shed geometry, the advantage is to split both shed bottom and top layers and reduce yarn density locally.
- Shed profile may be compared to cam shed profiles of cam motions where profile set motion law. Because of constant rotation motion in standard Jacquard weaving, hooks and cords motion cannot be changed, which is not the case anymore with UNIVAL technology. Available profiles are comparable to standard cam profiles P0 to P3 (c.f. Table 1) but can reach P4 to P7 profile (Fig. 61) in order to leave as much time as needed for carrier motion and set very dynamic yarn motion for yarn disentanglement and crossing. However, high profile should be used cautiously with sensitive yarns (small count or carbon and glass yarns). Due to high speed and acceleration during shedding, yarns are crossing each other more easily in the short term but such high profile tend to increase mid-term or long-term wear.

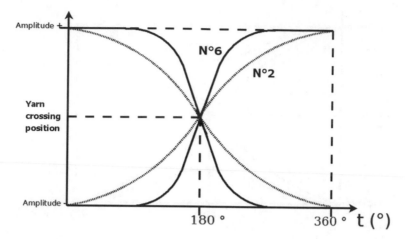

Fig. 61 Example of P2 and P6 UNIVAL shed profiles

- Shed phase difference corresponds to tagger indexes used for cams motion indexing. UNIVAL technology offers a shed phase difference within a range going from a 45° lead to a 45° delay. Such a parameter may have a significant use for high warp density technical application where yarns disentanglement can be difficult as it enables a local and progressive density reduction. Figure 62 illustrates this principle by depicting between t_0 and t_4 a setting a progressively decrease lead from first cords to last cords, for a profile $P_i > P_0$.
- Close shed profile parameter comes from centre close shed dobbies. Thanks to close shed profile, warp ends kept in upper or lower position for float weaving can move near shed centre and come back to its initial position. The UNIVAL technology allows a great flexibility as it can be set any value between 0 and 100% of half shed amplitude. Figure 63 illustrates the effect on yarn cinematic for a 40% and 90% close shed profile. Main advantages for this parameter is tension homogenisation during shedding and a lower pic tension as the reed is beating-up weft yarn. Another interesting aspect, especially for high density fabric

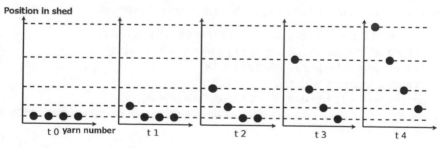

Fig. 62 Illustration of lead shed phase difference

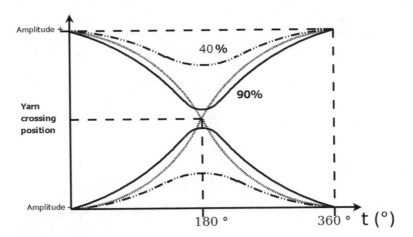

Fig. 63 Illustration of close shed profile effect

with a small values of close shed profile may be a better disentanglement at the beginning of motion.

A different configuration of these four parameter can be set for each pick in a row, in a column, hook by hook, for different harness areas or even by specific design areas: the possibility to overlay a .bmp file over weave design gives the opportunity to set a parameters configuration per each .bmp file colour. The adequate parameters levels can be adapted accurately according to the specificities of the design.

5.4 Exercise

A furnishing fabric needs to be woven with a warp density of 62 yarns/cm, selvages density is 100 yarns/cm, both at read location before templets. Pattern width is 40 cm, fabric width without selvages is 180 cm and total width is 182 cm. Shed amplitude is 100 mm, the load applied on a hook is 700 cN and loom speed is 420 rpm. According to these information, the following parameters need to be defined:

– number of Jacquard hooks and warp yarn needed for the fabric,
– number of Jacquard hooks and warp yarns needed for selvages,
– total number of Jacquard cords and hooks,
– optimal Jacquard machinery,
– definition of A.L.S. and choice of appropriate module,
– number of unused hooks, definition of hooks, harness configuration and number of pattern repeat.

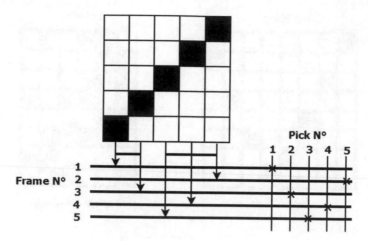

Fig. 64 Drawing-in-draft proposition for the exercise

6 Exercises Solutions

6.1 Shed Geometry and Tension Exercise

– A straight draw might not be the ideal solution as the tension difference between 1st and 5th frame could create visual defects.
– A more fitting solution for a homogeneous and small strain gap between yarns may be a pointed skip draw (Fig. 64), although other drafts may be also used.

6.2 Cam Indexing Exercise

– Pick repeat: 6
– List of cams:
– $\frac{21}{21}$, $R = \frac{6}{4} = 1.5$
– $\frac{3}{3}$, $R = \frac{6}{2} = 3$
– $\frac{2}{4}$, $R = \frac{6}{2} = 3$
– Order of rods for index positioning (see Fig. 65): 2, 3, 4, 6, 4, 6, 6, 5
– Cams will be placed facing the indexing table (B1 indexing code).

Jacquard fabric setting:

– Number of yarns and hooks for the fabric

$$N_{Yf} = \text{yarndensity} \times \text{fabricwidth} = 62 \cdot 180 = 11160 \text{ yarns}$$

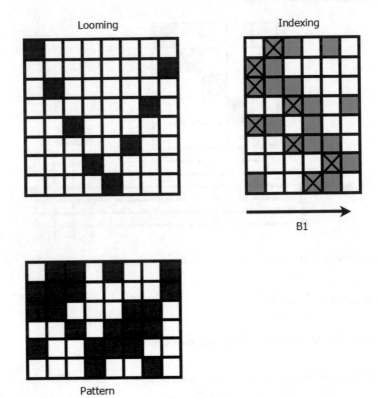

Fig. 65 Cam indexing solution

$$N_{Hf} = \text{yarndensity} \times \text{patternwidth} = 62 \cdot 40 = 2480\,\text{hooks}$$

– Number of yarns and hooks for the fabric

$$N_{Ys} = \text{yarndensity} \times \text{selvagewidth} = 100 \cdot (182 - 180) = 200\,\text{yarns}$$

$$N_{Hs} = \text{yarndensity} \times \text{selvagewidth} = 100 \cdot 2 = 200\,\text{hooks}$$

– Total number of cords (yarns) and hooks

$$N_Y = 11160 + 200 = 11360\,\text{cords}$$

$$N_H = 2480 + 200 = 2680\,\text{hooks}$$

– Optimal Jacquard machinery

2688 hooks SX Stäubli or Ji2 Bonas would be the most fitting machineries.

Unused	Selvage	Fabric	Selvage	Unused
4	100	11360	100	4

Fig. 66 Definition of hooks

- Definition of A.L.S. and choice of appropriate module,

$$A.L.S. = \text{amplitude (mm)} \times \text{load (daN)} \times \text{speed(rpm)}$$
$$= 100 \cdot 0.700 \cdot 420 = 29400$$

- Definition of hooks, harness configuration and number of repeat.

$$N_{unused} = N_{total\,Jacquard\,hooks} - N_H = 2688 - 2680 = 8\,hooks$$

Fabric is generally centred (Figs. 66 and 67).

$$N_{path} = \frac{N_Y}{N_H} = \frac{11360}{2680} = 4.2\,repeats$$

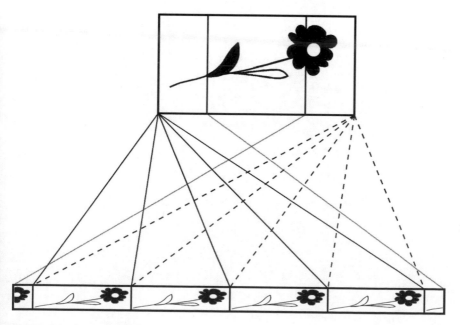

Fig. 67 Harness configuration

References

1. Adanur, S. (2001). *Handbook of weaving*. CRC Press.
2. Charlin, J.-C. (2003) The story of the Jacquard machine: From its origins to the present day. Stäubli: France.
3. Choogin, V., Bandara, P., & Chepelyuk, E. (2013). Mechnisms of Flat Weaving Technology. 1st edn. Woodhead Publishing. ISBN 978085097859.
4. Dantzer and Dantzer. (1908). Traité élémentaire de tissage. Beranger, Paris, France.
5. De Prat, D. (1921). Traité de tissage Jacquard. Charles Béranger, Paris, France.
6. Gandhi, K. (2020), Woven Textiles. 2nd edn. Woodhead Publishing. ISBN 9780081024980.
7. Lord, P. R., & Mohamed, M. H. (1982). *Weaving: Conversion of yarn to fabric* (2nd ed.). Darlington, UK.
8. Marks R., & Robinson, A. T. C. (1976). Principles of weaving. The Textile Institute, UK.
9. Morton, W. E., & Hearle, J. W. S. (2008). *Physical properties of textile fibres*. Woodhead Publishing in Textiles.
10. Neogi, S. K. (2015). *Role of yarn tension in weaving*. Woodhead Publishing India in Textiles.
11. Ormerod, A., & Sondhelm, W.S. (1998). Weaving: Technology and Operations. The Textile Institute, Manchester, UK.
12. Reh, F. (1928). *Traité de tissage mécanique* (3rd ed.). SAPI.

Non-uniformity and Irregularities in the Weaving

Alexander Buesgen

Abstract If a woven fabric has passed fabric inspection, it will look faultless and uniform at first glance. However, an accurate inspection of a woven fabric discloses a multitude of non-uniformities and irregularities. These are amongst others local crimp differences of warp and weft yarns, bow of weft threads, different frequency, distribution and size of openings at interlacing points and uneven spacing of warp- and weft threads. The following chapter explains, how thread geometry and weave can be used to increase the uniformity of woven structures. Manner and scale of woven irregularities are discussed in detail and how they are come into being by an ordinary weaving process. Especially but not exclusive for weaving technical textiles, it is important to increase evenness and to reduce quantity and size of irregularities. The options to improve the uniformity of woven structures at a weaving machine, e.g. optimization for timing and geometry of shed formation, hold-down-devices or full-width temples will be discussed.

1 Introduction

Most woven fabrics seem to have a uniform structure viewed without magnifier. They have passed fabric inspection successfully, which detects faults to count and classify and to remove them. So, most woven fabrics fulfil correctly the requirements for applications particularly in the traditional clothing and home textile markets.

Since technical textiles reached a significant portion of the textile market, new and more specific requirements are added to woven fabrics leading to a more precise and detailed consideration of the woven structure. What looks to be uniform with naked eye presents a multitude of unevenness and irregularities under Macro- and Microscopic devices. Width and spacing of single woven threads are spreading across a wide range. Openings, which are created by interlacing between weft and warp yarn show uneven frequency and a large variation in size. Crimp of treads changes

A. Buesgen (✉)
Faculty of Textile and Clothing, Hochschule Niederrhein—University of Applied Sciences, Mönchengladbach, Germany
e-mail: Alexander.Buesgen@hs-niederrhein.de

© Springer Nature Switzerland AG 2022
Y. Kyosev and F. Boussu (eds.), *Advanced Weaving Technology*,
https://doi.org/10.1007/978-3-030-91515-5_4

167

significantly from left to right side of the fabric. Weft thread orientations are curved and not rectilinear.

Non-uniformity and irregularities in the weaving downgrade quality and functionality of woven fabrics. For technical textiles, uniformity of woven structure becomes most essential. Airbags, filtration fabrics, forming fabrics, fabrics for sun shading and woven composite reinforcements are examples. Fabrics with increased requirements to structural uniformity are sometimes called precision fabrics [5, 13, 23]. However, precision fabrics are not always technical textiles. Other and so far traditional fabrics join this group. Quill pens of downy feathers may not penetrate tickings and down jacket fabrics which leads to high uniformity requirements for the woven structure. Mite dust tight fabrics surpass these uniformity demands realizing smallest or zero spaces to keep back particles of 1 to 3 μm.

Non-uniformity and irregularities are not necessarily faults of the weaving machine set-up. Often, they are created by state-of-the art and correct adjusted machine processing parameters. In the past, the non-uniformity of threads has been addressed repeatedly for non-uniformities of woven fabrics. But thread quality is only a part of the problem. This chapter focusses on woven structures if nearly perfectly even yarns are used.

Some parameters can be changed to improve quality of fabrics. Numerous suggestions and solutions are demonstrated after R&D projects of research institutions. However, some of them have not been taken over by the market so that industry has no benefit of these ideas.

2 Character and Cause of Unevenness in Woven Fabrics

2.1 Non-Uniformity of Thread Interlacing Structures

Non-Uniformity does not have to be regarded to be an irregularity of the fabric because it is a regular distribution of uneven areas caused inevitably by interlacing of threads. Nonetheless it is necessary to address this effect. The knowledge about it is very helpful to improve the evenness and quality of fabrics. And it has an influence to the extent of irregularities.

Investigations about the non-uniformity character of interlaced structures were done e.g. for primary backings of tuftings [22] and for woven reinforcements of printed circuit boards (PCB). The electronics industry called the non-uniformity of fabrics Fiber Weave Effect (FWE) and investigated its impact on signal quality coming to the conclusion, that reinforcing glass fibre fabrics should have flat and broad yarn cross sections like zero twist yarns for optimized signal integrity [9, 10].

Whenever a thread interlaces, it needs to displace two others by opening a space and penetrate between them to change from one side to the other side of the fabric. The size of this space corresponds more or less to the thickness of the penetrating thread. Figure 1 shows a unit cell of a plain woven structure. It demonstrates, that each

Fig. 1 Plain woven repeat
with different areas f_1, f_2 and
f_3 caused by interlacing

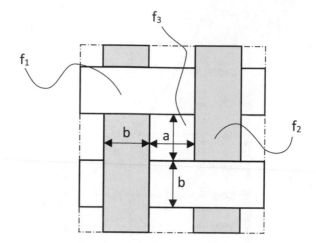

woven fabric has places, where two threads are laying above each other, places where
one single thread runs between two intersecting other threads and places where no
thread is present:

- (f_1) two threads above each other
- (f_2) one single thread
- (f_3) open space between threads.

The frequency distribution of these three areas depends on the cross-sectional
geometry of threads as well as on the weave. It is possible to reduce this structural non-
uniformity of woven fabrics by designing it to get a maximum share of one area type.
Calculation procedures, which can be applied to a common and standard spreadsheet
software are helpful to assess and improve the uniformity of the interlaced structure.

As a first option, the cross-sectional geometry of warp-ends and weft will be
considered. If warp-ends and weft threads have the same cross-sectional geometry,
the proportion of areas f_1, f_2 and f_3 in Fig. 1 can be calculated, e.g. the share of area
f_1 in a plain weave repeat is:

$$f_1 = \frac{4b^2}{(2a + 2b)^2} = \frac{4b^2}{4a^2 + 8ab + 4b^2} \text{with a = thickness, b = width} \quad (1)$$

A diagram like Fig. 2 can provide a better view to fabric design parameters and
their effect on fabric uniformity. To evaluate the influence of thread geometry, the
proportions of f_1, f_2 and f_3 are expressed versus the ratio of thickness divided by
width of woven threads. As an example, the proportion of area f1 (two threads above
each other) is calculated:

$$f_1 = \frac{4b^2}{4b^2(\frac{4a^2}{4b^2} + \frac{8ab}{4b^2} + 1)} = \frac{1}{\frac{a^2}{b^2} + 2\frac{a}{b} + 1} \quad (2)$$

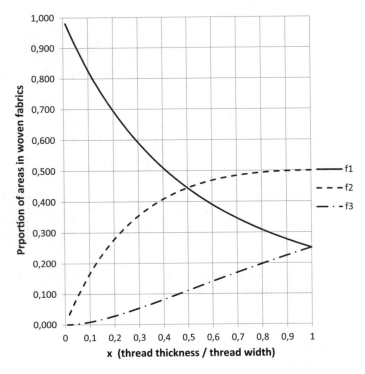

Fig. 2 Homogenization of woven structures by adjusting cross-sectional geometry of threads (plain weave)

$$x = \frac{a}{b} \Rightarrow f_1(x) = \frac{1}{x^2 + 2x + 1} = \frac{1}{(x+1)^2} \tag{3}$$

f_2 and f_3 are calculated analogous:

$$f_2 = \frac{2x}{(x+1)^2} \tag{4}$$

$$f_3 = \frac{x^2}{(x+1)^2} \tag{5}$$

The diagram Fig. 2 quantifies in which way the uniformity of woven structures can be improved by using threads of minimized thickness and maximized width. For extremely uniform fabrics, x is close to zero and the fabric is having a f_1 proportion close to 100%. So called "non crimp fabrics" are designed with maximum uniformity of the interlaced structure by threads with very large width and very small thickness but motivated by another reason: the minimal thickness of threads is minimizing crimp as well. On the other hand, threads having circular cross sections like typical monofils or like highly twisted yarns result into the highest degree of non-uniformity. 50% of the woven fabric consists of one thread, 25% of two threads and 25% of gaps.

A second option to increase the uniformity of woven structures offers the selection of another weave than plain. Each interlacing thread is pushing other threads apart, creating gaps and single thread areas. Floatings on the other hand enable closing or minimizing of openings. Length, number and arrangement of floatings can be used to increase the uniformity of fabrics by increasing the proportion of two threads in the fabric.

In the past, a calculation method has been published to determine the extension of openings between threads [25]. This calculation is based on the assumption, that threads can be moved more easy to approximate each other if a floating is present. Accordingly, the surrounding area of each floating in a weave repeat is analysed resulting into six different surrounding conditions. These six conditions are expressed by six movability factors of threads.

It is interesting that these movability factors—if they are taken for building weave tightness factors [18]—are proved to correlate much better to practical test results than other and more common weave tightness factors [21]. As an example, twill 2/2 has a movability factor of 0,25 (which means, threads are not sliding easily to get more close to each other) whereas basket 2/2 has a factor of 0,75. So basket 2/2 floatings can move together a great deal better than those of twill 2/2 which is confirmed by practical experience of weavers all over the world.

Calculation of uniformity including influence of weave starts in Fig. 3 for basket 2/2 with the large repeat on left side before move of floatings. The share of area f1 is:

$$f_{1basket} = \frac{16b^2}{(4a^* + 4b)^2}$$

with a^* = thread thickness before moving of loatings   (6)

It is still assumed, that a spacing is created by interlacing and that it has the size of thread thickness. Taking the moveability factors of floatings into account, which is 0,75 for basket 2/2, the size of spacing a will be reduced to 1/4 of a*. The upper right part of Fig. 3 presents basket 2/2 after floatings moved together. The calculation of area f_1 is then:

$$f_{1basket} = \frac{16b^2}{(a + 4b)^2}$$

with a = thread thickness after moving of loatings   (7)

$$f_{1basket} = \frac{16b^2}{\frac{a^2}{b^2} + \frac{8ab}{b^2} + \frac{16b^2}{b^2}} = \frac{1}{\frac{a^2}{b^2} + 8\frac{a}{b} + 16} \qquad (8)$$

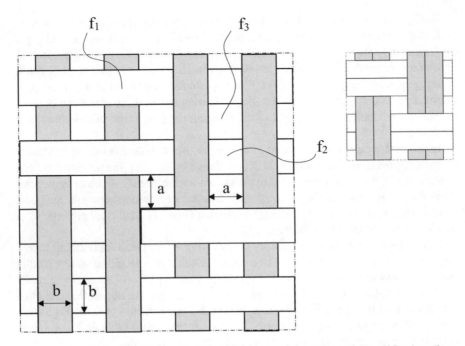

Fig. 3 Basket 2/2 repeat with areas f_1, f_2 and f_3 before applying the movability of floatings (large drawing on left side) and after moving threads to compact the woven structure (right upper corner)

$$x = \frac{a}{b} \Rightarrow f_{1basket}(x) = \frac{16}{x^2 + 8x + 16} = \frac{16}{(x+4)^2} \tag{9}$$

$f_{2\,basket}$ and $f_{3\,basket}$ are calculated analogous:

$$f_{2basket} = \frac{8x}{(x+4)^2} \tag{10}$$

$$f_{3basket} = \frac{x^2}{(x+4)^2} \tag{11}$$

Figure 4 presents the uniformity of a basket 2/2 woven structure after integration of the movability factor for this weave based on the work of Skliannikov [25]. If threads having a circular cross section like monofils are taken for weaving, only 4% of the fabric is not covered by threads, 32% is covered by one thread and 64% are covered by two threads, one upon each other.

It becomes clear, that in comparison to plain weave (Fig. 2) a basket 2/2 woven structure offers a significant higher structural uniformity. Selection of more flat yarns leads quicker to a dominating proportion of two-thread areas. For example, if flat yarns are woven having thickness/width ratio (x) of 0,1 the proportion of two threads

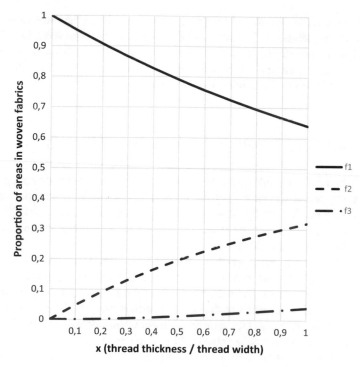

Fig. 4 Homogenization of woven structures by adjusting cross-sectional geometry of threads (basket 2/2 weave)

increases to 95,2%. In comparison to basket 2/2, this ratio only achieves 82,6% for plain woven cloth.

According to the previous calculations, the structural uniformity of every weave can be determined and predicted in combination with the cross-sectional geometry of woven threads. Table 1 provides an example of four different weaves woven with a ratio of 0,5 for thickness/width of threads [21]. It becomes clear, that basket 2/2 provides the best uniformity of this comparison, even better than satin 1/4 (2).

The conclusion of Sect. 2.1 is that interlaced structures like woven fabrics inevitably have uneven macroscale character built by areas of two threads, areas of one thread and openings between threads. Uniformity of woven structures is given,

Table 1 Uniformity comparison of weaves using 0, 5 as thickness/width ratio (x) for threads [21]

Proportion of area in %	Plain 1/1 with x = 0, 5	Twill 2/2 z with x = 0, 5	Basket 2/2 with x = 0, 5	Satin 1/4 (2) with x = 0, 5
Two threads (%)	44	53	79	64
One thread (%)	45	40	20	32
Openings (%)	11	7	1	4

when those areas are dominated by one type reducing the other two to minimum values. A first possibility to increase the fabric uniformity is the selection of thread cross-sectional geometry to flat and broad yarns. Another possibility is the selection of a weave having floatings, which allow pushing together of threads. The effect of both can be analysed and predicted by calculation.

2.2   Scattering of Thread Width, Spacing and Porosity

In contrast to most other automated machine processes, parameters of woven threads cannot be controlled within a very satisfactory tolerance. Yarn material and yarn design parameters frequently have a high degree of unevenness contributing to unevenness of woven structures (Fig. 5). This effect is investigated and discussed frequently in the past [11, 26, 30] and is not to be subjected here.

The focus of this chapter are those irregularities, which are mainly influenced by machine and processing parameter variations of weaving. To reduce the influence of yarn imperfections as much as possible, the following examples are using synthetic man-made fiber yarns having mostly endless fibers and sometimes in addition untwisted. These yarns offer significant lower CVm values than spun fiber yarns [15, 17], and thus enable a better observation of machine effects.

A good example, which has been designed to a uniform woven structure according to Sect. 2.1 by creating nearly 100% of double thread area (f_1) is shown in Fig. 6. This fabric is woven in plain weave with fine untwisted polyester yarns having a density of approximately 62 picks and 68 ends per cm. An analysis of the presented 70 interlacing points shows, that warp-ends as well as weft threads are changing in width about 20%, indicated in Fig. 6 by double arrows. Furthermore, yarn and fiber angles in the area of interlacing points unevenly deviate from 90°. As an example,

Fig. 5 Imperfections of the woven structure having fine but uneven cotton yarns (warp-ends: 45dtex, picks: 52 dtex)

Fig. 6 Different widths and different angles of picks and warp-ends

the orientation of weft at interlacing point (a) is 105° while it is 85° at interlacing point (b).

Another example for the non-uniformity of a woven structure, which is using warp- and weft yarns having a high degree of evenness is given in Fig. 7. The fabric samples have been woven with polyester filament yarns, warp-end count was Nm 100, 40 ends/cm, weft count was Nm 80/2, 36 picks/cm. The analysis was executed with a Keyence VHX-600 digital microscope [7].

The variation of spacing and opening sizes is clearly visible. First, the vertical (dark) warp-ends are arranged in pairs revealing the space requirement of reed dents during beat-up. Second, interlacing is creating a broad distribution of gaps and of gap sizes between threads.

It may be expected, that the occurrence and frequency of gaps in the woven structure depends on weft density (Fig. 8). The chosen weave however and the selection of weft and warp count has a more significant impact. Four samples analysed for

Fig. 7 Different gaps at interlacing points [7]

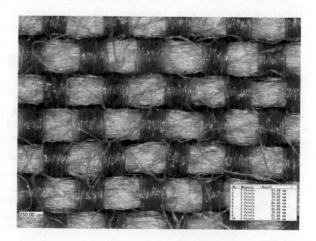

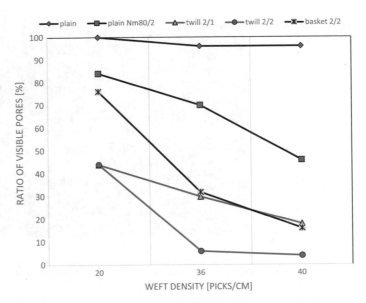

Fig. 8 Effect of weave and weft density to frequency of gaps in woven fabrics [7]

Fig. 8 have been woven with 40 ends/cm (Nm 100) and 20, 36 and 40 picks/cm having a count of Nm 106. One sample (plain Nm 80/2), woven using a coarser weft count of Nm 80/2 demonstrates that the woven structure can be tightened by selecting different counts for warp and weft. Balanced plain weave presents even under maximum cover (40 picks/cm) gaps at 96% of interlacing points [7].

The other possibility for evening the structure is to avoid gaps at interlacing points as far as possible. A good example for this is twill 2/2. It shows an almost closed structure having gaps at only 6% (woven with 36 Picks/cm) and 4% (woven with 40 picks/cm) of interlacing points [7].

In theory, it could be concluded that there are two possibilities for decreasing irregularities in the woven structure: creation of openings at every interlacing point (plain weave) or avoiding openings completely at interlacing points (twill weave). Unfortunately, this is wrong.

A further investigation reveals, that plain 1/1 woven fabric with balanced warp-end and weft counts is not as uniform as previously presumed. It may have 96% of porosity at interlacing points but the sizes of these gaps are very different. Figure 9 presents 25 interlacing points, which have very small gaps of 6000 μm and large ones up to 33,700 μm. This high degree of non-uniformity is independent from cover factor and present for low, medium and high weft densities [7].

If the yarn is provided with a very low CVm value, it cannot be made responsible primarily for the non-uniform woven structure. The weaving process must be taken into consideration. And here, a multitude of weaving machine effects reducing fabric uniformity can be observed. Timing and geometry of shed formation (Chap. 2.3), displacements of the cloth fell (Chap. 2.4), force-free picks at selvages (Chap. 2.5)

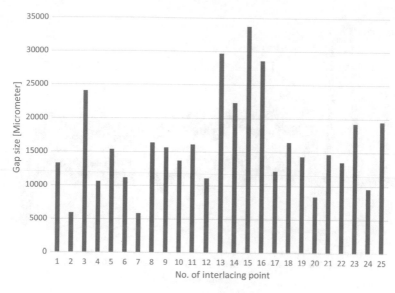

Fig. 9 Distribution of gap size in a plain woven structure [7]

and tension differences of weft ends at selvages in air-jet machines (Chap. 2.6) prevent woven threads from being positioned and treated identically one like the other. All these effects explain variations between different warp-ends.

But even if only one and always the same warp-end is studied a very large scattering of tension values over time can be observed. Figure 10 presents the tension of one single warp-end for only 50 weft insertions of a modern air jet weaving machine running not faster than 650 rpm [14]. The scattering of warp-end tension is remarkable high. Differences between maximum and minimum tensions are above 40% of the maximum values although a uniform polyester filament yarn has been taken for warp-ends.

Scattering of warp-tension is smaller for heald frames in the rear than in front of the machine. And it is smaller at fabric centre than at selvages [14].

Responsible for differences in warp-end tension are

(1)   length and tension differences of warping/sectional warping and
(2)   effects of the weaving process.

First, warping and sectional warping are commonly done using creels with mechanical tensioning devices. They adjust a pre-tensional level of ends without control or readjust of the actual tension value, if it becomes different during warping. Yarns which are delivered from the rear part of the creel do have much more length and more yarn guiding elements than yarns which are positioned in the front creel. Consequently, yarns get different elongation during warping/sectional warping [12]. In addition, fast unwinding of random packages shows individual and significant yarn tension differences especially caused by snagging of coils during the unwinding of patterning or ribboning zones ([16]: pp. 422).

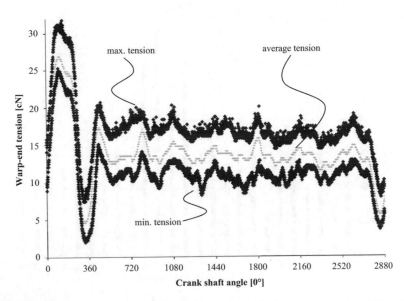

Fig. 10 Scattering of tension forces of one single warp-end after 50 cycles (front positioned heald frame, right side of the loom, sateen weave 1/7)

Second, tension scattering of single warp-ends can be traced back to dynamically swinging and vibrating elements of the weaving machine, which can be observed by modern high-speed video recording. Vibration of machine parts such as back-rest, heald frames, heddles and machine frame, friction of yarn-to-yarn and yarn-to-steel between reed dents and in the area of up-and-down moving heddles, uneven positioning of longitudinal moving yarn inside the eyelet of oscillating heddles and clasping of fibres or broken filaments, which are torn up by shed opening are contributing to the great extent of scattering warp-end tension.

Warp-end tension is a key parameter of weaving having a great influence to the woven result. Crimp of warp and weft threads, width and thickness of warp-ends and friction forces at interlacing points are affected by actual warp-end tension values. Whenever a warp-end presents an increased tension value, the interlacing weft threads are forced to higher crimp locally coming under increased tension also.

It should be mentioned, that many woven fabrics, which are similar to the above presented examples, are fulfilling quality requirements by 100%. A few applications however require more uniformity of the woven structure. Since the development of functional and technical fabrics, the number of those fabrics is constantly growing.

As a conclusion of Sect. 2.2 it becomes clear, that not only thread unevenness cause irregularities in woven fabrics but also a multitude of machine and operational parameters. The appearance and frequency of openings between interlacing threads can be very uneven. Weave design enables significant increase or decrease of openings. As openings can be extremely different in size, those weaves avoiding openings lead to more uniform structures. Because warp tension values can show a high scattering for each single warp-end during weaving even fabrics having no or minimized openings can get many structural irregularities.

2.3 Warp and Weft Tension Differences by Shed Formation

Systemic tension differences between one warp-end and the other are related to shed formation principles. The majority of weaving machines open shed geometries according to the clear shed principle (Fig. 11 above). The term clear shed refers to the front shed, which is set up to position warp ends of all heald frames having identical gradient angles in side view. The first heald frame creates the smallest shed and each following heald frame opens a bigger shed. Consequently, shed opening applies an increasing level of elongation and tension to warp-ends from the first to the last heald frame. The majority of warp-end materials provide enough elongation to tolerate these tension differences without any visible quality reductions. And because clear shed geometries allow to minimize open shed heights, it is a standard for set-up of weaving machine shed geometry.

The magnitude of clear shed tension differences between warp-ends which are lifted by different heald frames depends on front shed space requirement and number of heald frames. A geometrical model has been published to prepare a numerical simulation of the weaving process [33]. It is an example for the additional length

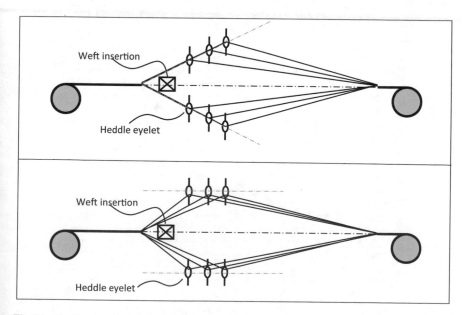

Fig. 11 Shed geometries of clear shed formation (above) and of unclear or parallel shed formation (below)

required from warp-ends between closed shed and open shed (Table 2). However, this additional length requirement is not identical to warp-end elongation, because warp-ends may stretch behind closed rear shed up to the back rest or furthermore up to warp beam.

Another example is given in Fig. 12, where a symmetrical shed geometry (upper shed = lower shed) has been adjusted for 16 heald frames weaving an E-Glass fabric with rigid rapier weft insertion. As the free elongation length of ends is open to other set-up parameters, Fig. 12 presents the length in mm, which is additional required from warp-ends during shed opening. The additional length requirement has a non-linear character and grows increasingly for heald frames in the rear. Therefore, tension differences of warp-ends are more critical for machines using a large number of heald frames.

The other and more seldom used shed geometry is called unclear or parallel shed (Fig. 11 below). All heald frames are moving into the same height position for upper as well as for lower shed formation. This must be a very high position to ensure that weft insertion has enough space even for the last heald frame. Consequently -with the exception of the last heald frame—every heald frame is elongating and tensioning

Table 2 Required additional warp-end length for shed opening using clear shed set-up of heald frames [33]

Heald frame No	1	2	3	4	5	6	7	8
Upper shed [%]	0,65	0,766	0,890	1,021	1,159	1,305	1,459	1,622
Lower shed [%]	1,049	1,061	1,078	1,101	1,128	1,158	1,193	1,232

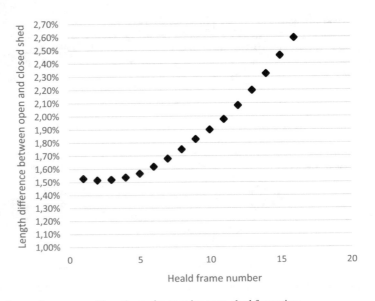

Fig. 12 Increasing warp-end length requirement by open-shed formation

warp-ends more than in clear shed geometries. Processing of most common warp yarns with parallel shed formation leads to increased machine downtime caused by broken warp-ends.

The advantage of parallel shed formation is the uniformity of warp-end tensions by shedding. If the optical appearance of woven fabrics reveals uneven crimp of warp-ends or if warp materials have very low or no elasticity (e.g. like many high performance fibers) parallel shed formation is recommended for set-up.

The synchronisation of open and closed shed formation with beat-up is another area of activity to create or even out warp-end tensions between ends, which are lifted by different heald frames. Standard weaving machines are set-up to reduce tensions and friction damage as much as possible. For this reason, beat-up takes place leaving heald frames in their actual position, if the weave requires them to be there again for the next weft insertion. This minimizes frictional damage of warp-ends by shed formation for every weave which is different from plain weave.

The disadvantage of open shed beat-up is, that different warp-end tension levels are present when the weft is beaten into the close fell (Fig. 13). To increase retaining forces of warp-ends fixing a beaten-up weft better at the cloth fell, upper and lower shed heights often are adjusted different in that way, that one (mostly the upper shed) is small and the other one is large. An easy way to adjust the relation of upper and lower shed height offers the vertical setting of back rest position.

Different shed openings result in (a) high tensioned warp-ends being in lower shed position, (b) medium tensioned warp-ends being in upper shed position and (c) low tensioned warp-ends being on their way from upper to lower or lower to upper shed when beat-up takes place. If warp material is very sensible, very fine or not providing sufficient elasticity, the woven structure becomes visible irregularities—e.g. visible crimp differences between one warp-end and the other.

A countermeasure can be change of warp-end drafting. A skipping order of entered warp-ends may distribute warp-ends with higher tension and embed them between lower tensioned ends. This is taken to reduce visible crimp differences of warp-ends

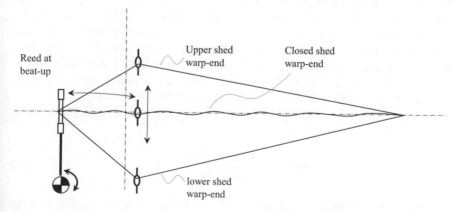

Fig. 13 Open shed beat-up creating three different warp-end stress conditions

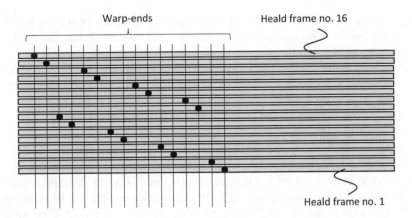

Fig. 14 Skipping entered warp-ends to reduce visibility of warp tension differences

which sometimes occurs on straight drafting, if a large number of heald frames is used. Figure 14 shows an example.

The other—and more effective—possibility to beat-up at even warp-end tensions is shed formation according to the so called closed shed principle. Every heald frame is moved into closed shed position after each weft insertion and before beat-up. Closed shed formation is independent from weave so that each warp-end has the same tension for beat-up.

The homogenized stress level of ends however is comparatively low and that does not fit for every set-up. In addition, wear of warp-ends is maximized in the area of heald frames due to more frequently up-and-down moving heddles. Each vertical heddle motion causes friction to contacting adjacent warp-ends and to the entered warp-end inside the eyelet because of longitudinal back-and forth motion of this end during shed opening and closing.

Weft tension and weft crimp is affected by shed formation timing also. The effect depends on the relaxation properties of weft material. If shed closing takes place early e.g. at 310° of the crank shaft, some weft materials do not relax until they are fixed by shed closing and interlacing with warp-ends. Their residual weft tension can contract the fabric at selvage areas. Late shed closing, e.g. 350° of the crank shaft angle enables better relaxation of weft threads providing better uniformity of the woven structure [5]. Unfortunately, weave efficiency as well as woven structural parameters are affectd by late shed closing in most cases unfavourable.

The time gained by delaying shed closing for 1° of the crank shaft angle can be calculated to:

$$\text{time} = \frac{60.000}{\text{rpm} * 360} \left[\frac{s\,10^{-3}}{1°} \right] \tag{12}$$

Thus, a delay of shed closing from 310° to 350° for an air-jet machine weaving with 1200 rpm will create an extra time for weft relaxation of 5,5 ms.

Concluding Sect. 2.3 needs to note, that shed geometry can either cause high differences in warp-end tension (clear shed) or can keep tension forces even but on a high level (unclear or parallel shed). As warp-end tension rises for clear shed from heald frames to heald frame non-linear, a large number of heald frames is critical in particular. Beat-up at open shed supports fabric unevenness, closed shed beat-up provides even tension level for warp-ends but increases friction and wear of them.

2.4 Displacement of Cloth Fell—Vertical and Horizontal Weft Bow

Bow and skew of weft are well known fabric faults in industry. The origin of bow is discussed controversial. In industry, the tension differences during warping causing different stretch of warp-ends has been addressed. Other reasons are named in literature are weave design like twill 2/2 or twill 3/1 and processing after weaving like finishing ([2] Chap. 6.7.2). Moreover, the weaving machine itself is able to create significant bow extent by creating two kinds of cloth fell arching. These effects are discussed here.

In theory, the cloth fell of a weaving machine should be a straight line oriented 90° to warp-end direction. In practice, the cloth fell can show arching effects. Vertical and horizontal arching changes spacing and angles between warp and weft from left to right side of the woven fabric.

If usual ring temples are taken the free length of cloth fell is able to conduct vertical motions between temples. Vertical arching of cloth fell is caused by warp sided or by weft sided weaves. Warp sided weaves have a majority of warp-ends in upper shed position exerting higher upwardly forces. The cloth fell is pulled up creating a bow between clamping temples (Fig. 15). Weft sided weaves pull the cloth fell down.

Pulling up and pulling down forces can be determined by front shed angles and warp-end tension. Figure 16 presents front shed tension forces of upper shed (F_{US}) and of lower shed (F_{LS}) and effective vertical force components (Fv_{US}, Fv_{LS}).

The resulting force, pulling the cloth fell up or down is:

vertical warp-end forces

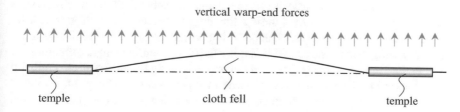

temple   cloth fell   temple

Fig. 15 Vertical arching cloth fell (majority of warp-ends in upper shed position)

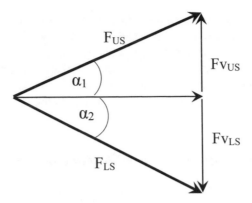

Fig. 16 Vertical forces of upper and lower warp-end tension

$$Fv = \sum Fv_{OS} - \sum Fv_{LS} = n_{OS} * F_{US} * \sin(\alpha_1) - n_{LS} * F_{LS} * \sin(\alpha_2) \quad (13)$$

with: α_1 = upper shed angle; α_2 = lower shed angle,
n_{OS} = number of upper shed warp-ends, n_{LS} = number of lower shed warp-ends.

An example shall demonstrate the magnitude of vertical arching forces. A weaving machine processes 7000 warp-ends, the angle α_1 of upper shed warp-ends is 22° and of lower shed 24°. Warp-end tension is 24 cN per end. Weave is satin 4/1 (2) resulting in 5600 ends in upper shed and 1400 ends in lower shed position. Vertical arching forces are 503 N – 137 N = 366 N pulling the cloth fell up.

There are two possibilities to avoid vertical arching at all. First, a full width temple keeps the cloth fell into a determinable vertical position. As warp-ends are under high tension inside a full width temple, machine stops may lead to plastic elongation for synthetic warp materials causing visible horizontal bars of increased weft spacing.

The other possibility is a simple hold-down device preventing the cloth fell from being pulled up, but not clamping it. Figure 17 shows an L-shaped steel bar which is fixed between the ring temples. It should be noted, that the hold the down device has to be mounted into the lowest vertical cloth fell position as it only prevents the cloth fell from being pulled up.

Beside vertical displacement, horizontal displacement creates so called weft bow. The majority of books and articles blame tension differences of the warping process and inaccurate fabric treatment during finishing to be responsible for weft bow. However, the weaving machines can contribute considerably to horizontal displacement as well. The machine caused effect is related to clamping length differences of warp-ends. Those warp-ends, which are clamped by temples are fixing the cloth fell at the forward position of the reed. Between temples, warp-ends have an increased elasticity because they are able to stretch with more length between back rest and take-up (Fig. 18).

The additional stretch of fabric and warp-ends between temples can pull weft threads to build an arching line. For cloth and home textiles, the extent of bow is

Fig. 17 Hold down device to prevent the cloth fell from being pulled up during weaving

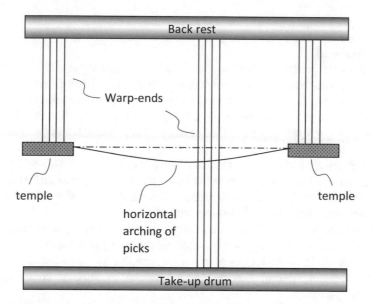

Fig. 18 Different clamping length of warp-ends and horizontal arching of the cloth fell

mostly small, e.g. 0 to 3 percent. For woven technical textiles and processing of synthetic man-made fibres bow can go up to 20%.

$$\text{Fabric bow [\%]} = \text{weft distance to cloth fell/fabric width} \times 100 \quad (14)$$

Horizontal bow is not accepted and needs to be corrected by clamping and tensioning frames or by special machines like the terry cloth bow and skew control machine [1]. Even if bow is not visible at weaving machines, the differences of

Fig. 19 Weft displacement caused by ring temples

warp-end stretch nevertheless are present if the warp material is able to elongate under stress at all.

Although full width temples can avoid horizontal arching at all, most common are ring temples, because they enable the adjustment of different stretch values. A fabric passing ring temples is guided partially over the cylinder surface and requires additional length for it. The resulting weft displacement is shown in Fig. 19

An interesting future prospect is the development of a new kind of temple, which is claimed to avoid or minimize weft bow [24]. This temple is designed to keep up the tension on cut ends of weft threads at selvages by using a two-folded device fixing the selvage between transportation belts and keeping up tension by the control of step motors. It may be helpful to reduce the bowed course of warp-end tensions also, which is the topic of the following section.

The conclusion of Sect. 2.4 is that there is a two-folded fabric arching, which can take place during weaving. Vertical arching caused by unbalanced weave can be suppressed by hold-down devices or full-width temples. Horizontal arching is more difficult to control. Among other reasons it is caused by different clamping length between ring temples.

2.5 Bowed Course of Warp-End Tensions Across Fabric Width

A comprehensive systemic irregularity of woven fabrics is created by force free selvages which are common for all shuttleless weaving machines. Force free selvage means that picks are ending up at selvages with a cut end, leading to a drop down of longitudinal weft tension to zero. This entails a plurality of consequences for the woven structure. Warp-ends and weft threads get another crimp in the middle part of

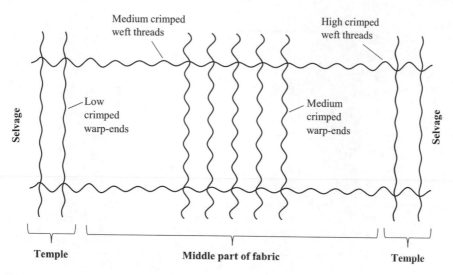

Fig. 20 Warp and weft crimp differences in woven fabrics

the fabric than at selvages. Warp and weft tension as well as warp and weft spacing are different at selvages leading to major differences in the macroscopic fabric structure.

Figure 20 presents the basic character of crimp differences caused by force free selvages. Crimp of warp-ends and of weft threads are balanced between present tensions of both at the central fabric body. Warp-ends are commonly set-up to higher tension than weft threads but weft threads are fixed by static friction of interlacing points and therefore both weft and warp are crimping with medium magnitudes in the middle. At selvages, there is no force keeping weft threads under tension because they end up with cut ends. Low tension of weft leads to increased crimp of weft threads. On the other hand, warp-ends do not have a counterpart for balancing out forces. Therefore, warp-end crimp is reduced at selvages leading to an increased length compared to central warp-ends.

The determination of crimp differences according to international standards like ISO 7211/3 or ASTM D3883 is labour intensive but instructive. A simple method for rapid monitoring of warp-end length differences at weaving machines is marking warp-ends by colour, when they just leave the warp-beam, start the weaving machine and check the distribution of marks when they arrive in the cloth fell area ([8] p.156).

A consequence of crimp differences between the middle part of the fabric and the selvage areas is a substantial non-uniformity of warp-end tension. Nearly every weaving machine shows a significant increase of warp-end tension at the central fabric and a lower stress level at both sides. This phenomenon has been investigated repeatedly in the past and it is discussed in literature until today ([3, 6, 19, 20, 29, 31]).

Figure 21 shows typical stress differences of a polyester awning fabric during air-jet weaving [14]. It is one of the most reliable diagrams about this effect with high correlation of upper and lower shed values and small standard deviations. The

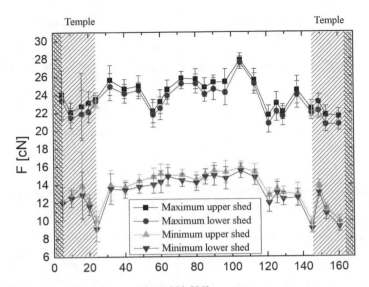

Fig. 21 Warp-end tension versus machine width [14]

large difference between minimum and maximum values looks strange but this is found to be characteristic by other researchers [20]. An analysis shows, that warping and sizing create stress differences of warp-ends having a reversed arch form [31]. This means, that warping and sizing reduce tension in the central part and cannot be blamed for the uneven stress distribution of Fig. 21.

If the warp stress distribution of machines having one single warp beam is compared to those machines weaving two pieces with two separated warp beams it becomes clear that force free selvages are mostly responsible for the effect ([29] pp. 905).

But crimp differences are not the only irregularities for woven structures in combination to the effect of Fig. 21. In addition, a distribution of thread densities can be found. Figure 22 presents warp and weft spacing measured in a sheeting cover fabric, which has been woven tightly with very fine cotton yarns. The measuring points 1 to 6 were evenly distributed across 1,7 m of fabric width. At selvage areas, weft threads have less densities and warp-ends have higher densities than at the central part of the fabric.

Figure 23 illustrates the irregularities of woven structures at selvage areas. The picture is taken from glass fiber fabric at a rapier machine. Spacing differences of warp and weft yarns are added by noticeable deviations of picks from the intended 90° orientation.

Options to reduce the uneven stress distribution of warp-ends are discussed controversal. Symmetric shed geometry is recommended [31] as well as high positioning of back rest leading to non-symmetrical shed geometry [29]. Likewise is the rating of full-width temples. A small equalization effect can be determined for using full-width temples instead of ring temples ([19] pp 149, [31] pp. 164, [29]). It is recommended

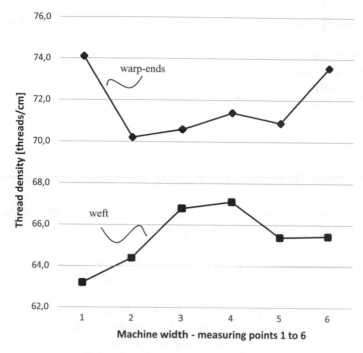

Fig. 22 Distribution of thread spacing across fabric width

Fig. 23 Non-uniform spacing and curved orientation of threads at selvage areas

as well as late shed closing. Another option is weaving with low tension levels but this affects the downtime of weaving machines significantly and therefore it should not be adjusted with care.

As mentioned before, a device to keep weft threads under tension at selvage areas has been developed in 2016 [24]. This will definitely reduce crimp differences of weft and warp between center and selvage of fabrics. If the market takes over this technology.

Concluding Sect. 2.5 needs to make clear, that the differences of warp-end tension across the fabric width are closely connected to force free selvages of shuttleless weaving machines. The fabric structure suffers under a very uneven crimp distribution for both weft and warp-ends. Spacing and orientation of warp-ends also differ between centre and selvage of fabrics.

2.6   Different Treatment of Weft Threads on Air-Jet Machines

Especially air-jet weaving can modify a weft yarn across the fabric width. After leaving the main nozzle, a weft thread gets additional blows from relay nozzles until it reaches the arrival side. Thus, the maximum blow intensity takes place for the initial length of the pick which has to travel through the whole weft insertion channel passing by all relay nozzles. The last part of the pick length receives no or very little blow from relay nozzles. Because air flow is untwisting yarn, yarn construction can be changed by air jet weft insertion from left to right side of the weaving machine ([28] pp. 977). The weft yarn untwisting can be detected for fiber yarns ([27] pp. 829). Filament yarns seem to be opened more intensely on the right side also ([5] pp. 324).

The other difference of air-jet machines is a stretching nozzle on the right (receiving) side, which keeps the weft thread more long under tension by air flow. Thus, the weft is treated different at both selvages. On the left side, the weft is cut and moves free of forces towards the fabric selvage. Therefore, the weft crimp of Fig. 20 will be different for air-jet machines: it should be lower on the right side than on the left side in theory.

The following studies proof, that there is a change of the woven structure produced on air-jet machines between left and right selvage areas. The structural variations can be detected by air permeability tests and by an analysis of weft crimp.

The air permeability of left side, fabric centre and right side is shown in Fig. 24 for air-jet woven fabrics. A polyester warp yarn of Nm 100 is interlaced in plain weave with three different cotton weft counts (Nm 36, Nm 50, Nm 80) having 17 picks/cm each. The differences of fabric structures are clearly visible. They show highest permeability on the left side, medium permeability on the right (receiving) side and lowest permeability in the fabric centre. Left and right selvage values are in contradiction to results of the above mentioned literature.

Both, air permeability (Fig. 24) and weft crimp (Table 3) show clearly, that the fabric structure is changed. On left side, selvage crimps of weft threads are higher than on right side. As differences of blow intensities could not be identified [4], air-jet woven cotton weft yarn rather seems to confirm the theoretical stretching nozzle effect.

In theory, weave design and weft densities may have an impact on this effect but this has not been identified in practice. Plain weave, twill 2/2 and sateen 1/7 (3) is changing the outcome just as little as small, medium or large weft spacing [4].

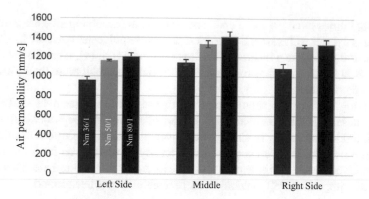

Fig. 24 Air permeability of air jet woven fabrics with cotton threads [4]

Table 3 Weft crimp of air jet woven fabrics with cotton weft yarn [4]	Fabric	Place of measurement		
		Left side	Centre	Right side
	Fabric Nm 36/1 Crimp [%]	1,96	0,99	1,72
	Fabric Nm 50/1 Crimp [%]	1,18	0,94	0,99
	Fabric Nm 80/1 Crimp [%]	2,34	1,38	2,06

A parameter which has a significant impact on the structural differences between left, centre and right side of air-jet fabrics is weft material and weft yarn structure. Figure 25 demonstrates the influence of texturized polyester as weft material on woven structure. The highest air permeability is present now on the right (receiving) side [4].

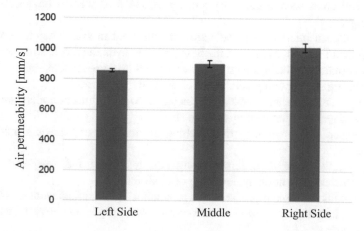

Fig. 25 Air permeability of air jet woven fabrics with texturized PES weft threads [4]

Air jet fabrics having a more open structure on the receiving side are confirmed for other polyester weft yarns. Because a more coarse polyester twist of Nm 70/s had the same influence like a texturized 76 dtex polyester yarn [4], it can be concluded, that material is affecting the structural fabric variations much more than yarn count and yarn construction.

Summing up Sect. 2.6 should point out that air-jet machines treat weft threads different at selvages. The suction nozzle on the receiving side keeps up a certain level of tension and that seems to change the woven structure, as air permeability tests show. The postulated difference of blow intensity of weft yarns from left to right side cannot be verified with experiments presented or referred to here. However, a noteworthy impact of weft material to air permeability has been found.

3 Conclusions and Recommendations

Non-uniformity and irregularities are present in woven fabrics even if yarns are very even. First, a system of interlaced threads suffers from structural non-uniformity caused by areas having two, one or no thread. Second, the usual process for interlacing yarns at weaving machines is responsible for significant tension differences of warp and weft threads during fabric formation. Inter alia unevenness of crimp, spacing and porosity are the consequences. Especially, the differences between fabric centre and selvages areas are critical.

This chapter points out the influence of weaving techniques and operations to the frequency and extent of non-uniformity and irregularities in the woven structure. Recommendations to minimize these disadvantages are:

- Structural non-uniformity of interlaced threads can be influenced by alteration of thread cross section geometry and by selection of weaves having movable floatings.
- Gaps between interlacing threads are very different in size. Therefore, weaves which avoid openings at all are the best option to even the fabric structure.
- Unclear (parallel) shed geometry of heald frames lead to even tensioning of warp-ends during shed opening better than clear shed geometries.
- Beat-up at closed shed enables interlacing of woven threads under more even warp-end tension than beat-up at open shed.
- Unbalanced weave cause vertical arching of cloth fell and needs a hold-down device.
- Ring temples lead to different elongation length of the warp-end—fabric continuum causing horizontal arching of cloth fell.
- Weft threads are force free at selvages producing significant unevenness of crimp resulting into uneven warp-end tensions and uneven spacing between centre and selvage areas.
- Air-jet machines keep weft threads under tension at the receiving side and thus produce different fabric structures at right and left side of the fabric.

References

1. Adanur, S. (2001). *Handbook of weaving* (p. 365). Technomic Pubishing Comp. Inc.
2. Ahmad, S., Rasheed, A., Afzal, A., & Ahmad, F. (2017) (ed) *Advanced textile testing techniques,* CRC Press, Taylor and Francis Group, FL/USA, chapter 6.7
3. Badve, N. P., & Bhattacharya, U. (1961). *Study of warp tension on Roper and Sakamoto automatic warp let-off motions,* part IV. *Textile Trends, 3,* 27–32.
4. Barner, S. (2016). *Luftdurchlässigkeit von Air-Jet-Geweben.* Niederrhein University of Applied Sciences, Faculty of Textile and Clothing, Mönchengladbach.
5. Bauder, H.-J., Julia, M., Helmut, W., & Heinrich, P. (2005). *In welcher Fertigungsstufe werden die Weichen für ein Präzisionsgewebe gestellt?,* Melliand Textilberichte, 2005. *Heft, 5,* 324–327.
6. Blanchonette, I. (1996). Tension measurements in weaving of singles worsted wool yarns. *Textile Research Journal, 66,* 323–328.
7. Benrui, T. (2019). *Optimizing of woven fabric tightness using weave design and weave construction,* Bachelor thesis, Niederrhein University of Applied Sciences, Faculty of Textile and Clothing, Mönchengladbach.
8. Choogin, V., Bandara, P., & Chepelyuk, E. (2013). *Mechanisms of flat weaving technology.* Woodhead Publishing Ltd.
9. Dudek, R., Goldman, P., Kuhn, J. (2007). *Advanced glass reinforcement technology for improved signal integrity,* HyperTransport™ Technology Developers Conference, October 2007.
10. Dietz, K. (2003). *Fine lines in high yield (Part XCVIII): Advances in reinforcement structures,* TechTalk in *CircuiTree,* November 2003.
11. Harlova, M. (2014). Detection of fabric structure irregularities using AirPermeability measurements. *Journal of Engineered Fibers and fabrics, 9*(4), 157–164.
12. Hübner, S. (1983). *Vorbereitungstechnik für die Weberei.* Springer Verlag, Berlin, Heidelberg.
13. Huschke, R. (2005). *Präzisionsgewebe - Eigenschaften, Herstellung und Anwendung,* Die Bibliothek der Technik (Band 280), SZ Scala GmbH.
14. Keilmann, L. (2006). *Analyse der Fadenbelastung an modernen Webmaschinen,* Niederrhein University of Applied Sciences, Diploma Thesis, Faculty of Textile and Clothing, Mönchengladbach.
15. Kretzschmar, S. D., Furter, R. (2008). *Process optimization in a filament yarn plant,* Pakistan Textile Journal (PTJ), September 2008, pp. 44–46.
16. Lawrence, C. (2003). *Fundamentals of spun yarn technology.* Taylor and Francis Inc.
17. Mamun, R., Repon, R., Jalil, M. A., & Uddin, A. J. (2017). Comparative study on card Yarn properties produced from conventional ring and compact spinning. *Universal Journal of Engineering Science, 5*(1), 5–10.
18. Milasius, V. (2000). An integrated structure factor for woven fabrics Part I: Estimation of the weave. *The Journal of the Textile Institute, 91*(2), 268–276.
19. Neogi, S.K. (2015). *Effects of temple,* Chapter 7.9 in: *Role of warp tension in weaving,* Woodhead Publishing India Pvt. Ltd., New Delhi, pp.148–151.
20. Obolenski, B., & Wulfhorst, B. (1993). Influence of various temple systems on the running characteristic of weaving machines and on fabric quality. *Melliand Textilberichte, 74,* 25–29.
21. Ratovo, K., Büsgen, A., Brunke, T., & Knein-Linz, R. (2017). New approach to increase the evenness of woven technical fabrics, Aachen-Dresden-Denkendorf International Textile Conference, November 30-December 1, 2017.
22. Rhodes, T.M. (1963). *Tufted pile fabric comprising a flat woven synthetic plastic backing,* US 3,110,905, 19.11.1963.
23. Scorl, H.-D., & Oguzlu, N., Weissenberger, W. (2003). Precision technical fabrics—a challenge for high-speed weaving machines. *Melliand International, Nr., 01,* 049.
24. Stark U., Bauder H-J., Gresser T. (2016). *Entwicklung eines neuartigen Breithaltesystems für schussgerades Weben von mittelschweren und schweren Geweben für Technische Textilien,* Research Project IGF 17845 N, Final report.

25. Skliannikov, V.P. (1974). *Optimisation of Structure and Mechanical Properties of Man-made-fibre Fabrics*, Liogkaya Industriya, Moscow, USSR.
26. Uster Technologies AG (2005). *Uster Tensojet 4 – Application Report – Successful yarn buying in the meaving mill by using modern high-performance tensile testing*, Uster, Switzeland.
27. Wahoud, A. (1987). *Ein Beitrag zum Schusseintragsverhalten von Garnen im Luftstrom*, phd thesis, Faculty of Mechanical Engineering, Institute of Textile Technology (ITA), RWTH Aachen.
28. Weinsdörfer, H., et al. (1989). *Auswirkungen des Luftstrahls beim Schusseintrag auf die Gewebequalität. Textil Praxis International, 9*, 972–977.
29. Weinsdörfer, H., Wolfrum, J., & Stark, U. (1991). The distribution of warp tension over the warp width and how it is influenced by the weaving machine setting. *Melliand Textilberichte, 72*, 903–907.
30. Weissenberger, W. (1993). *Prozessübergreifende Qualitätssicherung aus der Sicht vom Gewebe zum Garn*, Melliand Textilberichte, Vol 74, Nr. 4.
31. Wilbrand-Ludwig, H., Gries, T. (2003), *Maßnahmen zur Reduzierung der Spannungsbogigkeit an Webmaschinen*, Melliand Textilberichte 84, H. 3, 162–165.
32. Wolrum, J. (2005). Einflüsse auf den Schrägverzug von Geweben. *Melliand Textilberichte, 03*(2005), 145.
33. Wolters, T. (2002). *Verbesserte Webmaschineneinstellung mittels Simulationsrechnung*, phd thesis, University of Aachen, Faculty of Mechanical Engineering, Aachen.

Design of Weave Patterns

Radostina A. Angelova

Abstract This chapter aims to provide the basic knowledge on the weave patterns, used in the manufacturing of woven structures from one set of warp threads and one set of weft threads. The fabric construction, weave patterns, and weave diagrams are presented in details. The basic characteristics of a weave pattern: repeat, overlaps, and displacement number, as well as the designation of a specific weave by the repeat fraction, are explained. The elementary weaves: plain weave, twill weave, satin and sateen weaves and their derivatives are presented and discussed. The design of patterns for stripe effects, cell-like effects, rib effects, diagonal effects, zig-zag effects, curved effects, and 3D effects is presented. Original weave diagrams support each weave. Some new methods for the design of weave patterns, are also discussed.

Keywords Woven fabric · Weave patterns · Weave diagrams · Plain · Twill · Satin · Sateen

1 Fabric Construction, Weave Patterns and Weave Diagrams

A weave pattern is called the way of interlacing the warp threads (warp set) and weft threads (weft set) during the weaving of the fabrics. The weave pattern also represents the law of the warp threads motion (up and down), when forming the shed.

Most of the fabrics for clothing and household textiles, as well as a majority of the technical textiles, have simple woven structures [1, 2], as they are made of two sets of threads that interlace perpendicularly (Fig. 1):

- Warp set (ends): the vertical threads alongside the fabric's selvedge that are wound onto the warp beam.
- Weft set (picks): the horizontal threads that are inserted in the open shed to interlace with the warp set.

R. A. Angelova (✉)
Technical University of Sofia, 8 Kliment Ohridski Blvd, 1000 Sofia, Bulgaria
e-mail: joy_angels@abv.bg

© Springer Nature Switzerland AG 2022
Y. Kyosev and F. Boussu (eds.), *Advanced Weaving Technology*,
https://doi.org/10.1007/978-3-030-91515-5_5

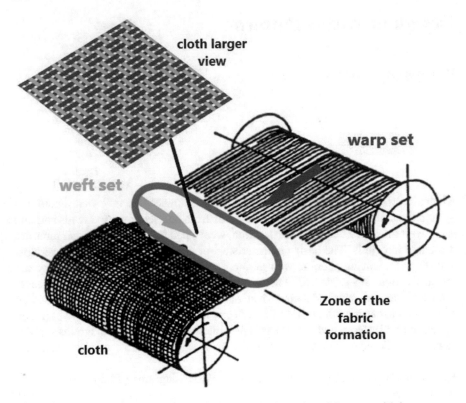

Fig. 1 Warp set of threads and weft set of threads in the formation of the woven fabric

When interlacing with the weft threads, the warp thread could have only two positions:

- To be over the weft thread when is in the up position in the shed—Fig. 2a. It is called a *warp overlap*.

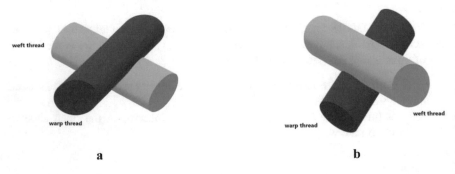

Fig. 2 The interlacing of a warp and a weft yarn: **a** warp overlap—the warp yarn is over the weft yarn; b/ a weft overlap—the weft yarns is over the warp yarn

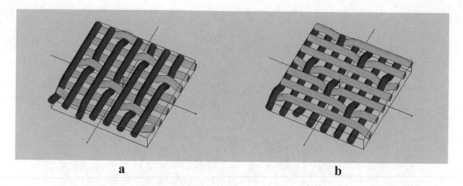

Fig. 3 Floating of yarn: **a** warp float; **b** weft float

- To be under the weft thread when is in the up position in the shed—Fig. 2a. It is called a *weft overlap*.

When a warp thread moves over two or more weft yarns, it forms a *warp float*. Figure 3a shows an example of a satin weave, where all warp threads float over 6 weft yarns, before crossing with the seventh one. When a weft thread moves over two or more warp yarns, it forms a *weft float*. Figure 3b is an example of a sateen weave: every weft thread floats over six warp threads, interlacing with the seventh one.

Is several warp yarns float together, following the same low for interlacing, it is called a *group warp float*. By analogy, the *group weft float* consists of several consecutive weft floats with the same low for interlacing. Figure 4 represents a basket weave with six warp and six weft threads in the repeat, where the group warp float over the weft yarns and the group weft flow over the warp yarns are visible.

Three specific characteristics can describe each weave [3]:

- The *repeat* (R) involves the smallest number of warp and weft yarns (or warp and weft overlaps) that defined the whole fabric. The repeat is practically multiplied several times (hundred and thousands) alongside the warp and weft yarns to form the peace of cloth. Figure 5 shows the cloth of the basketweave repeat from Fig. 4.

The repeat in the warp and weft directions of the cloth can be equal or different. When the *warp repeat* R_{wp} is the same as the *weft repeat* R_{wf} the weave repeat is called *square*.

- The *number of overlaps* (n) is the sum of the warp overlaps n_{wp} in the weave repeat and weft overlaps n_{wf}. The proportion of the warp and weft overlaps in the repeat defines the face of the fabric:

  – if $n_{wp} > n_{wf}$ the result is a warp-faced fabric.
  – if $n_{wf} > n_{wp}$ the result is a weft-faced fabric.
  – if $n_{wf} = n_{wp}$ the view of the fabric is preconditioned by both warp and weft yarns.

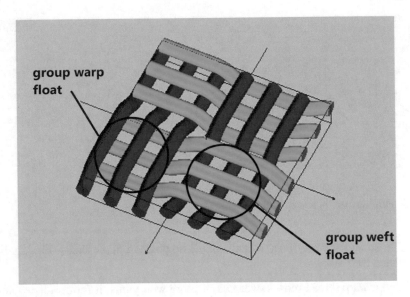

Fig. 4 Group float

- The *displacement number* (S) shows the position of the next overlap in the weave repeat concerning the previous one. It can be:

  - Positive integer number (+S):

    if the next overlap in the direction of the warp threads (S_{wp}) shifts upward;
    if the next overlap in the direction of the warp threads (S_{wf}) shifts rightward;

  - Negative integer number (-S):

    if the next overlap in the direction of the warp threads (S_{wp}) shifts downward;
    if the next overlap in the direction of the warp threads (S_{wf}) shifts leftward;

Figure 6 shows examples of the determination of the displacement number S. When a warp thread is considered (Fig. 6a, b), the upward position of the next overlap leads to positive S_{wp} ($S_{wp} = +2$, Fig. 6a), while the downward position of the next overlap leads to negative S_{wp} ($S_{wp} = -3$, Fig. 6b). When a weft thread is considered (Fig. 6c and d), the rightward position of the next overlap leads to positive S_{wf} ($S_{wf} = +4$, Fig. 6c), while the leftward position of the next overlap leads to negative $S_{wp=f}$ ($S_{wf} = -2$, Fig. 6d.)

The weave patterns are presented in the form of weave diagrams. The space between two vertical lines (a column) represents a warp thread. The space between two horizontal lines (a row) represents a weft thread. Usually, the weave diagram is presented in a grid of squares. The use of a square grid means that the density of the fabrics in both warp and weft directions is equal (equal numbers of ends and pick per dm). When the weave diagram has to represent a cloth with a larger number of

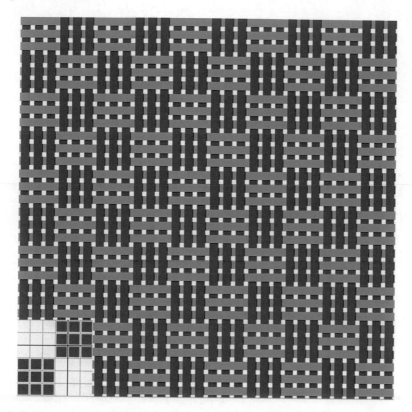

Fig. 5 Basketweave cloth: the repeat of the weave is multiplied several times on the direction of the warp and weft set

ends than picks (or vice a versa), then a grid with rectangles is more appropriate to be used (the short side of the rectangle coincides with the higher density).

The intersection of the warp and weft thread in the weave diagram represents an overlap. It can be:

- a warp overlap: in this case, the square is filled in with a mark. The mark can be of any type.
- a weft overlap: in this case, the square is left blank.

Thus, the designer has to mark only the warp overlaps in the weave diagram, filling the squares with a mark or colour.

Usually, the weave diagram represents only one repeat of the weave pattern. However, to visualize better the weave, more repeats could be marked in both horizontal and vertical directions. It is a good practice to highlight the repeat R, using different colours or symbols (Fig. 5).

The number of warp overlaps n_{wp} is a fraction of the total number of overlaps n in the weave repeat R. The number of weft overlaps n_{wf} is also a fraction of the total

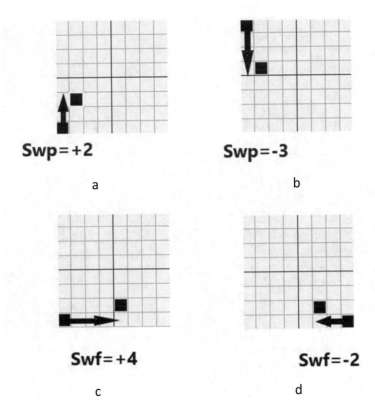

Fig. 6 Displacement number in warp and weft direction: **a** S_{wp} positive; **b** S_{wp} negative; **c** S_{wf} positive; **d** S_{wf} negative

number of overlaps n in the weave repeat R. For a single warp or weft yarn in the repeat, it is valid that:

$$R = n_{wp} + n_{wf}.$$

Therefore, the sequence of the warp and weft overlaps over a single warp thread in the repeat could be represented by a *repeat fraction* [3], e.g.:

To show which type of overlap should be marked first, dots are inserted in the repeat fraction, as follows:

- $\frac{n_{wp.}}{.n_{wf}}$, which means that a warp overlap(s) is marked first, followed by weft overlap(s).
- $\frac{.n_{wp}}{n_{wf.}}$, which means that a weft overlap(s) is marked first, followed by warp overlap(s).

The sequence of the warp and weft overlaps could also be marked over a single weft thread in the repeat, using the same repeat fraction. In this case, instead of a line that divides the repeat fraction, an arrow, showing the direction of the weft threads, is used, e.g.:

$$\frac{n_{wp}}{n_{wf}} >$$

2 Elementary Weaves

There are three *elementary weaves,* that give rise to all others: plain weave, twill and sateen/satin.

The elementary weaves have common features:

- They have a square repeat, i.e. the warp repeat R_{wp} is equal to the weft repeat R_{wf}, or

$$R = R_{wp} = R_{wf}$$

- The displacement number S is constant within the whole repeat.
- There is only one interlacing point between each warp and each weft thread within the repeat, i.e.:
  - there is only one single warp overlap alongside the yarn (warp or weft), and the rest are weft overlaps, or
  - there is only one single weft overlap alongside the yarn (warp or weft), and the rest are warp overlaps.

3 The Plain Weave and Its Derivatives

3.1 Plain Weave

The plain weave has the smallest repeat of all weaves: $R = 2$. It consists of 2 warp and two weft overlaps. The weave pattern could have only to appearances (Fig. 7), depending on the beginning of the first warp thread.

The plain weave has two warp threads in the warp repeat with two different motion laws: when the first thread is up, the second is dawn or vice versa. Therefore, only two harnesses are enough to produce cloth in a plain weave. However, when the fabric density is big, four, six or eight harnesses could be used for decreasing the friction between neighbour yarns and for decreasing the weight of the harnesses.

Well-known fabrics in the plain weave are percale, chiffon, taffeta, organza.

The plain weave is very simple, and it preconditions only two derivative weaves: the *rib weave* and the *hopsack weave.*

Fig. 7 The two possible
views of the plain weave

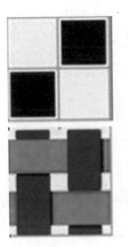

3.2 Rib Weave

The rib weave has two versions: *warp rib* and *weft rib*.

3.2.1 Warp Rib

The warp rib is obtained from the plain weave when the single overlap in the repeat fraction is multiplied by a number, different than 1 and 0. The additional overlaps are placed alongside the warp yarns.

The warp repeat of the weave is $R_{wp} = 2$. The weft repeat is a sum of the warp and weft overlaps in the repeat fraction, i.e. $R_{wf} = n_{wp} + n_{wf}$.

Figure 8 shows four samples of warp rib: (a) warp rib $\frac{2}{1}$; (b) warp rib $\frac{1}{4}$; (c) warp rib $\frac{2}{2}$; (d) warp rib $\frac{4}{2}$. The repeats in the fabric view are highlighted.

When $n_{wf} = n_{wp}$ in the repeat fraction, the warp rip is called *regular*. The horizontal ribs, formed on the fabric surface, are of equal width (Fig. 8a).

When $n_{wp} > n_{wf}$ (Fig. 8a) or $n_{wf} > n_{wp}$ (Fig. 8b, c), the width of the horizontal ribs is different, and the rib weave is called *irregular*.

3.2.2 Weft Rib

The weft rib is obtained from the plain weave when the single overlap in the repeat fraction is multiplied by a number, different than 1 and 0, and the new overlaps are added in the direction of the weft threads.

The warp repeat of the weave is $R_{wp} = n_{wp} + n_{wf}$. The weft repeat is $R_{wf} = 2$.

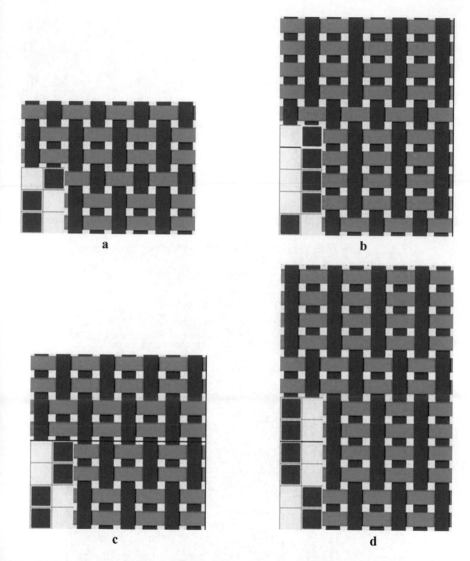

Fig. 8 Warp rib weaves: **a** warp rib $\frac{2}{.1}$; **b** warp rib $\frac{1}{.4}$; **c** warp rib $\frac{2}{.2}$; **d** warp rib $\frac{.4}{2.}$

Figure 9 shows two samples of weft rib: weft rib $\frac{2}{.2} >$ (Fig. 9a) and weft rib $\frac{.4}{2.} >$ (Fig. 9b). The fabric view and the repeats are illustrated. The weave in Fig. 9a is a regular weft rib, while the weave in Fig. 9b is an irregular weft rib.

The rib weave, being a derivative of the plain weave, could be produced using two harnesses only. In practice, four, six or eight harnesses are used.

Figure 10 presents a simulation of the fabric, woven with the warp rib from Fig. 8c. The horizontal ribs of the same size are visible, but depending on the yarns' density, the fabric could give the impression to be similar to the plain weave.

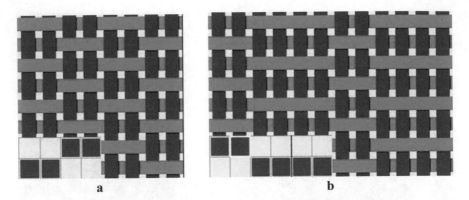

Fig. 9 Weft rib weaves: **a** weft rib $\frac{2}{2}$ >; **b** weft rib $\frac{4}{2}$ >

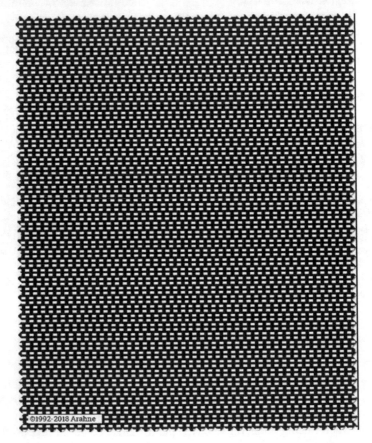

Fig. 10 Fabric simulation of warp rib $\frac{2}{2}$

Fig. 11 Fabric simulation of weft rib $\frac{.4}{2.}$ >

Figure 11 illustrates the fabric from Fig. 9b with vertical ribs of different size.

3.2.3 Complex Rib Weave

The complex rib can be either warp rib or weft rib. The repeat fraction contains more than one figure for both the warp overlaps n_{wp} and the weft overlaps n_{wf}. The weave pattern allows designing a cloth with more than two horizontal or vertical ribs of different width.

Figure 12 shows the repeat and the colour view of the complex warp rib $\frac{2.4.}{.3.5}$ with $R_{wp} = 14$ and $R_{wf} = 2$. The warp set is white, and the weft set is dark grey (the same colour palette is used from now on).

Figure 13 presents the repeat and colour view of the complex weft rib $\frac{.2.1.5}{6.3.4.}$ >. The warp repeat is $R_{wp} = 2$, the weft repeat is $R_{wf} = 21$.

Fig. 12 Complex warp rib $\frac{2.4.}{.3.5}$

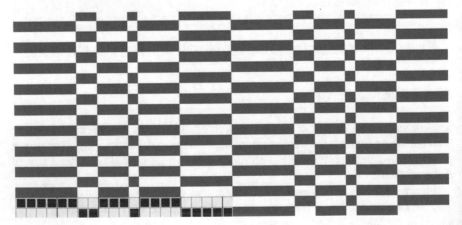

Fig. 13 Complex weft rib $\frac{.2.1.5}{6.3.4.} >$

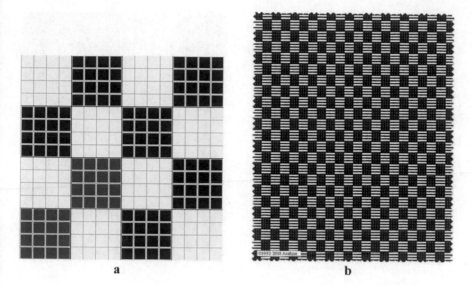

Fig. 14 Hopsack weave $\frac{4}{.4}$: **a** weave diagram with highlighted repeat; **b** fabric simulation

3.3 Hopsack Weave

The hopsack weave is obtained from the plain weave after multiplying the single warp and weft overlap in the repeat fraction by a number, different than 1 and 0. The new overlaps are marked in both warp and weft directions. When the multiplier is an equal number, i.e. $n_{wf} = n_{wp}$, the hopsack weave is regular and leads to the formation of texture with squares of warp and weft view. When the multiplier is different for the warp and weft overlaps, i.e. $n_{wf} \neq n_{wp}$, the hopsack weave is irregular and leads to the formation of texture with squares and rectangles of warp and weft view.

The repeat of the weave is square: $R = R_{wp} = R_{wf} = n_{wp} + n_{wf}$.

Figures 14 and 15 show the weave diagram and the fabric simulation of regular hopsack weave $\frac{4}{.4}$ and irregular hopsack weave $\frac{3}{.5}$, respectively.

4 The Twill Weave and Its Derivatives

4.1 Twill Weave

The twill weave leads to a formation of clear diagonals on the surface of the fabric. The effect is obtained due to the application of a constant displacement number of $S = \pm 1$.

Two types of diagonals are formed:

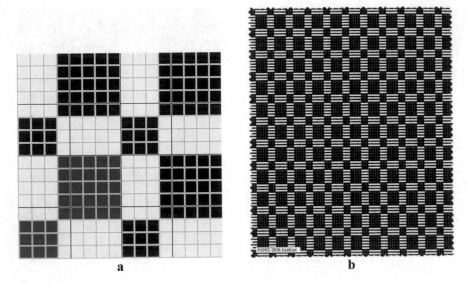

a   b

Fig. 15 Hopsack weave $\frac{3}{3}$: **a** weave diagram with highlighted repeat; **b** fabric simulation

- Right-handed or Z diagonal: when the next single overlap is placed with a displacement number of:
  - $S = S_{wp} = S_{wf} = +1$, or
  - $S = S_{wp} = S_{wf} = -1$.

- Left-handed or S diagonal: when the next single overlap is placed with a displacement number of:
  - $S_{wp} = +1$ and $S_{wf} = -1$, or
  - $S_{wp} = -1$ and $S_{wf} = +1$

For clarity, the direction of the twill diagonal could be mentioned after the repeat fraction: Z or S.

The twill weave has a square repeat: $R = R_{wp} = R_{wf} \geq 3$. The elementary twill weave has only one warp overlap or only one weft overlap in the repeat fraction, i.e.:

- $n_{wp} = 1$ and $n_{wf} = R - 1$, or
- $n_{wf} = 1$ and $n_{wp} = R - 1$.

The fabric with an elementary twill weave is warp-faced, when $n_{wp} > n_{wf}$, and weft-faced when $n_{wf} > n_{wp}$.

Figure 14 shows four examples of elementary twill weave with $R = 5$: a weft-faced, right-handed twill $\frac{1}{.4}Z$ (Fig. 16a); a warp-faced, right-handed twill $\frac{4.}{.1}Z$ (Fig. 16b); a weft-faced, left-handed twill $\frac{1.}{.4}S$ (Fig. 16c); and a warp-faced, left-handed twill $\frac{4.}{.1}S$ (Fig. 16d).

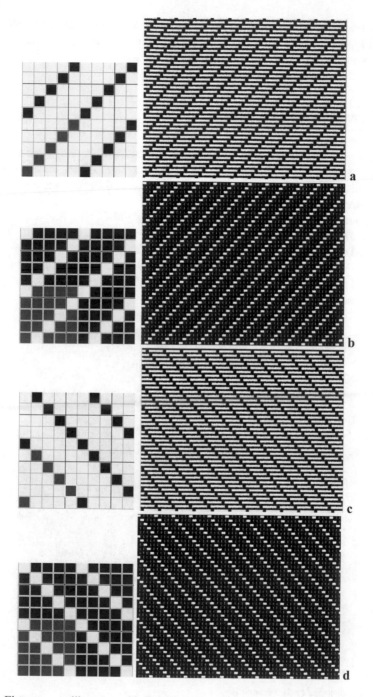

Fig. 16 Elementary twill weaves with $R = 5$; weave diagrams and fabric simulations: **a** $\frac{1}{.4}Z$; **b** $\frac{4}{.1}Z$; **c** $\frac{1}{.4}S$; **d** $\frac{4}{.1}S$

All warp threads in the repeat of the elementary twill weave have different law of motion (up and down). Therefore, the weaving of the fabric requires at least as many harnesses as the warp repeat of the weave pattern R_{wp}.

4.2 Strengthened Twill Weave

This derivative of the elementary twill weave is obtained when the single overlap in the repeat fraction is multiplied by a number, greater than 1. Thus, the single-thread diagonal disappears. Fabrics with diagonals of equal or different size are produced with the strengthened twill weave.

The weave has a square repeat: $R = R_{wp} = R_{wf} \geq 4$.

Figure 17 shows the repeat and the colour view of the strengthened twill $\frac{3}{3}Z$ with warp-faced and weft-faced diagonals of equal width.

Figure 18 shows the repeat and the colour view of the strengthened twill $\frac{4}{2}S$ with warp-faced and weft-faced diagonals of different width. As $n_{wp} > n_{wf}$, the result is a warp-faced fabric.

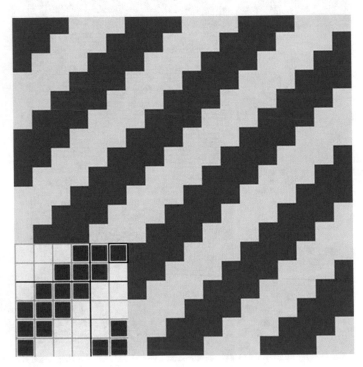

Fig. 17 Strengthened twill weave $\frac{3}{3}Z$

Fig. 18 Strengthened twill weave $\frac{4;}{.2}S$

4.3 Complex Twill Weave

This derivative of the elementary twill had at least two changes from warp to weft overlaps and vice versa alongside one thread in the weave repeat. The repeat fraction involves at least 4 figures: two for the fraction of the warp overlaps n_{wp} and two for the fraction of the weft overlaps n_{wf}.

The weave has a square repeat: $R = R_{wp} = R_{wf}$.

Figures 19 and 20 visualize the repeat and the colour view of the complex twill weaves $\frac{4.1.}{.2.3}S$ and $\frac{5.2.4}{.1.3.2}S$, respectively.

4.4 Zig-Zag Twill Weave

The zig-zag or wave twill requires a change in the displacement number S of the weave within the repeat. It can be obtained using either elementary twill or its derivatives—strengthened and complex twill.

The zig-zag effect appears when the basic weave changes its diagonal after certain threads. It could be done in the direction of the warp threads or the direction of the weft threads.

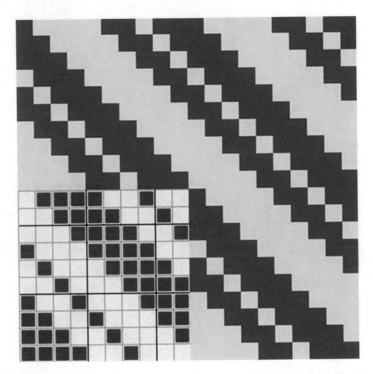

Fig. 19 Complex twill weave $\frac{4.1.}{.2.3} S$

When the diagonal direction changes from Z to S or vice-versa, a pick is formed in the weave. It could involve one thread (one-thread pick) or two threads (two-thread-pick).

4.4.1 Zig-Zag Twill in the Direction of the Weft Set

It is also called *horizontal zig-zag twill weave*. The rules for its design are:

- A basic weave is selected (elementary twill, strengthened twill or complex twill).
- The number of the warp threads, after which the displacement number S changes, is determined. Usually, it is equal to the warp repeat R_{wp} of the basic weave and leads to the formation of a zig-zag twill with *normal teeth*. If the number of the warp yarns after which the direction of the diagonal changes is lower than R_{wp}, a zig-zag twill weave with *reduced teeth* would appear. If the number of the warp yarns is greater than R_{wp}, a zig-zag twill weave with *enlarged teeth* would be designed.
- The type of the pick is determined: one-thread or two-thread.
- The repeat of the horizontal zig-zag twill weave is calculated in function with the repeat of the basic weave R_b and the above design choices.

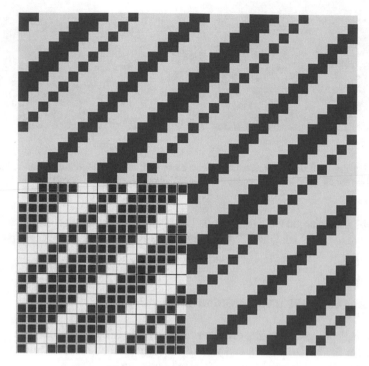

Fig. 20 Complex twill weave $\frac{5.2.4}{.1.3.2}S$

The warp repeat R_{wp} for a zig-zag twill weave with one-thread pick and normal teeth is:

$$R_{wp} = 2R_{b,wp} - 2$$

where $R_{b,wp}$ is the warp repeat of the basic weave.

The warp repeat R_{wp} for a zig-zag twill weave with two-thread pick and normal teeth is:

$$R_{wp} = 2R_{b,wp}$$

The weft repeat R_{wf} is equal to the weft repeat $R_{b,wf}$ of the basic weave.

If the design choice is to use a greater or lower number of warp yarns to design reduced or enlarged teeth, then the warp repeat R_{wp} could be calculated, applying the same expressions. Instead of the warp repeat of the basic weave $R_{b,wp}$, the number of the warp threads, after which the direction of the diagonal changes, should be used.

Figure 21 shows a horizontal zig-zag twill from a basic twill $\frac{2}{.4}Z$ with a one-thread pick (Fig. 21a) and a two-thread pick (Fig. 21b).

Figure 22 demonstrates the repeat and fabric simulation of a one-pick zig-zag weave in weft direction from a basic twill $\frac{4:}{.3}Z:$ with reduced (Fig. 22a), normal

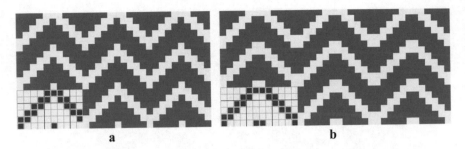

Fig. 21 Zig-zag twill weave (horizontal) from a basic twill weave $\frac{2}{4}$ **Z**: **a** with a one-thread pick; **b** with a two-thread pick

(Fig. 22b) and enlarged teeth (Fig. 22c). The change in the diagonal direction is after 4 warp threads (Fig. 22a), after 7 warp threads (Fig. 22b) and after 10 warp threads (Fig. 22c).

4.4.2 Zig-Zag Twill in the Direction of the Warp Set

It is also called *vertical zig-zag twill weave*. The basic rules for its design are similar to these for the horizontal zig-zag weave. The difference is that a number of weft threads are selected after which the diagonal direction changes.

The warp repeat R_{wp} is equal to the warp repeat $R_{b,wp}$ of the basic weave.

The weft repeat R_{wf} for a zig-zag twill weave with one-thread pick and normal teeth is:

$$R_{wf} = 2R_{b,wf} - 2$$

where $R_{b,wf}$ is the weft repeat of the basic weave.

The weft repeat R_{wf} for a zig-zag twill weave with two-thread pick and normal teeth is:

$$R_{wf} = 2R_{b,wf}$$

Figure 23 shows a vertical zig-zag twill with a one-thread (Fig. 23a) and a two-thread pick (Fig. 23b) using twill $\frac{3}{3}$ *S* as a basic weave. The fabric view is visualized together with the weave repeat.

4.4.3 Diagonal Zig-Zag Weave

The diagonal zig-zag weave is obtained when the number of threads in Z and S diagonal is different. The rules for its design are:

Fig. 22 One-pick zig-zag weave in weft direction from a basic twill $\frac{4}{3}$ Z—weave repeat and fabric simulation: **a** reduced teeth; **b** normal teeth; **c** enlarged teeth

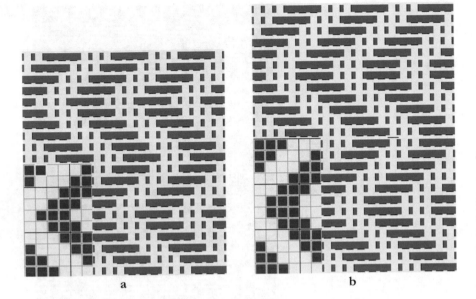

Fig. 23 Zig-zag twill weave (vertical) from a basic twill weave $\frac{3}{3}S$: **a** with a one-thread pick; **b** with a two-thread pick

- A basic weave is selected (elementary twill, strengthened twill or complex twill).
- The number of the threads, after which the displacement number S changes, is determined. The second number of threads (different from the first one), after which the displacement number S changes again, is selected.
- The type of the pick is determined: one-thread or two-thread.
- The direction of the zig-zag is also determined: towards the warp or the weft threads.

Figure 24 shows a diagonal zig-zag weave from a basic twill weave $\frac{3.1}{.1.3}Z$. The weave is with a one-thread pick, developed in the horizontal direction of the cloth.

4.4.4 Diamond Twill Weave

The diamond twill weave has both Z and S diagonals in the repeat. The following rules for its design are applied:

- A basic weave is selected (elementary twill, strengthened twill or complex twill).
- After a certain number of threads (usually equal to the repeat of the basic weave) the diagonal is changed from Z to S or vice versa (depending on the diagonal of the basic weave) in a horizontal direction.
- Using the same number of threads, the diagonal is changed in the vertical direction as well.

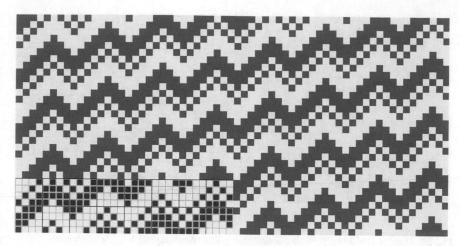

Fig. 24 Diagonal zig-zag twill from a basic twill weave $\frac{3.1.}{.1.3}Z$

The diamond twill weave usually has a one-thread pick. Therefore, the repeat R can be calculated as:

$$R = R_{wp} = R_{wf} = 2R_b - 2$$

where R_b is the repeat of the basic weave.

If the design choice is to use a greater or lower number of yarns than the repeat of the basic weave R_b, then the repeat of the diamond twill could be calculated, applying the same expressions. Instead of R_b, the number of the threads after which the direction of the diagonal changes, should be used.

Figure 25 presents a diamond twill weave from a basic twill $\frac{2}{.3}Z$.

4.4.5   Shadowed Twill Weave

The shadowed twill is designed from an elementary twill weave, and in each step of the repeat's design, a new overlap is added or omitted. The following rules are applied:

- An elementary twill in a warp-faced or weft-faced view is selected.
- After a certain number of threads (usually equal to the repeat of the basic weave) a new overlap is added to the single overlap (warp or weft) in the basic weave. In the next repeat, one more overlap is added. The process continues until the negative view of the basic weave is obtained.
- In the next repeats, the added overlaps are omitted one by one. The final repeat should have the view of the basic weave plus one additional overlap.

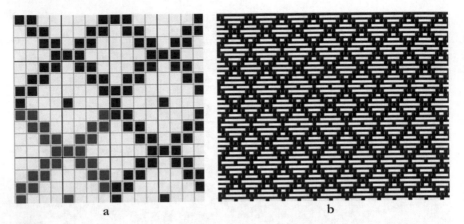

Fig. 25 Diamond twill weave from a basic twill $\frac{2}{3}$ **Z**: **a** weave repeat; **b** fabric simulation

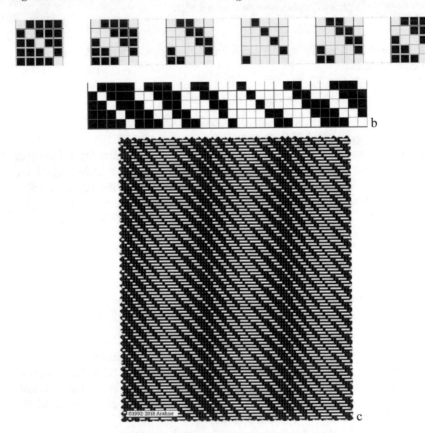

Fig. 26 Shadowed twill from a basic weave $\frac{4}{1}$ **S**: **a** the consecutive repeats in the design; **b** the whole repeat; **c** fabric simulation

Figure 26 presents the different repeats in the design of a shadowed twill from the basic weave $\frac{4}{.1}S$ (Fig. 26a) and the whole repeat (Fig. 26b). The fabric simulation is shown in Fig. 26c.

5 The Sateen and Satin Weaves and Their Derivatives

5.1 Satin and Sateen Weaves

The *satin* weave has a warp-faced repeat [4] with a single weft overlap alongside a single yarn. To enhance the effect of the weave, the fabric is produced with higher warp density than a weft density.

The *sateen* weave has a weft-faced repeat [4] with a single warp overlap alongside a single yarn. To enhance the effect of the weave, the cloth is produced with higher weft density than warp density.

Satin and sateen weaves have common features:

- The single overlaps have an even distribution within the weave repeat, and they do not contact (as it is in the case of the plain weave and twill weave).
- The weave repeat is $R \geq 5$.
- The displacement number S is determined as:

$$1 < S < R - 1$$

- The displacement number S could be applied either in the direction of the warp threads or in the direction of the weft threads.
- The displacement number S and the weave repeat R should not have a common divisor. It is recommendable S to be as closer as possible to R/2.

Following these requirements, it is easy to calculate that the 5-thread sateen weave $\frac{1}{.4}$ could have two possible displacement numbers S: $S = 2$ (Fig. 27a) and $S = 3$ (Fig. 27b). For comparison, a satin weave $\frac{4}{.1}$ with $S = 3$ is also presented (Fig. 27c).

The 6-thread satin or sateen cannot fulfil the requirements for calculation of S: the possible displacement numbers are 2, 3 and 4, but all they have common divisors with R (2 and 3). That is why the 6-thread satin and sateen have no constant displacement number and are called *irregular*. Figure 28 shows a possible version of the irregular 6-thread sateen weave.

Figure 29 presents the repeat and fabric view of two satin weaves: $\frac{7}{.1}$ with $S = 3$ (Fig. 29a) and $\frac{10}{.1}$ with $S = 6$ (Fig. 29b).

R. A. Angelova

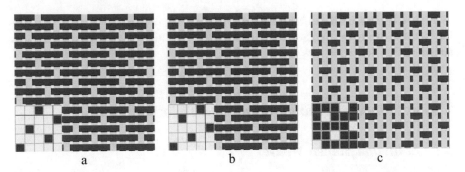

a   b   c

Fig. 27. 5-thread satin and sateen weaves: **a** sateen $\frac{1}{.4}$ with $S = 2$; **b** sateen $\frac{1}{.4}$ with $S = 3$; **c** satin $\frac{.4}{1.}$ with $S = 3$

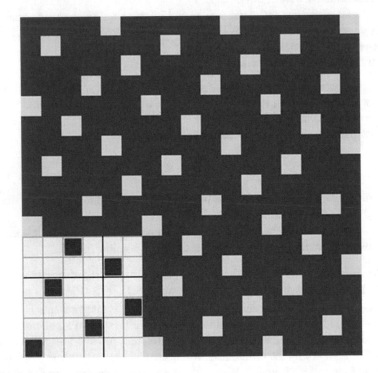

Fig. 28. 6-thread irregular sateen weave

5.2 Strengthened Satin Weave

This derivative of the elementary satin weave is designed by multiplying the single weft overlap by a number, greater than 1. The repeat of the basic weave does not

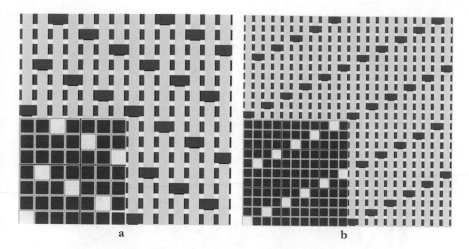

Fig. 29 Repeat and fabric view of satin weaves: **a** $\frac{7}{.1}$ with S = 3; **b** $\frac{10}{.1}$ with S = 6

change. The additional weft overlaps could be placed in the direction of the warp threads or the direction of the weft threads.

5.2.1 Strengthened Satin Weave in the Warp Direction

The additional one or more weft overlaps are placed above or below the single overlap in the basic satin weave. The position must be one and the same (e.g. only above the existing overlap) within the whole repeat.

Figure 30 shows strengthened satin weave with 2 additional weft overlaps from a basic satin $\frac{4}{.1}$ with $S = 2$.

5.2.2 Strengthened Satin Weave in the Weft Direction

The additional one or more weft overlaps are placed on the left or the right side of the single overlap in the basic satin weave. The position must be the same within the whole repeat.

Figure 31 demonstrates strengthened satin weave with 3 additional weft overlaps from a basic satin $\frac{6}{.1}$ with $S = 3$.

5.2.3 Strengthened Sateen Weave

This derivative of the elementary sateen weave is designed by analogy with the strengthen satin weave.

Fig. 30 Strengthened satin weave with 2 additional weft overlaps in warp direction from a basic satin $\frac{4.}{.1}$ with $S = 2$

5.2.4 Strengthened Sateen Weave in the Warp Direction

The additional one or more warp overlaps are placed above or below the single overlap in the basic sateen weave.

Figure 32 illustrates strengthened sateen weave with 1 additional warp overlap from a basic sateen $\frac{1.}{.7}$ with $S = 5$.

5.2.5 Strengthened Satin Weave in the Weft Direction

The additional one or more weft overlaps are placed on the left or the right side of the single overlap in the basic sateen weave.

Figure 33 shows strengthened sateen weave with 4 additional warp overlaps from a basic satin $\frac{1.}{.6}$ with $S = 4$.

Fig. 31 Strengthened satin weave with 3 additional weft overlaps in weft direction from a basic satin $\frac{6}{.1}$ with $S = 3$

5.2.6 Shadowed Satin Weave

The shadowed satin is obtained, adding a new weft overlap in each stage of the repeat's design.

The rules are very similar to the rules of the design of a shadowed twill weave:

- A basic satin weave is selected.
- After a certain number of threads (usually equal to the repeat of the basic weave) a new weft overlap is added to the single weft overlap of the basic weave. In the next repeat, one more overlap is added. The process continues untill the negative view of the basic weave is reached.
- In the next repeats, the added overlaps are omitted one by one. The final repeat should have the view of the basic satin weave plus one additional weft overlap.

Figure 34a shows the different repeats in the design of a shadowed satin from a satin weave $\frac{6}{.1}$ with $S = 3$. The whole repeat (Fig. 34b) and fabric simulation (Fig. 34c) are also presented.

Fig. 32 Strengthened sateen weave with 1 additional warp overlap in warp direction from a basic satin $\frac{1}{.7}$ with $S = 5$

5.2.7   Shadowed Sateen Weave

The shadowed sateen is designed in the same way as the shadowed satin weave. The only difference is that the basic weave is sateen and additional warp overlaps are added or omitted in each consecutive repeat of the shadowed sateen.

Figure 35 demonstrates the design of a shadowed sateen from a sateen weave $\frac{1}{.6}$ with $S = 3$. This basic weave is the negative view of the basic weave in Fig. 34, used for the design of a shadowed satin weave. The consecutive repeats (Fig. 35a), the whole repeat (Fig. 35b) and fabric simulation (Fig. 35c) are shown.

The comparison between the results in Figs. 34 and 35 shows that using different basic weaves a cloth with the same weave effect is produced.

6   Weave Patterns for Stripe Effects

The stripe effects on the cloth are a result of the combination of at least two different weave patterns. The following rules are applied:

Fig. 33 Strengthened sateen weave with 3 additional warp overlaps in weft direction from a basic satin $\frac{1}{6}$ with $S = 4$

- Two or more basic weaves are selected. They could be elementary weaves and their derivatives, or other weaves that have well distinct texture.
- The repeats of the basic weaves should be small. They had to be numbers that are multiples of each other or have a common divisor.
- When possible, the first thread of the next weave should be negative to the last thread of the previous weave. If this is not possible, a small dividing contrast strip could be used (e.g. in plain or rib weave).

6.1 Horizontal Stripe Effects

Two or more weave patterns may be combined.

The warp repeat R_{wp} could be found as the least common multiple (LCM) of the warp repeat of the basic weaves R_{wpi}:

$R_{wp} = LCM(R_{wpi})$.

The weft repeat R_{wf} is the sum of the weft threads in each basic weave n_{wfi} (as more than one repeat in weft direction may be applied):

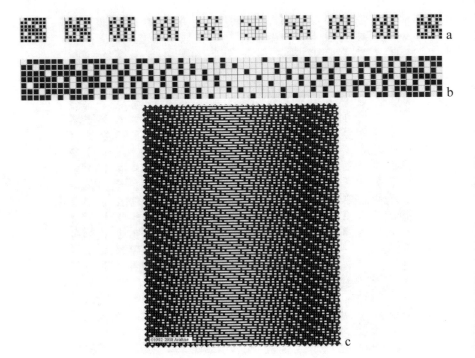

Fig. 34 Shadowed satin weave from a basic weave $\frac{6}{.1}$: **a** the consecutive repeats in the design; **b** the whole repeat; **c** fabric simulation

$$R_{wf} = \sum_{i=1}^{k} n_{wfi}$$

Figure 36 shows the repeat and fabric view of a pattern for horizontal stripe effects from two basic weaves: satin $\frac{4}{.1}$ with $S = 2$ and sateen $\frac{1}{.4}$ with $S = 2$.

6.2 Vertical Stripe Effects

Two or more weave patterns may be combined. However, the size of the warp repeat R_{wp} should be monitored, as it may become too large and require a jacquard weaving machine instead of a dobby.

The warp repeat R_{wp} could be calculated as the sum of the warp threads in each basic weave n_{wpi}:

$$R_{wp} = \sum_{i=1}^{k} n_{wpi}$$

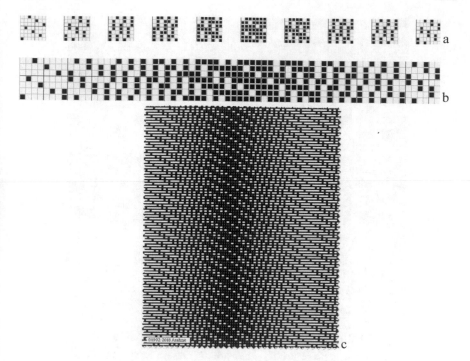

Fig. 35 Shadowed sateen weave from a basic weave $\frac{6}{.1}$: **a** the consecutive repeats in the design; **b** the whole repeat; **c** fabric simulation

The weft repeat R_{wf} is the least common multiple (LCM) of the weft repeat of the basic weaves R_{wfi}:

$$R_{wf} = LCM(R_{wfi}).$$

Figure 37 shows the repeat and fabric view of a pattern for vertical stripe effects from three basic weaves: plain weave (two repeats), warp rib $\frac{2}{.2}$ (two repeats) and a hopsack weave $\frac{2}{.2}$.

7 Weave Patterns for Cell-Like Effects

7.1 Square Weaves

The weave effects are squares or rectangles of similar or different size. The rules are as follows:

- Two or more basic weaves are selected with equal repeat or repeats that are multiples of each other.
- The cell may be of square or rectangular shape.

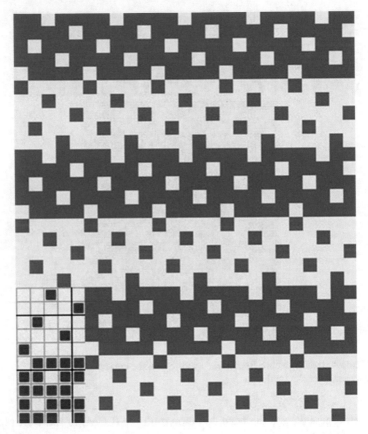

Fig. 36 Horizontal stripe effects from satin $\frac{4}{1}$ with $S = 2$ and sateen $\frac{1}{4}$ with $S = 2$

- When possible, the first thread of the next weave should be negative to the last thread of the previous weave.

The easiest way to design a square weave is to arrange two different basic weaves in a checkerboard pattern. In this case, the warp repeat R_{wp} could be calculated as

$$R_{wp} = 2n_{wp}$$

where n_{wp} is the number of the warp threads in a single cell.

The same is valid for the weft repeat R_{wf}:

$$R_{wf} = 2n_{wf}$$

where n_{wf} is the number of the weft threads in a single cell.

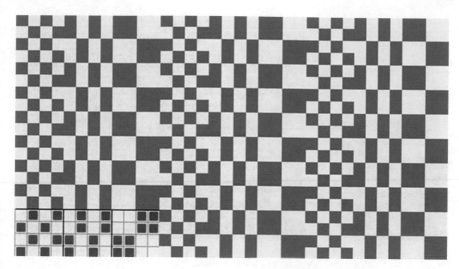

Fig. 37 Vertical stripe effects from plain weave (two repeats), warp rib $\frac{2}{2}$ (two repeats) and a hopsack weave $\frac{2}{2}$

A square weave from two basic twill weaves: $\frac{1}{3}Z$ and $\frac{3}{1}S$, is shown in Fig. 38. The repeat of the weave and the fabric simulation are demonstrated.

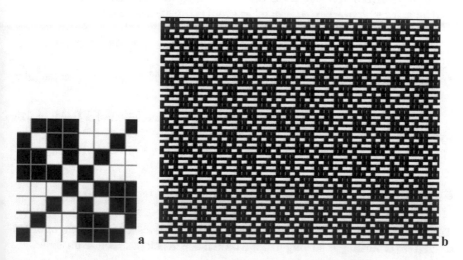

Fig. 38 Square weave from two basic twill weaves $\frac{1}{3}Z$ and $\frac{3}{1}S$: **a** repeat; **b** fabric simulation

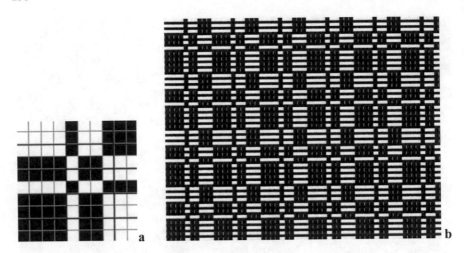

Fig. 39 Fancy basket weave $\frac{4.2.}{.1.3}$: **a** repeat; **b** fabric simulation

7.2 Fancy Basket Weave

The design of the fancy basket weave requires the completion of the following rules:

- The repeat fraction should involve at least two figures for warp overlaps and two figures for weft overlaps. It is good to avoid floats greater than 5.
- The presence of a single overlap is advisable.
- The repeat fraction is marked alongside both the first warp thread and first weft thread.
- The repeat of the weave is square and involves cells with a square and rectangular shape.

Figure 39 demonstrates the pattern and the fabric simulation of a fancy basket weave $\frac{4.2.}{.1.3}$.

8 Weave Patterns for Rib Effects

These patterns are used instead of warp or weft rib weaves when the length of the desired rib or floating is so long that it would make the cloth structure unstable.

8.1 Tightened Rib Weave

The structure of the traditional rib weave could be stronger if additional overlaps or threads in a different weave are added in the repeat of the rib pattern.

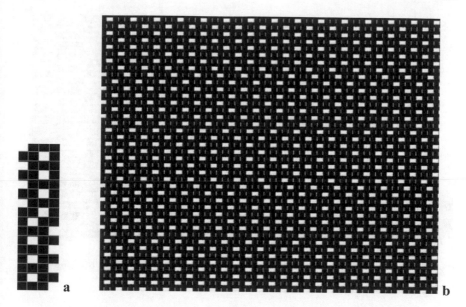

Fig. 40 Soleil weave from a basic warp rib $\frac{8.}{.8}$, tightened with additional warp overlaps in a plain weave: **a** repeat; **b** fabric simulation

8.1.1 Soleil Weave

This is a tightened rib weave, where additional overlaps are placed alongside the long yarn floats. The following rules are applied:

- A basic rib weave with yarn floats bigger than 6 is selected.
- A tightened weave with a small repeat is chosen: a plain weave and its derivatives or twill weave.
- If the basic weave is a warp rib, the new warp overlaps are added alongside the weft floats. If the basic weave is a weft rib, the new weft overlaps are added alongside the warp floats.

Figure 40 shows a soleil weave from a basic warp rib $\frac{8.}{.8}$, tightened with additional overlaps that follow the interlacing of a plain weave. The weave repeat (Fig. 40a) and the fabric simulation (Fig. 40b) are presented.

Figure 41 illustrates a soleil weave from a basic weft rib $\frac{6.}{.6} >$, tightened with additional overlaps that follow the interlacing of a twill weave $\frac{.2}{1.} Z$. Both the weave repeat (Fig. 41a) and fabric simulation (Fig. 41b) are shown.

8.1.2 Tightened Rib Weave with Additional Threads

Similar rules to the rules, used in the soleil weave, are applied:

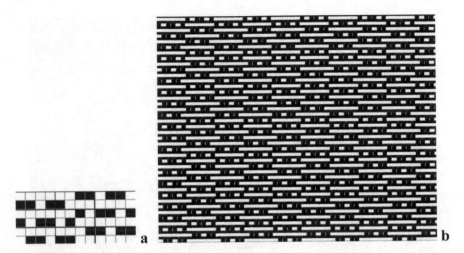

Fig. 41 Soleil weave from a basic weft rib $\frac{6}{.6}$ >, tightened with additional weft overlaps in a twill weave $\frac{2}{1.}$ Z: **a** repeat; **b** fabric simulation

- A basic rib weave with yarn floats bigger than 6 is selected.
- A tightened weave with a small repeat is chosen: a plain weave and its derivatives or twill weave.
- A ratio of alternation between the threads of the basic rib weave and the threads of the tightened weave is selected: 2:1 or 4:1.

Figure 42 shows a tightened rib weave from a basic warp rib $\frac{8}{.8}$, tightened with warp threads in the plain weave, with a ratio 4:1. Figure 43 presents a tightened rib

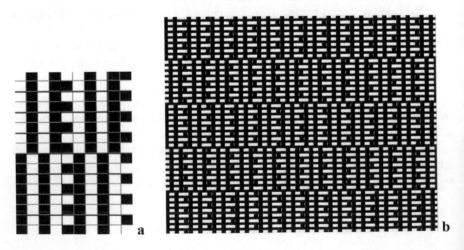

Fig. 42 Tightened rib weave from a basic warp rib $\frac{8}{.8}$, tightened with warp threads in the plain weave, with a ratio 4:1: **a** repeat; **b** fabric simulation

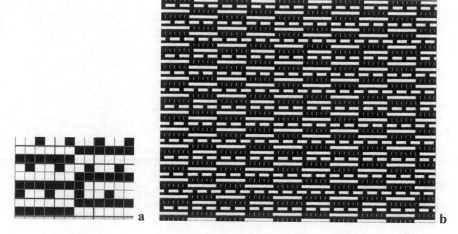

Fig. 43 Tightened rib weave from a basic weft rib $\frac{6:}{.6}$ >, tightened with additional weft threads in a twill weave $\frac{1:}{.2}Z$ with a ratio 2:1: **a** repeat; **b** fabric simulation

weave from a basic weft rib $\frac{6:}{.6}$ >, tightened with additional threads from a twill weave $\frac{1:}{.2}Z$ with a ratio 2:1.

8.2 Tightened Twill Weave with Additional Weft Threads

The reason to tighten the twill weave is to obtain a more stable fabric structure when longer twill repeats are used. The rules for designing are:

- Elementary twill weave with repeat greater than 8 is selected to be a basic weave.
- The plain weave is used as a tightened weave.
- A ratio of alternation between the threads of the basic twill and the plain weave is chosen: 2:2 or 4:2.

The application of thick weft yarns increases the texture effect. The ribs in plain weave constrict the cloth, thus creating a 3D visual effect.

Figure 44 illustrates a tightened twill weave with additional weft threads from a basic twill $\frac{1:}{.11}Z$ with alternation between the threads in twill and the threads in plain weave 4:2.

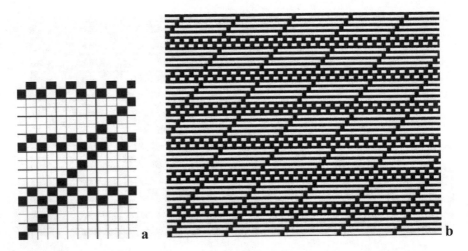

Fig. 44 Tightened twill weave with additional weft threads from a basic twill $\frac{1}{.11}$ Z and alternation with the plain weave 4:2: **a** repeat; **b** fabric simulation

8.3 Tightened Satin or Sateen Weave with Additional Weft Threads

The design rules are the same as for the tightened twill weave with additional weft threads. The difference is that the basic weave is satin or sateen weave with $R \geq 8$.

Figure 45 shows a tightened satin weave with additional weft threads from a basic sateen $\frac{1}{.9}$ with $S = 3$. The alternation with the threads in plain weave is 2:2.

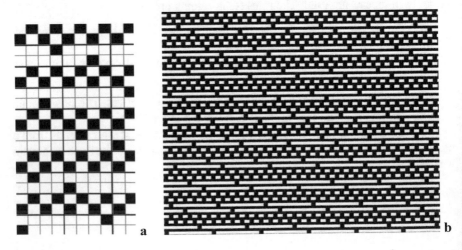

Fig. 45 Tightened sateen weave with additional weft threads from a basic sateen $\frac{1}{.9}$ with $S = 3$ and alternation with the plain weave 2:2: **a** repeat; **b** fabric simulation

9   Weave Patterns for Diagonal Effects

These patterns create diagonal effects on the cloth, which is different than the classical effect of the twill weave and its derivatives. Some of the weaves, however, are based on the twill weave again.

9.1   Elongated Twill Weave

The basic effect is diagonal, which angle is different than 45°. The rules for designing this weave are the following:

- A basic twill weave is selected: elementary, strengthened or complex twill.
- The desired inclination of the diagonal is chosen:

  - <45° to design *reclined twill;*
  - >45° to design *steep twill.*

- A displacement number $S \neq 1$ is selected. S should not break the twill diagonal.
- The overlaps of the steep twill are marked in the direction of the warp yarns, using the selected displacement number S.
- The overlaps of the reclined twill are marked in the direction of the weft yarns, using the selected displacement number S.

The warp repeat of the steep twill R_{wp} is:

$$R_{wp} = \frac{R_{wp,b}}{S}$$

where $R_{wp,b}$ is the warp repeat of the basic weave.

The weft repeat R_{wf} of the steep twill is equal to the weft repeat of the basic weave $R_{wf,b}$.

The warp repeat R_{wp} of the reclined twill is equal to the warp repeat of the basic weave $R_{wp,b}$.

The weft repeat of the steep twill R_{wf} is:

$$R_{wf} = \frac{R_{wf,b}}{S}$$

where $R_{wf,b}$ is the weft repeat of the basic weave.

Figure 46 illustrates elongated steep twill from a basic twill $\frac{5}{.4}Z$ with $S = 3$. The repeat (Fig. 46a), and fabric simulation (Fig. 46b) are shown.

A reclined elongated twill from a basic twill $\frac{4.1.}{.2.5} > Z$ with $S = 2$ is shown in Fig. 47: the repeat (Fig. 47a) and fabric simulation (Fig. 47b).

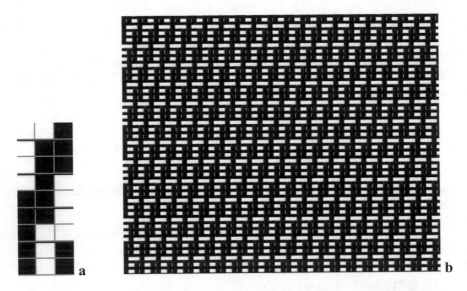

Fig. 46 Elongated steep twill from a basic twill $\frac{5}{4}Z$ with $S = 3$: **a** repeat; **b** fabric simulation

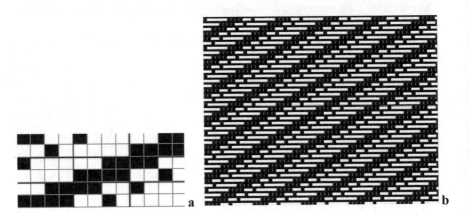

Fig. 47 Elongated reclined twill from a basic twill $\frac{4.1.}{.2.5} > Z$ with $S = 2$: **a** repeat; **b** fabric simulation

9.2 Diagonal Rib Weave

The basic rib weave creates diagonals on the cloth. The following rules are used:

- A basic weave: warp or weft rib, is selected.
- An elementary weft-faced twill is selected. Its warp overlaps show the beginning of the whole repeat of the basic rib weave. Therefore it is good the repeat of the twill to be equal to the greater repeat of the basic rib weave.

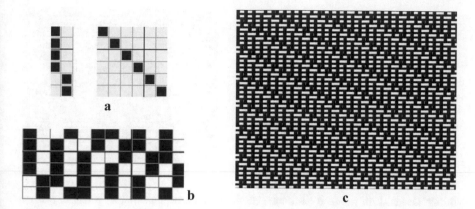

Fig. 48 Diagonal rib weave: **a** basic warp rib $\frac{4}{2}$ and a twill weave $\frac{1}{5}S$; **b** repeat; **c** fabric simulation

The calculation of the repeat depends on the type of the basic rib weave.

If the basic weave is a warp rib, the warp repeat of the diagonal rib R_{wp} is:

$$R_{wp} = 2R_t$$

where R_t is the repeat of the twill weave.

The weft repeat R_{wf} of the diagonal rib is equal to the weft repeat of the basic rib weave $R_{wf,b}$.

If the basic weave is a weft rib, the warp repeat of the diagonal rib R_{wp} is equal to the warp repeat of the basic rib weave $R_{wp,b}$. The weft repeat R_{wf} of the diagonal rib is

$$R_{wf} = 2R_t$$

Figure 48 presents the repeat (Fig. 48b) and the fabric simulation (Fig. 48c) of a diagonal rib weave from a warp rib $\frac{4}{2}$ and a twill weave $\frac{1}{5}S$ (Fig. 48a).

Figure 49 shows a diagonal rib weave from a weft rib $\frac{4}{3} >$ and a twill weave $\frac{1}{6}Z$ (Fig. 49a). The repeat (Fig. 49b) and the fabric simulation (Fig. 49c) are also presented.

9.3 Fancy Diagonal Twill

This weave creates fancy diagonal of changeable width, due to the insertion of basic twill weaves with the same direction of the diagonals. The applied rules are as follows:

- Two basic twill weaves are selected with the same diagonal direction. Three or more patterns could also be used.

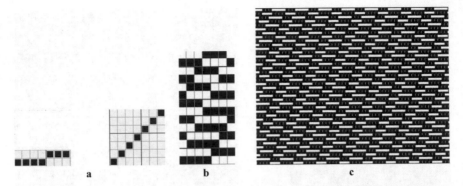

Fig. 49 Diagonal rib weave: **a** basic weft rib $\frac{4}{.3}$ > and a twill weave $\frac{1}{.6}Z$; **b** repeat; **c** fabric simulation

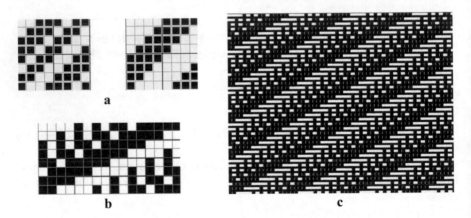

Fig. 50 Fancy diagonal twill weave after warp alternation of the basic twill weaves: **a** basic weaves: twill $\frac{2.3.}{.1.2}Z$ and twill $\frac{4.}{.4}Z$; **b** repeat; **c** fabric simulation

- A ratio of alternation between the threads of the basic weaves is selected; usually, it is 1:1.
- A direction of the alternation is chosen: in the direction of the warp yarns or weft yarns.

Figure 50 shows the fancy diagonal twill from two basic weaves: complex twill $\frac{2.3.}{.1.2}Z$ and strengthened twill $\frac{4.}{.4}Z$ (Fig. 50a). A warp alternation 1:1 is applied. The repeat of the new weave (Fig. 50b) and the fabric view (Fig. 50c) are presented.

The same basic twill weaves are used in Fig. 51, but a weft alternation 1:1 is applied to obtain the repeat (Fig. 51a) of the fancy diagonal twill. The fabric view is presented in Fig. 51b.

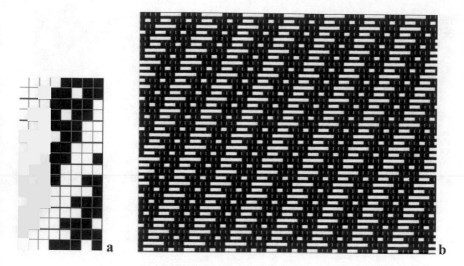

Fig. 51 Fancy diagonal twill weave after weft alternation of the basic twill weaves: **a** basic weaves: twill $\frac{2.3.}{.1.2}$ Z and twill $\frac{4.}{.4}$ Z; **b** repeat; **c** fabric simulation

9.4 Broken Twill

The broken twill has two diagonals in the repeat, but they do not form a zig-zag effect. The diagonals are located opposite in the repeat. The following rules could be applied for the weave's design:

- An elementary twill weave with a single-overlap diagonal is used as a basic weave.
- Part of the single-overlap diagonal is left as it is, while the other part is rotated to 90°. It is not necessary for the diagonal to be divided into two equal parts.

The repeat of the broken twill remains the same as the repeat of the basic twill.

Figure 52a presents the repeat and colour view of a broken twill from a basic twill $\frac{1.}{.6}$ Z, where the diagonal rotates after half of the repeat. Figure 52b shows the repeat and colour view of a broken twill from a basic twill $\frac{.7}{1.}$ Z, where the diagonal rotates after 5 threads, forming diagonals with different length in the left and right direction.

10 Weave Patterns for Zig-Zag Effects

10.1 Zig-Zag Rib Weave

The creation of the zig-zag rib is quite similar to the diagonal rib weave, but the whole rib repeat follows the displacement pattern of a zig-zag twill. The rules are:

- A basic weave: warp or weft rib, is selected.

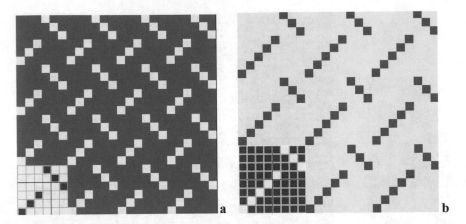

Fig. 52 Broken twill weave from: **a** a basic twill $\frac{1}{.6}Z$; **b** a basic twill $\frac{7}{.1}Z$

- A zig-zag twill, based on elementary twill weave (with a single warp or weft overlap) is chosen; it could be with one-thread or two-thread pick:

  o   If the basic weave is a warp rib, a horizontal zig-zag twill is selected.
  o   If the basic weave is a weft rib, a vertical zig-zag twill is selected.

Figure 53 shows a zig-zag rib weave from a basic warp rib $\frac{3}{3.}$ and a two-thread horizontal zig-zag twill from an elementary twill $\frac{1}{.5}Z$ (Fig. 53a). Both the repeat of the zig-zag rib weave (Fig. 53b) and the fabric simulation (Fig. 53c) are presented.

Figure 54 shows a zig-zag rib weave from a basic weft rib $\frac{4.}{.2}$ and a one-thread vertical zig-zag twill from an elementary twill $\frac{1}{.3}Z$ (Fig. 54a). The repeat of the zig-zag rib weave is presented in Fig. 54b, and the fabric simulation—in Fig. 54c.

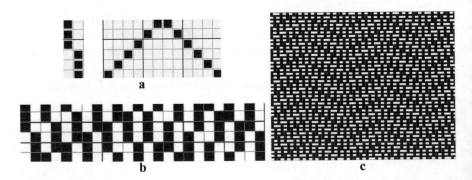

Fig. 53 Zig-zag rib weave: **a** basic weaves: warp rib $\frac{3}{3.}$ and a two-thread horizontal zig-zag twill from an elementary twill $\frac{1}{.5}Z$; b repeat; **c** fabric simulation

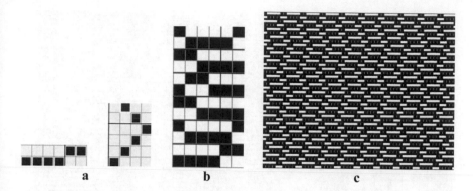

Fig. 54 Zig-zag rib weave: **a** basic weaves: weft rib $\frac{4}{.2}$ and a one-thread vertical zig-zag twill from an elementary twill $\frac{1}{.3} Z$; **b** repeat; **c** fabric simulation

10.2   Herringbone Twill Weave

The herringbone twill weave is similar to the zig-zag twill weave. However, not only the direction but the view of the diagonal changes from warp to the weft or vice versa. The rules for the composing the weave are:

- A twill weave with equal warp and weft overlaps in the repeat is selected as a basic weave.
- The direction of the zig-zag is selected: alongside the warp threads (vertical zig-zag) or alongside the weft threads (horizontal zig-zag).
- The number of the threads, after which the displacement number S and the view of the diagonal will change, is determined. Usually, it is equal to the repeat of the basic weave, but could be lower or greater number.

A herringbone twill from a basic twill $\frac{4}{.4} Z$ is shown in Fig. 55. The diagonal change is done after a whole repeat (8 threads) in the direction of the weft yarns. The repeat of the weave (Fig. 55a) and the fabric simulation (Fig. 55b) are presented.

Figure 56 illustrates the same basic twill as in Fig. 55, but the diagonal changes in the direction of the warp yarns. Both the repeat of the weave (Fig. 56a) and the fabric simulation (Fig. 56b) are shown.

11   Weave Patterns for Curved Effects

11.1   Curved Twill Weave

The curved effect is obtained from a tightened twill weave as a basic weave. Curved diagonals are formed in the direction of the warp (*longitudinally curved twill weave*)

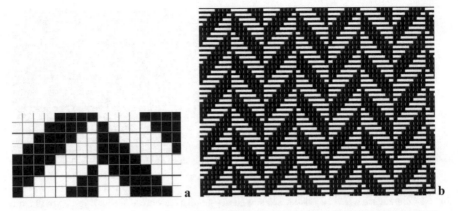

Fig. 55 Herringbone twill weave in a weft direction from a basic twill $\frac{4}{4}$ **Z**: **a** repeat; **b** fabric simulation

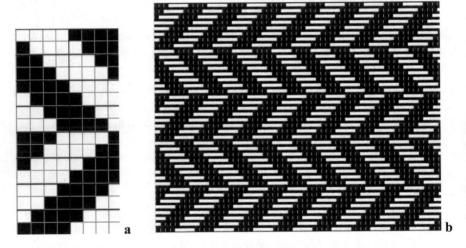

Fig. 56 Herringbone twill weave in a warp direction from a basic twill $\frac{4}{4}$ **Z**: **a** repeat; **b** fabric simulation

or the weft threads (*transversally curved twill weave*). The rules that are applied for the design are as follows:

- A strengthened twill weave, with a warp-faced diagonal, is selected. The diagonal has to be with at least three warp overlaps.
- After a certain number of threads (usually equal to the repeat of the basic weave, but could be lower or higher), the displacement number S increases with 1 (e.g. from $S = 1$ to $S = 2$).

- The increment of S in each consequent repeat continues until $S_{max} = n_o - 1$, where n_o is the number of the warp overlaps in the diagonal. The idea is not to break the diagonal.
- After the repeat with S_{max}, the displacement number S starts to decrease in each consequent repeat, until reaching $S = 2$.

The maximal curved effect is obtained when the displacement number changes from 1 to S_{max}. A lower displacement number than S_{max} could also be applied to obtain a lighter curved effect.

Figure 57 presents the different repeats in the design of a curved twill weave in the direction of the warp threads from a basic weave $\frac{4}{2}Z$ (Fig. 57a) and the whole repeat (Fig. 57b). The fabric simulation is shown in Fig. 57c.

The same basic weave $\frac{4}{2}Z$ is used for the design of curved twill weave in the direction of the weft threads. Figure 58a shows the repeats with changeable displacement number. The whole repeat (Fig. 58b) and fabric simulation (Fig. 58c) are also shown.

11.2 Curved Zig-Zag Twill Weave

The effect on the cloth is obtained, after adding the mirror image of the repeat of a curved twill weave. The same rules are applied, as for the curved twill weave:

- A strengthened twill weave, with a warp-faced diagonal with at least three warp overlaps n_o, is selected.
- After a certain number of threads (usually equal to the repeat of the basic weave, but could be lower or higher), the displacement number S increases with 1 (e.g. from $S = 1$ to $S = 2$).
- The increment of S in each consequent repeat continues until reaching $S_{max} = n_o - 1$.
- After the repeat with S_{max}, the displacement number S starts to decrease in each consequent repeat, until $S = 1$.
- The mirror image of the obtained repeat is added to design the whole repeat. One-thread or two-thread pick could be formed in the contact zone between the designed repeat and its mirror version.

Figure 59 presents the repeat (Fig. 59a) and the colour view (Fig. 59b) of a curved zig-zag twill in the direction of the warp threads from a basic weave $\frac{4}{2}Z$.

Figure 60 presents the repeat (Fig. 60a) and the colour view (Fig. 60b) of a curved zig-zag twill in the direction of the weft threads from a basic weave $\frac{4}{2}Z$.

Fig. 57 Curved twill weave (in warp direction) from a basic weave $\frac{4}{2}$ **Z**: **a** the consecutive repeats in the design; **b** the whole repeat; **c** fabric simulation

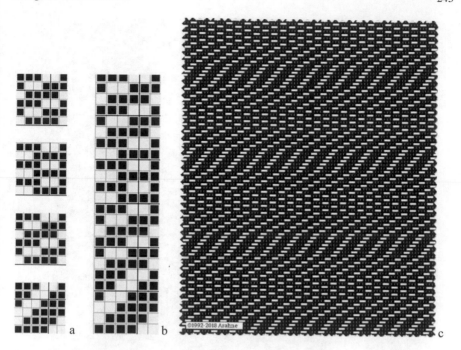

Fig. 58 Curved twill weave (in weft direction) from a basic weave $\frac{4.}{.2}$ **Z**: **a** the consecutive repeats in the design; **b** the whole repeat; **c** fabric simulation

Fig. 59 Curved zig-zag twill weave (in warp direction) from a basic weave $\frac{4.}{.2}$ **Z**: **a** the repeat; **b** colour view

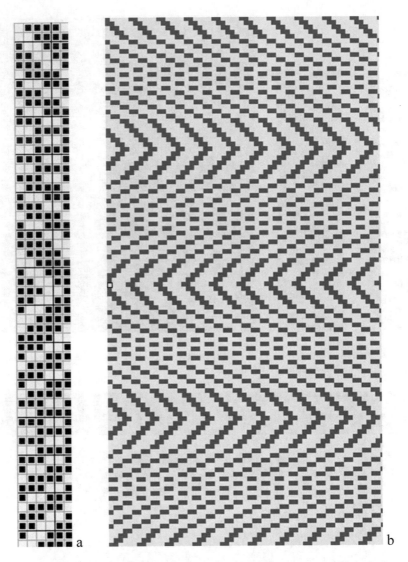

Fig. 60 Curved zig-zag twill weave (in weft direction) from a basic weave $\frac{4}{2}$ Z: **a** the repeat; **b** colour view

12 Weave Patterns for 3D Effects

These weaves are characterized by the formation of cells on the cloth with raised sides and a concave centre. The cells can have different shapes: square, rhombus, rectangle. The concave zones are formed by long warp or weft floats, and the raised ones are obtained due to the more intensive interlacing of the two systems of threads. Because of its view, these weave patterns are also called *waffle* weaves.

The rules for designing a waffle weave are:

- A basic zig-zag twill weave is selected.
- A mirror version of the zig-zag twill weave is added to obtain a square repeat.
- In the formed rhomboids, additional warp or weft overlaps are marked in a checkerboard pattern.

Figure 61 shows the steps in designing a waffle weave from a basic twill $\frac{1}{.5}Z$ (Fig. 61). The one-thread pick zig-zag twill is shown in Fig. 61b and the whole repeat of the waffle weave—in Fig. 61c. Figure 61d presents the fabric simulation of the cloth.

Waffle weaves could be designed by using square motifs, as shown in Fig. 62a. Two repeats of the positive version of the motif, and two repeats of the negative version of the motif, are placed in a checkerboard order in the repeat (Fig. 62b) of the waffle weave. The fabric simulation of the weave is demonstrated in Fig. 62c.

Fig. 61 Waffle weave: **a** the basic twill weave $\frac{1}{.5}Z$; **b** the zig-zag twill weave; **c** the repeat; **d** fabric simulation

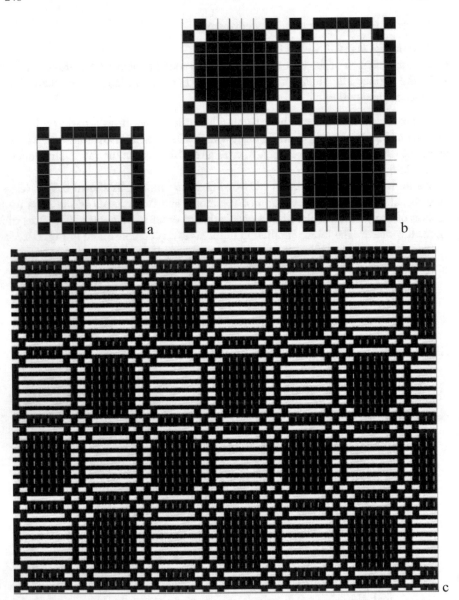

Fig. 62 Waffle weave: **a** motif; **b** the repeat; **c** fabric simulation

13 New Methods for the Design of Weave Patterns

There are several ways to design a new weave pattern, but most of the designed weaves are already known or similar to already known patterns. Some inspiration could be found in traditional motifs for weaving, knitting or embroidery in different cultures, as was done in [5–7]. Others look in the ornamentation of pottery and other crafts [8].

The design of new weave patterns using Boolean operations has been demonstrated in the papers of [9, 10]. The design is based on the presentation of the weave patterns in the weave diagrams as matrices. A software for automatic application of the Boolean operations was also developed [11].

The possibility for the conversion of musical notations in weave patterns has been proposed and researched in [12]. The idea is based on the visual similarity between the musical scores and the weave diagrams.

The relationship between texts and weave patterns has been investigated in [13]. Several possibilities for the transformation of words and sentences to weave diagrams were presented and discussed.

Another option for the creation of unusual weave patterns from the Braille alphabet has been demonstrated in [13].

In any case, the design of a new weave pattern is a challenging work, which could result in a beautiful and unique woven fabric.

14 Conclusion

The basic knowledge on the weave patterns, necessary to start designing well-known or new weave patterns has been provided.

The transformation of the elementary weaves into their derivatives was presented in details.

The design of different texture effects on the woven fabrics was presented.

The chapter could be effectively used to gradually and in detail enter the world of weave patterns, as well as to inspire the experts in patterns design for new experiments.

15 Recommendations

All geometric 3D view of woven patterns have been prepared using the software: Wisetex © [14].

All weave patterns have been designed with the CAD weaving softwares: DBweave © [15] and ArahWeave © [16].

All fabric simulations were done using CAD weaving software ArahWeave © [16].

References

1. Adumitroaie, A., & Barbero, E. J. (2011). Beyond plain weave fabrics - I. *Geometrical Model. Composite Structures, 93*, 1424–1432.
2. Dash, A. K., & Behera, B. K. (2018). Role of weave design on the mechanical properties of 3D woven fabrics as reinforcements for structural composites. *The Journal of The Textile Institute, 109*(7), 952–960.
3. Damyanov, G., & Chobanov G. (1988). *Proektirane i stroej na takanite [Design and construction of fabrics*, VMEI Lenin, Bulgaria.
4. Grosicki, Z. J. (2004). *Watson's advanced textile design, compound woven structures* (4th ed.). Reprinted by Woodhead Publ. Ltd.
5. Watt, M. (1998). Exploring pattern in woven design: A comparison of Two Seventeenth Century Italian textiles. In *Proceedings from Textile Society of America Symposium*, 446–455, New York, US.
6. Rikert, S. L., Harp, Sh. S., Horridge, P. E., & Shroyer, J. L. (1999). Documentation of Swedish patchwork quilts: 1830 to 1929. *Clothing and Textiles Research Journal, 17*(3), 134–143.
7. Huang, J., & Zhou, X. (2018). Oroqen decorative pattern design and research on the application of creation thought. In *IOP Conference Series: Materials Science and Engineering* (Vol. 371, No. 1, p. 012062). IOP Publishing.
8. Hurley, W. M. (1979). *Prehistoric cordage: identification of impressions on pottery* (No. 3). Taraxacum.
9. Griswold, R.E. (2002). Designing weave structures using Boolean operations, Part 1, Available from http://www.cs.arizona.edu/patterns/weaving/webdocs/gre_bol1.pdf
10. Angelova, R. A. (2005). Designing new weaving patterns using Boolean operations. In *Proceedings from X International conference of the faculty of power engineering, EMF'2005* (Vol. 2, pp. 137–142), Bulgaria.
11. Kyosev, J., & Angelova, R. A. (2005). Specialized software for generation and investigation of the woven pattern. In *Proceedings from Aachen textile conference*, Aachen, Germany.
12. Angelova, R. A. (2017). Design of weave patterns: When engineering textiles meets music. *The Journal of The Textile Institute, 108*(5), 870–876.
13. Angelova, R. A. (2019). *Design of new weave patterns*. CRC Press.
14. "WISETEX SUITE," [Online]. Available: https://www.mtm.kuleuven.be/Onderzoek/Composites/software/wisetex. Accessed 29 May 2020.
15. "DB-WEAVE," [Online]. Available: https://www.brunoldsoftware.ch/dbw.html. Accessed 29 May 2020.
16. "ArahWeave CAD", [Online]. Available: https://www.arahne.si/learn-support/software-demo/. Accessed 29 May 2020.

CAD for Weaving

TexGen

Louise P. Brown

Abstract This chapter gives an overview and introduction to the use of TexGen, open source software developed at the University of Nottingham as a pre-processor for 3D geometric modelling of textile structures. An overview is given of the modelling theory used in the software. There is a guide to creating automatically generated textile models using the built-in weave wizards and a detailed example of the method for creating a textile model using the functionality within the graphical user interface (GUI). An overview of the Python application programming interface (API) is given, illustrated by an example script, as well as information on how to use the Python functions to edit existing textiles. Finally there is an overview of the options for different meshing and export options used to prepare models as input for simulations.

Keywords TexGen software · Textile modelling · Python scripting · Meshing

1 Introduction

TexGen [3] is open source software developed at the University of Nottingham for 3D modelling of textiles and textile composites. It is used as a modelling pre-processor for simulation of a variety of material properties including textile mechanical properties and permeability [11].

This chapter aims to give a practical introduction to the use of TexGen for creating textile models and a guide to exporting them in a form which can be used for further analysis by finite element (FE) or computational fluid dynamics (CFD) packages. It starts with a brief overview of the modelling theory used in the software which will give a better understanding of the functions available both via the user interface and the Python scripting API. Utilising this functionality it is possible to create a range of textile models as shown in Fig. 1.

Supplementary Information The online version contains supplementary material available at (https://doi.org/10.1007/978-3-030-91515-5_6).

L. P. Brown (✉)
Composites Research Group, Faculty of Engineering, University of Nottingham, Nottingham, UK
e-mail: louise.brown@nottingham.ac.uk

© Springer Nature Switzerland AG 2022
Y. Kyosev and F. Boussu (eds.), *Advanced Weaving Technology*,
https://doi.org/10.1007/978-3-030-91515-5_6

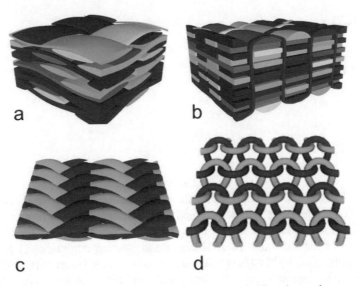

Fig. 1 Examples of TexGen models **a** Layered plain weaves **b** 3D orthogonal weave **c** Triaxial braid **d** Knitted

The software can be downloaded either as executables to run on Windows computers from https://sourceforge.net/projects/texgen/, or as source code for building on either Windows or Linux operating systems from https://github.com/louisepb/TexGen.

2 TexGen Modelling Theory

A good understanding of the TexGen modelling theory will enable the user to make the most effective use of its capabilities and give the knowledge required to enable modelling of a wide range of textile structures. This section gives a brief overview of the theory; a more comprehensive description is given in [8].

Using TexGen, textiles are built up from an assembly of yarns which are modelled as solid volumes. These are defined by a yarn centreline, specifying the smallest repeating section of the yarn, and cross-sections along that yarn. A domain can be specified which defines the area of the textile to be investigated. This is typically a unit cell for prediction of, for example, mechanical properties or permeability. For other simulations, for example impact, a larger domain which covers more repeats of the weave pattern may be required.

Fig. 2 Yarn with 3 master nodes showing **a** non-periodic and **b** periodic interpolation

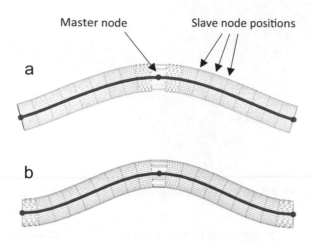

2.1 Yarn Path

Yarn paths are specified by a set of master nodes, given by a set of x, y, z coordinates, which define the centreline of the yarn. For a yarn with a repeating pattern the smallest repeating section of the yarn is specified. The exact path between those nodes is determined by an interpolation function: Bezier, natural cubic or linear spline. When the yarn is built (automatically by TexGen) the interpolation function calculates the yarn path at a set of intermediate positions, slave nodes, the number of which is determined by the yarn resolution parameter. Figure 2a shows a yarn with three master nodes and indicates the positions of the calculated slave nodes. If the yarn is to be repeated then the 'periodic' option should be selected so that the cross-sections at either end of the yarns are continuous if the yarn is repeated, as shown in Fig. 2b.

2.2 Cross-Sections

Cross-sections are specified along the length of the yarn. These are given as 2D sections which are then oriented in a plane perpendicular to the yarn tangent. By default, the cross-section will be constant along the length of a yarn but cross-sections may also be set at each master node or at positions along the yarn, selected by the user.

Several cross-sectional shapes are available: ellipse, lenticular, power ellipse, rectangle and polygon. There is also a hybrid section which can be specified using a combination of different shapes. The polygon section is not available in the user interface and can only be generated from a Python script. The hybrid and polygon shapes give the most flexibility which may be useful when creating yarns which are in contact with each other and therefore deform to non-standard cross-sections.

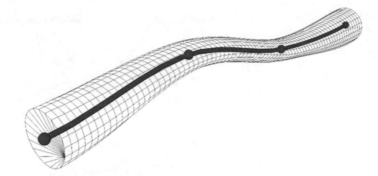

Fig. 3 Yarn with varying cross-section

The specified 2D cross-sections are then oriented in a plane orthogonal to the yarn tangent at their specified positions along the yarn. An interpolation function generates intermediate cross-sections at each slave node and then a surface mesh is generated by joining points around the edge of each cross-section with the corresponding points on the section at the adjacent slave nodes, shown in Fig. 3.

2.3 Domain and Yarn Repeats

Defining the master nodes for the smallest repeating section of a yarn will allow that specific length of the yarn to be built. In order to generate a larger section of the textile comprising several repeats of the weave pattern a set of repeat vectors need to be specified. These are a set of x,y,z vectors, one for each direction in which the yarn is to be repeated. Typically this would be one vector for a repeat in the x direction and another one for the y. If the material is sheared then it might include a vector with both x and y components.

At this point it is possible (in theory) to define an infinite textile. A domain is therefore specified to bound the region of the textile which is to be generated. This may include a number of repeats and will be the area of the model which is exported for further analysis. In many cases the domain will correspond to a unit cell, typically for analysis to predict mechanical properties or permeability. In some cases, for example ballistic simulations, a larger area of the textile is needed and a larger domain can be selected accordingly.

2.4 Yarn Orientations and Volume Fractions

When outputting textile models for simulation it is necessary to know material orientations and, in the case of composite materials, local yarn volume fractions. In order

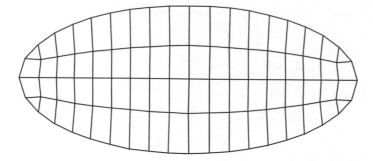

Fig. 4 Section mesh

to calculate these TexGen generates a mesh for each slave node section as shown in Fig. 4. Orientations are generated by finding the vector between centre points of corresponding elements on meshes at adjacent slave nodes.

Volume fractions are calculated automatically within TexGen using Eq. 1. Yarn properties are used to calculate the yarn fibre area and the area of the local yarn section is calculated by extracting the yarn cross-section at the required point on the yarn and then calculating its area.

$$VolumeFraction = \frac{FibreArea}{SectionArea} \tag{1}$$

3 Automatically Generating Textile Models Using the TexGen Weave Wizards

The simplest way to start using TexGen is to use one of the wizards to automatically generate textiles. These all use the CTextile class as a base class which contains yarn information defined in the format described in Sect. 2. Additional functionality is implemented in the inherited classes which automates the process of generating the master nodes and cross-sections. The class hierarchy is shown in Fig. 5.

3.1 3D Weave Wizard

The 2D wizard defines a grid which specifies whether the warp or weft is up at a given crossover point of the warp and weft yarns. In the weave wizard this information is input using the Pattern Dialog (Fig. 6). Master nodes are defined at each of the crossover points and their x, y coordinates are defined by the yarn widths and spacing defined in the initial dialog (Fig. 7). The z positions are defined by the value in

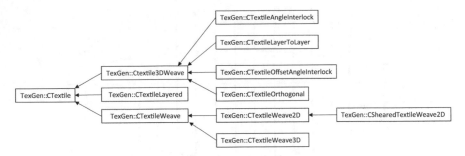

Fig. 5 Textile class hierarchy

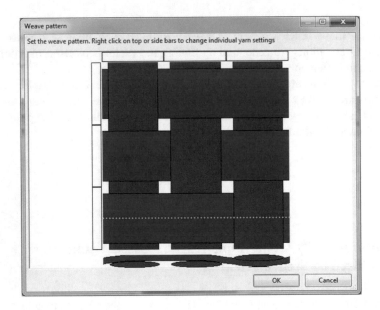

Fig. 6 2D weave pattern dialog

the grid and the thickness of the yarns specified. The yarns are assumed to have constant, elliptical cross-sections. Once the weave pattern (and hence the master node positions) is specified the textile is built in the way described in Sect. 2.

There is an option to create a sheared textile. The additional shear angle parameter is used for calculation of the master node positions and the option is given to create a sheared domain. In this case a parallelogram shaped domain is created with sides at the shear angle specified.

For both straight and sheared textiles there are refine options which interrogate the model for intersections and apply adjustments to the cross-sections to remove the intersections. In this case extra cross-sections are added to the model which use the polygon section. These are described more fully by Sherburn [8] and Zeng [12].

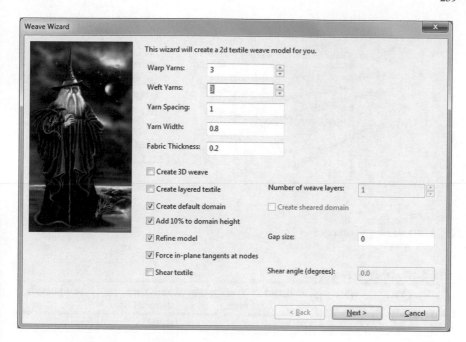

Fig. 7 2D weave wizard

Table 1 Yarn data for
Exercise 1

	Yarn width/mm	Yarn height/mm	Yarns/cm
Warp	0.9	0.3	10
Weft	0.7	0.2	13

Exercise 1: Create a 4 × 4 satin weave textile using the 2D weave wizard and the parameters in Table 1.

3.2 3D Weave Wizard

The 3D weave wizard builds a 3D grid of crossover positions as illustrated in Fig. 8. At each of these positions (i, j) it stores information as to whether it is a warp, weft or no yarn in a cell vector (see Fig. 8). The 3D weave dialogs allow entry of warp, weft and binder yarn information including numbers of yarns and their dimensions. The weave pattern dialog populates the grid and then uses the information to build the yarns for the textile.

The orthogonal, angle interlock and layer-to-layer options impose constraints in the pattern dialog using information about the known yarn configurations for each of these textiles. For example in an orthogonal weave the binder yarns must pass

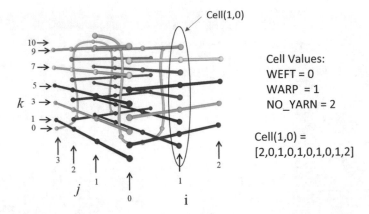

Cell Values:
WEFT = 0
WARP = 1
NO_YARN = 2

Cell(1,0) =
[2,0,1,0,1,0,1,0,1,2]

Fig. 8 3D weave grid structure

Table 2 Yarn data for Exercise 2

	Yarn width /mm	Yarn height /mm	Yarn spacing /mm	Yarn fibre diameter /mm	Fibres per yarn
Warp	3.6	0.35	3.8	0.007	5000
Weft	2.58	0.25	2.8	0.007	8000
Binder	1.375	0.16	1.4	0.007	3500

through all weft layers and cross over the weft yarns either above the top weft layer or below the bottom layer.

A refine option is available in the orthogonal weave which implements observed geometric deformations to give a more realistic textile model, adjusted to a given thickness. Yarn properties are required for this option as local volume fractions are calculated during the process to ensure that the compaction being modelled is realistic. An example of a refined orthogonal weave can be seen in Fig. 1b. A more detailed description of these refinements is given in [11]

Exercise 2: Create a 3D orthogonal weave textile with a total of 6 warp yarns, with a ratio of 2 warp stuffers to 1 binder yarn, and 4 weft yarn stacks. There should be 3 yarns in each weft stack and 2 in each warp stack. Textile thickness is 1.4 mm. Use the yarn parameters in Table 2.

4 Creating Custom Textiles and Editing Textiles Using the Graphical User Interface

A User Guide is available on the TexGen webpage [10] which gives information on the different features in the GUI. This section illustrates how to use these to create a TexGen model and highlights some of the features available. It is recommended that the reader opens the TexGen GUI and follows the steps given in order to gain practical experience in its use. Also, if the Python Output tab is selected in the Log window then it is possible to see the Python commands which have been executed in order to carry out the option selected. These can be related to both the theory described in Sect. 2 and the Python commands described in Sect. 5 and will give a better understanding of how TexGen works. User interface commands are given in *italics*. Also note that there is no Undo option in TexGen so it is advisable to save the model at frequent stages using the *File-> Save TexGen File* option.

First, ensure that the Controls, Outliner and Log windows are all open as shown in Fig. 9: Select the *Window* option from the main menu and ensure that all three window options are checked. The windows can be dragged and dropped into the preferred locations.

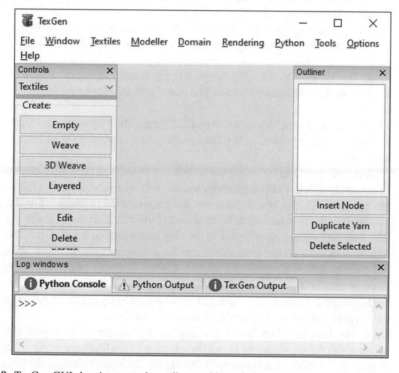

Fig. 9 TexGen GUI showing controls, outliner and log windows

4.1   Create a Simple Textile

1.   Create an empty textile: Either select *Create: Empty* from the Controls window or *Textiles-> Create Empty...* from the main menu. Enter a name for the textile to be created in the *Textile name* dialog box.

     The Render window in the centre of the GUI becomes black and shows a set of axes and a tab containing the name of the textile which was input. At this point a textile has been created and added to the textile list. Select *Modeller* from the *Controls* drop-down menu. This changes the options in the Controls window to reflect those in the *Modeller* menu option.

2.   Create a yarn: Either select *Create: Yarn* from the Controls window or *Modeller-> Create Yarn...* A dialog is displayed which allows for entry of the start and end points of the yarn and the number of nodes to be selected. For this example use the default settings.

     A single yarn is displayed in the Render window and a tree is displayed in the Outliner window showing that Yarn 0 has been created with Nodes 0 and 1.

3.   Select the Yarn by clicking on the yarn in the Render window or by selecting *Yarn(0)* in the Outliner window. The yarn will be displayed in grey when selected.

4.   Create a second yarn: Make a copy of the yarn, either by using the *Duplicate Yarn* button at the bottom of the Outliner window or by pressing *Ctrl-d* on the keyboard.

     A second yarn has been created which can be seen in the Outliner window. The Render window still apparently shows only one yarn, although the colour has changed. This is because the second yarn created is identical to the first and so only the second one to be rendered is visible. This is shown in Fig. 10.

5.   Move second yarn: Select *Yarn 1* in the Outliner window. A set of axis handles will be displayed on the yarn. Click and hold the green y-axis arrow using the left mouse button. Keeping the button pressed, the yarn can now be dragged in the y direction. Drag the yarn until the *Position: Y* text box in the Controls Dialog has the value 5. The yarn can also be moved by selecting the yarn and then typing the value '5' into the *Position: Y* text box.

6.   Duplicate Yarn 0: Select Yarn 0 and duplicate as described in steps 4 and 5.

7.   Move node: Select *Node1* of *Yarn2* by selecting in the Outliner window, shown as greyed. Set values x $= 0$ and y $= 10$ using the *Position* text box in the Controls window.

8.   Duplicate Yarn 2: Select Yarn 2 and duplicate as described in steps 4 and 5.

9.   Move Yarn 3: As describe in step 6, either drag Yarn 3 type in position to move Yarn 3 to x $= 5$.

     At this point the textile should look as shown in Fig 11.

10. Insert nodes: Select node 1 in each yarn using the Outliner window. Multiple nodes can be selected using *Ctrl-Left Click*. Use the *Insert Node* button in the Outliner window to insert a node before the selected nodes. Note that inserted

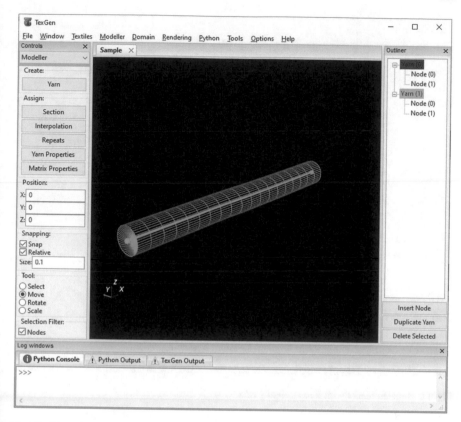

Fig. 10 Yarns created with two nodes

nodes will be positioned at the mid-point between the selected node and the previous node (ie will always be inserted before the selected node).

11. Move nodes: In order to create a woven pattern in the yarns the z coordinate of some of the nodes needs to be changed. Using the Outliner window select *Nodes 0 and 2 of Yarns 0 and 3 and Node 1 of Yarns 1 and 2*. Using the *Position:Z* text box in the Controls window change the z coordinate to '2'.

The resulting yarn configuration is shown in Fig. 12.

12. Select all yarns either by selecting the yarns in the Render window using *Shift-Left Click* to select multiple yarns or by selecting *Yarn 0* in the Outliner window and then *Shift-Left Click Yarn 3* to select all yarns between the two yarns. Note that in the Outliner window *Ctrl-Left Click* is used to make multiple discrete selections and *Shift-Left Click* is used to select a block between the two selections.

13. Assign yarn cross-sections: Either select *Assign: Section* from the Controls window or *Modeller-> Assign Section…* The *Select Yarn Section dialog* is displayed. In this case a constant cross-section is assigned for each yarn. Select the *Edit Section* button.

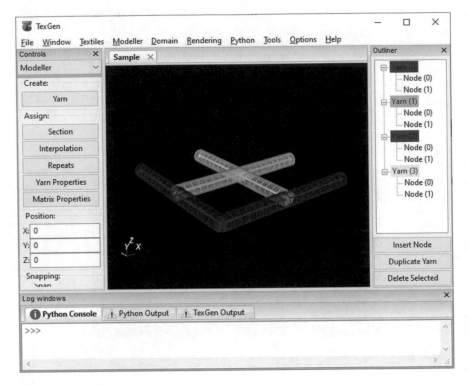

Fig. 11 TexGen model with four straight yarns

The *Select Cross-Section Shape dialog* is displayed as shown in Fig. 13a. Select *Lenticular* from the drop down menu. Enter *Width, 4 and Height, 2,* then select *OK* and also on the *Select Yarn Section dialog.*

The lenticular sections will now have been applied to all yarns as shown in Fig. 13b.

14. Assign domain: It can be seen that, at the moment just the single section of the yarns have been specified. The way in which the yarns are to repeat and the boundaries of the unit cell need to be specified. In this case the domain will be specified first using the *Domain- < Create Box...* option. In the *Box Domain dialog enter Maximum X and Y values, '10' and Minimum Z value, '-1' and Maximum Z value, '3'.*

If the *Rendering->Trim To Domain* option is selected the model will now look as shown in Fig. 14.

15. Add repeats: All of the yarns created have a length of 10. To create a continuous textile all of the yarns need to have repeat vectors of 10 specified in both the x and y directions. Select all of the yarns as described in step 13 and then either select *Assign: Repeats* from the Controls window or *Modeller-> Assign Repeats...* The *Repeats dialog* will be displayed and two vectors should be specified as shown in Fig. 15. Two repeat vectors are defined here; row 1 gives the repeat in the x direction and row 2 the repeat in the y direction.

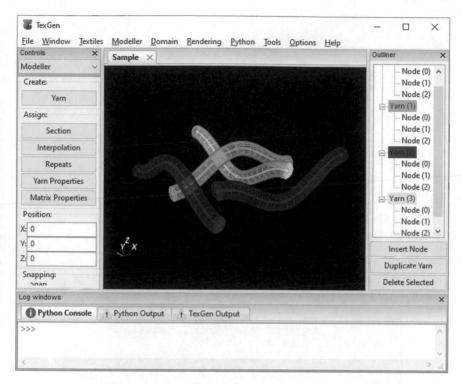

Fig. 12 TexGen model with four yarns and nodal z positions set

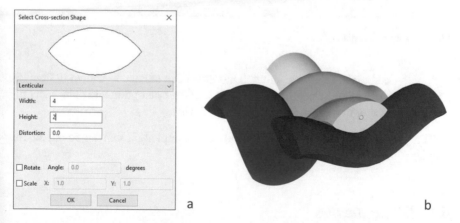

Fig. 13 **a** Cross-section shape dialog **b** TexGen model with lenticular sections

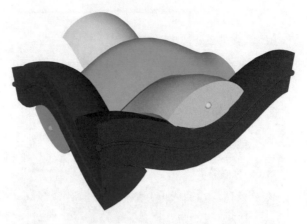

Fig. 14 TexGen model with yarns trimmed to domain

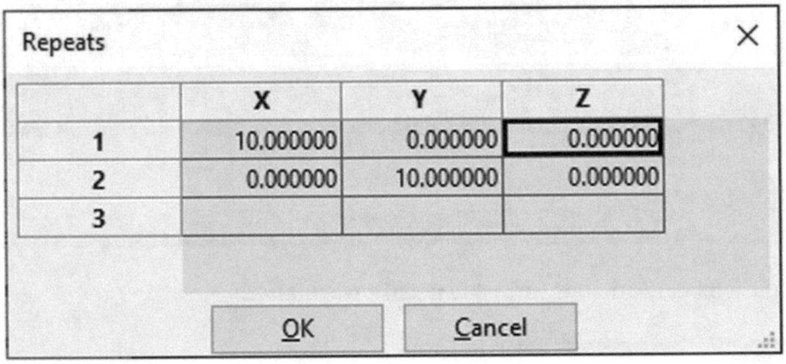

Repeats			✕
	X	**Y**	**Z**
1	10.000000	0.000000	0.000000
2	0.000000	10.000000	0.000000
3			

OK   Cancel

Fig. 15 Yarn repeat dialog

A complete, repeating unit cell has now been created as shown in Fig. 16.

16. Save the model: Select the *File-> Save TexGen File* option and then select the *Standard* option in the subsequent *Save dialog*.

The file for the final textile produced, GUISample.tg3, can be found in the supplementary material.

4.2 Editing Textiles

Once a textile model has been created, either by using the GUI or a Python script, it can be edited using the same methods that were used to create the textile. The yarns and nodes can be selected in the same way and then either dragged using the arrow widgets or by entering values into the *Position* text box in the Controls window.

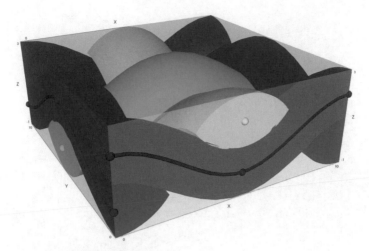

Fig. 16 TexGen model with both yarn repeats and domain set

One example of where changes may need to be made to a model is if there are intersections between the yarns, which are both unrealistic and may prevent FE simulations from running. The *Rendering-> Render Textile Interference* or *Rendering-> Render Textile Interference Depth options* can be used to give a visualisation of where intersections between yarns occur. The intersection depth is calculated at each surface mesh point. In the case of the latter option the depth of the intersection is reflected by the size of the points rendered as shown in Fig. 17 which shows the interference for the model created in Sect. 4.1. A message is given in the TexGen Output tab of the Log window stating the depth of the maximum interference in the model.

In this case the following steps could be taken to edit the textile and reduce interference:

1. It can be seen that there is interference where the two central yarns (yarns 1 and 3) cross in the centre of the model shown in Fig. 17. In this case the interference could be reduced by changing the cross-section at node 1. The model was created with constant cross-sections along the yarn so first it is necessary to change the interpolation to interpolate between nodes. This allows a different cross-section to be specified at each node point.

    Select *Yarn 1* either using the Outliner window or by *left-clicking* on the yarn in the Render window.
2. Select *Modeller-> Assign Section* from the main menu or the *Section button* in the *Modeller tab* of the Controls window.
3. Select *Interpolate between nodes* from the drop down menu in the *Select Yarn Section dialog*. The dialog will then display a list of the nodes for the yarn as shown in Fig. 18.
4. Click on *Section at Node 1* to select the node and then click on the *Edit Section button.*

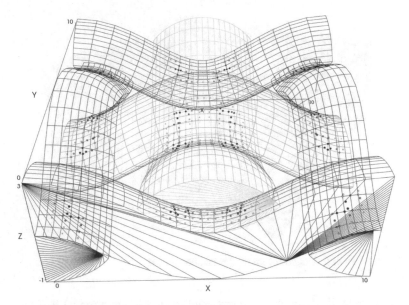

Fig. 17 TexGen model showing interference depths

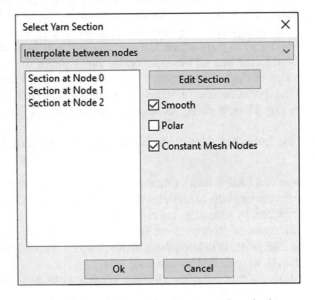

Fig. 18 Select Yarn section dialog with Interpolate between nodes selection

5. The Select Cross-Section Shape dialog is displayed, showing the information for the current cross-section at the selected node. In this case the first change that will be tried in order to reduce the intersections will be to change the Distortion of the lenticular section. This changes the vertical position of the maximum width of the section.

    Enter *Distortion, 0.1* and then select *OK* on this and the Select Yarn Section dialogs.

6. Select the *Refresh View* option in the Rendering menu. On inspection of the interference markers on the central section of Yarn 1 it can be seen that the intersection towards the edge is reduced but there is still some intersection towards the middle. A better solution might be a hybrid section which can be used to change the lower half of the yarn section.

7. Repeat steps 1–4 to again allow the section at Node 1 to be edited. In the *Select Cross-section Shape dialog* select *Hybrid* from the drop-down menu.

8. Click outside the shape shown in the centre of the dialog box twice to create two points to split the shape. The circular tags can then be used to drag the points to the desired points. In this case use them to create a horizontal line as shown in Fig. 19.

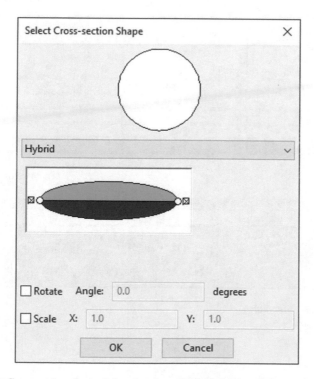

Fig. 19 Select Cross-section shape dialog showing split for hybrid section

9.   The top section will be set to be the same as the original section: *Click on the
     upper section,* then *Select Lenticular, Width = 4.0, Height = 2.0* and then click
     OK. To change the lower section *Click on the lower section*. In the subsequent
     Cross-Section Shape dialog *Select Lenticular, Width = 4.0, Height = 1.8* and
     then click *OK*.
10. Select the *Refresh View* option in the Rendering menu. It will be seen that there
     are now no intersections in the central region.

   This edited version of the textile is available in CorrectInterferenceYarn1.tg3. The
domain can be edited using the *Domain-> Edit Domain* option from the main menu
or the *Edit button* in the *Domain tab* of the *Controls window*. The *Domain Editor
dialog* is displayed as shown in Fig. 20. It can be seen that six planes are shown,
one for each face if a box domain was created. In order to change the domain planes
*Click on the row of one of the planes, enter the required values in the x, y, z and d
boxes and then select Replace Plane*. Planes can be added or deleted using the *Add
Plane* and *Delete Plane* buttons.

   Exercise 3: Edit the domain of the textile so that two repeats of the weave are
displayed with the edges of the domain running between the yarns.

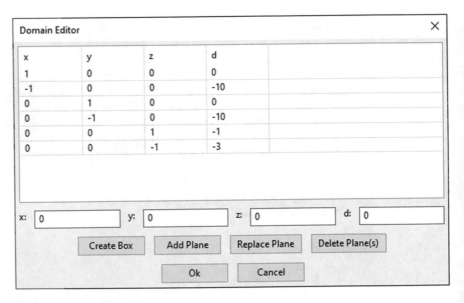

Fig. 20 Domain editor dialog

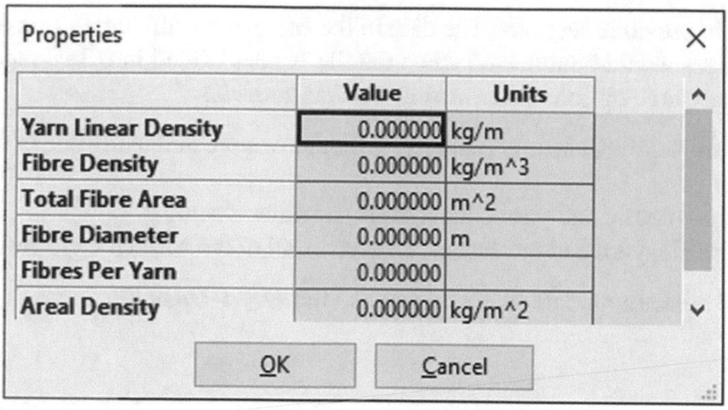

	Value	Units	^
Yarn Linear Density	0.000000	kg/m	
Fibre Density	0.000000	kg/m^3	
Total Fibre Area	0.000000	m^2	
Fibre Diameter	0.000000	m	
Fibres Per Yarn	0.000000		
Areal Density	0.000000	kg/m^2	v

Properties ✕

<u>O</u>K   <u>C</u>ancel

Fig. 21 Properties dialog

4.3 Adding Material Properties

Selected yarn properties can be set for the whole textile. If the *Modeller-> Yarn Properties* option is selected without selecting any yarns the *Properties dialog* allows the yarn linear density, fibre density, total fibre area, fibre diameter, fibres per yarn and areal density to be set for the textile. In the dialog, shown in Fig. 21, both the Value and the Units can be entered for any of the properties. Standard metric units can be used as well as imperial length units and the textile-related units of tex and denier.

If one or more yarns are selected then the *Properties dialog* will allow input of material properties, Young's Modulus, Shear Modulus, Poisson's Ratio and Alpha (coefficient of thermal expansion) which will be applied to the selected yarns.

For composite materials it is assumed that the space in the domain which is not occupied by yarns is matrix whose properties can be set using the *Modeller-> Matrix Properties* option. The Young's Modulus, Shear Modulus, Poisson's Ratio and Alpha can be set using the *Properties dialog*.

The yarn densities and areas are used in TexGen to calculate fibre volume fractions. These must be set if element fibre volume fractions are required in the exported files. The mechanical properties are used for populating the materials sections of the export files.

4.4 Loading and Saving TexGen Files

TexGen models can be saved using the *File-> Save TexGen File* option from the main menu. Models are saved with a .tg3 file extension. These files are in XML format and can be viewed and edited using a text editor. Examining one of these files in a text editor will show that the structure of the file follows the modelling format described in previous sections. The data in the file can be edited in a text editor and these changes would be reflected when the file is next loaded into TexGen.

There are three options when saving TexGen models:

- Minimal—textile data only (only use if model was created using one of the weave classes)
- Standard—textile and yarn data. This is the default setting
- Full—textile, yarn and mesh data. This saves all of the surface mesh information.

To load TexGen models in.tg3 format use the *File-> Open TexGen File* option.

5   Python Scripting API

TexGen software is written using the C++ programming language and is split into several modules as shown in Fig. 22. Use of the GUI has been described in earlier sections which allows easy generation of standard textile weaves but is more time consuming for non-standard textiles. For more complex textiles use of the Python scripting interface gives greater flexibility. The Python interface is automatically generated from the C++ code using the Simplified Wrapper and Interface Generator (SWIG) [1] and allows functions from the Core, Renderer and Export modules to be called from a Python script. All menu items in the GUI execute one or more of these Python commands and much can be learned by looking at the Python Output window which echoes the commands executed by the GUI.

There is a comprehensive scripting guide available on GitHub [2] and this section highlights a selection of the Python scripting capability rather than giving an in-depth guide.

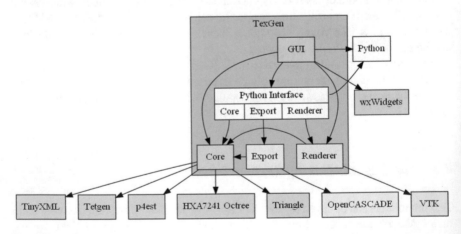

Fig. 22   TexGen modules

TexGen Python scripts can be used either as standalone scripts or can be run using the *Python-> Run Script* menu item in the TexGen GUI. In the latter case the textile will be visualised in the GUI after being added to the textile list within the script. Python commands can also be used to edit a textile already loaded into the GUI.

5.1 Creating a Textile Model Using a Python Script

To create a Python script a text file is created in an editor with a .py extension. In Windows Notepad++ (Notepad++, 2019) is a good option for doing this. Python code will be displayed as shown below:

```
# Python code
# Lines beginning with a # are comments
```

Using Python scripting a textile is created in the order described in the previous sections. First a textile is defined:

```
# Create a textile
Textile = CTextile()
```

Yarns are then created which are added to the textile. If more than one yarn is to be created these can be created as a Python list:

```
# Create a Python list containing 4 yarns
Yarns = [CYarn(), CYarn(), CYarn(), CYarn()]
```

Yarns contains 4 empty yarns which, as yet, are not associated with the textile which has been created. Data which defines the geometry of the yarns will be added to each yarn and then these yarns will be assigned to the textile. First, master nodes are added to each yarn to define the yarn path:

```
# Add nodes to the yarns to describe their paths
Yarns[0].AddNode(CNode(XYZ(0, 0, 0)))
Yarns[0].AddNode(CNode(XYZ(0.22, 0, 0.05)))
Yarns[0].AddNode(CNode(XYZ(0.44, 0, 0)))

Yarns[1].AddNode(CNode(XYZ(0, 0.22, 0.05)))
Yarns[1].AddNode(CNode(XYZ(0.22, 0.22, 0)))
Yarns[1].AddNode(CNode(XYZ(0.44, 0.22, 0.05)))

Yarns[2].AddNode(CNode(XYZ(0, 0, 0.05)))
Yarns[2].AddNode(CNode(XYZ(0, 0.22, 0)))
Yarns[2].AddNode(CNode(XYZ(0, 0.44, 0.05)))

Yarns[3].AddNode(CNode(XYZ(0.22, 0, 0)))
Yarns[3].AddNode(CNode(XYZ(0.22, 0.22, 0.05)))
Yarns[3].AddNode(CNode(XYZ(0.22, 0.44, 0)))
```

Nodes are specified using the CNode class which is constructed using an instance of the XYZ class giving the position of the node. The XYZ class is constructed using 3 parameters representing the x,y,z coordinates of the point. In the Python code above the number in square brackets after Yarns refers to a particular yarn in the list and then the Yarn class function AddNode is called with the nodes created using CNode.

So far the nodal positions have been specified for each yarn but there is no information about cross-sectional shape or interpolation between the nodes.

If the same information is to be assigned to each yarn then a loop can be used to iterate through each yarn in turn and assign the properties to each in turn. To loop through each yarn in the list:

```
# Loop over all the yarns in the list
for Yarn in Yarns:
```

When inside the loop the current yarn being changed is referred to using the Yarn variable. In Python the instructions within the for loop are defined by indentation.

First the interpolation between the nodes is defined by creating an instance of one of the classes derived from CInterpolation. In this case periodic cubic spline interpolation is used, defined by the class CInterpolationCubic:

```
# Set the interpolation function
Yarn.AssignInterpolation(CInterpolationCubic())
```

Next, cross-sections are defined using the AssignSection function. This takes an instance of a class derived from CYarnSection, in this case a constant cross-section is created using CYarnSectionConstant. This, in turn, takes an instance of a class derived from CSection as a parameter, in this case an elliptical cross-section is defined using CSectionEllipse using numerical parameters defining the width and height:

```
# Assign a constant cross-section along the yarn
# with elliptical shape
```

```
Yarn.AssignSection(CYarnSectionConstant(CSectionEllipse(0.18,0.04)))
```

Further information on creating different cross-sectional shapes and varying the cross-section along the length of the yarn can be found in the TexGen scripting guide [2].

The resolution of the surface and volume meshes created are defined using the `SetResolution` function. If one parameter is used then this gives the number of points around a cross-section and the number of points along the length of the yarn (defining the slave nodes) is calculated automatically so that the distance along the length is similar to that around the cross-section. If two parameters are provided the first specifies the number of slave nodes along the yarn and the second the number of points around the yarn. In this case one parameter is given:

```
# Set the resolution of the mesh created
Yarn.SetResolution(20)
```

The repeat vectors are defined as described in Sect. 6.2 using the function `AddRepeat` which takes an instance of the `XYZ` class as a parameter to define the x,y,z components of each repeat:

```
# Add repeat vectors to the yarn
Yarn.AddRepeat(XYZ(0.44, 0, 0))
Yarn.AddRepeat(XYZ(0, 0.44, 0))
```

This completes the definition of the yarn which may now be added to the textile:

```
# Add the yarn to the textile
Textile.AddYarn(Yarn)
```

This is the end of the `for` loop, marked by the end of indentation in the following code.

The last requirement for creation of the textile is specification of a domain to define the area of the textile being observed. The function `AssignDomain` takes an instance of a class derived from `CDomain`, in this case `CDomainPlanes` using the constructor which takes two `XYZ` points defining the lower left and upper right corners for a box shaped domain:

```
Textile.AssignDomain(CDomainPlanes(XYZ(0,0,-0.02),XYZ(0.44,0.44,0.07)))
```

Lastly `Textile` is added to the list of textiles so that it can be rendered in the TexGen GUI. Each textile is given a name which is either passed as a parameter to the `AddTextile` function or will be generated automatically if no parameter is given. This name will be shown at the top of the render window in the GUI:

```
            # Add the textile with the name "polyester"
            AddTextile("polyester", Textile)
```

The completed script is shown here:

```
# Create a textile
Textile = CTextile()

# Create a python list containing 4 yarns
Yarns = [CYarn(), CYarn(), CYarn(), CYarn()]

# Add nodes to the yarns to describe their paths
Yarns[0].AddNode(CNode(XYZ(0, 0, 0)))
Yarns[0].AddNode(CNode(XYZ(0.22, 0, 0.05)))
Yarns[0].AddNode(CNode(XYZ(0.44, 0, 0)))

Yarns[1].AddNode(CNode(XYZ(0, 0.22, 0.05)))
Yarns[1].AddNode(CNode(XYZ(0.22, 0.22, 0)))
Yarns[1].AddNode(CNode(XYZ(0.44, 0.22, 0.05)))

Yarns[2].AddNode(CNode(XYZ(0, 0, 0.05)))
Yarns[2].AddNode(CNode(XYZ(0, 0.22, 0)))
Yarns[2].AddNode(CNode(XYZ(0, 0.44, 0.05)))

Yarns[3].AddNode(CNode(XYZ(0.22, 0, 0)))
Yarns[3].AddNode(CNode(XYZ(0.22, 0.22, 0.05)))
Yarns[3].AddNode(CNode(XYZ(0.22, 0.44, 0)))

# Loop over all the yarns in the list
for Yarn in Yarns:

    # Set the interpolation function
    Yarn.AssignInterpolation(CInterpolationCubic())

    # Assign a constant cross-section all along the yarn of
elliptical shape
    Yarn.AssignSection(CYarnSectionConstant(CSectionEllipse(0.18,
0.04)))

    # Set the resolution of the surface mesh created
```

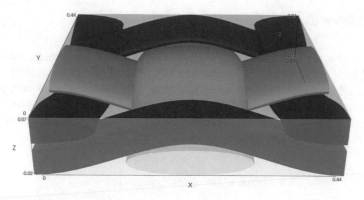

Fig. 23 Textile model generated using Polyester.py script

```
Yarn.SetResolution(20)

# Add repeat vectors to the yarn
Yarn.AddRepeat(XYZ(0.44, 0, 0))
Yarn.AddRepeat(XYZ(0, 0.44, 0))

# Add the yarn to our textile
Textile.AddYarn(Yarn)

# Create a domain and assign it to the textile
Textile.AssignDomain(CDomainPlanes(XYZ(0, 0, -0.02), XYZ(0.44, 0.44,
0.07)))

# Add the textile with the name "polyester"
AddTextile("polyester", Textile)
```

The resulting textile is shown in Fig. 23. The complete script Polyester.py can be found in the supplementary material.

This example shows the general process for creating a textile script. Much more complex textiles can be created by making use of the functions available with the TexGen Python API. A range of sample scripts for a variety of textiles are available at https://github.com/louisepb/TexGenScripts.

Exercise 4: Make the following changes to the Polyester.py script:

1. Change the cross-sections to lenticular.
2. Change the cross-section dimensions: Width = 0.2, height = 0.048
3. Change the height of the domain to suit the amended yarn dimensions so that it does not cut off any of the yarns.
4. Change the yarn resolution to 40.

5.2   Editing Textiles Using the Python Console

Textiles which are displayed in the TexGen render window, whether created using the GUI, by loading an existing TexGen model from a .tg3 file or by running a Python script are all created using the same set of classes. Because of this it is then possible to access the textile and yarn information via the Python console and use the same set of Python commands to edit the textile. Some examples of this are given below.

The most recent textile loaded can be retrieved using.

```
textile = GetTextile()
```

If more than one textile is loaded in the GUI then the required textile can be selected by passing a string parameter to the GetTextile function containing the name of the textile (as displayed on the render window tab).

A specific yarn can be accessed using the GetYarn function using the required yarn number as a parameter:

```
yarn = textile.GetYarn(1)
```

The API functions can then be used to edit the textile. For example the resolution could be set using:

```
yarn.SetResolution(20,40)
```

which would change the resolution of yarn 1 to have 20 slave nodes and 40 points around the outline of each yarn.

To change the position of a node in a yarn the ReplaceNode function can be used, again using the CNode class to create an instance of a node which is passed to the function:

```
yarn.ReplaceNode(1, CNode(XYZ(5,5,2.4)))
```

The first parameter of ReplaceNode gives the index of the node to be changed.

6   Pre-Processing Textile Models for Simulation

TexGen is mainly used as a pre-processor for simulations, typically finite element (FE) simulations for prediction of mechanical properties or computational fluid dynamics (CFD) simulations for permeability predictions. In order to generate input files it is necessary to create a mesh of the textile model, exported in a format which can be imported into third party software for analysis. Additional information is exported which includes yarn orientations and volume fractions for each mesh element.

ABAQUS® input files can be generated which include periodic boundary conditions [4] and steps for prediction of thermo-mechanical properties of a unit cell. Alternatively an input file can be created which just contains the mesh information, leaving the user to set up the rest of the input according to the analysis to be performed.

There are several options for exporting models within TexGen and the option selected may depend on the type of analysis to be carried out, the type of textile and whether a dry textile or a textile composite is to be simulated. The following sections describe the different export options available within TexGen and highlight advantages and limitations of these methods.

6.1 Dry Fibre Volume Mesh Export

This method can be used to create a conformal mesh of just the yarns. The mesh is created in two stages: first a two dimensional mesh is created at each slave node using a rectangular/triangular mesh generation technique as illustrated in Fig. 4 and then the mesh points between subsequent sections are joined to form hexahedral and wedge elements as illustrated in Fig. 24.

The mesh is exported as an ABAQUS® input (.inp) file as node and element sets. Options are available to specify deformations in the x, y and z directions and to apply compression plates. In dry fibre simulations contact between the yarns must be taken into account and either upper and lower yarn contact surfaces (suitable for woven fabrics) or whole yarn surfaces (more suitable for textiles such as knits) can be selected. The input file generated creates boundary conditions, contacts and steps for applying the specified deformations. Supplementary files are also created; a .ori file contains element yarn orientation information and a .eld file contains yarn volume fraction information for each element.

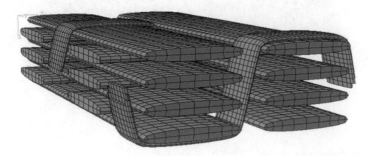

Fig. 24 Dry fibre export mesh

6.2   Tetrahedral Volume Mesh Export

To perform simulations on a composite material both the yarn and matrix need to be meshed. In a TexGen model the resin volume is assumed to be represented by the specified domain. In the volume mesh export option a tetrahedral mesh is created. A mesh showing both the whole volume and just the yarns is shown in Fig. 25. If the Periodic option is selected then matching triangulated surfaces will be generated for opposite boundaries of the domain, ensuring that nodes and elements match for use in the periodic boundary conditions created in the input file. It should be noted that this method of mesh generation is robust for 2D weaves but less successful for textiles containing vertical or near-vertical yarns. In this case the algorithm used to generate the tetrahedra can result in elongated elements.

If the Periodic Boundary Conditions option is selected then equations are set up in the input file according to those described by Li [4]. A set of steps are also specified which, when used with the dataHandling.py and effective-MatPropRVE.py scripts supplied with TexGen, allow material properties to be extracted. A tutorial for extraction of material properties using this method can

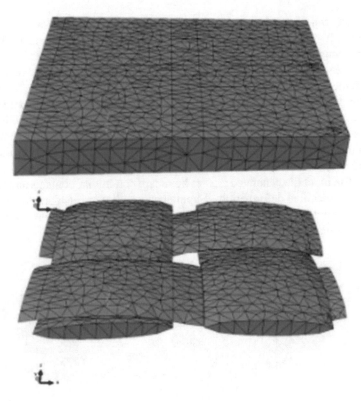

Fig. 25 Tetrahedral mesh export

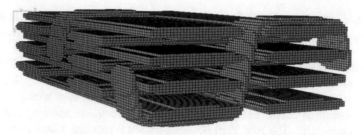

Fig. 26 Voxel mesh export

be found at http://texgen.sourceforge.net/index.php/Extraction_of_Material_Propert ies_using_Voxel_Meshing_and_Abaqus

An alternative method of generating a tetrahedral mesh in TexGen is available using the Tetgen Mesh option. This uses the tetgen library [9] and may be used to generate meshes for both 2D and 3D woven textiles. The algorithm is not robust for textiles with yarns in contact with each other. To use this method a TexGen model should be created which enforces small gaps between all yarns, a limitation of the method as this is not necessarily a realistic representation of the actual textile.

6.3   Voxel Mesh Export

It can be challenging to generate conformal meshes of both yarn and matrix regions. Where yarns come into contact with each other there are liable to be small, wedge-shaped areas which are difficult to mesh and often give rise to badly distorted elements. In many cases it has been shown [5] that a non-conformal voxel mesh may be adequate, particularly in the case of prediction of mechanical properties where damage mechanisms are ignored.

In TexGen the voxel mesh is generated by dividing the domain area into cuboids. The voxel is defined as belonging to either the yarn or matrix depending on whether the centre point of the cuboid (element) is located in the yarn or matrix. The information for the orientations and volume fractions of the elements are also based on the values at this centre point. Figure 26 shows a voxel mesh for a 3D textile with the matrix elements removed.

Periodic boundary conditions are set up as described in Sect. 6.2 and the analysis for extracting material properties is carried out using the same method.

6.4   Geometry Export

An alternative approach to preparing models for simulation can be to export the geometry and then use third-party software to mesh the model. The disadvantage of this method is that it doesn't allow the yarn orientations and volume fractions to be exported, although it is be possible to write a script to interrogate the TexGen model using the centre points of the meshed elements.

TexGen uses the OpenCASCADE [7] library to create geometry in STEP and IGES formats. The algorithm is more robust for the Faceted option which creates a geometry which corresponds to the model surface mesh (as seen when the X-Ray rendering option is used). The Smooth option is less reliable and may not always generate a geometry. The Join Yarn Sections option forces one shape to be created for each yarn, rather than a separate shape for each yarn repeat, but may be slow to generate the geometry.

The Surface Mesh option exports the surface mesh in VTK .vtu format or either binary or ASCII stl format.

7 Conclusion

This chapter has aimed to give the reader a working knowledge of the TexGen software, giving them the knowledge necessary to create a wide range of textile structures using either the GUI or the Python API. For further information the reader should visit the TexGen webpage, www.texgen.sourceforge.net, which has a range of documentation including a user guide, usage examples and a list of publications which have used TexGen models for analysis. There is also an active user forum, http://texgen.sourceforge.net/phpBB3/index.php, which is both a source of information and a place to ask for help with creating TexGen models.

Acknowledgements Continued development of TexGen software has been supported by Engineering and Physical Sciences Research Council RSE Fellowship [Grant number: EP/N019040/1].

Appendix

Exercise 1 Solution

1.   Either select *Textiles-> Create Weave…* from the main menu or *Weave* from the *Textiles* tab of the Controls menu.
2.   Set the number of warp and weft yarns to 4 (to create a 4 × 4 satin weave).
3.   Use the warp yarn data to set the yarn spacing and yarn width (as these parameters are different for the warp and weft yarns the weft dimensions will be set later using the Pattern Dialog). The *Yarn Spacing* is set to 1.0 based on 10 yarns/cm.
4.   The combined heights of the yarns is 0.5 mm so this is selected as the *Fabric Thickness*. The wizard will automatically set both yarn heights to 0.25 mm

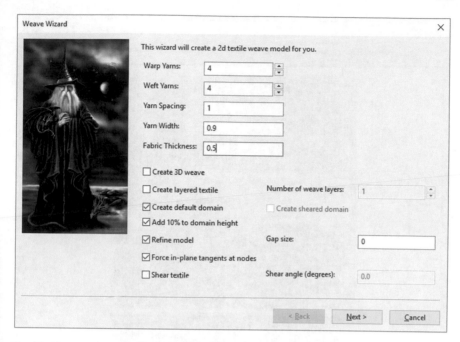

Fig. 27 2D weave wizard yarn parameters

(half the thickness) but, again, these can be adjusted in the Pattern Dialog. Steps 2–4 are shown in Fig. 27.

5. Click the *Next* button to move to the *Weave pattern dialog.*
6. *Shift-click* on the bars at the left side of the weave pattern to select all of the warp yarns.
7. *Right-click* on one of the selected side bars to show options to change yarn parameters for these yarns (Fig. 28)
8. Select *Set yarn height...* and enter the value 0.3 (Fig. 29)
9. Repeat steps 6–8 to select the bars at the top of the weave pattern and set the weft yarn width, height and spacing to 0.7, 0.2 and 0.769 respectively. (The spacing is calculated using 13 yarns/cm)
10. Click on the crossovers on the weave pattern to create the desired weave configuration (Fig. 30)
11. Select *OK*. The textile is created as shown in Fig. 31. The textile can also be loaded from the Ex2_9_1.tg3 file.

Exercise 2 Solution

1. Either select *Textiles-> Create 3D Weave...* from the main menu or *3D Weave* from the *Textiles* tab of the Controls menu.
2. Select the *Orthogonal* weave type and select *Next*.

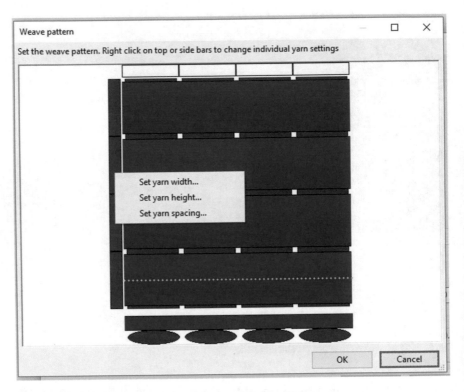

Fig. 28 Options to change selected yarns in Weave Pattern Dialog

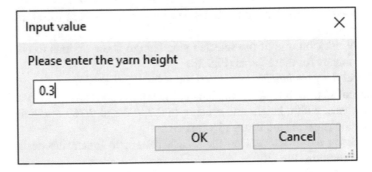

Fig. 29 Dialog to input yarn height for selected yarns

3. Fill in the weft, warp and binder data as shown in Figs. 32, 33 and 34. In the binder yarn window select the *Refine* option and set the *Target Thickness* to 1.4.
4. In the three subsequent windows set the yarn properties. This is shown for the weft yarn in Fig. 35.
5. In the final weave pattern window click on the points on the binder yarns to create the desired weave configuration (Fig. 36)

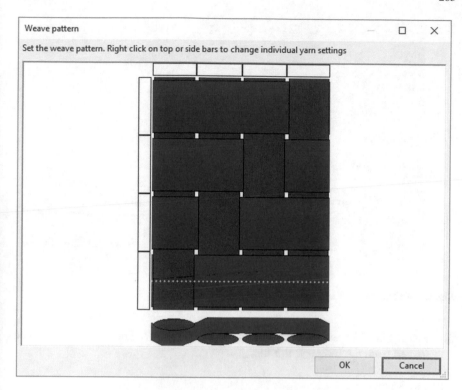

Fig. 30 Weave configuration for Exercise 1

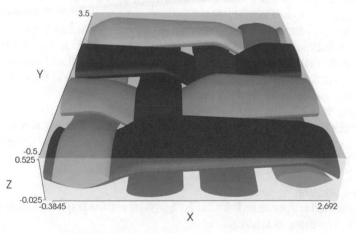

Fig. 31 Exercise 1, finished textile

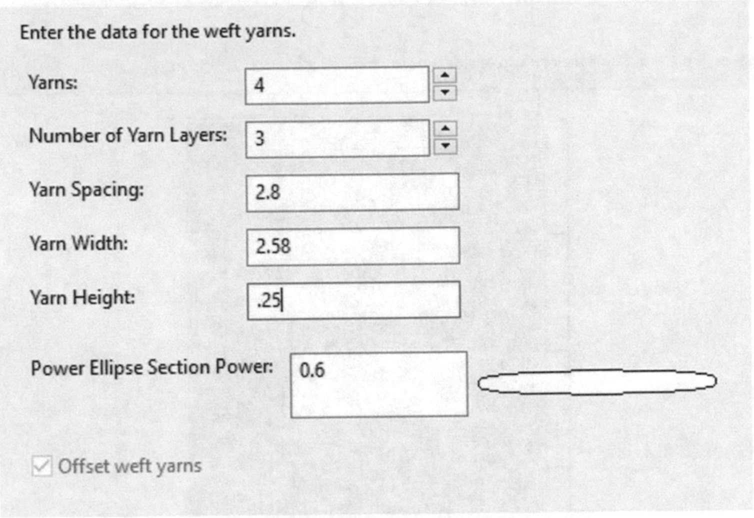

Enter the data for the weft yarns.

Yarns: 4

Number of Yarn Layers: 3

Yarn Spacing: 2.8

Yarn Width: 2.58

Yarn Height: .25|

Power Ellipse Section Power: 0.6

☑ Offset weft yarns

Fig. 32 Weft yarn input for Exercise 2

6. Select *OK*. The textile is created as shown in Fig. 37. The textile can also be loaded from the Ex2_9_2.tg3 file.

Exercise 3 Solution

1. Either select *Domain-> Edit Domain...* from the main menu or *Edit* from the *Domain* tab of the Controls menu.
2. The x and y dimensions of the domain are to be changed so the first four planes (ie the first four lines in the dialog) need to be changed. Select the first row. The values from this row will be displayed in the x, y, z and d text boxes. Set the *d* value to 2.5 and select *Replace Plane*. Note that the values in the main window are not updated until *Replace Plane* is selected.
3. Repeat step 2 for the next 3 planes as shown in Fig. 38.
4. Select *OK*. The size of the domain will be updated and the newly specified region of the textile will be displayed as shown in Fig. 39. The updated textile can be found in Ex2_9_3.tg3.

Exercise 4 Solution

The amended script can be found in Ex2_9_4.py and the resulting textile is shown in Fig. 40. The changes are as follows:

1 & 2.
```
# Assign a constant cross-section along the yarn of lenticular
shape
Yarn.AssignSection(CYarnSectionConstant(CSectionLenticular(0.2,
0.048)))
```

3.
```
# Set the resolution of the mesh created
Yarn.SetResolution(40)
```

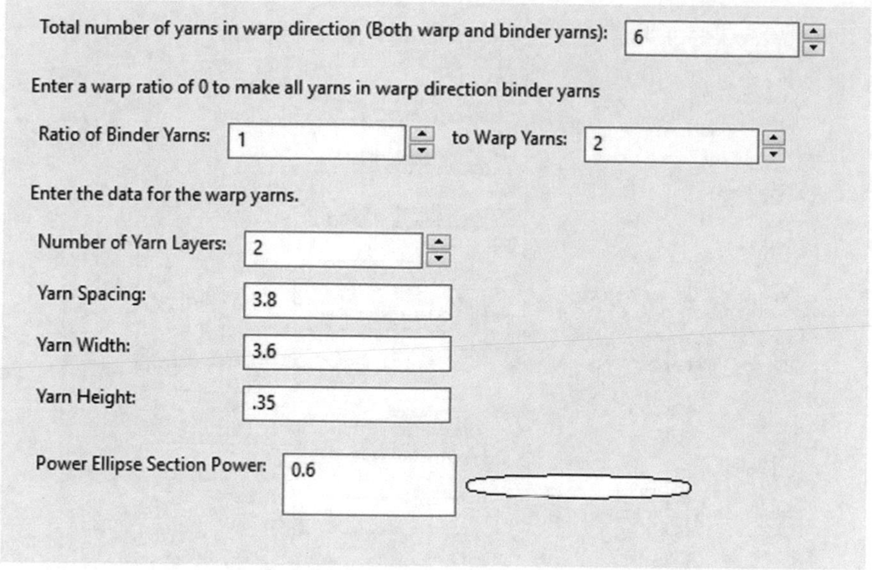

Total number of yarns in warp direction (Both warp and binder yarns): 6

Enter a warp ratio of 0 to make all yarns in warp direction binder yarns

Ratio of Binder Yarns: 1   to Warp Yarns: 2

Enter the data for the warp yarns.

Number of Yarn Layers: 2

Yarn Spacing: 3.8

Yarn Width: 3.6

Yarn Height: .35

Power Ellipse Section Power: 0.6

Fig. 33 Warp yarn input for Exercise 2

4.
```
# Create a domain and assign it to the textile
Textile.AssignDomain(CDomainPlanes(XYZ(0, 0, -0.024), XYZ(0.44,
0.44, 0.074)))
```

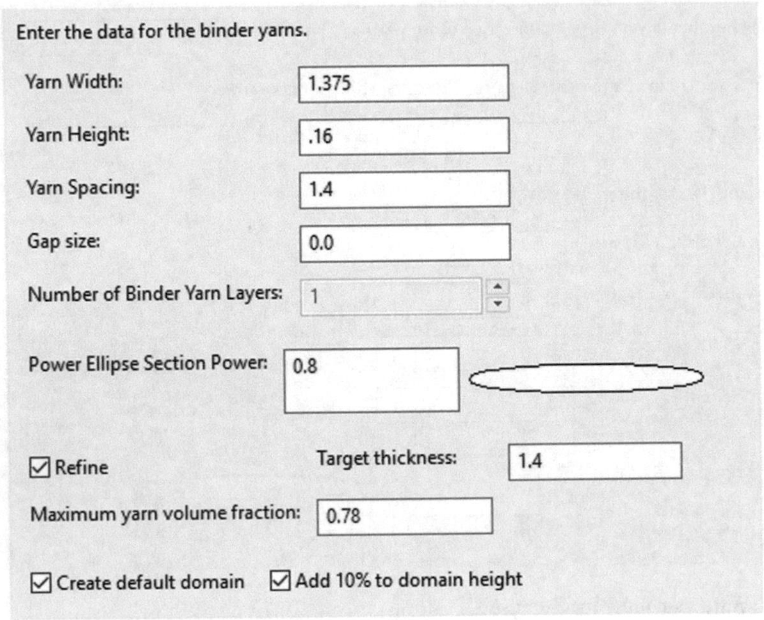

Enter the data for the binder yarns.

Yarn Width: 1.375

Yarn Height: .16

Yarn Spacing: 1.4

Gap size: 0.0

Number of Binder Yarn Layers: 1

Power Ellipse Section Power: 0.8

☑ Refine   Target thickness: 1.4

Maximum yarn volume fraction: 0.78

☑ Create default domain   ☑ Add 10% to domain height

Fig. 34 Binder yarn input for Exercise 2

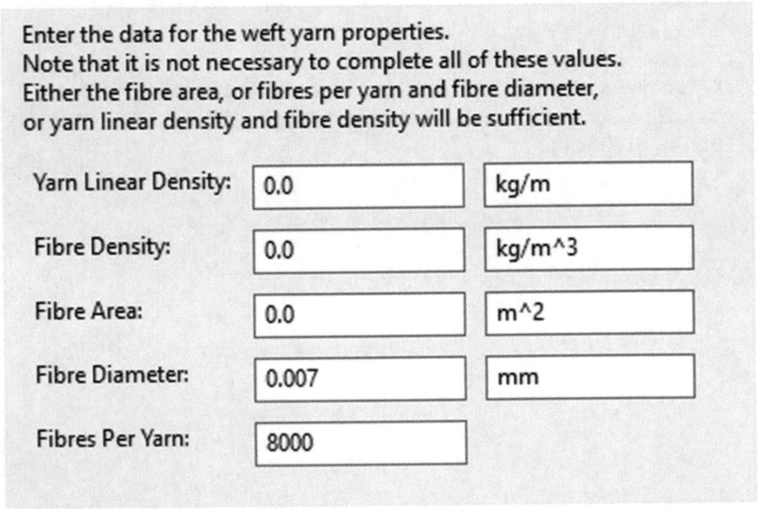

Enter the data for the weft yarn properties.
Note that it is not necessary to complete all of these values.
Either the fibre area, or fibres per yarn and fibre diameter,
or yarn linear density and fibre density will be sufficient.

Yarn Linear Density: 0.0   kg/m

Fibre Density: 0.0   kg/m^3

Fibre Area: 0.0   m^2

Fibre Diameter: 0.007   mm

Fibres Per Yarn: 8000

Fig. 35 Weft yarn properties for Exercise 2

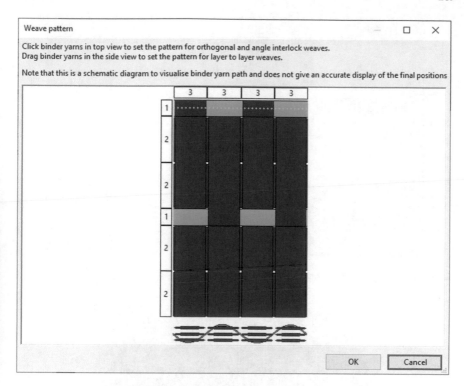

Fig. 36 Weave configuration for Exercise 2

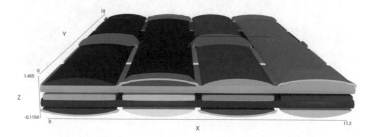

Fig. 37 Exercise 2, finished textile

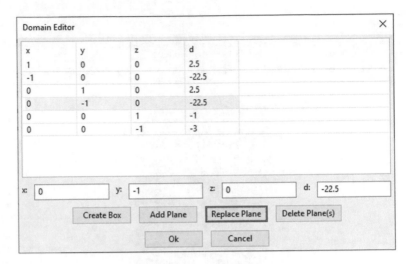

Fig. 38 Domain planes for Exercise 3

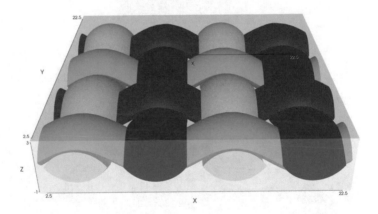

Fig. 39 Exercise 3, textile with domain specified for two repeats

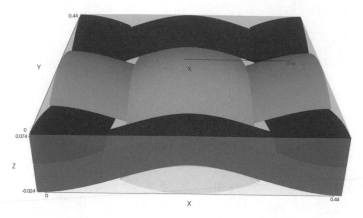

Fig. 40 Exercise 4, finished textile

References

1. Beazley, D. & Matus, M. 2007. *SWIG* [Online]. Available: www.swig.org. Accessed 26 Sep 2019.
2. Brown, L. P., Gommer, F., Long, A. C., Matveev, M., Zeng, X. & Yan, S. 2019. *TexGen Scripting Guide* [Online]. Available: https://github.com/louisepb/TexGenScriptingGuide. Accessed 1 Oct 2019.
3. Brown, L. P. & Sherburn, M. 2019. *louisepb/TexGen: TexGen v3.11.0 (Version 3.11.0). Zenodo* [Online]. Available: https://doi.org/10.5281/zenodo.3241493. Accessed 21 Oct 2019.
4. Li, S., & Wongsto, A. (2004). Unit cells for micromechanical analyses of particle-reinforced composites. *Mechanics of Materials, 36*, 543–572.
5. Matveev, M. Y., Long, A. C., & Jones, I. A. (2014). Modelling of textile composites with fibre strength variability. *Composites Science and Technology, 105*, 44–50.
6. *Notepad++* [Online]. Available: https://notepad-plus-plus.org/. Accessed 3rd Jul 2019.
7. OPENCASCADE. (2019). *OpenCASCADE* [Online]. Available: www.opencascade.com. Accessed 26 Sep 2019.
8. Sherburn, M. (2007). *Geometric and Mechanical Modelling of Textiles*. University of Nottingham.
9. SI, H. (2015). TetGen, a Delaunay-Based Quality Tetrahedral Mesh Generator. *ACM Trans. on Mathematical Software, 41*(2).
10. TEXGEN. (2019). *TexGen* [Online]. Available: http://texgen.sourceforge.net. Accessed 3 Jul 2019.
11. Zeng, X., Brown, L. P., Endruweit, A., Matveev, M., & Long, A. C. (2014). Geometrical modelling of 3D woven reinforcements for polymer composites: Prediction of fabric permeability and composite mechanical properties. *Composites Part A, 56*, 150–160.
12. Zeng, X., Endruweit, A., Brown, L. P., & Long, A. C. (2015). Numerical prediction of in-plane permeability for multilayer woven fabrics with manufacture-induced deformation. *Composites Part A: Applied Science and Manufacturing, 77*, 266–274.

WiseTex—A Virtual Textile Composites Software

Stepan Lomov

Abstract A virtual textile composites software WiseTex builds geometrical models of woven (2D and 3D), braided (bi- and tri-axial) and warp-knitted multiaxial (non-crimp) fabrics. The chapter describes the models of woven fabrics. The models can be based on full geometrical data, provided by the user, or can calculate the equilibrium crimp state of the yarns and their deformed cross section dimensions using minimisation of the bending deformation energy and experimental data of compressibility of the yarns. The yarn interlacing topology is described using a coding system, applicable both to 2D and 3D woven fabrics. The mechanical model of the fabric repeat allows analytical calculations of the fabric resistance to flat compaction, shear and biaxial tension. The software has GUI and a command-line instruction set which allows creating interface with software for upstream (weaving) and downstream (mechanical performance of textiles and textile composites) of the textile production/performance chain.

Keywords Weaves · Woven geometry · Mechanical properties · Software

1 Introduction

A virtual textile composites software WiseTex has been developed in early 2000s by the author of the present chapter [22, 23, 31, 41], is in continuous development since then [18, 29, 40] and is widely used by industry and academy around the globe. WiseTex has its origin in the previous work of its author in St-Petersburg in 1990–1995, resulted in CETKA software [16, 21, 25, 26].

WiseTex is a meso-level textile processor (MLTP) [29], i.e. a numerical tool, which accepts information of a textile parameters, overall fibre volume fraction (FVF) in a composite, deformation of the reinforcement (shear, compression, etc.), and creates a geometrical model of the textile (Fig. 1a). The geometrical model can be further processed by different models to produce such parameters as permeability of the

S. Lomov (✉)
Department of Materials Engineering, KU Leuven, Leuven, Belgium
e-mail: stepan.lomov@kuleuven.be

© Springer Nature Switzerland AG 2022
Y. Kyosev and F. Boussu (eds.), *Advanced Weaving Technology*,
https://doi.org/10.1007/978-3-030-91515-5_7

293

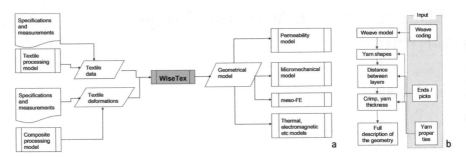

Fig. 1 WiseTex data flow: **a** input/output and links to the upstream and downstream software; **b** geometry calculation sequence

textile, stiffness matrix, thermal conductivity or dumping parameters of the consolidated composites, etc. Moreover, the unit cell geometrical model can be transformed into a 'general purpose' meso-level finite element (FE) model of the composite's unit cell, allowing further in-depth simulation of the composite/reinforcement properties and behaviour.

Textile input data is organised in a hierarchy of data levels for fibres, yarns and the fabric (Appendix A). The easy and open way of implementing this hierarchy is use of XML (Extensible Markup Language), that defines a set of rules for encoding documents in a format that is both human- and machine-readable.

The internal calculation sequence for building of a geometry model of a fabric repeat (unit cell) is shown in Fig. 1b. It starts from the definition of the weave topology, based on the weave coding. Then the yarn geometry is created, based on the weave density information (ends/picks count) and yarns mechanical properties. Calculation of the geometry is based either on the user-defined yarn crimp and user-defined dimensions of the yarn cross-sections, or on the calculations of the yarn crimp balance under the warp/weft interaction, defined by the yarn bending rigidity and their compression resistance. The full description of the geometry contains information of all yarns trajectories and their cross section dimensions. Yarns mechanical interaction is recalculated if the fabric is under compression, biaxial tension or shear deformation, resulting in the deformation resistance forces and the deformed geometry of the repeat. The following sections describe these calculations in more details.

2 Weave Coding

The weave structure coding based on section of the weave in the warp direction. Figure 2 depicts a fabric with $L = 7$ weft layers. The warps in intersections with the weft can occupy $L + 1$ levels, level 0 being the fabric face and level L being the back. Each warp is coded as a sequence of level codes and the entire weave is coded as a set of these warp sequences. In composite reinforcements warps are often layered as shown in Fig. 2. Warp paths are coded as warp zones, identifying sets of warp

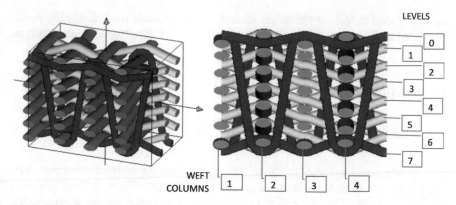

Fig. 2 Weave coding in WiseTex

yarns layered over each other. Hence the weave shown in Fig. 2, is coded as shown in Table 1.

The coding of weaves with the warp intersections codes allows an analysis of their topology. First, the warp intersections sequences gives definition of the *crimp intervals* of the warps, which are yarn segments extending between two crossovers, with identification of the weft yarns supporting the warp crimp at the ends of the interval and the positions of the supporting wefts in relation to the warp yarn: below or

Table 1 Coding of the weave shown in Fig. 2—warp intersection levels

Warp zone	Warp number in the zone	Weft column				Warp colour in Fig. 2
		1	2	3	4	
1	1	0	1	0	1	Red
2	1	0	7	0	7	Red
3	1	7	6	7	6	Red
4	1	1	0	1	2	Grey
	2	2	1	2	3	Grey
	3	3	2	3	4	Grey
	4	4	3	4	5	Grey
	5	5	4	5	6	Grey
	6	6	5	6	7	Grey
5	1	7	0	7	0	Red
6	1	1	2	1	0	Grey
	2	2	3	2	1	Grey
	3	3	4	3	2	Grey
	4	4	5	4	3	Grey
	5	5	6	5	4	Grey
	6	6	7	6	5	Grey

above it. Based on this identification, the weft crimp intervals are defined. The reader
is referred to [31] for more detailed description of the weave topology calculations.

3 Fabric Geometry

3.1 Yarn Centre-Line Shape in a Crimp Interval

The topology and waviness (crimp) of interlacing yarns are set by the weave pattern.
The term *crimp* also characterises the ratio of the length of a yarn to its projected
length in the fabric. Crimp is caused primarily by out-of-plane waviness; composites
reinforcements do not feature significant in-plane waviness because of the flat nature
of the tows. Typical crimp values in 2D composite reinforcements are below 1%; for
through-the-thickness yarns in 3D fabrics (e.g. warps 2 and 5 in Table 1) the crimp
values will be much higher.

Consider a wavy shape of the warp yarn in the crimp interval AB, shown in
Fig. 3a. p and h are distances between points A and B along the warp direction x and
thickness z, with p being the weft yarn spacing in the fabric. The distance h is called
the crimp height. The shape of the warp middle line $z(x)$ between A and B satisfies
the following conditions:

$$z(0) = \frac{h}{2}; z'(0) = 0; z(p) = -\frac{h}{2}; z'(p) = 0 \tag{1}$$

If the yarns cross section dimensions d_1 and d_2 are known (see Fig. 3a), then the
shape $z(x)$ can be defined, using two variants of the assumptions on compressibility
of the yarns.

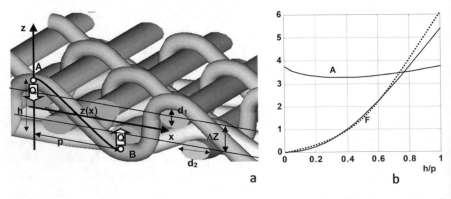

Fig. 3 Description of the fabric geometry in WiseTex: **a** parameters describing the yarn shape; **b**
characteristic functions

In the first case, cross section shapes are defined by the curved shape of interlacing yarns. This is acceptable when fibres move easily in the yarns, for example with untwisted continuous fibres. In the second case cross section shapes are fixed and define the curved shape of interlacing yarn. This is the case for monofilaments or consolidated yarns with high twist and heavy sizing.

With the first approach one can consider yarn crimp in isolation, find an elastic line satisfying boundaries represented by (1) and minimising the bending energy:

$$W = \frac{1}{2} \int_0^p B(\kappa) \frac{(z'')^2}{\left(1 + (z')^2\right)^{5/2}} dx \tag{2}$$

where $B(\kappa)$ is the yarn bending rigidity which depends on local curvature κ. A further simplification using an average bending rigidity leads to the well known problem of the elastica for which a solution can be written using elliptical integrals. Calculations are made easier with an approximation of the exact solution:

$$\frac{z}{h} = \frac{1}{2}(4\bar{x}^3 - 6\bar{x}^2 + 1) - A\left(\frac{h}{p}\right)\bar{x}^2(\bar{x} - 1)^2\left(\bar{x} - \frac{1}{2}\right); \bar{x} = \frac{x}{p} \tag{3}$$

where function $A(h/p)$ is shown in Fig. 3b; the value $A = 3.5$ provides a good approximation in the range $0 < h/p < 1$. The first term of Eq. (3) is the solution to the linear minimisation problem. This cubic spline very closely approximates the yarn line.

As the yarn shape defined by Eq. (3) is parameterised with the dimensionless parameter h/p, all properties associated with the bent yarn centreline can be written as a function of this parameter only. This allows introduction of a characteristic function $F(h/p)$ for the crimp interval, used to define the bending energy of the yarn W, transversal forces at the ends of the interval Q and average curvature κ, Fig. 3b:

$$W = \frac{1}{2} \int_0^p B(\kappa) \frac{(z'')^2}{\left(1 + (z')^2\right)^{5/2}} dx = \frac{B(\bar{\kappa})}{p} F\left(\frac{h}{p}\right) \tag{4}$$

$$Q = \frac{2W}{h} = \frac{2B(\bar{\kappa})}{ph} F\left(\frac{h}{p}\right) \tag{5}$$

$$\bar{\kappa} = \sqrt{\frac{1}{p} \int_0^p \frac{(z'')^2}{\left(1 + (z')^2\right)^{5/2}} dx} = \sqrt{\frac{2}{p} F\left(\frac{h}{p}\right)} \tag{6}$$

If a spline approximation of (3) is used, then

$$W \approx \frac{1}{2} B(\kappa) \int_0^p \left(z'' \right)^2 dx = 6B(\kappa) \frac{h^2}{p^3}$$

and $F \approx 6(h/p)^2$ (a dotted line in Fig. 3b).

If the second approach to the definition of the yarns interactions is used, then the yarn cross sections have pre-defined shapes. In *WiseTex* a mixed model is used, where average characteristics of crimp intervals are obtained from (3–6), taking advantage of the single parameter h/p involved in these equations, while yarn shapes over crimp intervals are defined by Peirce-type models [34] for an assumed cross-section shape.

3.2 Crimp Balance

The full geometry of the yarn in the fabric unit cell is created using the principle of the minimum energy of banding deformation of the warp and weft yarns and the experimental data on the compressibilityof the yarns. Throughout the section, subscript i, $i = 1..NWa$ designates a warp yarn, subscripts $j = 1..NWe$, $l = 1..L$ designate a weft yarn, NWa is number of warps in the fabric, NWe is the number of wefts in each layer of the fabric and L is the number of weft layers. The distribution of yarns over the yarn zones (see Sect. 2) is not represented by this numbering (the warps have a throughout numbering).

The following input data are given:

1. Fabric weave, given by a set of the warp intersection level sequences.
2. Compression and bending behaviour of warp and weft yarns (all the yarns in warp and weft can be different).
3. Spacing of warp and weft yarns (which can be non-uniform).
4. Shift between the weft layers in the warp direction. This is defined by the weft insertion and battening process. Two typical cases are zero shift (weft yarns are one above another) and shift of 50% of the weft spacing (weft yarns of the upper layer in between yarns of the lower layer).

Analysis of the weave matrix allows determining sets of crimp intervals on warp and weft yarns. Subscript k designates the interval number, for warp yarns $k = 1..NWe$, for weft yarns $k = 1..K_{jl}$, the number of crimp intervals on different weft yarns K_{jl} may be different. The ranges of the subscripts i,j,k are not given explicitly in the formulae below. For each interval the support yarns at the interval ends and signs of the yarn position relative to them are known from the analysis of the weave. If we consider a crimp interval k on a warp yarn i, then the indices of the weft yarn (and crimp intervals of it) supporting the warp at the ends of the interval are designated as j',l',k' and j'',l'',k''. If we consider a crimp interval k on a weft yarn (j,l), then the indices of the supporting warp yarns and intervals on them are designated by i',k' and i'',k'' (Fig. 4a).

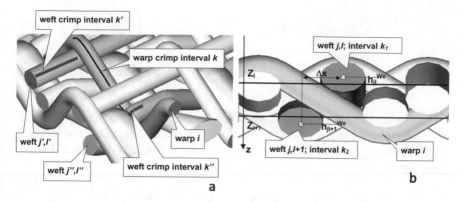

Fig. 4 Internal geometry of a woven fabric: **a** warp and weft crimp intervals; **b** tight packing of the layers

We assume that all the crimp intervals of a weft yarn (j,l) have the same crimp height $h_{jl}{}^{We}$. The weft yarns deviate at the ends of the crimp intervals in z direction from the average planes of the weft layers by $h_{jl}{}^{We}/2$. These average planes have z co-ordinates Z_l, $Z_0 = 0$. Dimensions of cross-sections can differ along a yarn. These dimensions at the ends of crimp intervals of warp and weft yarns are designated as $d_{1,ik}^{Wa}$, $d_{2,ik}^{Wa}$, $d_{1,jlk}^{We}$, $d_{2,jlk}^{We}$.

The parameters listed above allow building a fill description of the internal geometry of the fabric. Vertical positions of the centres of weft yarns at the ends of weft crimp intervals are given by

$$z_{jlk}^{We} = Z_l + h_{jl}^{We} P_{jlk}^{We} \tag{7}$$

where P gives a relative position of the supporting yarns: $P_{jlk}^{We} = +1$, if the weft yarn lies over the supporting warp, and $P_{jlk}^{We} = -1$ otherwise. The same notation is used for warp crimp intervals.

Vertical positions of the warp yarns at the end of crimp intervals are given then as

$$z_{ik}^{Wa} = z_{(il)'k'}^{We} - \left(\frac{d_{1ik}^{Wa}}{2} + \frac{d_{1j'l'k'}^{We}}{2} \right) P_{ik}^{Wa} \tag{8}$$

Positions of the ends of the crimp intervals in x and y directions are determined by the yarn spacing. With dimensions of the yarns and position of the yarns at the end of crimp intervals and support yarns defined, the models, described in the previous section are employed to generate fill paths of the yarns in the repeat.

To calculate the unknowns h_{jl}^{We}, Z_l, d_{1ik}^{Wa}, d_{2ik}^{Wa}, d_{1jkl}^{We}, d_{2jkl}^{We}, equations expressing the minimum of the total bending energy of the yarns in the repeat, compression of the yarns and the tight packing of the yarn layers should be constructed.

Yarn dimensions are defined by the laws of the yarn compression:

$$d_{1ik}^{Wa} = d_{10ik}^{Wa} \eta_{1i}^{Wa} \left(\frac{Q_{ij'l'}}{d_{2ik}^{Wa}} \right), d_{2ik}^{Wa} = d_{20ik}^{Wa} \eta_{2i}^{Wa} \left(\frac{Q_{ij'l'}}{d_{2ik}^{Wa}} \right)$$

$$d_{1jlk}^{We} = d_{10jlk}^{We} \eta_{1jl}^{We} \left(\frac{Q_{i'jl}}{d_{2jlk}^{We}} \right), d_{2jlk}^{We} = d_{20jlk}^{We} \eta_{2jl}^{We} \left(\frac{Q_{i'jl}}{d_{2jlk}^{We}} \right) \tag{9}$$

Here and below subscripts with a prime designate the corresponding indexes of crimp intervals and yarns, which are the support yarns for the crimp interval under consideration. Q_{ijl} in these equations are transversal forces of interaction if the warp yarn I and weft yarn (j,l), calculated with (5):

$$Q_{ijl} = \frac{1}{2} \left(\frac{B_i^{Wa}}{p_{ik'}^{Wa} h_{ik'}^{Wa}} F \left(\frac{h_{i'k'}^{Wa}}{p_{i'k'}^{Wa}} \right) + \frac{B_i^{Wa}}{p_{ik'+1}^{Wa} h_{ik'}^{Wa}} F \left(\frac{h_{ik'+1}^{Wa}}{p_{ik'+1}^{Wa}} \right) \right)$$
$$+ \frac{1}{2} \left(\frac{B_{jl}^{We}}{p_{jlk'}^{We} h_{jl}^{We}} F \left(\frac{h_{jlk'}^{We}}{p_{jlk''}^{We}} \right) + \frac{B_{jl}^{We}}{p_{jlk''+1}^{We} h_{jl}^{We}} F \left(\frac{h_{jl}^{We}}{p_{jlk''+1}^{We}} \right) \right) \tag{10}$$

Relation between crimp heights of the warp crimp intervals and the weft crimp is given by a constraint expressing tight contact between the interwoven yarns:

$$h_{ik}^{Wa} + \frac{1}{2} \left(h_{j'l'}^{We} + d_{1j''l''}^{We} \right) = Z_{l''} - Z_{l'} + \frac{1}{2} \left(d_{1ik}^{Wa} + d_{1ik+1}^{Wa} + d_{1j'l'k'}^{We} + d_{1j''l''k''}^{We} \right) \tag{11}$$

The coordinates of the weft layer planes Z_l are calculated from the condition of tight packing of the woven structure in z direction. Consider two weft yarns—(j,l) and $(j,l + 1)$ in two consecutive weft layers (Fig. 4b). k_1 and k_2 are indices of crimp intervals on these yarns which either have warp i as a support or are supports for the warp i at a certain crimp interval k'. z-coordinates of the centres of the cross-sections of the weft yarns are given by

$$z_{jlk1} = Z_l + h_{jl}^{We} P_{jlk2}^{We}; z_{j,l+1,k2} = Z_{l+1} + h_{j,l+1}^{We} P_{j,l+1,k2}^{We} \tag{12}$$

where P are the position codes of the weft yarns derived from the weave analysis. The distance Δx between the centres of the weft yarns in the warp direction is defined by spacing of weft and possible shift between the weft layers. Consider a distance Δz between these centres. As the yarns are packed closely, then this is the distance defined by the condition: the distance between contours is equal to $d^{Wa}{}_{ik'}$. The solution of such a problem, designated as Δz_{tight} depends on the shapes of the yarns, their dimensions and Δx. The value of Δz_{tight} defines the distance between the layers. As we assumed existence of a common middle plane of a weft layer, then

$$Z_{l+1} = Z_l + \max_{j,k} \left(\Delta z_{tight} \left(shape_{jl}^{We}, shape_{jl+1}^{We}, d_{1jlk1}^{We}, d_{2jlk1}^{We}, d_{1j,l+1,k2}^{We}, d_{2j,l+1,k2}^{We}, d_{1i'k'}^{Wa} \right) \right.$$
$$\left. - h_{jl+1}^{We} P_{j,l+1,k2}^{We} + h_{jl}^{We} P_{j,l,k2}^{We} \right) \tag{13}$$

With the condition $Z_1 = 0$, this equation defines all the weft layers positions.

Finally, *weft crimp heights* $h_{jl}{}^{We}$ are independent variables in the minimum problem:

$$W_\Sigma = \sum_{ik} \frac{B_{ik}^{Wa}}{p_{ik}^{Wa}} F\left(\frac{h_{ik}^{Wa}}{p_{ik}^{Wa}}\right) + \sum_{jlk} \frac{B_{jlk}^{We}}{p_{jlk}^{We}} F\left(\frac{h_{jl}^{We}}{p_{jlk}^{We}}\right) \to min \qquad (14)$$

where h^{Wa} are related to h^{We} with (11). The minimum is reached when

$$\frac{\partial W_\Sigma}{\partial h_{jl}^{We}} = 0 \qquad (15)$$

The system of equations, which defines the parameters of the fabric internal geometry, is now complete. It is solved by an iteration procedure, checking the convergence by the convergence of weft crimp height values. The reader is referred to [31] for more detailed description of the geometry calculations.

3.3 Full Description of the Fabric Geometry

Figure 5 illustrates the description of the configuration of the yarns in a unit cell of the woven fabric. In this description the unit cell can be sheared, with an angle α between the warp and weft directions (see Sect. 5).

The midline of a yarn is given by the spatial positions of the centres of the yarn cross-sections O: $r(s)$, where s is the coordinate along the midline, r is the radius-vector of the point O. Let $t(s)$ be the tangent to the midline at point O. The cross-section of the yarn, normal to t, is defined by its dimensions $d_1(s)$ and $d_2(s)$ along axis $a_1(s)$ and $a_2(s)$. These axes are "glued" with the cross-section and rotate around $t(s)$, if the yarn is twisted along its path (such a twist can be the result of fabric shearing). Because of this rotation the system $[a_1 a_2 t]$ may differ from the natural coordinate

Fig. 5 General representation of yarns geometry in WiseTex

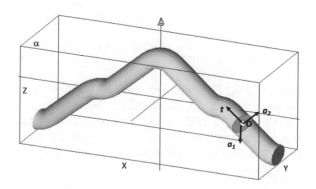

system along the spatial path $[nbt]$ $(t = \frac{dr}{ds}, n = \frac{dt}{ds}, b = t \times n)$. The shape of the cross-section can be assumed elliptical, lenticular etc. The shape type does not change along the yarn, but dimensions d_1 and d_2 can change because of different compression of the yarn in the contact zones and between them. The definition of a yarn with a given cross-section shape consists therefore of the five periodic functions: $r(s)$ (then $[nbt]$ vectors can be calculated), $a_1(s)$, $a_2(s)$, $d_1(s)$, $d_2(s)$.

When used in numerical calculations, the yarn description is given as arrays of values for a set of cross-sections S_i along the yarn midline:

- Co-ordinates of the cross-section centre point $O = (x,y,z)$
- Tangent, normal and bi-normal to the yarn centre-line t, n, b
- Vectors of (orthogonal) axis of the contour a_1, a_2
- Dimensions in the directions of the axis d_1, d_2

This description fully defines volumes of the yarns in a unit cell. The format is the same for orthogonal and non-orthogonal (angle α) unit cells. The in-plane dimensions of the unit cell X, Y are given by the repeat size of the textile structure, the thickness Z—as difference between the maximum and minimum z-coordinates of the cross-sections of all the yarns in the unit cell.

The geometry description is stored in binary, ASCII or XML format (Appendix A).

4 Nested Laminates

A module of *WiseTex* suite, *LamTex* creates a geometrical model of a nested laminate [30]. In order to build a geometrical model of a laminate, consider first a model of one layer of the fabric. As described above, the *WiseTex* model provides positioning of all yarns in the volume of the fabric unit cell (repeat). The placement hence defines the *surface profile functions* of the face and back surface of the laminate. The functions $h_f(x,y)$ and $h_b(x,y)$ define the distance of the points on the fabric face (subscript f) and back (subscript b) surface from the middle surface of the fabric:

$$
\begin{aligned}
h_f(x, y) &= \frac{Z}{2} - z_f(x, y) \\
h_b(x, y) &= z_b(x, y) + \frac{Z}{2}
\end{aligned}
\tag{16}
$$

where x,y,z are Cartesian coordinates, with the centre of the coordinate system in the centre of the unit cell, Z is the fabric thickness, z_f and z_b are coordinates of the face and back surface of the fabric, defined as maximum and minimum z-coordinate of yarns at (x,y) position. Equation (16) applies if there is a point (x,y,z) inside the yarns or fibrous plies for the given (x,y). If no such point exist, then

$$h_f(x, y) = h_b(x, y) = Z/2$$

Functions h_f and h_b are actually computed on a finite mesh. Note that this definition applies for skewed unit cell, as well (which is the case for sheared fabrics), as the *WiseTex* geometrical model treats orthogonal and non-orthogonal unit cells in a similar way.

With the surface profile functions defined, it is easy to calculate the nesting of the layers. Consider two identical layers of the laminate, with one layer shifted relative to another by dx and dy in x and y directions. To define the nested position, we calculate the distance h^* between centre planes of the layers, when the yarns in the layers are just touching one another and there is no inter-penetration of the yarns. When nesting is zero, $h^* = Z$. Consider certain distance between centre planes h. The distance (depending of the (x,y) position) between the back surface of the upper layer and the face surface of the lower layer is

$$\delta(x, y; h) = h - Z + h_f(x, y) + h_b(x - dx, y - dy)$$

Defining

$$\delta*(h) = \min_{x,y} \delta(x, y; h)$$

we can compute the nesting distance h^* as a solution of the equation

$$\delta*(h) = 0$$

As the surface depth functions are defined on a finite mesh, these calculations are easily performed if shifts dx, dy are done in integer mesh units. In this case the calculations involve $O(Nx*Ny)$ comparisons (Nx and Ny being the mesh size in x and y directions).

For a laminate of L layers, the set of shifts dx_l, dy_l, $l = 1...L$ is defined, and the algorithm is applied to one layer after another. When the nested positions of layers are defined, the descriptions of the yarns and fibrous plies of the original one layer fabric are copied and positioned according to the in-plane shifts and vertical placement of the layers.

Figure 6 illustrates the models built with *LamTex* algorithm. Figure 6a shows a non-nested laminate of a plain weave glass fibre fabric. The thickness of the stack is 4.82 mm (indicated by the red dashed lines) and the fibre volume fraction is 35.0%. When layers have been shifted randomly and nested (Fig. 6b), the thickness of the stack becomes 45.3 mm, and the fibre volume fraction 39.1%. Finally (Fig. 6c), if the maximum nesting is sought, the thickness becomes 3.95 mm and the fibre volume fraction 42.7%. Fo a balanced fabric this corresponds to regular shifts of the layers. Note that the configurations shown in Fig. 6 is subject to the translational symmetry transformation in x–y plane, and the apparent voids in the unit cell volume are actually filled by the yarns belonging to the adjacent repeatable units.

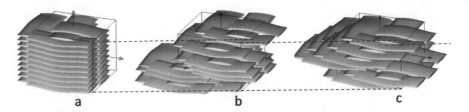

Fig. 6 Building nested laminates with LamTex: 10-layers laminate of a glass fibre plain weave fabric: **a** no nesting; **b** random shift of the layers; **c** maximum nesting. Red dashed lines indicate the thickness of the stacks

5 Fabric Deformations: Flat Compaction, Shear and Bi-Axial Tension

A model of compression of woven fabrics implemented in *WiseTex* [17, 27] accounts for two physical phenomena associated with the fabric compression: change of the yarns crimp and compression of the individual yarns. Internal structure of laminated preforms after compression is affected also by relative shift and nesting of the layers in lay-up [30].

Biaxial tension of a woven fabric [17, 28] is characterised by change of the fabric dimensions in warp (x-axis) and weft (y-axis) directions $X = X_0(1 + e_x)$, $Y = Y_0(1 + e_y)$, where X and Y are sizes of the fabric repeat, subscript "0" designates the undeformed state, e_x, e_y are technical deformations of the fabric. As discussed in Sect. 3, the internal structure of the fabric is described based on weft crimp heights and weft and warp cross-section dimensions at the intersections and. These values change after the deformation. Tension of the yarns induces transversal forces, which compress the yarns, changing their dimensions. The same transversal forces change the equilibrium conditions between warp and weft, which leads to a redistribution of crimp and change of crimp heights. When the crimp and the yarn dimensions values in the deformed configuration are computed, the internal geometry of the deformed fabric is built as explained in Sect. 3.

Model of shear of woven fabrics [17, 28] accounts for the following mechanisms of the yarns deformation, determining the shear resistance: friction,(un)bending; lateral compression; torsion; vertical displacement of the yarns. The geometrical model of the sheared fabric is similar to the model of a bi-axial braid (non-orthogonal unit cell), with the additional complication of the change of yarn cross section dimensions induced by the lateral compression of the yarns during the fabric shear. When a fabric is sheared, the deformation is resisted by friction between yarns, bending and compression of the yarns. Friction forces are estimated in the model using normal forces of the yarn interaction, tension being a pretension normally employed in the shear test. The transversal forces are increased by the internal pressure, developed inside yarns due to their lateral compression in the sheared structure. This is taken into account using the experimental compression diagrams of the yarns. Resistance due to bending is estimated using the difference in bending energies in deformed and

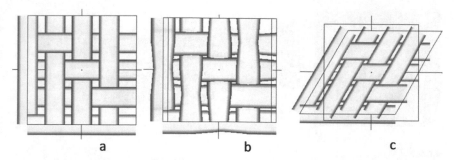

Fig. 7 Deformation of a 5H satin glass fibre woven fabric [32], calculated with WiseTex: **a** undeformed fabric; **b** uniaxial tension in warp direction, elongation 10%; **c** shear by 30°

undeformed configurations, the latter computed with algorithms for non-orthogonal structures.

The fabric geometry after deformation is described using the same data as shown in Sect. 3.3. The dimensions X, Y, Z and the angle α of the unit cell corresponds to the fabric parameters after the deformation. Figure 7 shows the results of WiseTex calculations of a deformed geometry of a 5H satin glass fibre woven fabric [32].

Figure 8 illustrates the calculations of tension of a two-layer PES safety belt with a 3D weave structure [15]: the input data, WiseTex models and the calculated and experimentally measured tensile diagrams. The initial elongation of the fabric without significant resistance is explained by compaction of the multifilament yarns. The calculation correctly predicts the stiffness of the belt; the rupture of the belt is not simulated well, most probably because of the difference between the strength of the highly crimped yarns in the weave and the tensile strength, measured by the tension on a straight yarn.

The reader is referred to [17] for a detailed description of WiseTex deformability models.

6 Scripting

WiseTex geometrical modelling and the micromechanical, permeability etc.calculations (see Fig. 1) can be controlled by direct user input via GUI or by commands in a script, which allows programming of complex calculation tasks and integration into wider modelling environment, with script commands generated by other modules of the software. Scripts in *WiseTex* and in micromechanical (*TexComp*) and permeability (*FlowTex*) components of the *WiseTex* package use the syntax of DOS command line, as shown in Appendix B.

There are different possibilities for use of the scripts. The simplest are serial calculations with textile input data sets with varying parameters (for example, changing ends/picks spacing of a woven fabric or braiding angle for braids)—the result of

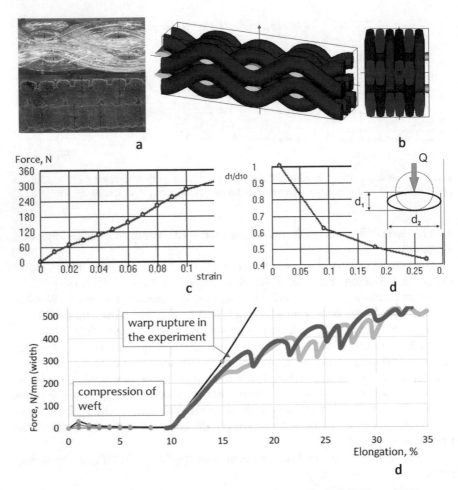

Fig. 8 WiseTex modelling of a narrow weave tension (two-layer safety belt, PES yarns) [15]: **a** cross sections of the belt, **b** WiseYex models; **c** tension and **d** compression diagrams of a multifilament PES yarn, 480 tex; **e** experimental (coloured lines) and calculated (black line) tensile diagrams

such calculations will be a set of GEO models and the corresponding reinforcement permeability and/or stiffness of the impregnated composite. Look-up tables can be built with this technique for the following use in macro-level flow modelling or structural analysis of composite parts. As an example, the script for creation of a look-up table *permeability vs fibre volume fraction* for a woven preform can be schematised as follows:

```
for p = 0.001, 0.002, 0.05, 0.01, 0.02, 0.05, 0.1, 0.2, 0.5 // MPa
    WiseTexCL "woven.XML" "woven_compressed.XML" compress:p
    extract VF value from "woven_compressed.XML"
    WiseTexCL "woven_compressed.XML" "woven_compressed.vox" voxels:100,100,20
    FastFDFlowTex  -W "woven_compressed.vox" WX
    extract permeability value Kx from FlowTex results file
    print p, VF, KX
end
```

Scripting allows also integration of the meso-level simulations in macro-analysis. For example, consider the problem of structural analysis of a 3D part, with shear angles of the textile reinforcement different from point to point because of draping of the reinforcement over the mould. Assume that a draping simulation software generates values of the reinforcement shear angle GAMMA for all finite elements (i is the element number) in the part model. Then the script for calculation of the stiffness of the composite in all the element will be as follows:

```
for all elements
    WiseTexCL "unsheared.xml" "sheared.xml" in_plane:element.GAMMA
    TexCompCL sheared.xml epoxy.xml C.csv
    transform C matrix into the local coordinate system of the element
    store C in element
end
```

7 Transfer of a WiseTex Model to Finite Element Tools

The downstream integration of MLTP models requires a tool, which (a) automatically translates the geometry into finite element (FE) software, (b) creates solid models and performs the automatic corrections of yarn volumes, and (c) assigns properties and sets contact surfaces.

Such a tool should read the WiseTex geometrical model and organise the geometric primitives in accordance to the data structures of the FE package. This can be done, as an example, as follows (Fig. 9a–e).

Every yarn is treated as a separate part. Each part represents an object, which contains a set of attributes and sub-objects. In the context of textile modelling, the starting point for creation of FE model are (a) the section sketches, which contain the planar geometry of yarn cross-sections, (b) the coordinate planes in 3D space where the sketches are projected to. The sketch sections are discretized. They come up from WiseTex as ellipses and they are then transformed to polygons connecting a user-defined number of points on the ellipse contour. Once the cross-sections are projected, the solid volumes are constructed by connecting these sections through a loft operation (in the case of high curvature of the yarn more complex algorithms may be needed). The result is the solid model of a yarn built as a set of segments connecting the cross sections. Every segment has its own fibre coordinate systems and, if necessary, a set of material properties defined by intra-segment fibre volume fraction. The part instances are then combined in an assembly object representing all the yarns in a unit cell model.

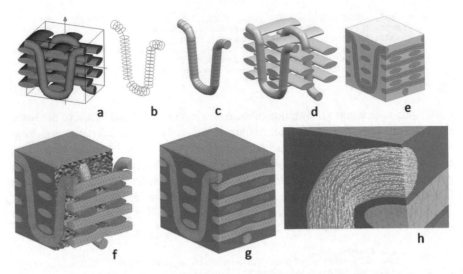

Fig. 9 Transformation of WiseTex internal geometry into finite element models, stages (produced in Virtual Materials Characterisation Tool of Simcenter, courtesy A. Matveeeva): **a** WiseTex model; **b** cross sections of a yarn; **c** solid volume of a yarn; **d** solid model of all yarns in the unit cell; **e** the yarns are placed inside a box and matrix volume added; (**f, g**) the mesh; **h** detail of the mesh: local fibre directions

The meshing of the resultant FE model presents the major difficulty of inter-penetration of the yarn volumes, created by the geometric pre-processor [10, 24]. Even if the interpenetrations are avoided, as they are in Fig. 9f–i, the meshing still is difficult, because of appearance of highly distorted elements. The mesh continuity demands an ideal geometrical match between the volumes of matrix and yarns. In a textile model adjacent yarns can either contact each other directly or be separated by a thin layer of matrix. This very thin layer inevitably leads to highly distorted elements and poor quality of stress approximation in that region. Various numerical schemes are used to resolve the problem of matrix pocket meshing. For instance, within the domain (mesh) superposition (a.k.a. embedded elements) yarn and matrix meshes share the same space. Independent meshes interact through the integration points of the elements. The yarn stiffness is reduced to account for the combined action of these volumes. Hence, the matrix can be meshed with regular elements. In [36–38, 42] embedded elements are used in conjunction with the WiseTex geomet-rical modelling. Voxel-based meshes is another way to radically solve the meshing problems [8], WiseTex has an option to output a voxel-style discrete representation of the fabric model volume.

Table 2 shows the existing links between *WiseTex* and FE packages. Figure 10 illustrates these links, showing models created based on WiseTex models for:

– hybrid carbon/self-reinforced polypropylene laminate in Abaqus [38]
– 3D woven fabric in Composites Dream

Table 2 Links between WiseTex and FE software tools

Software	Producer	Comment	References
Abaqus	Dassault Systèmes Simulia Corp	The Python script for creation an Abaqus model is included in *WiseTex* package	[1, 5, 10, 12, 18, 29, 36, 38, 41, 42]
ANSYS	ANSYS Inc	An *FETex* application of *WiseTex* package outputs an ANSYS APDL script for creation of a FE model	[11, 19]
Composites Dream (SACOM)	ITOCHU Techno-Solutions Corporation	Incorporates *MeshTex* tool for transformation of the *WiseTex* geometry in a FE model which is solved by SACOM FE solver or can be transferred to Abaqus or LS Dyna	[20, 24, 35]
Digimat	e-Xstream engineering, part of Hexagon Manufacturing Intelligence	Digimat includes "WiseTex engine"—i.e. WiseTex algorithms for creation of woven structures, which are further used for homogenisation of mechanical properties (fast mean-field method, FE with conforming tetrahedral or voxel meshes or FFT over regular grids; could be exported to external FE tools)	[6, 7, 33]
Simcenter	Siemens	Virtual Material Characterization (VMC) ToolKit imports *WiseTex* XML files or voxel files which are further used for homogenisation of mechanical properties (fast mean-field method or FE with conforming or voxel meshes)	[2, 3, 9]

(continued)

Table 2 (continued)

Software	Producer	Comment	References
TexMind	TexMind	*WiseTex* files can be read in all *TexMind* textile generator packages and then transferred to Abaqus, LS Dyna, ANSYS APDL, Impact FEM, VTMS, Obj, STL, x3d, vrml, TexGen Python script. TexMind *Weaver* can save the topology information for the woven structuers as WiseTex XML file	[13, 14]

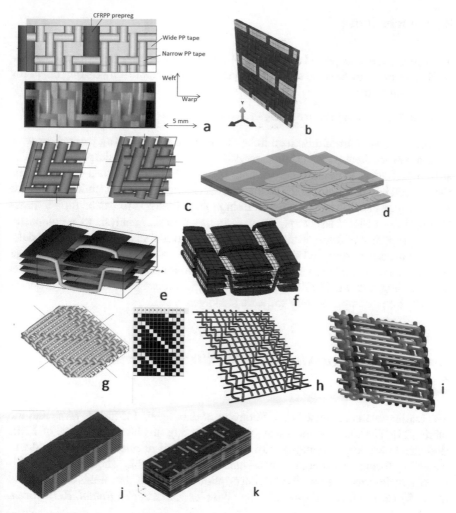

Fig. 10 Transformation of WiseTex internal geometry into finite element models, examples: (**a**, **b**) hybrid self-reinforced polypropylene/carbon tows laminate, WiseTex and the actual fabric image (**a**)—ABAQUS (**b**) [38],(**c**, **d**) sheared twill 2/2 fabric: WiseTex and LamTex (**c**)—Simcenter (**d**); (**e**, **f**) 3D woven glass/epoxy composite, WiseTex (**e**)—SACOM/CompositeDream (**f**) [20],(**g**, **h**, **i**) narrow satin weave WiseTex (**g**)—weave structure in TexMind (**h**)—LS DYNA (**i**) (courtesy Y. Kyosev); (**j**, **k**) WoxTex woven structures in Digimat: (**j**) a solid model; (**k**) voxel mesh (courtesy L. Adam)

- a laminate of sheared 2/2 twill fabeic in Simcenter
- narrow sateen fabric in LS DYNA via TexMind.

8 Conclusions

The open data exchange and scripting in *WiseTex* allow an easy integration of meso-modelling in custom textile simulation systems and in the full modelling chain for textile composites:

> textile process – composite processing – composite material – composite part

The upstream part of this chain links manufacturing/processing parameters with the internal architecture of the fabric, treated locally in the composite part. The downstream part refers to models transforming the structural, geometrical description of the unit cell into properties of the composite: mechanical, transport (fluid, thermal) or physical (e.g. electrical conductivity). Being local, these results can be further transferred to macro-level structural, thermal, impregnation,... analysis of a composite part as properties of discretisation elements (most commonly, FEs) of these models.

The data organisation in *WiseTex* is suitable to be a part of a future open exchange format for meso-level textile modelling, which would cover the needs of all levels of modelling and would allow quick and simple exchange between the different competitive or complementary software packages.

Appendix A. Textile Objects in WiseTex and Their Parameters

The textile data is organised in a hierarchy of data levels for fibres, yarns and the fabric [23]. The easy and open way of implementing this hierarchy is use of XML (Extensible Markup Language) [43], that defines a set of rules for encoding documents in a format that is both human—and machine-readable. The XML data fields are briefly described below. The full description of the TEX data fields can be found in the *WiseTex* software documentation (*WiseTex v3.0 User's Manual, KU Leuven, 2012*).

The fibre data include fibre geometry and transversely orthotropic elastic constants for the fibre material.

The yarn data include general yarn parameters such as linear density, shape and dimensions of the cross sections, data on the fibrous structure of the yarn and also description of the yarn mechanical behaviour in compression, bending and tension. The latter group of the data is not compulsory. For example, if the compression data is omitted, the sections of the yarn will retain the specified dimensions in the fabric; if the bending data is omitted, then crimp of the yarns in the fabric must be specified by the user, as crimp balance calculation will not be possible etc. Note that the yarn data describe the yarn per se—some of the specified parameters can be modified after the fabric geometry model has been built, for example, the yarn cross section dimensions in the fabric can be different because of the yarn compression. See Sect. 3 for description of the use of the yarn parameters in the geometry model.

The woven fabric data contain data on the yarns placement density (the distance between the yarn centre lines p), the weave pattern and the placement of different yarns in warp and weft fabric. The *Weave* data section contains a weave code sequences (see Sect. 2). The *ModellingParameters* data section holds information on the computational parameters used for building the geometrical model.

The XML TEX data can be modified either manually, using an XML editor or directly the text representation of the XML file, or via a custom program, hence open for integration in custom simulation software. The freeware and open source libraries for manipulation of XML files are available for all programming languages, for example [39]. The TEX XML data is also open in a sense that it can be augmented for different MLTPs, but the presence of already defined fields will ensure compatibility of the formats with *WiseTex*.

The generic description of the fabric geometry (see Sect. 3.3) is stored as arrays of values for a set of cross-sections S_i along the yarn midline.

Once the geometrical model is generated by *WiseTex*, the textile inpt data in principle is not needed any more for use in subsequent processing. However, it could be beneficial for some cases to keep specific textile data together with the geometry data created for them.

Appendix B. Command Line Components of *WiseTex* Package

Table 3 shows the functionality and command syntax of the command line versions of *WiseTex* package components. The syntax of the commands in Table 2 is simplified for the needs of the discussions in this paper; the full syntax definition can be found in the software documentation.

Table 3 Functionality and command syntax of command line components of WiseTex package

Software	Function	Command syntax	Explanation
MLTP *WiseTex*	Generic command syntax	WiseTexCL <in-file> <out-file> <mode> [: <options>]	The command creates a geometrical model of the <out-file> based on data in the <in-file> . <ode> specifies the function and/or deformations applied to the unit cell; <options> are the model and computation parameters
	Create a geometrical model of relaxed fabric	WiseTexCL <in-file> <out-file> build	
	Create a geometrical model of compressed fabric	WiseTexCL <in-file> <out-file> compress: <p> [, <p>]...	p is the applied pressure in MPa; if more than one p-parameter is given, then a set of GEO files is generated with p values added to the file names
	Create a geometrical model of the fabric, deformed in plane	WiseTexCL <in-file> <out-file> in_plane: <gamma> , <epsX> , <epsX> [; <gamma> , <epsX> , <epsY>]...	*gamma* is the shear angle in degrees, *epsX* and *epsY* are tensile deformations in X and Y directions
	Write a voxel file	WiseTexCL <in-file> <voxel file> voxels: <Nx> , <Ny> , <Nz> [:D]	Voxel file: the name of the text file, which will contain description of the voxels; Nx, Ny, Nz: number of divisions of the unit cell in three directions; if D is specified, then local permeabilities for Brinkmann calculations are written to the voxel file
Micro-mechanics *TexComp*	Calculate a stiffness matrix of the composite	TexCompCL <WiseTex file> <matrix data> <stiffness file> [CS: <phi>] [VF: <vf>]	matrix data: XML file containing values of Young modulus and Poisson coefficient for the matrix; stiffness file contains the result of the calculation: stiffness and compliance matrices and engineering constants in the co-ordinate system CS defined by angle *phi* of the in-plane rotation; *vf* is the fibre volume fraction if different from the one defined by the WiseTex file data

(continued)

Table 3 (continued)

Software	Function	Command syntax	Explanation
Permeability *FlowTex*	Calculate permeability of the fabric	FastFDFlowTex -W <VOX file> WX [WA <angle>]	WX specifies the flow direction (could be also WY or WZ) WA specifies the angle of the unit cell of the GEO model The results of the calculation (permeability values) are written in the ASCII file "FlowTex_results.txt"

Notation:

<...> text string to be specified.

[…] parameter can be omitted.

… possible repeat of the previous element.

References

1. Bedogni, E., Ivanov, D. S., Lomov, S. V., Pirondi, A., Vettori, M., & Verpoest, I. (2012). Creating finite element model of 3D woven fabrics and composites: semi-authomated solution of interpenetration problem. In *15th European Conference on Composite Materials (ECCM-15)*. Venice: electronic edition, s.p.
2. Cortelli, S. (2019). "Virtual Material Characterization: enabling efficient materials engineering for composites." Retrieved 03.07.2019, from https://community.plm.automation.sie mens.com/t5/Simcenter-Blog/Virtual-Material-Characterization-enabling-efficient-materials/ba-p/579569.
3. Cortelli, S. (2019). "VirtualCT: realistic composite material modeling using micro-CT-based voxel approach." Retrieved 26.11.2019, from https://community.plm.automation.siemens.com/t5/Simcenter-Blog/Virtual-Material-Characterization-enabling-efficient-materials/ba-p/579569.
4. Daggumati, S., Van Paepegem, W., Degrieck, J., Xu, J., Lomov, S. V., & Verpoest, I. (2010). Local damage in a 5-harness satin weave composite under static tension: Part II - Meso-FE modelling. *Composites Science and Technology, 70*(13), 1934–1941.
5. Daggumati, S., Voet, E., Van Paepegem, W., Degrieck, J., Xu, J., Lomov, S. V., & Verpoest, I. (2011). Local strain in a 5-harness satin weave composite under static tension: Part II - Meso-FE analysis. *Composites Science and Technology, 71*(8), 1217–1224.
6. Digimat (2019). Digimat, users' and example manuals—e-Xstream engineering, part of Hexagon Manufacturing Intelligence.
7. Digimat and Sonaca (2019). Simulation of 3D woven materials in aerospace industry. (electronic resource, www.e-Xstream.com).
8. Doitrand, A., Fagiano, C., Irisarri, F. X., & Hirsekorn, M. (2015). Comparison between voxel and consistent meso-scale models of woven composites. *Composites Part A: Applied Science and Manufacturing, 73*, 143–154.
9. Farkas, L., Vanclooster, K., Erdelyi, H., Sevenois, R., Lomov, S. V., Naito, T., Urushiyama, Y., & Van Paepegem, W. (2016). Virtual material characterization process for composite materials: An industrial solution. In *ECCM17 - 17th European Conference on Composite Materials Munich*. electronic edition, s.p.
10. Ivanov, D. S., & Lomov, S. V. (2015). Modelling the structure and behaviour of 2D and 3D woven composites used in aerospace applications. In P. E. Irving, & C. Soutis (Eds.), *Polymer composites in the aerospace industry* (pp. 21–52). Cambridge, Elsevier - Woodhead Publishers.
11. Ivanov, D. S., Lomov, S. V., Baudry, F., Xie, H., Van Den Broucke, B., & Verpoest, I. (2009). Failure analysis of triaxial braided composite. *Composites Science and Technology, 69*, 1372–1380.
12. Iwata, A., Inoue, T., Naouar, N., Boisse, P., & Lomov, S. V. (2019). Coupled meso-macro simulation of woven fabric local deformation during draping. *Composites Part A, 118*, 267–280.
13. Kyosev, Y. (2017). Möglichkeiten der TexMind Software für die Generierung von textilen Strukturen für FEM Simulationen und CAD Anwendunge 9th Saxon Simulation Meeting, Chemnitz: electronic edition.
14. Kyosev, Y. (2019). Automated virtual development and analysis of structures for textile wearables. In *AUTEX-2019—19th World Textile Conference on Textiles at the Crossroads*, Ghent, Belgium: electronic edition.
15. Kyosev, Y., Lomov, S. V., Kuster, K. (2016). The numerical prediction of the tensile behaviour of multilayer woven tapes made by multifilament yarns. In Y. Kyosev (Ed.), *Recent developments in braiding and narrow weaving* (pp. 69–82). Springer.
16. Lomov, S. V. (1995). Prediction of geometry and mechanical properties of woven technical fabrics with mathematical modelling. Dept. Mechanical Technology of Fibrous Materials. St.-Petersburg, SPbSUTD. Thesis: Doctor of Technical Sciences (Dr Habil.): 486 (in Russian).
17. Lomov, S. V. (2007). Virtual testing for material formability. In A. Long (Ed.), *Composite forming technologies* (pp. 80–116). Woodhead.

18. Lomov, S. V. (2015). From a virtual textile to a virtual woven composite. In M. H. Aliabadi (Ed.), *Woven Composites* (Computational and Experimental Methods in Structures, vol. 6). London, Imperial College Press: 109–140.
19. Lomov, S. V., Bernal, E., Ivanov, D. S., Kondratiev, S. V., & Verpoest, I. (2005). Homogenisation of a sheared unit cell of textile composites: FEA and approximate inclusion model. *Revue Européenne de Mécanique Numérique (former Revue européenne des éléments finis), 14*(6–7), 709–728.
20. Lomov, S. V., Bogdanovich, A. E., Ivanov, D. S., Hamada, K., Kurashiki, T., Zako, M., Karahan, M., & Verpoest, I. (2009). Finite element modelling of progressive damage in non-crimp 3D orthogonal weave and plain weave E-glass composites. In *2nd World Conference on 3D Fabrics*, Greenville, SC: CD edition.
21. Lomov, S. V., & Gusakov, A. V. (1995). Modellirung von drei-dimensionalen gewebe Strukturen. *Technische Textilen, 38*, 20–21.
22. Lomov, S. V., Gusakov, A. V., Huysmans, G., Prodromou, A., & Verpoest, I. (2000). Textile geometry preprocessor for meso-mechanical models of woven composites. *Composites Science and Technology, 60*, 2083–2095.
23. Lomov, S. V., Huysmans, G., & Verpoest, I. (2001). Hierarchy of textile structures and architecture of fabric geometric models. *Textile Research Journal, 71*(6), 534–543.
24. Lomov, S. V., Ivanov, D. S., Verpoest, I., Zako, M., Kurashiki, T., Nakai, H., & Hirosawa, S. (2007). Meso-FE modelling of textile composites: Road map, data flow and algorithms. *Composites Science and Technology, 67*, 1870–1891.
25. Lomov, S. V., & Primachenko, B. M. (1992). Mathematical modelling of two-layered woven fabric under tension. *Technologia Tekstilnoy Promyshlennosty, 1*, 49–53 (in Russian).
26. Lomov, S. V., & Truevtzev, N. N. (1995). A software package for the prediction of woven fabrics geometrical and mechanical properties. *Fibres & Textiles in Eastern Europe, 3*(2), 49–52.
27. Lomov, S. V., & Verpoest, I. (2000). Compression of woven reinforcements: A mathematical model. *Journal of Reinforced Plastics and Composites, 19*(16), 1329–1350.
28. Lomov, S. V., & Verpoest, I. (2006). Model of shear of woven fabric and parametric description of shear resistance of glass woven reinforcements. *Composites Science and Technology, 66*, 919–933.
29. Lomov, S. V., Verpoest, I., Cichosz, J., Hahn, C., Ivanov, D. S., & Verleye, B. (2014). Meso-level textile composites simulations: Open data exchange and scripting. *Journal of Composite Materials, 48*, 621–637.
30. Lomov, S. V., Verpoest, I., Peeters, T., Roose, D., & Zako, M. (2002). Nesting in textile laminates: Geometrical modelling of the laminate. *Composites Science and Technology, 63*(7), 993–1007.
31. Lomov, S. V., Verpoest, I., & Robitaille, F. (2005). Manufacturing and internal geometry of textiles. In A. C. Long (Ed.), *Design and manufacture of textile composites* (pp. 1–60). Woodhead.
32. Long, A. (2000). Process modelling for textile composites. In *International Conference on Virtual Prototiping EUROPAM 2000* (pp. 1–17). Nantes.
33. Melchior, M. A., Duflot M., Gerard J. S., Melchior S., Hebert P., Adam L., & Assaker, R. (2015). End-to-End FE Based Homogenization of Woven Composites. In *Proceedings of the American Society for Composites: Thirtieth Technical Conference* (pp. 39–46).
34. Peirce, F. T. (1937). The geometry of cloth structure. *Journal of the Textile Institute, 28*(3), T45–T96.
35. Sakakibara, T., Yamamoto, T., Yamada, G., Imaoku, A., Ohtagaki, R., & Zako, M. (2018). Recent developments in "COMPOSITES DREAM" for meso-FE modelling of advanced materials. In *IOP Conf. Series: Materials Science and Engineering (13th International Conference on Textile Composites (TEXCOMP-13)* (Vol. 406, p. 012027).
36. Tabatabaei, S. A., & Lomov, S. V. (2015). Eliminating the volume redundancy of embedded elements and yarn interpenetrations in meso- finite element modelling of textile composites. *Computers & Structures, 152*, 142–154.

37. Tabatabaei, S. A., Lomov, S. V., & Verpoest, I. (2014). Assessment of embedded element technique in meso-FE modelling of fibre reinforced composites. *Composite Structures, 107,* 436–446.
38. Tabatabaei, S. A., Swolfs, Y., Wu, H., & Lomov, S. V. (2015). Full-field strain measurements and meso-FE modelling of hybrid carbon/self-reinforced polypropylene. *Composite Structures, 132,* 864–873.
39. Thomason, L. (2012). "TinyXML." 2012, from http://www.grinninglizard.com/tinyxml/.
40. Vanaerschot, A., Cox, B. N., Lomov, S. V., & Vandepitte, D. (2016). Multi-scale modelling strategy for textile composites based on stochastic reinforcement geometry. *Computer Methods in Applied Mechanics and Engineering, 310,* 906–934.
41. Verpoest, I., & Lomov, S. V. (2005). Virtual textile composites software Wisetex: Integration with micro-mechanical, permeability and structural analysis. *Composites Science and Technology, 65*(15–16), 2563–2574.
42. Vorobiov, O., Tabatabaei, S. A., & Lomov, S. V. (2017). Mesh superposition applied to meso-FE modelling of fibre reinforced composites: Cross-comparison of implementations. *International Journal for Numerical Methods in Engineering, 111,* 1003–1024.
43. W3C. (2008). "Extensible Markup Language (XML) 1.0 (Fifth Edition)." 2012, from http://www.w3.org/TR/REC-xml/.

Fiberworks—A Loom Simulator for the Shaft Loom

Robert Keates

Abstract Fiberworks is a design aid for the shaft loom with a maximum of 64 shafts and treadles in the Silver level version and 9996 warp and weft threads. It permits easy editing of threading, tieup, treadling, liftplan and colour sequences to generate weave plans which can be view in conventional modes, or more specialized modes for warp or weft faced weaves and double weave. Fiberworks Silver also includes a simple graphic editor for fabric analysis and free form design. Additional features allow for pattern repetition and substitution of weave structures into profile drafts.

Keywords Weaving simulator software · Draft editor · Shaft loom

1 Representation of Woven Fabric in Two Dimensions

A woven fabric consists of orthogonal sets of threads, warp threads along the length of the fabric and weft threads across the width. To give the fabric integrity, warp and weft threads must interlace such that each warp thread passes over some weft threads and under others. The loom provides a means to raise or lower warp threads collectively through a shedding mechanism. Such mechanisms can be simple, where alternating lifts raise the all the odd warp threads and then all the even threads, or complex, where each warp thread is controlled independently as in the Jacquard loom. Between these extremes, the multishaft loom permits complex patterning that follows simple rules, based on which warp threads are connected to each shaft and the order of raising shafts, and this is readily simulated in the computer [1].

A weaving draft (Fig. 1a, b) is a schematic representation of the action of the multishaft loom, showing how pressing a treadle is linked to raising of a selection of shafts, and in turn how the raised shafts determine which warp threads will pass over a given weft thread. The complexity of pattern is a function of the number of shafts and the number of treadles available. For looms with 8 shafts or more, shafts may be controlled independently either through the dobby mechanism or with individual

R. Keates (✉)
K&B Fiberworks, P.O. Box 649, Salt Spring Island, BC V8K 2W2, Canada
e-mail: info@fiberworks-pcw.com

© Springer Nature Switzerland AG 2022
Y. Kyosev and F. Boussu (eds.), *Advanced Weaving Technology*,
https://doi.org/10.1007/978-3-030-91515-5_8

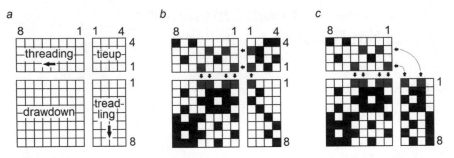

Fig. 1 **a** Layout of a weaving draft in the North American convention, with threading oriented right to left, shaft 1 at the bottom and treadling oriented top down, treadle 1 on the left. **b** Relationship of the draft to the woven fabric. The shaded square at the top of the treadling shows pedal 1 pushed; the tieup shows that treadle 1 is linked to shafts 1 and 3, which rise. Shafts 1 and 3 raise warp threads 1,2,4 and 5 so that weft will pass under these warp threads, and over threads 3,6,7 and 8. **c** The liftplan is an alternative representation for dobby looms or lever-controlled table looms, where each shaft can be controlled independently rather than being linked to a treadle

levers to raise each shaft. This can be depicted as a *liftplan* which shows the shafts to lifted for each weft row (Fig. 1c), and which replaces the treadling. The advantage of the liftplan is the increased number of combinations of shafts permissible for each lift; 254 for an 8-shaft loom and 65,534 for a 16-shaft loom. By comparison, an 8-shaft treadle loom generally only has 10 treadles.

Weaving drafts can be drawn by hand on gridded paper, but weaving design computer programs now available make the task easier and less tedious (Lourie, 1973).

The advantages of the computer include ease of correction and modification, including insertion or deletion of threads, and automation of repetitive tasks. For example, most weaving software will generate the drawdown that represents the woven textile and only require manual entry of threading, tieup and treadling, or threading and liftplan. Threads can also be represented in full colour, and some programs allow variation of thread spacing or thickness.

Weaving software simulates the structure and appearance of the woven fabric, allowing the weaver to predict the outcome of a particular loom setup. Fiberworks is one such suite of software that simulates the action of the shaft loom, and is targeted at the handweaver and small design studio. It has certain constraints; each weft thread is continuous across the width of the fabric and warp threads are assumed to be parallel without intersecting or crossing over other warp threads.

2   Fiberworks Screen Layout and Drawing Tools

Fiberworks opens with a ready-to-use empty drawdown frame. The number of shafts and treadles can be preset in Preferences to a maximum of 64 shafts and 64 treadles in the Silver level version. (File menu, Windows or Fiberworks menu, Macintosh).

One can also preset it to open in Liftplan format. The threading and treadling are intended to be open-ended; the actual limit is 9996 warp ends and 9996 weft picks, a limit which is rarely encountered. Scroll bars allow off-screen portions of a design to be brought into view.

A toolbar displays icons for commonly used tools. The most important of these are the drawing tools that control mouse action (Fig. 2). When one of the red-outlined tools is selected, the mouse will draw directly into the threading, tieup and treadling drafts; this is indicated by making the mouse cursor take the shape of a pencil. The blue rectangle is the selection tool that allows a rectangular area of threading, tieup or treadling to be selected. When this tool is active, the mouse cursor appears as the conventional arrowhead pointer.

While it is possible to fill grid squares by clicking into them individually, Fiberworks takes advantage of a common feature of weaving drafts, that sequences often run in diagonal lines. The straight draw tool makes parallel diagonals, and the point draw tool makes opposing diagonals that come to points (Fig. 3). One does not "draw" in the sense of tracking the path of the actual threading with the mouse cursor; the object is to drag out a rectangle of the desired height and width, and the rectangle fills itself with the chosen pattern. Figure 3b shows the actual path of the mouse cursor when drawing the point threading shown.

Drawing over any part of an existing threading or treadling draft replaces the original contents of the drawing rectangle; the program is constrained to allow only one thread per heddle and only one treadle per pick. However, tieup and liftplan allow multiple entries per treadle or per pick.

Fig. 2 The drawing tool icons from the toolbar (windows version shown). The outline around the straight draw tool indicated that it is the currently selected tool. Icons with a red outline draw into threading, tieup or treadling drafts

□ selection

✎ colour pickup

▨ straight draw

▨ point draw

▨ line draw

▨ freehand

▨ network draw

Fig. 3 Drawing action of the mouse tools. The drawing process involves clicking and dragging out a rectangle in the grid area; the rectangular space fills itself with **a** parallel diagonals with the straight draw tool or **b** opposing diagonals with point draw. In **b** the dashed line shows the path followed by the mouse

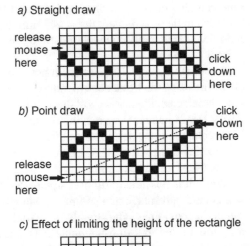

a) Straight draw

b) Point draw

c) Effect of limiting the height of the rectangle

3 Editing and Corrections

Fiberworks was designed to allow a weaving draft to be modified as easily as modifying text with a text editor. One can draw over or type over to replace any part of any existing draft. An Undo function (type Ctrl + Z) stores 64 levels of undo. The Insert key (Windows) or spacebar (Mac) or toolbar button can create a gap to allow one to add extra threads anywhere in a draft; the delete key or toolbar button removes unwanted threads. To add or remove a block of threads, one can outline the block with the selection tool; insert creates a gap the size of the outlined block and delete removes the contents of the outlined block, and closes the gap. To empty a block without closing the gap, double click and drag with any drawing tool, or move the insertion point to the problem grid square and type the minus key.

Where single click draws by marking grid squares, double click unmarks, making a black square white again. This is important in tieup and liftplan.

Not all weaving drafts are amenable to these drawing tools. In cases with irregular sequences, it is easier to type in the shaft numbers than to click each grid square with the mouse for example Fig. 4. Typing action takes place at the *insertion point*, represented on screen as a blinking grid square (analogous to the vertical bar when typing in a text editor). The insertion point advances automatically as one types, as in a text editor. In the threading draft, the direction of advance is right to left (in the North American convention) and in the treadling draft, the direction of advance is top down.

For multidigit shaft or treadle numbers, Fiberworks uses the next row down on the keyboard to represent 11 – 20 (q w e r t y u i o p on an English keyboard), and

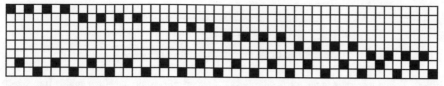

828 18 28 17 27 17 27 16 26 1 626 1 52 51 52 51 42 41 42 41 32 31 32 31

Fig. 4 An example of a sequence that is easier to enter by typing at the keyboard than by clicking with the mouse. This is commonly the case with tied weaves

the row below that to represent 21–30. This allows each thread to be marked with a single keystroke. There is an option in Preferences to adjust for foreign keyboards; if the French keyboard is selected, numbers 11 - 20 are represented by a z e r t y u i o p.

The insertion point always goes to the last location clicked with the mouse, and this is a convenient way to position the insertion point in preparation for typing.

4 Adding Colour and Manipulating Colours

The right edge of the Fiberworks drawdown frame (Windows) contains a colour palette with 82 colour tiles (Fig. 5); on Mac, the palette floats independently of the drawdown window. Two squares at the top of the palette indicate the colours currently selected for single click and double click. To select a colour, click or double click on its tile.

To apply the selected colour, click into the warp or weft colour bar with any drawing tool selected (the mouse cursor should appear as a pencil). The warp colour bar lies immediately above the threading draft (Fig. 6), and the weft colour bar lies to the right of the treadling draft or liftplan. To create a band of colour, drag over the

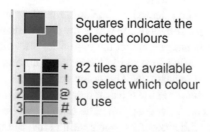

Squares indicate the selected colours

82 tiles are available to select which colour to use

Fig. 5 A portion of the colour palette. The upper left **square** at the top indicates the currently selected colour applied by single mouse click; the lower right square is the colour selected to apply with a double click. Below the squares there are 82 **tiles** showing the colours available for selection. Next to each tile is a character that indicates the keystroke that applies a colour by typing

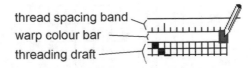

Fig. 6 To apply colours in the warp, one clicks the pencil cursor into the **warp colour bar**, which lies immediately above the threading draft. The pencil cursor will display if any drawing tool has been selected in the toolbar. Weft colours can be applied by clicking into the **weft colour bar**, which lies immediately right of the treadling draft or liftplan. width

required number of warp or weft threads. To add a different colour, click on its tile in the palette to select the new colour.

Complex colour sequences can sometimes be entered more easily by typing. The symbol of an associated keystroke shows next to each colour tile in the palette. To type a colour sequence, place the insertion point in the colour bar, then type. The insertion point can be placed in the colour bar by pressing the Tab key and arrow keys, or by clicking the mouse pencil or arrowhead in the desired location. The insertion point advances as one types, right to left in the warp colour bar and downwards in the weft colour bar.

Colours may or may not show in the drawdown area depending on the current display mode. Fiberworks distinguishes between *Structure* displays and *Colour* displays. Structure displays show the orientation (warp or weft) of a thread without regard to its colour. A *warp drawdown* displays warp over weft as a black square and weft over warp as white. A *weft drawdown* displays weft over warp as black and warp over weft as white. The colour displays are *Normal cloth* (a colour drawdown that shows thread colour regardless of whether it is warp or weft, useful for colour and weave effects), *Interlacement* which shows both colour and thread direction, which is the default display mode. In addition, there are more specialized display modes: *Warp-faced/Rep, Weft-faced, Bound weave* and *Double Weave*. These additional display modes do not convert an arbitrary draft into the required form, but if the drafts have been designed as warp-faced, weft faced or double weave, will display the result more accurately.

The display mode can be selected in the Cloth menu, or by clicking in the status bar at the bottom of the drawdown frame where the current display mode is indicated.

If a required colour is not present in the colour palette, any existing colour that is not in use can be modified to provide the desired colour. To do so, first select the target colour by clicking its tile. The color will appear as the current selected colour in the indicator squares at the top of the palette. Click on the indicator, and the *Modify Colour* panel (Fig. 7) will appear. The colour can be changed by adjusting the position of the three slider controls representing Hue, Saturation and Brightness. The red, green and blue content of the new colour are also shown at the bottom of the panel, and RGB values can be entered here to regenerate a colour with known RGB content.

If the colour being modified is already present in the current drawdown, the changes made will appear simultaneously in the drawdown. This allows the designer

Fig. 7 The Modify Colour panel allows a given colour to be changed by adjusting the Hue, Saturation and Brightness slider controls, or by entering RGB values. This control can act on any tile in the current palette, including those already in use in the displayed drawdown

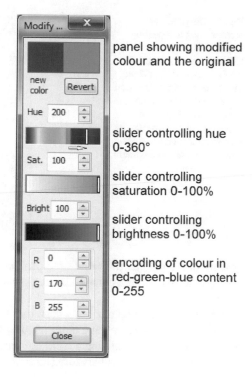

panel showing modified colour and the original

slider controlling hue 0-360°

slider controlling saturation 0-100%

slider controlling brightness 0-100%

encoding of colour in red-green-blue content 0-255

to assess how the newly created colour interacts with the remaining colours in the sample.

Colour palettes are automatically saved in the drawdown file provided that at least one thread has been coloured other than the default white warp and blue weft. A palette can be extracted from one drawdown file and applied to another drawdown. To extract a palette, open the file, and choose Select Drafts in the Edit menu. Place a mark only against the Palette item in the panel, and click Copy (Fig. 8). Then go to the new design and click Paste.

The colour pickup tool (the eyedropper icon in Fig. 2) allow one to reselect a colour that already appears in an existing draft. To reselect, click on the Color Pickup toolbar icon, and the cursor changes to an eyedropper. Click in the required colour in the warp or weft colour bar, and the "picked up" colour becomes the selected colour in the upper indicator square of the colour palette. When there are several similar

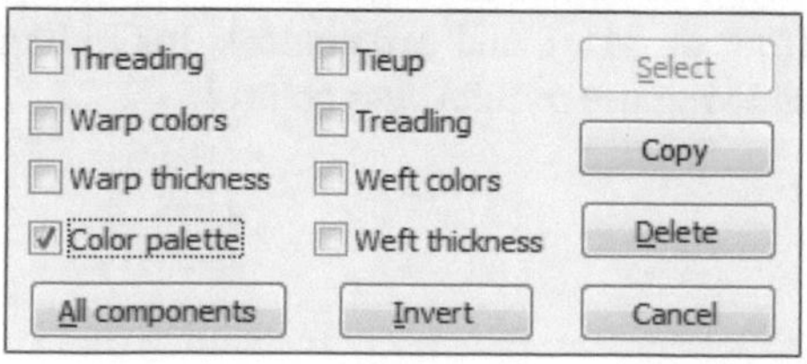

Fig. 8 This panel (Edit menu, Select All) allows a palette to be copied from one file to paste into another drawdown

Fig. 9 The panel for
creating simple repeating
sequences of colours.
Colours can be selected from
the full range of the palette
by sliding the scroll bar until
the desired colour in the
centre patch. A preview
appears at right to display the
pattern effect and to assess
how the two selected colours
work together

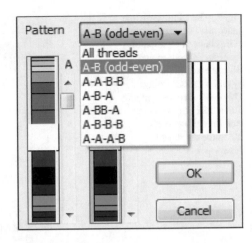

colours in the palette, this solves the problem of deciding which one was actually
used previously. After a colour has been picked up, the cursor reverts to its previous
form.

Simple repetitive colour sequences can be created automatically (Warp menu,
Colors… or Treadling menu Colors…). The Colors panel (Fig. 9) has a dropdown
menu with a selection of single solid colour or simple patterns of two colors (four
colours on Mac), and two bars for selecting the colours.

5 Varying Thread Spacing and Thickness

The band above the warp colour bar or to the right of the weft colour bar controls
thread spacing or thickness (Fig. 6). A mouse click in this band doubles the displayed
width of a thread; double click halves the displayed width. This variation may represent actual thread thickness or spacing; threads displayed wide may actually represent
normal threads that are widely spaced apart, and threads displayed narrow may actually be normal width but packed closely together, particularly for warp and weft
faced weaves.

In the majority of cases, thickness or spacing variations occur in regular repeating
patterns which can be created automatically (Warp menu, Thickness… or Treadling
menu Thickness…). The Thread thickness panel (Fig. 10) has a dropdown menu
with a selection of single uniform thickness or simple patterns of two thicknesses
(four on Mac), and two controls for setting the values. See Fig. 11 for an example
showing use of thickness control.

Fig. 10 The panel for creating simple repeating sequences of thickness or spacing. The numerical values are relative numbers with 4 representing an average thread. This allows 2 for half width, 1 for quarter width, 8 for double width and so on

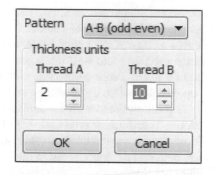

Fig. 11 Example of thick and thin in a rep rug weave. The thickness variations allow the sections of different warp colours to be emphasized or de-emphasized to create the block design

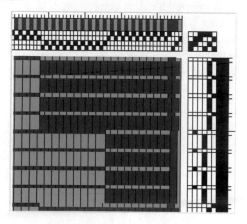

6 Selection, Cut, Copy and Paste, Transforms and Repeats

The selection tool (blue rectangle icon in Fig. 2) allows the mouse to outline rectangular areas in threading, tieup, treadling or liftplan, as well as warp and weft color bars. When the selection tool is active, the mouse cursor appears as the standard arrowhead instead of the pencil. To dismiss the selection tool, click on one of the drawing tools.

The selected area is indicated by a blue outline and can be cut (leaving a gap), deleted (closing the gap), copied, transformed or repeated. The Insert key or toolbar button opens up a gap the width of the selected block by moving displaced content, left in the case of warp items or downwards in the case of weft items, so no content is lost.

To copy an area that extends beyond the current screen, choose the selection tool and click on one corner of the required selection. Then scroll to the diagonally opposite corner and Ctrl + click; the blue selection rectangle will expand to include both corners.

Cut or copied items can be pasted. The pasted item will appear at the location of the current insertion point, which marks the lowest shaft number, lowest warp end, lowest treadle number or lowest pick number of the pasted block. The pasted block retains its blue outline, and can be dragged to a new location if needed. Ctrl + drag leaves a copy of the pasted block in place while moving to a new location. Warp items can paste into weft items and vice versa. The paste will fail if the insertion point is not in a zone compatible with the copied item, so one can't paste a block of copied colours into the threading and vice versa.

A block outlined with the selection tool can be transformed; in the Edit menu choose Transform or type Ctrl + T. The transform panel (Fig. 12) appears in two forms; (*a*) for linear sequences (each position in the sequence can only hold one value, e.g. one shaft per warp thread in the threading), and (*b*) for multivalued components tieup and liftplan.

A selected block can be repeated; choose the Repeat… item in Warp or Treadling menus. If no selection has been made, the entire warp or treadling sequence will be repeated. The Repeat panel (Fig. 13) allows different styles of repeat and options to repeat selected components of threading or treadling. The Apply Now button allows the repeat effect to be previewed without leaving the Repeat panel.

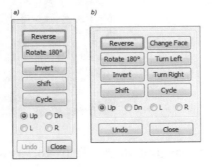

Fig. 12 The Transform panels: **a**, applicable to linear sequences threading, treadling and colour bars; **b**: applicable to area selections such as tieups and liftplans. Change face means to replace white grid squares with black and vice versa. If the entire tieup is so changed the upper and lower faces of the woven cloth exchange places

Fig. 13 The Repeat panel includes a dropdown menu (shown expanded at right) for different styles of repeat. At left, check boxes select the components to repeat. If a prior selection has been made in the threading or treadling draft, the box "Repeat selected block" is marked by default. The item is grayed out if no selection was made, in which case the entire threading or treadling is repeated

7 Profile Drafts and Block Substitution

A profile draft lays out the design elements of a textile without considering the structure of the woven areas. The profile has the appearance of a normal weaving draft, with threading, tieup and treadling or threading and liftplan (Fig. 14). The profile draft is best represented in one of the structure views, usually warp drawdown, in which case black squares in the profile drawdown represent pattern weave and white squares represent ground weave. The drawdown area may appear to contain long warp or weft floats and lack coherence for weaving, but this will be corrected when the appropriate weave structure is added to each area. Pattern and ground structure elements provide contrast when inserted into the grid squares. The substitution expands the threading and treadling, since each unit of the profile represents multiple threads of weave structure, and may also increase the number of shafts and treadles required.

Any draft can be used as a profile in Fiberworks. More elaborate profiles may depict images of objects (Fig. 15), however this may only be possible if large numbers of shafts are available to the weaver. Objects that are symmetrical need fewer shafts because of repetition across the axis of symmetry. The number of treadles is not an issue if woven as a liftplan on a dobby or table loom.

The Block Substitution Panel contains a catalog of about 90 weave structures that can be used to substitute for units of the profile draft (Figs. 16 and 17). The substitution replaces one grid square of the profile with one unit of the weave structure. For example, a single unit of Double weave needs four shafts, four warp threads, four treadles and four picks. This will quadruple the number of shafts and treadles originally in the profile, and quadruple the number or warp and weft threads.

After choosing a weave structure, click the Apply button to see a preview of the resulting fabric, for example Fig. 18.

The preview panel offers additional controls (Fig. 19); buttons at the top allow a change of magnification, and switch between colour and structure views, needed because the normal controls are not accessible while the preview panel is visible and because the threading and treadling have been expanded. The button *Change Face of Profile* exchanges the position of pattern and ground. The button *Change Face of Cloth* exchanges the orientation of front and back of the textile. Some weave

Fig. 14 Example of a simple profile representing the layout of design elements, for example for a placemat. The long floats on picks 4 and 8 will be corrected after adding the ground structure

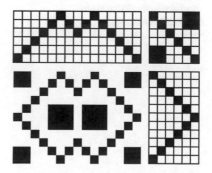

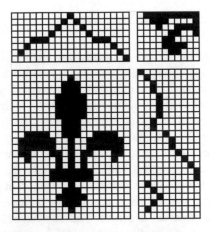

Fig. 15 A profile incorporating imagery. The inclusion of one empty row and one empty column in the tieup allows for horizontal and vertical bands that will be all ground in the final textile. These empty rows and columns allow for borders and for separation of the image motifs if they are to be repeated multiple times. In designing imagery, the number of shafts and treadles in the profile depend on the number of unique columns and unique rows in the profile drawdown

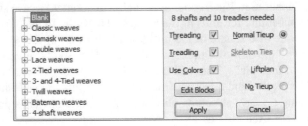

Fig. 16 The initial Block Substitution panel provides a range of categories of weave structure. Click the [+] symbol next to a category to expand the list of structures included in that category. The panel also allows one to choose which component will be expanded and a choice between tieup and liftplan mode. The Edit button allows an existing weave structure to be modified, or if the Blank option is chosen, to create one's own series of blocks. Once a weave structure has been selected, the panel shows how many shafts and treadles will be needed

Fig. 17 The Double Weave
category expanded to show
the weave structures
available in that category

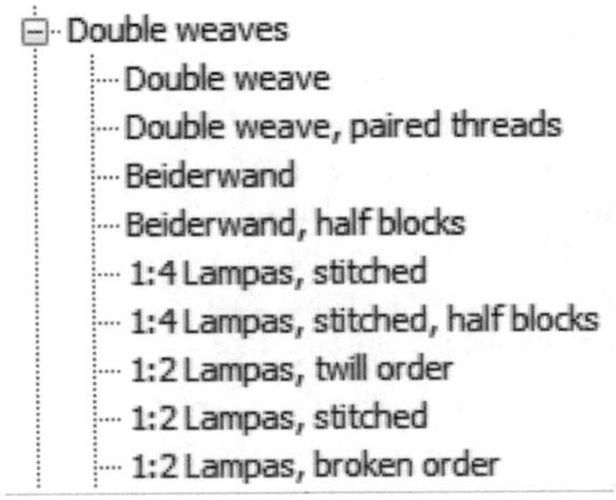

- Double weaves
  - Double weave
  - Double weave, paired threads
  - Beiderwand
  - Beiderwand, half blocks
  - 1:4 Lampas, stitched
  - 1:4 Lampas, stitched, half blocks
  - 1:2 Lampas, twill order
  - 1:2 Lampas, stitched
  - 1:2 Lampas, broken order

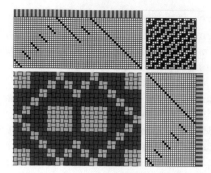

Fig. 18 The result of the substitution process with double weave for the simple profile. The six profile shafts and treadles have expanded to 24 shafts and treadles, and the 15 warp blocks and 11 weft blocks have expanded to 66 warp threads and 44 weft threads. Colors were added after the substitution was complete and the result is shown in double weave display mode

Fig. 19 The Block Substitution Preview panel allows one to review the result of block substitution, with options to Try Another weave structure or to OK the result obtained

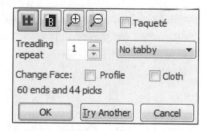

structures may require tabby to be added (not the case for double weave). The button *Try Another* restores the original profile draft and the list of structure in order to continue to experiment. If satisfied with the result, click OK, and the substituted textile is created in a new window while the original profile still exists in its own window.

8 Pattern Lines and Network Drafting

Conventional weaving drafts create patterns that are generally based on rectilinear boundaries, parallel to warp and weft. Even when imagery is used for block substitution, the end result shows the rectangular boundaries of the blocks. Otherwise, threading or treadling sequences based on curves have the potential for long floats because of placement of adjacent threads on the same shaft or treadle. The work of Brandon-Guiguet [2] and elaborated by Roussel and Masson [3] demonstrated how to displace adjacent threads on the same shaft to a new location that would avoid creating floats.

The process starts by drawing an arbitrary curve in the threading called a *pattern line*. The freehand drawing tool in the toolbar (Fig. 2) is provided for this purpose. The points on the original pattern line are then promoted to the next higher location that lies on a virtual grid formed by tiling the threading with a unit called the network initial. The most commonly used initial is a simple 4-end, 4-shaft straight draw directed either to left or right (Fig. 20). The original curve is then redrawn so every warp end lies on a network position (Fig. 21a).

If the pattern line lies too close to the top of the threading, there may not be a network position above, in which case the displaced end *wraps* around to the bottom of the threading to find the first available network position (Fig. 21). If wrapping is to be avoided, the highest point on the pattern line should be on shaft $S - H + 1$ (Schlein's Shaft Rule, [4], where S is the total number of shafts and H is the height of the initial. For an 8-shaft draft with a 4-end straight initial, the pattern line should not go above shaft 5. Wrapping may sometimes be permitted if the number of shafts is divisible by the height of the initial.

In addition, the tieup rule requires that there is no run of more than $H - 1$ successive shafts on a single treadle or successive treadles linked to the same shaft. A similar rule applies to liftplans. An initial of height 3 results in smother curves but limits the possibilities in the tieup; an initial of height 5 gives more possibilities in the tieup but increases the jagged appearance of curves. In practice, the initial of height 4 is usually the best compromise.

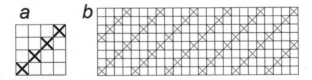

Fig. 20 Conversion of a freehand curve to lie on a grid based on a 4-shaft initial. The gray grid squares in the middle frame depict the network formed by tiling the initial. The frame on the right shows the pattern line redrawn to lie on the network. The general form of the original curve is retained, but the network imposes jaggedness

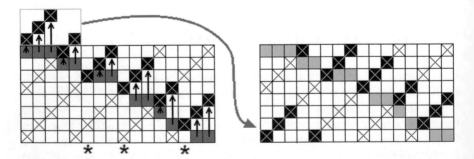

Fig. 21 Illustration of wrapping; when the pattern line lies above any network position, it wraps around and seeks the first network position from the bottom of the draft. Wrapping should only be allowed if the number of shafts in the grid is exactly divisible by the height of the network initial

Fig. 22 The panel controlling redraw on network (Warp and Treadling menus). The upper half has the controls for shaft reduction, and the lower half for setting the height of the network initial and the style. Left or right straight initials are the most commonly used. The Apply button allows the result to be previewed, and the Done button accepts the result and closes the panel

To transform a pattern line to network draft, choose the menu item Redraw on Network, present in both Warp and Treadling menus. The menu item brings up a control panel shown in Fig. 22.

In some cases, drawing a pattern line on more shafts than are available to weave allows for more control. The original pattern line can then be reduced to the final number of shafts by processes called telescoping and digitizing (Fig. 23).

Telescoping takes the section that lies above the available number of shafts and drops the whole section down to lie within the allowed range of shafts. Telescoping should be avoided if the final number of shafts is not divisible by the height of the initial. Telescoping also produces effects that are distinct from digitizing, as seen in Fig. 24.

Digitizing squeezes the pattern line to fit in the allowed space. Digitizing can be done without wrap, with the highest shaft in the pattern line follows the Shaft Rule, or if the final number of shafts is exactly divisible by the height of the initial, with wrap.

9 Fabric Analysis

Fiberworks Silver includes a module called the Sketchpad. This has two main functions—fabric analysis and free-form design. The sketchpad presents itself as a plain grid, initially 80×80, but can be re-sized as needed. Fabric analysis involves taking a woven sample of unknown structure, and by inspection under a magnifying lens, observing each thread intersection in turn to distinguish warp over weft from weft over warp. Each warp over weft is marked as a black square on the grid, and weft over warp remains white. When one or more repeats of the sample have been recorded on the grid, the grid contents can be analysed to determine the threading, tieup and treadling that was used to weave the sample, with the result displayed in the normal drawdown frame. Each unique column in the grid represents an additional shaft required, and each unique row represents an additional treadle.

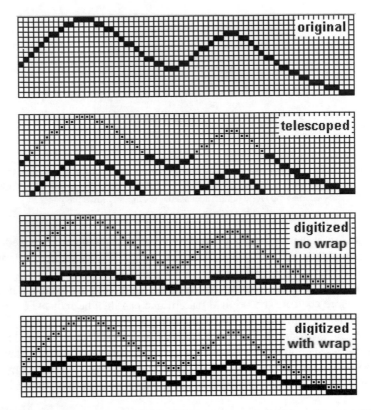

Fig. 23 Shaft reduction by telescoping, digitizing no wrap and digitizing with wrap. The dotted squares show the original pattern line before shaft reduction

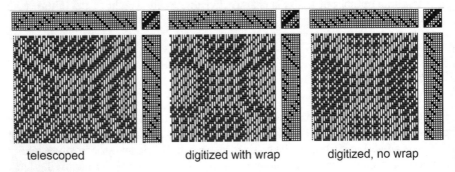

telescoped   digitized with wrap   digitized, no wrap

Fig. 24 Effect of the different shaft reduction methods. Telescoping produces multiple contour lines that "echo" the original pattern line. Digitizing with wrap on a turtle line produces hollow circular objects. Digitizing without wrap creates solid circular objects

If the sample lacks selvedges, it may be difficult to predetermine which axis is warp and which is weft. If necessary, it is possible to turn the sketchpad contents before analysis, or to turn the drawdown after analysis.

The sketchpad can also be used to draw items in free-form, and then analyze the result to generate a threading, tieup and treadling. Since free-form design may generate long floats, the results can be treated as a profile draft and a structurally stable weaving derived by block substitution. The images in Figs. 14, 15 and 16 show examples.

The sketchpad will also accept complex bitmap imagery and re-pixellate it on a coarser grid. For coloured images, there are tools for reducing the colour content. The result may not be suitable for a conventional shaft loom, but could be used to generate cartoons for tapestry weaving or drawloom weaving.

10 Exercises

A copy of Fiberworks Silver is required to complete these exercises. A free trial version can be downloaded from the Fiberworks website http://www.fiberw orks-pcw/download.htm (Windows) or from http://www.fiberworks-pcw/mac.html (Mac). The free version is unable to Save files, print or operate a dobby loom, but all other program functions are available for use.

1. (a) Draw the Threading and Tieup Shown in Fig. 25; Try to Minimize the Number of Mouse Actions.
   (b) Use the selection tool to outline the threading, and copy it. Click the selection tool at the beginning of the treadling to locate the insertion point and paste into the treadling.
   (c) Experiment with variations in the tieup.
2. (a) Draw the Threading and Tieup Shown in Fig. 26.
   (b) Copy the threading and paste into the treadling as described in 1b above.
   (c) Position the insertion point on pick 33 in the treadling and click the Insert button 8 times.
   (d) Into the gap, add the sequence 9,10, 9,10,9,10, 9,10 (this can be done with one stroke of the mouse.
   (e) Choose Colors… from the Warp menu and set up an odd–even pattern of two colours.
   (f) Choose Thickness… from the Treadling menu and set up an odd–even pattern of thin and thick.
   (g) View the result as Warp Drawdown, as Interlacement and as Rep/Warp faced.
   (h) Choose the eye dropper icon in the toolbar and click on one of the colours in the warp colour bar (the cursor should take on the shape of the eye-dropper). This should make it the selected colour at the top of the colour palette. Click on the large square at the top of the palette to bring up the Modify Color panel. Vary the settings for hue, saturation and brightness.

3.1   Windows version

(a)   Choose New Sketchpad and draw a free-form image such as the owl in Fig. 27. For best results, keep the image simple and symmetrical.

(b)   In the Analysis menu, choose Include Border.

(c)   In the Analysis menu, choose Make Drawdown. This will create a drawdown with a profile threading, tieup and treadling.

3.2   Mac version

(a)   Draw a single point threading with the point facing down as in 'V'.

(b)   In the Tieup menu choose Convert to Liftplan, and choose Warp Drawdown in the toolbar.

(c)   Draw a free-form image by clicking directly into the drawdown area (double click to erase an unwanted black square). Align the centre of a symmetrical image with the point of the 'V'. When done, reduce the number of shafts to eliminate unused shafts, keeping one unused shaft for a border.

       For both Windows and Mac: The result of steps a-c should generate a profile draft.

(d)   In the Tools menu, choose Block Substitution.

(e)   Choose a category, for example Twill weaves, and choose 1:3 Turned twill. Click on Preview.

(f)   In the Preview panel, click on Try Another to revert to the Profile draft and to explore the different weave structures.

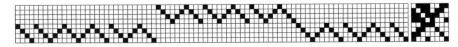

Fig. 25 A simple threading and tie-up that can be drawn with the mouse Point-Draw tool

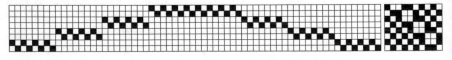

Fig. 26 A more complex threading and tie up that can be drawn with the mouse straight-draw tool

Fig. 27 An example of free-form design to use as a profile draft. Draw in the Sketchpad of the Windows version or draw as a lift plan in the Mac version of Fiberworks. Use this profile to experiment with Block Substitution

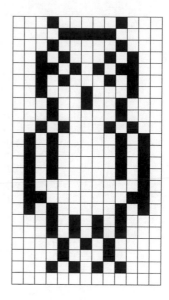

References

1. Lourie, J. R. (1973) *Textile graphics/computer aided*. Fairchild Publications, NY, p. 21–32.
2. Brandon-Guiguet Methode des Initiales, Joannès Desvigne & Cie, Lyon, (1938)
3. Olivier Masson, François Roussel haft Weaving and Graph Design, Editions en Bref, Montréal (1988)
4. Schlein, A. N. (1994). *Network drafting: an Introduction*. Bridgewater Press.

3D and Multi-axis Weaving: Product and Process

Process Based Method for Pattern Development of Narrow Woven Complex Profiles

Yordan Kyosev

Abstract This chapter presents a practical method for development of the weaving pattern for tapes with complex cross section. The method is based on analysis and reproduction of the weaving process and can be learned and applied for development of woven tapes even if developer has less knowledge of the classical methods of the woven pattern development. The chapters starts with the development of hollow woven structures and fold structures. The development steps for complex profiles are demonstrated with step by step example, beginning form the principal idea, going through the weft insertion planning and ending with the arrangement of the warp yarns and their positions during the different weft inserts. At the end of the chapter are given the basic ideas for using machines with multiple shedding positions and multiple weft insertion devices (needles or shuttles).

1 Introduction

The modern narrow weaving technique is mostly used for production of belts, tapes, ribbons, labels and similar products with flat cross section, where the pattern development is based on classical methods and theories. New, specialized products in the area of composites, medical and other specific applications introduces already woven products with complex cross sections. For such products, the application of the classical pattern design methods is not really efficient and can error-capable. This chapter presents a process based method for creation of the weaving pattern for such profiles. It is based on the natural following of the weft yarn paths and the required positions of the warp yarns in the corresponding sheds. The methods helps significantly in the development of woven profiles with tubular filled or hole cross section and more complex profiles. The method can be applied both using all kinds of machines—shuttle, needle, rapier or projectile.

Y. Kyosev (✉)
Technische Universität Dresden, Dresden, Germany
e-mail: yordan.kyosev@tu-dresden.de

Fig. 1 Possibility for creation of narrow weaving products

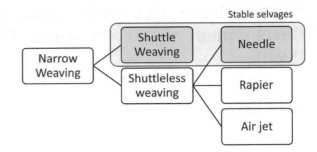

2 Terminological Remarks

Narrow textiles is a collective term for all kind of stable textile structures with limited up to 30–50 cm width [1]. As stable structure is interpreted any structure, based on interlacement or interloping of yarns (or other materials) which keep remains the topologically equivalent state after influence of external (mechanical) loads. Practically it means, that the yarns are not starting to fray and go out of the fabrics and thus stable selvages are required. Wide fabrics, just cut to some tape with the width to less than half meter does not count to the narrow fabrics, but if their selvages become fixed during the cutting (for instance by melting of the cut areas as used for labels), then they represents stable product and are classified as narrow fabrics, too.

The narrow fabrics include narrow woven fabrics—with all kind of tapes and belts, braided products—as ropes and laces, crochet knitting products and pillow and bobbin laces (which are actually a subclass of the braiding products [2].

The narrow weaving itself can be separated into several subclasses, depending on the type of the weft insertion into classical narrow **shuttle** weaving and the several modern ways of narrow **shuttleless** weaving (Fig. 1).

The shuttleless weaving include the needle narrow weaving, used for production of the most kind of tapes and belts; the rapier weaving, which come in application for some wider complex tapes or for the weaving of sensible materials and the air jet weaving where normally labels for clothing at high speed with several colours in each weft line are produced at large width and cut directly on the machine into smaller tapes.

3 Pattern Development

The development of the main part of the pattern for the narrow weaving products is described in several books for classical wide weaving [3, 4], but there are several specific issues, which the "wide"-weavers do not consider, ignore and just do not know. One of the main difference is the preparation of the selvages. These very small areas of less than 1 mm width have to be stable, aesthetic, soft, resistance to friction and abrasion and their production requires both high level of engineering in the placement and development of the mechanisms for the production and good understanding of the mechanics of the yarn tension and weaving pattern.

One of the classical books in this area was written by Walter Kipp in German language [5], and translate in English [6] too, but actually can be found only as used book. The company Jakob Müller published in 2005 both in English and German very good book with the details about the narrow weaving [7]. In these books already is described already the principle of the construction of the weaving pattern based on step by step analysis of the relative positions of the warp yarns to the weft yarn at different steps for production of hollow tubes and hollow or "half-hollow" selvages. This technique was applied and explained in this chapter as main method for the development of products with complex cross section.

4 Hollow and Solid Woven Structures with Circular Diameter

The hollow structures have wide application in the narrow weaving. In the most case the people connect these with the tubes for fire brigade only, but tubes are used as medical implanters, for filters, cable covering. In the last twenty years, Jakob Müller AG, Swiss developed specialised high productivity needle weaving machine NH2M 53/NC2M (https://www.mueller-frick.com/produkte-systeme/sei lwebsysteme/nh2m-53-/-nc2m) for woven ropes with diameter up to 8 mm (https:// seilereiherzog.ch), which cover is a hollow woven structure (Fig. 2).

Each curtain tape has normally two or three hollow areas, where additional threads or braids are placed and used for building regular, controlled wrinkles. Each high quality label in the clothing is produced as a complete piece and the selvages there are hollow structure, too. They are produced normally by 16–20 thin filament yarns,

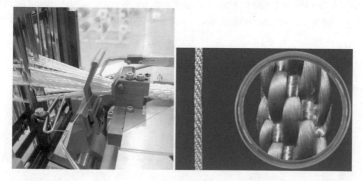

Fig. 2 Needle weaving machine for production of solid ropes with circular cross section and view of such product. With kind permission of Jakob Müller AG Frick, Switzerland. *Source* https://www.mueller-frick.com/fileadmin/storage/Datenimport/Leaflets/WM/NH2M53__ NC2M/NC2M_NH2M_61146_en_07_15_.pdf

which build a nice rounded selvage which feel softer than the standard selvages and do not scratch on the neck.

There is a classical, weaving pattern based way for creating pattern and then the control carts for the weaving of hollow structures. It require good understanding of the weaving pattern and following several rules. The presented here method can be used for development of more complex structures even of people with less weaving pattern knowledge.

4.1   Hollow Tubular Weave

Let a single tube structure has to be developed and for the initial case a shuttle weaving machine is used. The same principle can be applied for needle weaving, actually there not all variations are possible, because the weft yarn loop has to be fixed at each weft insertion cycle and this condition disturbs in the creation of some of the geometries.

For simplification in the current tape it will be considered, that it is based on plain weave (Fig. 3a and d). For the weaving of plain weave are required two yarn groups, each inserted in a separated heald frame. For the weaving of the two layers then are required four yarn groups (Fig. 3b and c). Figure 1e demonstrates a cross section through the warp yarns and possible way of the weft yarn. Normally, there is only **one** weft yarn for the both cycles, but in the figure two different line types (solid and dashed) are used in order to make the identification of the weaving cycles more simple.

The following development of the pattern is based on the following the weaving process. Let the first weft insertion build a interlacement in the upper layer of the structure (Fig. 3). During this weft insertion all warp yarns (visualised as circles), which are **below** the weft line have to be moved to the upper position of the shed, and all other warp yarns have to stay down. Each warp yarn, which stays in the upper position receive a marked square in the corresponding row of the pattern (Fig. 4).

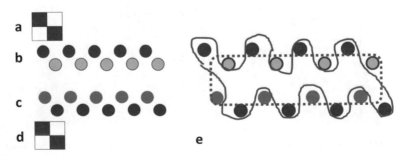

Fig. 3 Principle setup for weaving hollow tube. **a** Plain weave for the upper layer. **b** Plain weave for the bottom layer. **c** Two yarn systems for the upper layer and **d** two yarn systems for the bottom layer. **e** Ppossible cross section and path of the weft yarn

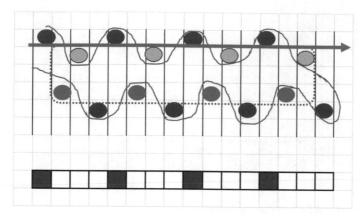

Fig. 4 First row of the upper layer of the woven tubular structure

In the same principle, the first weft insertion for the bottom layer can be prepared (Fig. 5). At this case, all warp yarns, except the purple, have to be set to upper position.

In the same principle the second weft insertions for the upper and bottom layer have to be prepared, following the already prepared path, so that the proper weave type is build (Fig. 6). The complete sequence of all four rows is presented in Fig. 7. Here has to be taken into account, that for this kind of structure the complete technological data set has to be transferred properly on the machine, and not only the pattern through the control chart. The complete information consists additionally of the initial position and the moving direction of the shuttle.

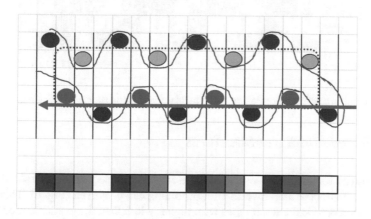

Fig. 5 First row of the bottom layer of the tubular woven structure

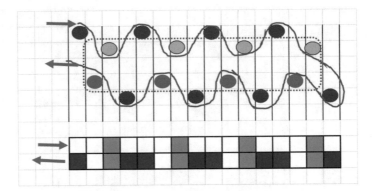

Fig. 6 Second weft for upper for the bottom layer

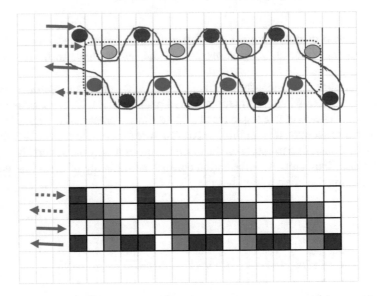

Fig. 7 Complete sequence for weaving of the tubular structure on shuttle weaving machine

4.2 Folded Multilayer Structures

If the pattern for tubular weave is applied in different sequence, an folded, open structure with double larger width after unfolding can be woven (Fig. 6). This effect is used in the practical case, if tapes with higher width than the available reed on the machine has to be produced (Fig. 8).

Theoretically, there is no limitation of the width of the final structure, if the fabrics is woven in folded state, as visualised in Fig. 8. For the weaving of one complete repeat (two complete weft insertions) of the represented there four layers will be required eight weft cycles at all, which makes the process less productive then if the

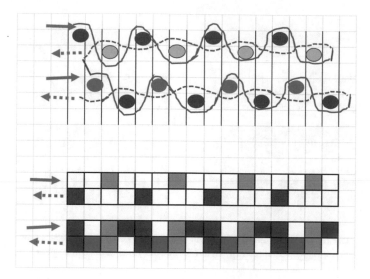

Fig. 8 The same pattern for folded fabrics with double width

fabrics is woven on machine with the suitable width. For this reason this method make sense for production of structures, which are deployed in folded state and which has to change their state later. Such one example can be an airbag for the car sits, which is integrated into the safety belt, named Airbelt. More information about it can be found in the web pages of wired.com (https://www.wired.com/2010/12/air-bag-containing-seatbelt-coming-to-luxury-cars/ and https://www.wired.com/2009/11/ford-inflatable-seatbelts/ and in youtube under the following link https://www.youtube.com/watch?v=qePgg1Goq_4).

There is actually a technological limit for this production (Fig. 9b). With the increasing of the number of the parallel layers, which have to be simultaneously woven, are increasing the number of the yarns, which have to be taken through the same reed opening, in order to obtain the same density of the fabrics. In the case of four layers there are then eight yarns for instance in one reed (or twelve, if three yarns per layer are going into one opening). This increases the friction between the yarns and the friction between the yarns and the reed during the weaving makes the realisation from some critical number of yarn not more possible.

5 Profiles

The cross section of the beams determine significantly its mechanical behaviour. With well selected orientation and material distribution can be achieved high stiffness and strength at lower weight of the composite parts. The weaving process allow production of various cross sections at one process, which reduces the costs for

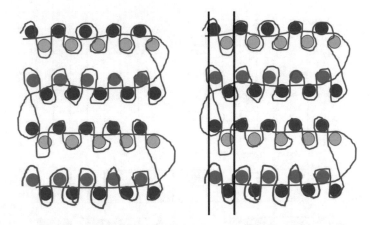

Fig. 9 Example for cross section of folded flat fabrics for weaving. Increased number of layers reduce the weaving speed and increases the number of the yarns in the same opening of the reed

the production compared to the production of flat pieces and following connection by stitching. The following example demonstrate the sequence of creation of the weaving pattern for commonly used "omega" cross section. The sequence can be applied for any "weaveable" type of cross section.

5.1   Cross Section Preparation

A long profile with cross section as presented in the Fig. 9a has to be produced on a shuttle weaving machine. The final form of the profile is not really suitable for weaving, because the classical weaving machines have all warp yarns distributed in one plane. In order to make the sample weavable, the diagonal edges have to be flatten and deformed to be in a parallel position as the main edge of the cross section (Fig. 10b). If each side edges becomes split to two parts, these can be fold to outside (Fig. 10b) or to the inner side (Fig. 9c) of the cross section. The folding to the inner part results into building four (4!) layers, which have to be woven at the same space of the reed, which can cause problems. Alternative solution can be created after rotation of the cross section to 90° (Fig. 10d). then its upper and bottom part rotated again (Fig. 10e) and then the middle vertical lines fold to outside (Fig. 10f) or inside (Fig. 10 g). These both variants require weaving of four layers at the same reed position, too. Analysing all variants, it seems, that the configuration on the Fig. 9b is the best, because it has the lowest number of parallel layers and will be used for the next steps.

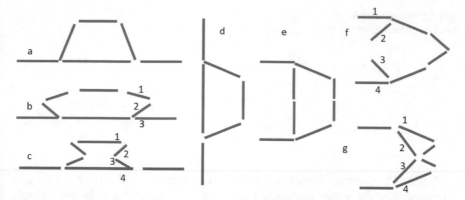

Fig. 10 Possible orientation of omega-profile for preparation for weaving. **a** Horizontal, **d** vertical, the other images represents variants for decomposition of the cross section into subsections

5.2 Shuttle Motion Sequence

As next step, the shuttle motion sequence has to be developed. In this task it is assumed, that all parts of the cross section needs to have the same weft density and the shuttle have to pass them all the same number of time for one complete repeat.

Figure 11 represents a first trial for the shuttle motion sequence, starting with motion from left to right through the complete width (Fig. 11a) (1), then moving to the left, but interlacing only the most right hand piece (2), then moving to the right (3), all upper parts (4), to the right (5) and left (6). The complete cycle is demonstrated summarized in Fig. 11b, and there is well visible, that the shuttle will work on the side edges twice often than the upper part of the cross section. This will results to double more weft yarns in this areas, which is not wished.

Fig. 11 First trial for shuttle motion sequence

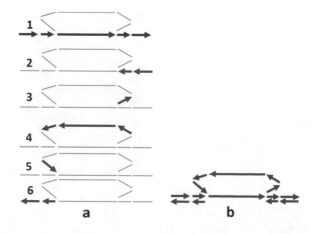

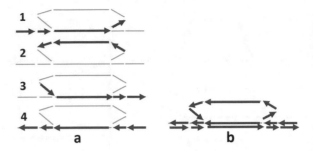

Fig. 12 Second trial for the weft motion sequence

Figure 12 represents modification of the first one, with trial to create motion sequence with the same number of passing through all sections. In this variant, during the first motion step the shuttle does not goes to the end of the bottom edge, but starts to move to the upper leg (Fig. 12a-part 1). Then moves to the left (Fig. 12a-2), then right (Fig. 12a-3)until the end and then back (Fig. 12a-4). The summarized all motions in Fig. 12b demonstrates some improvement—the middle bottom line is passed two times, too, but the upper part of the cross section will still have the half of the weft density, related to the remaining parts.

Learning from the both previous sequences, the next variant is created (Fig. 13). There after moving to the right (Fig. 13a-1), and then partial left (Fig. 13a-2), the upper section lines are passed (Fig. 13a -3, 4, 5) but with the last line there again the middle bottom part is passed (Fig. 13a-5), and the shuttle makes one additional step through the upper line segments (Fig. 13a-6, 7) and then ends with motion to the left hand side (Fig. 13a-8). Putting all motion steps on one figure (Fig. 13b) demonstrates, that this sequence will lead to the same number of weft yarns in all

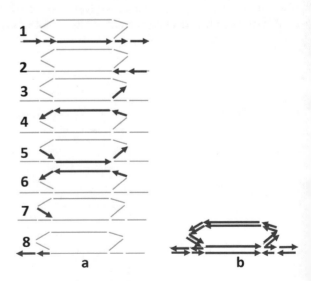

Fig. 13 Final sequence for the shuttle motion

Fig. 14 Motion of the shuttle and position of the warp yarns

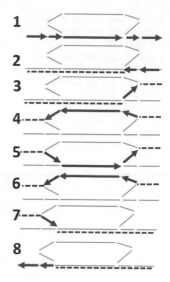

segments. The complete pattern required eight shuttle motions and create two weft inserts, it means, that per weft row in the final product are required four cycles as mean value.

The weft yarn does not interlace with the warp yarns in the segments during all cycles in the selected sequence. For instance, at the step 2, it has to interlace only with the warp yarns on the right hand side segment and then has to pass under (or over) all remaining yarns (Fig. 14–2) until gets it end position on the left hand side of the machine. During the next step (Fig. 14-3) it has to pass back through the same shed under (or over) the yarns, until the shuttle reaches the working area, where its yarn has to start interlacement. These places, where the shuttle goes through the machine without interlacement have to be marked on the picture, too, in order to be considered during the pattern creation.

5.3 Warp Yarns Distribution

The profile segmentation from the previous step allows the distribution of the warp yarns. Each line segment (Fig. 15a) has to receive enough warp yarns in order to build interlacement and to reach its desired size. For the pattern development, it is enough to represent only one repeat of the pattern. In the current case for simplicity, plain structure is selected, and two yarns pro line segment are required. In the middle segments are represented more repeats (Fig. 15b), so that the principle of the yarn distribution into the reed openings to be more clear (Fig. 15c). The warp yarns, which need to have close position at the end profile, have to be in the same (or close to each other) reed openings. The areas with the three layers have to have yarns in same reed openings, in order to receive the woven area there dense enough.

Fig. 15 Distribution of the warp yarns **a** line segments, **b** minimal number of warp yarns per segment, **c** possible (but not always the best) distribution of the yarns in the reed

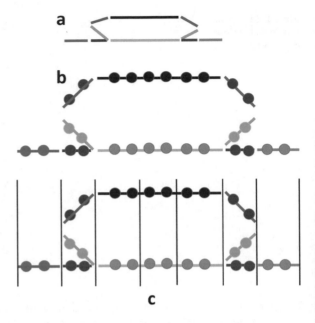

5.4   Warp and Weft Yarn Interlacement

After the warp yarns (at least one repeat of each segment) are prepared and placed on the picture, the interlacement with the weft yarn can be noted (Fig. 15a). For each shuttle motion step, a curve going over and under warp yarn cross section, following the wished pattern (in the current case plain weave) has to be drown. Figure 15b represents the motion in the second shuttle step, where after interlacing with the warp yarns in the right edge, the shuttle moves to the left part without crossing the yarns. This drawings have to be prepared for each shuttle motion steps in the same way.

5.5   Pattern Drawing

The figures with the interlacements (Fig. 16a and b) for each step are the basis for the drawing the woven pattern, which from other side is used for creation of the control sequence for the machine.

The pattern is created for each weft line following simple rule:

All warp yarns, which are placed OVER the weft yarn, are send to the upper position, and their corresponding cells in the pattern have to be filled out (Fig. 17a and b). The yarns, which remains under the weft, remains not filled. For the correct pattern creation, per year warp yarn a separated column on the figure is required. This step can be performed with a little bit training with Excel or Open Office Calc

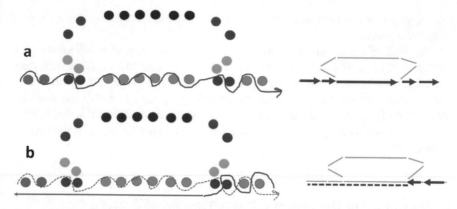

Fig. 16 Planning of the weft and warp yarn interlacement

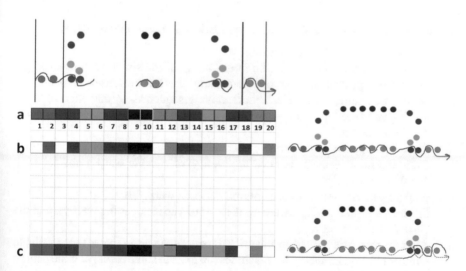

Fig. 17 Drawing the pattern based on the figures with the interlacements

using paining. The colours have not real meaning in this case, but they helps in the orientation of the user- to recognize to which yarn group the cells are corresponding.

5.6 Number Shuttles

After the pattern is ready, it can be counted, how many different yarn groups are allocated. In the upper case, there are ten line segments, with two yarn groups per segment (for plain structure) leads to twenty independent yarn groups. This means, that the structure can be woven only if the dobby machine has 20 heald frames, or if jacquard is used.

If instead of plain, twill 1–3 has to be implemented, then at all 40 groups will be there and obviously only jacquard machine will be able to be used in such case. For some pattern, after careful analysis and design can be found yarn groups which perform the same motion. These can be then merged into one group and threaded into one and the same heald frame. During the development process it is better, if the user concentrate on the pattern creation and check such possibility after the process is finished.

6 Summary of the Process Based Method of Pattern Development

The complete development step can be summarized in the following steps:

1. Create a sketch of the cross section
2. Try different orientations and different folding of the structure in order to get the minimal possible number of layers, parallel to the weft insertion plane
3. Define the shuttle (or needle, rapier) moving sequence, so that each segment line is passed the same number of times with the shuttle as all other (if the same density is required)
4. Place one repeat of warp yarns in each separated segment of the cross section
5. Draw the interlacement between the weft yarn and the warp yarns for each shuttle moving step
6. Prepare pattern sheet with one column per each warp yarn
7. Fill the pattern, marking each cell, if its warp yarn is over the weft yarn at the current shuttle motion
8. After the pattern is created, evaluate if this pattern can be produced with available number of heald frames on the machines (up to 22 in some cases) or if the pattern needs directly jacquard weaving machine.

7 Needle Weaving with Two Needles

Weaving tapes with two or more layers requires multiple weft insertions for each density line and is takes time. Several pattern can be produced with two needles (shuttles, rapiers) which provides then double higher productivity compared to the weaving with single needle. In order two simultaneous needles to be used on one machine, it has to be able to hold the warp yarns at three positions—upper (H), lower (L) and Middle (M) (Fig. 18). In such case two sheds gets opened and the two weft insertion elements can go there and place their yarns or loops.

Figure 19 represents one weft insertion step where two yarns (or loops) are inserted on the same time, as cross section through the warp yarns. The circles demonstrated

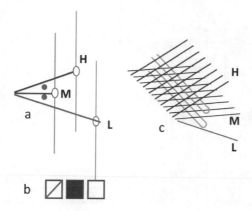

Fig. 18 Shed confirmation for weaving with two needles—**a** additionally to High and Low position, the warp yarns are able to hold in the Middle place. **b** Possible coding in the weave pattern. **c** Side view

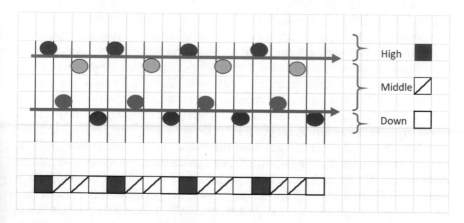

Fig. 19 One weft insertion step with two weft insertion elements for production of two layers

the warp yarns and the two thick arrows—the weft yarns. For the pattern creation the warp yarns are separated into three groups:

- warp yarns over the both wefts—these goes in upper position (H)
- warp yarns between the two wefts—these remains in the middle position (M)
- warp yarns under the both wefts—these stay in low position (L).

In Fig. 20 this pattern is extended with a second weft insertion step, so that two layers with plain structure get created. In this way, more complex pattern can be woven too, as for instance two layers angle interlock or other two layer structures with connections, too.

Using two needles requires not only special shaft or jacquard machine (with the option for stopping in the middle position, in the German Language this machines

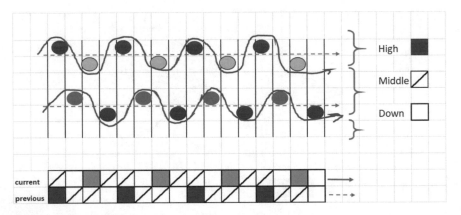

Fig. 20 Second weft step for two layer plain structure

are named TMH-Tief-Mitte-Hoch—translated _ Low-Middle-High machine). The developed in the last years shading machines, especially those with individual drives have possibilities to stop at different positions, but this process requires high forces for holding the warp yarns at these positions. The forces in the electromagnetic heald frame drives are normally caused by electromagnetic field, which means, that for high forces more energy is required and the forces remain still limited mainly in the area of light up to middle heavy tapes. One efficient way to "convert" classical heald frame machine with only two positions into machine with option for middle position is using of heald frame wires with different positions of the eyelets (openings) (Fig. 21). If one frame have to provide only two positions (for instance upper and middle), shifting the eyelet higher from the middle position can provide the required solution. Several producers provide the wires with two openings, so that these can be used both as normal shaft (central opening) and for upper or lower position. This solution does not provide enough flexibility for all pattern, especially if some frame (which means group of warp yarns) have to be moved to all three positions, for instance in order to connect the both layers. Cost effective and space saving solution for such case is to

Fig. 21 Different types of heald frame wires for weaving with multiple sheds

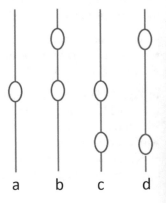

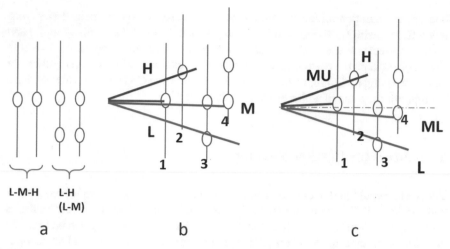

Fig. 22 Possible combination of different types of heald frame wires for the arrangement of different combinations for high, middle and low position of the warp yarns

have (mechanical) frame drive with three positions for 2 or 4 frames only (Fig. 22a) and the remaining frames with normal drives with upper and lower position, adjusted with suitable wires.

Additionally to the heald frame drives the multiple needle machines require specialized equipment for the selvage preparation. There are several types of systems for fixing such selvages available, known as "Z"-system or for multiple needles as 3N, 4N etc. [7]. The principle possibilities varies between—working with each needle independently (Fig. 23 a)—where the two layers stay not connected at the selvage or getting two needles together (Fig. 23b), so that the two layers gets connected at the selvage line.

This short overview about the frame drives and the selvages demonstrates, that the pattern designer has to consider the available machine options before finally decide how to develop the pattern for some profile. Pure "pattern making theory" without direct connection to the machine in the area of the narrow weaving is useless,

Fig. 23 Two (from multiple) possibilities for selvage holding when using needle waving with two needles

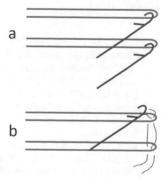

because the variations of the machines types (weft insertion type), frame drives (how many frames with Low-High and how many frames with "Low-Middle-High") and selvage systems is large and their total number of combinations are more than 200. This fact explains the reason, why there is no one existing suitable CAD system for narrow weaving design for technical products—such one CAD has to cover all these configurations, some of which are used only once.

8 Sources for Further Reading

The readers could find interesting results about relations between the woven pattern and properties in [8]. Applications for conveyor belts are reported in [9], for ballistics for instance in Abtew et al. [10]. Good review about the 3D weaving can be found in [11] and Kovačević and Schwarz [12]. The application for Wisetex [13] for computation of the mechanical properties of woven tapes is reported in [14]. Another book for narrow weaving, which the author did not succeed to access and evaluate at the current time is published in 2013 by Albert Thompson [15].

9 Conclusions

This chapter provides step by step explanation for a process based method for pattern creation of narrow woven fabrics with tubular or complex cross section. The method can be applied for people without deep knowledge in the weaving pattern design, but require systematic painting or at least good visualization of the different steps. As the pattern development of the complex profiles is dependent on the machine parameters, like type of the weft insertion system, possible positions of the frames, type of the selvages, is clear, that this process can be performed correctly only if the data about the available machines is provided.

References

1. Engels, H. (1996). Handbuch der Schmaltextilien. Die schützenlose Bandwebtechnologie: [Festschrift anlässlich der Vollendung des 80. Lebensjahres von Jakob Müller]. Frick: Fachhochschule Niederrhein (Mönchengladbach).
2. Kyosev, Y. (2015). *Braiding technology for textiles*. Amsterdam: WP Woodhead Publ./Elsevier (Woodhead Publishing series in textiles, 158).
3. Adanur, S. (2020). *Handbook of weaving*. CRC Press.
4. Gandhi, K. L. (2012). The fundamentals of weaving technology. In: *Woven Textiles* (pp. 117–160)Elsevier.
5. Kipp, H. W. (1988). *Bandwebtechnik*. Aarau: Sauerländer.
6. Kipp, H. W., & Greenwood, K. (1989). Narrow fabric weaving [Engineering, electronics, textile technology]. Aarau etc. Heidelberg: Sauerländer; distribution other countries Melliand.

7. Essig, E. (2005). Needle weawing technique. Frick, Switzerland: Jakob Müller Institute of Narrow Fabrics.
8. Boussu, F., Picard, S., & Soulat, D. (2018). Interesting mechanical properties of 3D warp interlock fabrics. In Y. Kyosev, B. Mahltig, & A. Schwarz-Pfeiffer (Eds.), *Narrow and smart textiles* (pp. 21–31). Springer International Publishing.
9. Berbig, I., Holschemacher, D., Kern, C., & Michael, M. (2016): Narrow woven fabrics applied for conveyor facilities. In: Y. Kyosev (Ed.), *Recent developments in braiding and narrow weaving* (1st ed, pp. 121–127). Cham, S.l.: Springer International Publishing.
10. Abtew, M. A., Boussu, F.., Bruniaux, P., Loghin, C., & Cristian, I. (2019). Engineering of 3D warp interlock p-aramid fabric structure and its energy absorption capabilities against ballistic impact for body armour applications. *Composite Structures, 225*, 111179.
11. Gokarneshan, N., & Alagirusamy, R. (2009). Weaving of 3D fabrics: A critical appreciation of the developments. *Textile Progress, 41*(1), 1–58.
12. Kovačević, S., & Schwarz, I. (2015). Weaving complex patterns-from weaving looms to weaving machines. In: C. Volosencu (Ed.), *Cutting Edge research in technologies* (pp. 93–111). IN TECH.
13. Verpoest, I., & Lomov, S. V. (2005). Virtual textile composites software WiseTex: Integration with micro-mechanical, permeability and structural analysis. *Composites Science and Technology, 65*(15–16), 2563–2574.
14. Kyosev, Y., Lomov, S., & Küster, K. (2016). The numerical prediction of the tensile behaviour of multilayer woven tapes made by multifilament yarns. In: Y. Kyosev (Ed.), *Recent developments in braiding and narrow weaving* (1st edn, pp. 69–80). Cham, S.l.: Springer International Publishing.
15. Thompson, A. (2013). *Narrow fabric weaving.* Read Books Ltd.

3D Orthogonal Woven Fabric Formation, Structure, and Their Composites

Abdel-Fattah M. Seyam

Abstract This chapter deals with new class of 3D woven structures, which is also known as non-crimp preform because the warp and weft yarns are not interlaced. The structure is formed from three yarn systems x-, y-, and z-yarns (or weft, warp, and binding yarns or through the thickness yarns) that are orthogonal to each other. The structure can be formed with desired thickness by controlling the number of x- and y-layers and yarns linear density to form fairly thick structure that is suitable as preforms for composite. These unique structures can be formed using recently invented 3D orthogonal weaving or traditional weaving technology. The 3D orthogonal weaving process and its structures possess numerous advantages over traditional 3D structures that are formed using traditional single phase weaving (referred here as 2D weaving) and those formed by stacking layers of 2D woven preforms. The features (advantages and drawbacks) of the 3D orthogonal weaving process and structures are covered. Additionally, applications of 3D orthogonal structures in composites are addressed. The chapter also covers innovative dobby and Jacquard shedding systems that are capable of forming 3D orthogonal woven preform with greater flexibility compared to traditional 2D weaving process. Part of the chapter is devoted for the development of a generalized geometric model to predict analytical (thickness, areal density, x-, y-, z-yarns fiber volume fraction) and mechanical properties. The models predict the properties of jammed and non-jammed 3D orthogonal woven preforms from flat and circular yarns in terms of their construction parameters.

Keywords 3D orthogonal preforms · Composite · Fiber volume fraction · Single-phase weaving · Multi-phase weaving · Geometric modelling · Dobby · Jacquard

A.-F. M. Seyam (✉)
Charles A. Cannon Professor of Textiles, Wilson College of Textiles, North Carolina State University, Raleigh, NC, USA
e-mail: aseyam@ncsu.edu

© Springer Nature Switzerland AG 2022
Y. Kyosev and F. Boussu (eds.), *Advanced Weaving Technology*,
https://doi.org/10.1007/978-3-030-91515-5_10

1 Introduction

In its simplest form, single layer (2D) woven fabric is formed from two sets of orthogonal threads (or warp and weft yarns) that are interlaced in a certain fashion (weave design or structure) to provide desired performance and/or aesthetic quality. While this class of 2D fabrics can be manufactured with numerous construction parameters, namely warp and weft fiber type (including blends), warp and weft thread density, warp and weft linear density, and weave structure, they possess inherent limitations. These include low thickness and finite number of maximum threads that can be woven with given yarn and weave (known as weavability limit or maximum construction). Requiring more thread density and thicker fabrics than what the 2D fabrics can accommodate is driven by the need for higher thermal insulation and mechanical performance. This need may be achieved by stacking 2D woven fabrics to form thick structure with desired thickness and performance for apparel, home textiles, and technical textiles, including composites. However, composites from stack of 2D preforms are susceptible to in-plane shear since the layers are only held together by brittle matrix material that are easy to crack and delaminate that leads to premature failure of the composite structure. To prevent delimitation, stack of 2D preform are z-pinned or over stitched manually or using sewing machine post weaving, which lacks accuracy and increase in the manufacturing cost. Z-pinning or over stitching causes significant preform distortion and damage yarns [1]. Additionally, stack of 2D fabric requires more fabric production.

To overcome such drawbacks mentioned above, specialty structures were developed, including warp and weft backed fabrics, double fabrics, and ply fabrics (more than two layers). Double and ply fabrics are classified as multilayer or 3D fabrics. Unlike staking of 2D fabric, a 3D fabric is formed by utilizing special woven stitching techniques to produce an integrated structure using single-phase weaving machine (known as 2D weaving machine), in which one weft yarn is inserted at a time. The early use of backed, double, and treble fabrics was limited to thermal performance and increased weight for apparel and home textiles. More recently, 3D fabrics found their ways in technical textiles' markets such as medical textiles, fiber reinforced composite markets (automotive, space, aerospace, marine, sports, defense, civil, etc.), and industrial hoses (fire, oil, water, protective sleeves).

As mentioned above, the formation of the traditional 3D fabrics are multilayer fabrics where the layers are integrated by woven stitches. Each layer is formed by interlacing a set of warp yarns and a set of weft yarns and each layer is connected with other layer(s) by woven stitches that can be conducted by several methods. For example, treble fabric is formed from three layers with each layer woven from two sets of orthogonal yarns (i.e. three sets of warp yarns and three sets of weft yarns). To integrate the three layers (face, center, and back layers) together, one or more of the following stitching techniques may be utilized:

1.   Raising back ends over center picks (to stitch back and center layers) and raising center ends over face picks (to stitch center and face layers).

Fig. 1 Yarn strength
contribution in the test
direction due to crimp

$\cdot F_{yarn} \cos\theta$

F_{yarn}

2. Lowering face ends under center picks (to integrate face and center layers) and lowering center ends under back picks (to integrate center and back layers).
3. Raising back ends over center picks (to integrate back and center layers) and lowering face ends under center picks (to stitch face and center layers).
4. Raising center ends over face picks (to integrate face and center layers) and lowering center ends under back picks (to integrate center and back layers).
5. Using extra warp yarns devoted for stitching. In this case, the extra warps will weave with each of the three layers to integrate them into one structure.

As a result of interlacing warp and weft yarns and stitching the layers by one or more techniques mentioned above, each warp and weft yarn are bent around each other and as such the path of each yarn is wavy. This is known as crimp due to weave interlacing and stitching, which reduces the strength contribution of the yarns to the test principle direction (warp or length and weft or width directions). The reduction of strength is approximately proportional to Cos θ, where θ is the maximum angle of thread axis to the principle (test) direction. Figure 1 shows simple force analysis of crimped yarn and its contribution in one of the principle directions (warp or weft), where F_{yarn} is yarn strength. Another issue caused by the crimp is the reduction of initial modulus since the yarn crimp removal requires low force. The actual yarn modulus will be reflected in the structure after crimp removal. High initial modulus is essential requirements in most of composites applications.

It is obvious that traditional 2D and 3D structures' mechanical performance (strength and modulus) is reduced by the crimp due to weave interlacing and stitching to integrate layers in 3D structures. For this reason 3D structures are formed from unidirectional layers, where a unidirectional layer is formed from a single set of yarns to avoid interlacing and produce non-crimp structures. Unidirectional layers can be laid in principle directions (0° and 90°) and may be also laid at different directions depending on the performance requirement. Preforms from 3D structures from unidirectional layers are commonly used in composites. While these structures exhibit high in-plane mechanical properties compared to their counterpart from 3D traditional woven preforms, they are susceptible to delamination and premature failure due to lack of presence of through the thickness binding elements. To avoid the drawbacks of 3D preforms from unidirectional layers and 3D woven preforms, unique structures were developed that is 3D orthogonal woven structures. The process to form these structures was invented by Mohamed and Zhang [2].

2 Structure and Formation of 3D Orthogonal Woven Preforms

2.1 Structure

Figure 2 shows 3D orthogonal preforms of different yarn compactions and thicknesses. In Fig. 2a, the number of y-yarn layers is four and the number of x-yarn layers is five. In Fig. 2b the number of y-yarn layers is seven and the number of x-yarn layers is eight. In 3D orthogonal woven structures, the number of x-yarn layers exceeds the number of y-yarn layers by one. The reason for this relationship is due to the way this class of preforms are formed, which will be explained later in this section. It can be noted from the figure that: (1) there is no interlacing between x-yarn and y-yarn layers, (2) x-yarns and y-yarns do not have crimp and lay straight in the x-direction and y-direction, respectively, (3) z-yarns binds and integrate the x-yarn and y-yarn layers, (4) the through the thickness component of z-yarns are vertical in case of maximum compact structures and tilted at an angle for other structures that are not woven to the maximum compaction, and (5) The z-yarns are divided to two groups that weave opposite to each other similar to plain weave in 2D fabric (other weave structures are possible). Like traditional 2D fabrics, each of the two z-yarn groups is assigned (or drawn-in) to a harness in case of plain weave (more harnesses are required for weaves other than plain weave). While the entire structure of Fig. 2 is so different from plain weave, the convection to identify the weave of 3D orthogonal structures follows 2D weave design. Figure 3 shows simulation of 3D orthogonal fabrics in different number of x- and y-layers and weave designs.

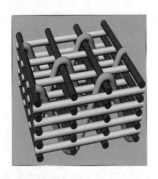   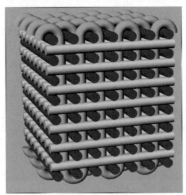

(a) Open 3D orthogonal preform   (b) Compact 3D orthogonal preform

● x-yarn (weft), ○ y-yarn (warp), ◯ z-yarn (binding yarns)

Fig. 2 Structures of 3D orthogonal woven preforms from yarns with round cross section [3]

(a) Plain weave, 2 y-layers, 3 x-layers

(b) 2x2 Basket weave, , 3 y-layers, 4 x-
layers

(c) 2x2 Twill weave, 4 y-layers, 5 x-layers
● x-yarn (weft), ◯ y-yarn (warp), ◉ z-yarn (binding yarns)

Fig. 3 3D Orthogonal preforms in different number of layers and weave structures from flat yarns
[3]

2.2 *Formation*

Figure 4 depicts 3D weaving machine that is specialized to manufacture 3D orthogonal preforms. In the setting shown, the y- and z-yarns are supplied from creels
(two in this case) with individual rotating package for each yarn. Rotating packages
maintain the yarns twist, whether it is zero twist for flat yarns or at certain level for
twisted yarns (spun or twisted continuous filament yarns). Addition of twist to flat
yarn reduces its strength, coverage, increase preform thickness, and reduces preform
fiber volume fraction (defined as volume of fibers divided by the total volume of

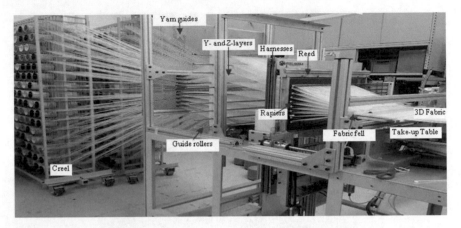

Fig. 4 Overview of 3D orthogonal weaving machine (donated by 3TEX, INC) forming preform of nine y-layers (ten x-layers) [4]

preform or composite). Each yarn is passed through tension device and guides and threaded through a dedicated closed ceramic guide to maintain position and prevent entanglement with neighbouring yarns. The ceramic guides are fixed in a Plexiglass vertical panel. Each y-layer (nine layers in the figure) is passed over or under a guide roller to maintain the layer position in the weaving zone (between fabric fell and harnesses in the figure). The z-yarns are guided to be in the middle of the y-layers. Each z-yarn is threaded through an eye of a heddle wire in a harness. Like traditional 2D weaving, the number of harnesses and the assignment of each z-yarn to a harness depend on the weave design. The y- and z-yarns are threaded through reed dents (spaces between reed wires) according to the denting (reed) plan. The denting plan for the setting in Fig. 4 is shown in Fig. 5, where the black vertical lines represent

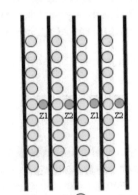

○ y-yarn (warp), ● z-yarn (binding yarns)
Z1 & Z2: z-yarns drawn in harness 1 and harness 2, respectively

Fig. 5 Denting plan of plain weave (2 repeats) for 9 y-layer

the reed wires. Different denting plans to control fiber volume fraction of z-yarn is possible. The denting plan in the figure is known as 1:1 z-yarn/y-yarn per layer ratio. Ratios such as 1:2 and 1:3 allow reduction in z-yarn density and increase in y- and x-yarn thread density, which increases the in-plane performance while maintaining the structure integrity. To accommodate the y-layers and z-yarn large motion during formation, the reed and harnesses' heights are higher than that used in 2D traditional weaving. Each x-yarn is supplied from individual package and all x-yarn packages are supported by creel (not shown) situated on one side of the machine (right side in this example). The fabric is collected at the delivery end of the machine by the take-up table. The traditional take-up mechanism used to collect the fabric on roll is not suitable due to the fairly high thickness and rigidity of the 3D orthogonal fabrics. The friction, pressure, and bending imposed on the fabric by traditional take-up system would cause damage to the high performance fibers that are vulnerable to such complex field of stresses. The table take-up system keeps the 3D orthogonal fabric flat till it is cut to the desired length then the table is reset to start weaving the next fabric length.

Like 2D, the 3D orthogonal weaving essential motions are: (1) shedding motion, (2) weft insertion motion, (3) best-up motion, and (4) warp and fabric control motion. Figures 6, 7 and 8 show the sequential events of the weaving essential motions to form 3D orthogonal fabric. The y-yarns are spaced apart to form required number of layers and do not move to form the sheds. They stay stationary throughout the entire weaving process. Only z-yarns, which are controlled by harnesses, move up to the top most position or down to the bottom most position to form the top and bottom sheds (Fig. 6). Each harness moves up or down according to the weave design and the assignment of yarns to each harness (drawing-in). With such positioning, the z-yarns form two layers and the number of sheds formed equals the number of y-layers plus

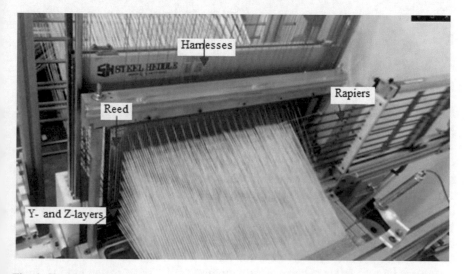

Fig. 6 Shed formation

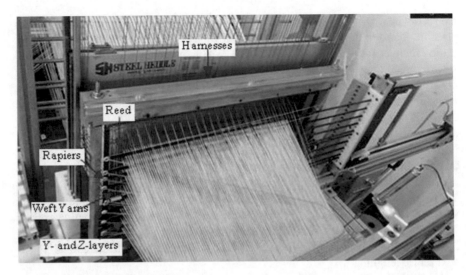

Fig. 7 Multi-insertion of x-yarns

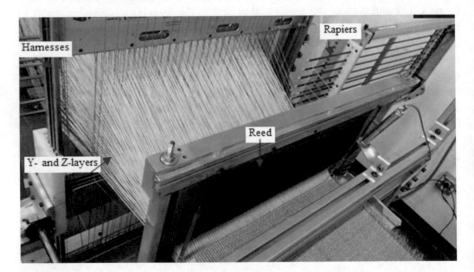

Fig. 8 Beat-up

one. Since the number of y-layers is nine as it can be seen from Fig. 4, the number of sheds is ten, which is equal to the number of x-layers. Following the shed formation, the rapiers (ten in this case) simultaneously insert (multi-weft insertion) the weft yarns as it can be seen in Fig. 7. The inserted x-yarns (weft yarns) are then beaten-up by the aid of the reed wires. In Fig. 8, the reed moved from its most back position (as seen in Fig. 6 and 7) to its most forward position to push the newly inserted weft yarns into the fabric to become the new fabric fell. The warp and fabric control motion controls the y- and z-yarns (warp sheets) tension and the preform take up speed that controls the x-yarn density (number of x-yarns/layer/unit preform length). The tension of each y- and z-yarns is controlled at desired level by individual tension devices for each yarn. The amount of fabric take-up is controlled by the take-up table, which is controlled by individual motor according to the x-yarn density.

While the setting of the 3D weaving machine in Fig. 4 provides ultimate flexibility in terms of hybridization and incorporation of functional yarns, it takes considerable preparation time to back wind y- and z-yarns and threading each yarn and layers through machine parts. It is more economic to supply y- and z-yarns on beams. Since the length of z-yarns required to form certain preform length is much higher than y-yarns, z-yarns must be supplied from different beam(s) than y-yarns.

3D orthogonal woven structures can be manufactured using traditional 2D weaving technology (single phase weaving or single weft insertion) where one shed is formed at a time. As mentioned in the introduction section, this technique was used to form traditional multilayer structures such as double and treble fabrics. Figure 9 shows the weave design, DID (Draw-In Draft or draw), and CP (chain plan) of minimum number of harnesses on square paper required to form four y-yarn layers 3D orthogonal preform using 2D weaving machine. Additionally, the figure depicts the denting plan. In this example, the z-yarns form plain weave structure. The corresponding simulation of 3D orthogonal fabric with four layers of y-yarns is shown in Fig. 2. In Fig. 9, y1–y4 are y-yarns of layers 1–4, z1 and z2 are two z-yarns of plain weave repeat, and H1–H6 are six harnesses, which is the minimum number

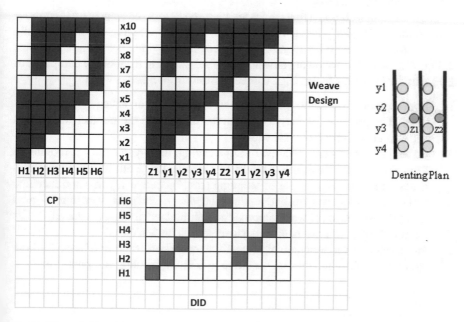

Fig. 9 Weave design, DID, CP of minimum number of harnesses, and denting plan of 4-yarn layers 3D orthogonal preform

required to form the structure. Harnesses H2–H5 are dedicated to the four y-layers (y1–y4) while harnesses H1 and H6 are devoted to z-yarns (z1 and z2). The 3D orthogonal woven structure can be interpreted as a 4-layer fabric stitched by extra yarns (z-yarns). The weave design of Fig. 9 can be developed in conjunction with Fig. 2 for better understanding. The x1–x10 weft yarns in Fig. 9 are the first two sets of five each in Fig. 2. The ten sheds for ten weaving cycles to form shed of:

1. x1 weft (top weft yarn in Fig. 2): only z1 yarns (H1) are raised
2. x2 weft: z1 yarns and y1 yarns (H1 and H2) are raised
3. x3 weft: z1 yarns, y1 and y2 yarns (H1–H3) are raised
4. x4 weft: z1 yarns, y1–y3 yarns (H1–H4) are raised
5. x5 weft: z1 yarns, y1–y4 yarns (H1–H5) are raised
6. x6 weft: only z2 yarns (H6) are raised
7. x7 weft: z2 yarns and y1 yarns (H6 and H2) are raised
8. x8 weft: z2 yarns, y1 and y2 yarns (H6, H2, and H3) are raised
9. x9 weft: z2 yarns, y1–y3 yarns (H6, H2–H4) are raised
10. x10 weft: z2 yarns, y1–y4 yarns (H6, H2–H5) are raised.

To situate the axis of each yarn in a set of x-yarns (five x-yarns in this example) in the same vertical plane, the 2D weaving machine should be equipped with a take-up system that can be set to stop taking up the fabric for four picks and move the fabric after each fifth pick by a preset distance based on desired x-yarn density. If the take-up continues to move a distance for each weft insertion, a structure with staggered x-yarns would be obtained.

For this example, ten weaving cycles are required to form one repeat of 3D orthogonal preform with four y-layers (five x-layers) using 2D weaving technology while this requires only two weaving cycles if 3D weaving machine, such as the one shown in Fig. 4, is used. The following advantages of 3D weaving technology compared to 2D traditional weaving technology can be realized from this example:

1. Higher productivity for the same weaving speed (number of weaving cycles/unit time): The productivity increases with the increase in the number of x-layers due to multi-insertion of weft yarns.
2. Lower stress on y- and z-yarns: In 3D weaving, the y-yarns are not moving to form sheds and z-yarns move much less frequently (1/5th in this example) a matter that reduces the rigor of weaving (friction between yarns and machine parts, impact from beat up, and repeated bending). Unlike 2D weaving, y-yarns strength loss due to 3D weaving process is eliminated or highly reduced, and as a result the preform mechanical performance is higher.
3. More suitable for weaving high performance fibers such as Glass, Carbon, Aramid, and Ceramic that are sensitive to friction, bending and pressure.

3 Features of 3D Orthogonal Preforms and Their Composites

The features (advantages and draw backs) of 3D orthogonal woven preforms and their composites are addressed in this section. The advantages are:

1. Composites from 3D orthogonal preforms are not susceptible to delamination due to the presence of through the thickness portions of the z-yarn that binds the x- and y-yarn layers as it can be noticed from Figs. 2 and 3.
2. Like composites from 3D woven structures made from stacking 2D woven fabric or unidirectional laminates, composites from 3D orthogonal preforms possess high strength to weight ratio compared to metals. However, composites from 3D orthogonal preforms possess higher strength to weight ratio than composites from stacking 2D preforms due to the fact that in 3D orthogonal preforms x- and y-yarns do not have crimp and strength loss due to weaving is very low if the preforms are formed by 3D orthogonal weaving process.
3. The x-, y-, z-yarns and total volume fraction can be controlled at broad range of levels because the x- and y-yarns are not crimped and z-yarn length can be easily calculated (see modeling section). Other 3D woven preforms have crimp within a layer and crimp due to woven stitches that are extremely difficult to model and predict due to the numerous parameters that affect the crimp. Additionally, crimp increases the preform/composite thickness that leads to reduction in fiber volume fraction. The, higher preform/composite total fiber volume fraction and hence higher performance can be achieved compared to any other 3D preforms.
4. Preforms possess high conformability due to extremely low inter-fiber and inter-yarn friction since there is no interlacing (no crimp) between x- and y-yarns. The low friction allows the x- and y-yarns to slip over each other when pressed against complex shape molds (spherical, double curvature, etc.) to form specific composite parts such as curved beams, tub, helmet, etc.
5. The straight path of x- and y- yarns creates channels that make the preforms acquire excellent ability to transfer resin, which allows shorter time for resin transfer and as a result the manufacturing cost is lowered.
6. Given the above points, composite from 3D orthogonal woven preforms are able to sustain damage tolerance, impact resistance and multi-hit ballistic protection capability than any other composites from other 3D preforms.
7. 3D orthogonal preform and their composites are good hosts for embedded sensors for applications such as structure health monitoring. For example, these structures are excellent hosts for embedded optic fibers (whether silica- or polymer-base). Optic fiber sensors can be laid straight (parallel to x- and/or y-yarns and hence their signal loss due to bending is avoided, which provide high dynamic range and better sensitivity [5].

The 3D orthogonal preforms are not without drawbacks. These preforms cannot be used in their soft form applications such as body armor since the inter-yarn friction is low and a bullet or shrapnel can easily penetrate through such structures. Like other 3D woven preforms and their composites, 3D orthogonal woven preforms and their composites are anisotropic structure with high performance in x- and y-directions and lower performance in other in-plane directions. 3D from 2D stack can be laid to have yarns in more than two directions. Despite these few drawbacks the advantages outweigh the disadvantages. The advantages of composites from 3D orthogonal woven preforms qualify them for numerous high performance applications, including aerospace, automotive, boats, military armor (such as tanks, helmets, etc.), wind blades, etc.

4 Innovation in Shed Formation System

While the 3D orthogonal weaving machine (Fig. 4) has advantages, the serious issue is the long time to thread the individual yarns in guides and the y- and z-yarn layers around the rollers prior to weaving. The long setting time is needed every time when switching to a different structure. Recently, weaving machine manufacturers developed innovative shedding systems that could be used to manufacture 3D orthogonal woven structures. These technologies are able to maintain the advantages of the 3D orthogonal weaving machine (Fig. 4) while eliminating the long setting time. The innovative shedding systems include dobby and Jacquard systems. Historically, dobby and Jacquard woven fabrics offer countless color, texture, and intricate designs and as such they established themselves in commercial applications such as high fashion dresses, decorative fabrics, drapery, sheeting, upholstery, matelasse, mattresses, table cover, wall décor, terry, and woven carpets. Jacquard weaving extended to non-traditional technical textile applications such as seamless airbag and awning. Other potential applications that are still being under research and development include electronic textiles, shaped seamless apparel, and 3D shaped preforms for fiber reinforced composites. The specific innovations that are relevant to 3D orthogonal weaving are individual control of harnesses and warp yarns in dobby and Jacquard weaving, respectively. Adapting such systems would allow the use of automatic drawing-in, tying-in, quick style change to significantly reduce the setting time to prepare y- and z-yarns for weaving. The focus of this section is to review the recent innovations in the dobby and Jacquard shedding systems and their benefits relevant to 3D orthogonal weaving.

4.1 Innovation in Dobby Shedding Systems

Dobby shedding motion controls the movement of large number of harnesses and as the number of harnesses increases so does the load on the motor that drives all moving parts connected to the weaving essential motions (shedding, weft insertion, beat up, and warp and fabric controls). When the weaving process is stopped for any reason (warp stop, weft stop, or other stop), the potential for start mark defect exists. The reason for many of these start mark defects stems from the heavy mass of the machine moving parts that the motor has to overcome their inertia upon resuming weaving and as a result the machine set speed is not reached and consequently the beat up force is weaker and the new pick is not forced enough distance and stays a farther distance from the cloth fell than preset spacing and hence start mark defect is developed, which is a major defect. As the number of harnesses increases, the possibility for start mark defect increases, an issue that caused limiting the number of harnesses and the design capability. The shed timing in traditional shedding motion is rigid and fixed for each weaving cycle. The individual control of harnesses made it possible to time each shed differently for each weaving cycle if so desired. The individual harness control dobby systems allow presetting shed timing independently from other weaving motions. Synchronizing all weaving events to work in harmony during a given weaving cycle is accomplished by wireless system using radio frequency. Early shed, which is required to weave high weft density, is not desired in general due to the dragging force of the weft in crossed warp yarns that cause abrasion and weakening the warp and weft yarns. The combination of early shed and normal time shed is useful when weaving variable weft density. This can be used to strategically form fabric with parts of variable performance in the same fabric rather than seaming different fabric pieces woven using traditional shedding motion.

The drawbacks of traditional shedding systems motivated shedding systems producers to develop dobby shedding systems with individual control of each harness. Today, these shedding systems are commercially available by major machine manufacturers. Staubli was the first to pursue such development; the UNIVAL 500, which is offered in up to 52 harnesses. This electronic dobby system is using powerful actuators to control harnesses' movement and each harness is controlled by an actuator. Currently, several machine manufacturers (Toyota, MAGEBA, Dornier, Picanol, etc.) provide electronic dobby systems with individual harness control. Additional benefit of these shedding systems is the elimination of electromagnets and numerous links are eliminated that allowed significant reduction in noise and vibration and higher weaving speed compared to the traditional dobby systems.

The individual harness control shedding systems can be used to manufacture 3D orthogonal woven preforms with higher flexibility than the setting of 3D orthogonal weaving depicted in Fig. 4. To produce 3D orthogonal woven fabric using a machine with individual harness control, each y-yarn layer is drawn in a harness and each z-yarn layer is drawn in a harness. For example, to weave 3D orthogonal preform of nine y-layers (ten x-yarn layers) with z-yarns forming plain weave, a total of eleven harnesses is required. To maintain the advantage that y-yarn layers are stationary throughout the entire weaving process, the nine y-yarns harnesses are raised at different heights (as shown in Fig. 4) and remain stationary. Only the z-yarns harnesses move according to the weave design. With such setting, individual harness control shedding systems with all yarns controlled by harnesses can be employed to form not only 3D orthogonal structures but also weave other 3D woven fabrics such as interlock, angle stitched, and cellular structures woven individually or combination of these in the same preform which provides design flexibility for different applications. Such 3D weaving machine requires reed and harnesses with enough height to accommodate the sheds formed by the y-yarns and z-yarns. Multi weft insertion motion is also required.

4.2   Innovation in Jacquard Shedding Systems

Until the 1980's, the mechanical Jacquard systems were most common in industry setting and they still exist today, mostly in developing countries. These systems have too many parts; pattern cards, cylinder, needles, needles' springs, hooks, neck cords, harness cords, heddle wires, and returning springs for each harness cord. The parts are working in harmony for selection of hooks (and hence warp yarns) to be raised or lowered for each shed according to the weave structure, which is registered on the punch pattern cards. The Jacquard head parts get their movement from a drive connected to the main weaving machine motor that added more complications due to the fact that each weaving machine has a different main drive configuration that requires different connection configuration to the Jacquard head. These too many parts of mechanical Jacquard and their heavy mass and complicated motions limited weaving speed to a degree that mechanical Jacquard shedding systems were not compatible with shuttleless high-speed weaving machines.

The first commercial electronic Jacquard was exhibited by Bonas Machine Company Ltd. at ITMA 1983 show held in Milan, Italy. The electronic Jacquard shedding systems eliminated many of the mechanical Jacquard parts. The pattern cards converted to an electronic file that is used to select hooks (hence warp yarns) according to weave design via electro magnets. As a result, the cylinder, needles and their springs were eliminated. The electronic selection of hooks and reduction in parts led to significant increase in Jacquard weaving speed. In fact electronic Jacquard shedding systems are capable of running at much higher speed than any of today's shuttleless weaving machine. It is fair to say that today's electronic Jacquard systems are ready for the next generation high-speed weaving machines.

The traditional Jacquard (whether mechanical or electronic) capacity (maximum number of hooks) was limited, which requires harness tie, and weaving several small/medium design repeats across fabric width. This posed challenges for design innovation and flexibility. The standard size or capacity of Jacquard machines ranged from 448 to 1,792 hooks. A given harness tie type (straight, center, boarder, or mixed harness tie) remains unchanged for long time since their construction takes fairy long time. For this reason a Jacquard designer has to fit patterns to exiting harness ties. The Jacquard machine manufacturers responded to the design capability challenge by introducing: (1) individual control of harness cords (or warp yarns) and (2) mega size capacity. In mega size Jacquard machines, the number of hooks may reach up to 18,432. The individual warp yarns control and mega size Jacquard machines permitted the creation of extremely intricate large size patterns; one large pattern or several small/medium size patterns across the fabric width may be created.

At ITMA 1999, Grosse and Staubli introduced entirely new Jacquard weaving concept. Grosse named their system UNISHED and Staubli termed their machine

UNIVAL 100. In these systems each wrap yarn is controlled by an actuator and both are considered electronic Jacquards that are communicating with the machine controller to form the shed from electronic pattern file, which is created by CAD software. Grosse's earliest version (exhibited at ITMA 1999 and termed UNSHED 1) was shown as a prototype. The machine was shown again at ITMA 2003 with little improvement. At ITMA 2007 the much improved version (UNISHED 2) machine was exhibited and about two years later the machine was available commercially. Figure 10 illustrates the shed formation principle of the UNISHED 2. The config-uration of the UNISHED led to the elimination of harness cords, magnets, hooks, pulleys, springs, and the gantry. AT ITMA 2011, the UNISHED 2 was shown at Tsudakoma booth on water jet weaving machine (ZW8100). This was the first time to combine water jet weft insertion with Jacquard shedding system at ITMA [6].

At ITMA 2003, the UNIVAL 100 was introduced in its commercial configuration. Figure 11 illustrates how the shed is formed. In the figure, an actuator (extremely small stepping motor termed JACTUATR) is shown while selecting its harness cord (warp yarn) for the upper shed (left image). The right image shows an actuator with its warp yarn selected for the bottom shed. In this system, magnets, hooks, knives,

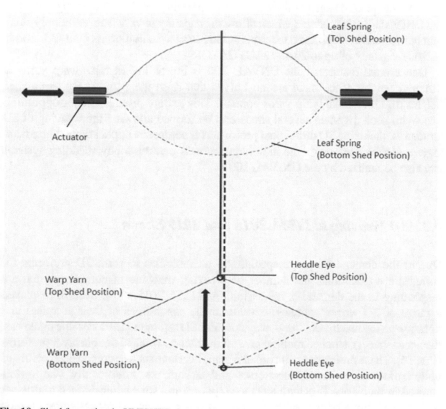

Fig. 10 Shed formation in UNISHED 2

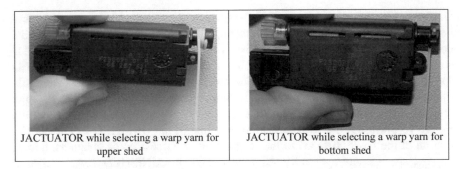

| JACTUATOR while selecting a warp yarn for upper shed | JACTUATOR while selecting a warp yarn for bottom shed |

Fig. 11 Shed formation in UNIVAL 100

and pulleys are eliminated. At ITMA 2003 a UNIVAL 100 machine with capacity of 7,920 was mounted on Picanol OMNIplus-6-J 250 weaving Mattress Ticking and Table Cloth (pattern was changed on the fly) at 1,025 picks/min and filling insertion rate 2,460 m/min. This rate of filing insertion is the highest in the history of Jacquard weaving. With this technology, weavers are capable of manufacturing intricate Jacquard designs at the speed of commodity fabrics. The configuration of the UNIVAL 100 is modular, which allow the capacity to vary from extremely small number for narrow fabric and extremely wide fabric. Staubli offers UNIVAL 100 in variable capacity of up to 20,480 JACTUATORs.

One unique feature of the UNVAL 100 is the ability to raise warp yarns at different heights since there are no knives that raise the yarns at the same shed height; true individual warp yarns control. This feature provides major opportunities. With such a system several sheds can be formed allowing the weaving of 3D structures, including 3D orthogonal preforms for composite applications. The advantages mentioned above on the individual harness control dobby shedding systems can also be realized by the UNIVAL 100.

4.3 3D Weaving at ITMA 2015 and 2019 Shows

Due to the demand for more specialized technologies to form 3D preforms for composite applications, the number of weaving machine manufacturers that are responding to the demand is increasing. At ITMA 2015 show, **Staubli** displayed a range of 3D woven samples to validate the capabilities of their Jacquard and dobby weaving machines. The display contains 3D orthogonal, 3D variable thickness, distance (spacer), stitched multilayer, and unstitched/stitched double layer preforms (Fig. 12). These preforms are intended for applications such as composites, soft/hard body armor, and embedded electronics and sensors, etc. Additionally, **Vandewiele** showed in their Face Brochure distance (spacer) and stitched/unstitched multilayer preforms. Figure 13 shows applications of distance fabrics in boat and transportation applications.

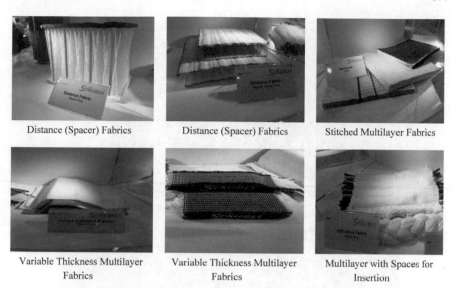

| Distance (Spacer) Fabrics | Distance (Spacer) Fabrics | Stitched Multilayer Fabrics |
| Variable Thickness Multilayer Fabrics | Variable Thickness Multilayer Fabrics | Multilayer with Spaces for Insertion |

Fig. 12 Delete3D samples exhibited by Staubli at ITMA 2015

At ITMA 2019, **VUTS** (Liberec, Czech Republic) exhibited an air jet weaving machine specialized in formation of 3D distance fabric with constant or variable distance (Fig. 14). Potential applications for the distance fabrics include inflatable boats of different type, dock, mats, flood protection, lifting bags, etc. Currently, the inflatable boat industry is using drop stitch technique to distance two fabrics. In this technique, numerous needles are used for joining two base fabrics formed by a traditional 2D weaving machine. The procedures of the drop stitching method are complex and lengthy. The process may take over 20 days to remove and replace the needles (http://www.nrs.com/tech_talk/dropstitch.asp). The distance fabrics produced by the weaving process are much faster (productive) and easier to design and form desired shape.

Optima 3D, new company formed in 2018, has designed and developed a range of 3D weaving machines for formation of preforms for the composite industry. The company showed its Optima 500/150, which uses shuttle system for weft insertion and a Jacquard shedding system with electronic user interface for structure parameters input that provide preforms design flexibility. While **Dornier** did not exhibit its new 3D weaving machine, the Company developed line of machinery for forming preforms for composite systems that included 3D weaving, which is used to manufacture multilayer preforms for composite applications. The features of the machine include the use of CAD system to program the weave structure, rigid rapier filling insertion system, individual warp yarn control Jacquard shedding system, creel for feeding warp yarns from wound packages, and an optional horizontal take-up system for thick preforms.

As indicated earlier, the multi-layer (3D) weaving techniques have been known for long time and used for the production of upholstery, hoses, tubular grafts, air bags,

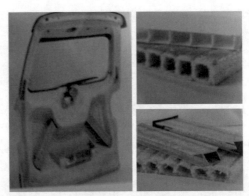

Transportation Applications

New 500 L Fiat Boat from Distance (Spacer) Fabric

Fig. 13 Applications of 3D multilayer and distance fabrics produced by Jacquard weaving machines. *Source* Van De Wiele's ITMA 2015 Face Brochure

Fig. 14 Constant and variable thickness distance fabrics produced by DIVA machine (*Source* VUTS brochure ITMA 2019)

etc. using one shed at a time. These products were developed before the invention of CAD systems using mechanical dobby and Jacquard shedding systems. With the development of high-speed electronic dobby and Jacquard system, the CAD systems are must to create digital weaves to control the movement of warp yarns to form individual sheds according to weave design. While these CAD systems can be utilized to develop 3D orthogonal, cellular, I-shape, and T-shape structures, it requires very skilled woven designer with deep understanding of the 3D fabric geometries. The need for CAD systems that is intuitive and easy to use to develop 3D fabrics is sorely needed. The software company **EAT** recently developed the 3D Weave Composite Software. The warp or the weft cross-section is drawn digitally on square paper along with information on warp and weft yarn size and color (type) of each warp layer via user interface. Then the system provides colored 3D visualization of the structure. The software then converts the design to traditional up and down motion (similar to Fig. 9) of each warp yarn to weave with one shed at a time weaving machine.

5 Geometric Modelling of 3D Orthogonal Preforms

The ultimate goal of a composite manufacturer is to develop structures that meet the performance requirements for specific end use while maintaining manufacturing cost at minimum. Trials and errors approach to achieve the goal is costly and it is impossible to cover the endless structures due to the numerous numbers of parameters involved in the formation of the preform and the resin infusing process. The number of structure parameters of 3D orthogonal woven preforms is very high. These include fiber type, fiber volume density, fiber blend, yarn volume density, yarn packing factor, yarn cross section shape, yarn linear density, weave structure, yarn spacing at weave intersection, yarn spacing under weave float, number of yarns/layer in the weave repeat, number of yarn layers, and length of yarn required per weave repeat. These parameters are repeated for each of x-, y-, and/or z-yarn. Add to this the fact that for each parameter there is a wide range of levels. For these reasons, geometrical modelling approach is used to develop geometrical relationships to predict analytical and mechanical properties in terms of structural parameters. The accuracy of the geometrical relationship depends on the assumptions used to facilitate deriving them. Regardless of how accurate the assumptions are, numerical solutions of the geometrical relationships permits quantitative direction of the effect of a parameter on a property and thus can be used as a guide to design a composite with desired performance. Several researchers dealt with the topic of geometrical modelling of 3D orthogonal woven preforms. Extensive critical review on the topic is covered by Ince [3]. The limitations of the previous research on the topic are briefly discussed below.

Early studies adopted a unit cell of jammed structure of x-, y-, and z-yarn and neglected the effect of number of layers and z-yarn crowns at crossover area with x-yarns that led to underestimation of the x- and z-yarn volume fraction and overestimation of the y-yarn fiber volume fraction [7, 8. Most published papers covered plain weave jammed geometry and assumed segmented straight portion of z-yarn, [9–14], which causes error in predicting z-fiber volume fraction as pointed in [15]. Other publications covered plain weave and few other weaves such as twill and stain [15–18]. There are numerous weaves that are not considered such as plain weave derivatives (warp rib, filling rib, and basket weaves), twill weave derivatives (such as extended and multiple twills), and satin weave derivatives (such as double and extended satins/sateen). The limitation of geometric modelling prompted undertaking approach to develop a generalized geometric model to predict analytical properties (thickness, areal density, x-, y-, z-yarns and total fiber volume fraction) of 3D orthogonally woven preforms in terms of yarns and preform construction parameters listed above. Several models were derived to predict the analytical properties for any weave structure (numerical parameter that accounts for the z-yarn interlacing and its through the thickness components) for jammed and non-jammed structures from spun yarns (or twisted yarns from continuous filament) as well as flat yarns (untwisted yarns from continuous filaments) [3], Seyam and Ince [3, 19]. Ince et al. [19] developed the following nomenclature and set of equations for flat yarns with rectangular cross section considering jammed and non-jammed 3D orthogonal woven preforms.

6 Nomenclature

k	$k = x$, y or z
ρ_{vk}	Volume density (g/cm^3)
$\rho_{\ell k}$	Linear density of yarn (g/km)
w_k	Yarn width (cm)
t_k	Yarn thickness (cm)
AR_k	Yarn aspect ratio (width to thickness ratio)
Φ_k	Yarn packing factor
p_x	X-yarn spacing for non-jammed structure (cm)
p_{xi}	X-yarn spacing at weave intersection for jammed structure (cm)
p_{xf}	X-yarn spacing under the z-yarn float for jammed structure (cm)
$\overline{p}_x$	Average x-yarn spacing for jammed structure (cm)
p_y	Y-yarn spacing for non-jammed structure (cm)
p_{y1}	Yarn spacing of y-yarn followed by z-yarn for jammed structure (cm)
p_{y2}	Yarn spacing of two successive y-yarns for jammed structure (cm)
$\overline{p}_y$	Average y-yarn spacing for jammed structure (cm)
N_k	Number of yarns/layer/weave repeat
i_z	Number of z-yarn intersections/weave repeat
M_z	Z-yarn weave factor ($=N_x/i_z$)
θ	Maximum angle of the z-yarn axis to plane of preform (radian)
n_y	Number of y-yarn layers
n_x	Number of x-yarn layers $= n_y + 1$
a	Arc portion length of z-yarn (cm)
b	Through-thickness part length of z-yarn (cm)

L_x Weave repeat width (cm)
L_y Weave repeat length (cm)
L_z Preform thickness (cm)
V_{repeat} Preform volume/weave repeat (cm^3)
ℓ_k Total length of a yarn/weave repeat (cm)
V_{fk} Fiber volume of a yarn/weave repeat (cm^3)
F_{fk} Fiber volume fraction of a yarn (%)
F_{fxyz} Total fiber volume fraction (%)
W_k Weight of a constituent yarn (g/m^2)
W_{xyz} Areal density of preform (g/m^2)

The following assumptions are considered to develop the geometrical relationships:

1.   x-, y-, and z-yarns are uniform rectangles and fiber packing along their length
2.   x-, y-, and z-yarns are completely flexible, inextensible, and incompressible
3.   x- and y-yarn spacing is uniform in case of non-jammed structures

Micrographs of composites from flat yarns showed that yarn cross section shape is rectangle (Figs. 22 and 23). Yarn width (w_k) and yarn thickness (t_k) of rectangular yarns can be calculated from:

$$w_k = \sqrt{\frac{\rho_{\ell k} AR_k}{\Phi_k \rho_{vk} 10^5}} \tag{1}$$

$$t_k = \sqrt{\frac{\rho_{\ell k}}{\Phi_k \rho_{vk} AR_k 10^5}} \tag{2}$$

Assumption 2 is reasonable for the purpose of predicting geometrical relationships.

The geometries showed in Figs. 15, 16 and 17 are developed based on the assumptions and to facilitate deriving the geometrical relationships.

Generalized 3D orthogonal preforms for non-jammed and jammed construction are shown in Figs. 16 and 17, respectively. In jammed preform, the x-, y-, and z-yarns are at their maximum thread density (maximum set) or minimum x- and y-yarn spacing. The structure parameters are labelled per the nomenclature.

From the geometry of Figs. 15, 16 and 17, the preform thickness and volume can be calculated from:

$$L_z = n_x t_x + n_y t_y + 2t_z \tag{3}$$

$$V_{repeat} = L_x L_y L_z \tag{4}$$

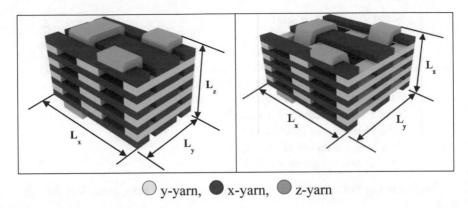

○ y-yarn, ● x-yarn, ● z-yarn

Fig. 15 Delete3D orthogonal woven preforms of jammed (left) and non-jammed of one repeat of plain weave

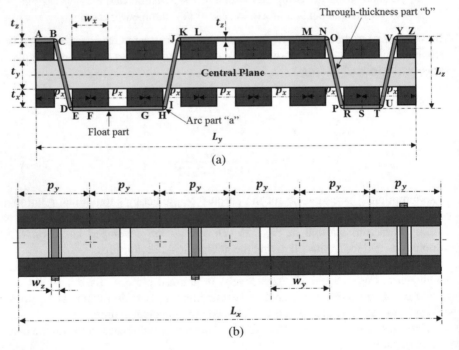

Fig. 16 One repeat of generalized non-jammed structure. **a** x-yarn cross section, **b** y-yarn cross section (x-yarn in red, y-yarn in yellow, and z-yarn in green)

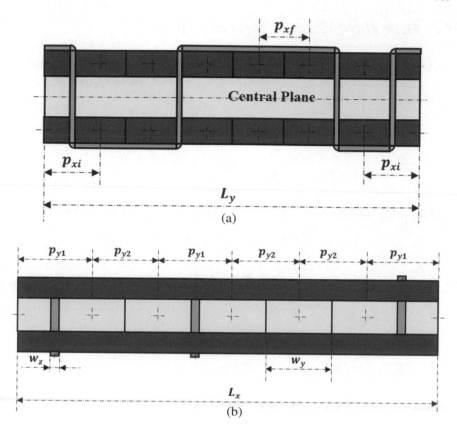

Fig. 17 One repeat of generalized jammed structure **a** x-yarn cross section, **b** y-yarn cross section

6.1 Generalized Model for Non-Jammed Preforms

From geometry of Fig. 2, preform repeat length and width for generalized non-jammed weave structure are given by:

$$L_y = N_x p_x \tag{5}$$

$$L_x = N_y p_y \tag{6}$$

6.2 Generalized Model for Jammed Preforms

From the geometry of Fig. 3, there are two different x-yarn spacing: the spacing at weave interlacing ($p_{xi} = w_x + t_z$) and the spacing under the float ($p_{xf} = w_x$).

Thus, the repeat length and average x-yarn spacing are:

$$L_y = N_x \overline{p}_x \tag{7}$$

$$\overline{p}_x = \frac{M_z w_x + t_z}{M_z} \tag{8}$$

Similarly, preform repeat width and average y-yarn spacing can be written as:

$$L_x = N_y \overline{p}_y \tag{9}$$

$$\overline{p}_y = \frac{N_y w_y + N_z w_z}{N_y} \tag{10}$$

6.3 FVF of x-, y-Yarn, and z-Yarns

Total x-yarn and y-yarn length per preform repeat are:

$$\ell_x = L_x N_x n_x \tag{11}$$

$$\ell_y = L_y N_y n_y \tag{12}$$

Fiber volume of a yarn is given by:

$$V_{fk} = \frac{\ell_k \rho_{\ell k}}{\rho_{vk} 10^5} \tag{13}$$

Fiber volume fraction of a yarn is the ratio of the fiber volume of the yarn/repeat to total volume per repeat, then

$$F_{fk} = \frac{V_{fk}}{V_{repeat}} 100 = \frac{\frac{\ell_k \rho_{\ell k}}{\rho_{vk} 10^3}}{L_x L_y L_z} \tag{14}$$

From Eqs. (3), (5), (6), (11) and (14), x-yarn FVF equation for non-jammed is:

$$F_{fx} = \frac{\frac{n_x \rho_{\ell k}}{\rho_{vx} 10^3}}{p_x (n_x t_x + n_y t_y + 2t_z)} \tag{15}$$

From Eqs. (3), (7)–(10), (11) and (14), x-yarn FVF equation for jammed structures is:

$$F_{fx} = \frac{\frac{n_x \rho_{\ell k}}{\rho_{vx} 10^3}}{(\frac{t_z}{M_z} + w_x)(n_x t_x + n_y t_y + 2t_z)} \tag{16}$$

From Eq. (3), (5), (6), (12) and (14), y-yarn FVF equation for non-jammed is obtained as:

$$F_{fy} = \frac{\frac{n_y \rho_{\ell k}}{\rho_{vy} 10^3}}{p_y (n_x t_x + n_y t_y + 2t_z)} \tag{17}$$

From Eqs. (3), (7)–(10), (12) and (14), y-yarn FVF equation for jammed structures is:

$$F_{fy} = \frac{\frac{N_y n_y \rho_{\ell y}}{\rho_{vy} 10^3}}{(N_y w_y + N_Z w_z)(n_x t_x + n_y t_y + 2t_z)} \tag{18}$$

Z-yarn consists of horizontal straight parts on top or bottom surface of the preform, x-yarn width halves, arcs at crossovers points with x-yarns, and through-thickness tilted (in non-jammed structure) or through-thickness vertical (in jammed structure) parts.

Hence, total length of z-yarn per repeat for non-jammed and jammed structures is:

$$\ell_z = \left[2ai_z + bi_z + w_x i_z + (N_x - i_z)p_x\right]N_z \tag{19}$$

$$\ell_z = \left[2ai_z + bi_z + w_x i_z + (N_x - i_z)w_x\right]N_z \tag{20}$$

Calculation of the arc length (a) and tilted length (b) for non-jammed structures requires calculation of the angle (θ). It can be shown that:

$$\theta = \tan^{-1}\left(\frac{-2(p_x - w_x)(n_x t_x + n_y t_y) - \sqrt{(2(p_x - w_x)(n_x t_x + n_y t_y))^2 - 4(t_z^2 - (p_x - w_x)^2)(t_z^2 - (n_x t_x + n_y t_y)^2)}}{2(t_z^2 - (p_x - w_x)^2)}\right) \tag{21}$$

Then (a) and tilted length (b) are:

$$a = \left(\frac{t_z}{2}\right)\theta \tag{22}$$

$$b = \frac{n_x t_x + n_y t_y + t_z \cos\theta}{\sin\theta} \, or \, \frac{p_x - w_x - t_z \sin\theta}{\cos\theta} \tag{23}$$

Detail of derivation of the angle (θ) is shown elsewhere [3].

For jammed structures the vertical length (b) is calculated from:

$$b = n_x t_x + n_y t_y \tag{24}$$

Therefore, the total z-yarn length (ℓ_z) can be calculated using Eq. (19) or (20). Similar to x- and y-yarn FVF, FVF of z-yarn in weave repeat (F_{fz}) is the ratio of the fiber volume of z-yarn (V_{fz}) in weave repeat to the total weave repeat volume (V_{repeat}). From Eqs. (3), (5), (6), (19) and (14) results in z-yarn FVF equation for non-jammed structures is obtained as:

$$F_{fz} = \frac{\frac{\left[\frac{2a}{M_z} + \frac{b}{M_z} + \frac{w_x}{M_z} + \left(1 - \frac{1}{M_z}\right)p_x\right] N_z \rho_{\ell k}}{\rho_{vk} 10^3}}{N_y p_y p_x (n_x t_x + n_y t_y + 2t_z)} \tag{25}$$

From Eqs. (3), (7)–(10), (20) and (14), z-yarn FVF for jammed structures is obtained as:

$$F_{fz} = \frac{\frac{[2a + b + w_x + (M_z - 1)w_x] N_z \rho_{\ell k}}{\rho_{vk} 10^3}}{(N_y w_y + N_z w_z)(M_z w_x + t_z)(n_x t_x + n_y t_y + 2t_z)} \tag{26}$$

Thus, the total fiber volume fraction is given by:

$$F_{fxyz} = F_{fx} + F_{fy} + F_{fz} \tag{27}$$

From Eqs. (15), (17), (25) and (27), the total FVF for non-jammed structures is:

$$F_{fxyz} = \frac{\frac{n_x \rho_{\ell k}}{\rho_{vx} 10^3}}{p_x (n_x t_x + n_y t_y + 2t_z)} + \frac{\frac{n_y \rho_{\ell k}}{\rho_{vy} 10^3}}{p_y (n_x t_x + n_y t_y + 2t_z)}$$
$$+ \frac{\frac{\left[\frac{2a}{M_z} + \frac{b}{M_z} + \frac{w_x}{M_z} + \left(1 - \frac{1}{M_z}\right)p_x\right] N_z \rho_{\ell k}}{\rho_{vk} 10^3}}{N_y p_y p_x (n_x t_x + n_y t_y + 2t_z)} \tag{28}$$

From Eqs. (16), (18), (26) and (27), the total FVF for jammed preforms is given by:

$$F_{fxyz} = \frac{\frac{n_x \rho_{\ell k}}{\rho_{vx} 10^3}}{\left(\frac{t_z}{M_z} + w_x\right)(n_x t_x + n_y t_y + 2t_z)} + \frac{\frac{N_y n_y \rho_{\ell y}}{\rho_{vy} 10^3}}{(N_y w_y + N_z w_z)(n_x t_x + n_y t_y + 2t_z)}$$

$$+ \frac{\frac{[2a+b+w_x+(M_z-1)w_x]N_z\rho_{\ell k}}{\rho_{vk}10^3}}{(N_yw_y + N_Zw_z)(M_zw_x + t_z)(n_xt_x + n_yt_y + 2t_z)} \qquad (29)$$

6.4 Preform Areal Density

Areal density of a constituent yarn is given by:

$$W_k = \frac{\ell_k\rho_{\ell k}}{10L_xL_y} \qquad (30)$$

Areal density of the preform is given by:

$$W_{xyz} = \frac{[\ell_x\rho_{\ell x} + \ell_y\rho_{\ell y} + \ell_z\rho_{\ell z}]}{10L_xL_y} \qquad (31)$$

From Eqs. (5), (6), (11), (12), (19) and (31) areal density of non-jammed preforms is obtained as:

$$W_{xyz} = \frac{\left[N_yp_yn_x\rho_{\ell x} + p_xN_yn_y\rho_{\ell y} + \left[\frac{2a}{M_z} + \frac{b}{M_z} + \frac{w_x}{M_z} + \left(1 - \frac{1}{M_z}\right)p_x\right]N_z\rho_{\ell z}\right]}{10N_yp_yp_x} \qquad (32)$$

From Eqs. (7)–(12), (20) and (31) areal density of jammed preform is obtained as:

$$W_{xyz} = \frac{(N_yw_y + N_Zw_z)M_zn_x\rho_{\ell x} + (M_zw_x + t_z)N_yn_y\rho_{\ell y} + [2a + b + w_x + (M_z - 1)w_x]N_z\rho_{\ell z}}{10(N_yw_y + N_Zw_z)(M_zw_x + t_z)} \qquad (33)$$

6.5 Numerical Results

The value of the generalized geometrical model is demonstrated in this section. The effect of z-yarn interlacing pattern (weave factor), and z-yarn linear density on preform thickness, areal density, constituent yarns' and total FVF are presented graphically. Range of weave factor are considered (which has not been previously explored) to illustrate the significance of the weave factor in determining the constituents' and total FVFs and areal density of preforms that in turn determine the in-plane and out-of-plane performance of the resultant composites. Table 1 shows the weaves selected for the numerical results. Three different z-yarn linear density

Table 1 Weaves and weave factors used for the numerical results

Weave	N_x	i_z	$M_z=N_x/i_z$
Plain or any filling rib weave	2	2	1.00
2 × 1 warp rib	3	2	1.50
3 × 2, 2 × 1, 1 × 1 twill	10	6	1.67
2 × 1, 2 × 2 warp rib	7	4	1.75
3 × 2, 1 × 2, 2 × 1 basket	11	6	1.83
2 × 2 warp rib	4	2	2.00
3 × 3, 2 × 1 warp rib	9	4	2.25
3 × 2 warp rib	5	2	2.50
3 × 3 warp rib	6	2	3.00
4 × 3 warp rib	7	2	3.50
4 × 4 warp rib	8	2	4.00
10-H. sateen	10	2	5.00

levels (500, 1,000, and 2,000 g/km) were considered to reveal its influence on the constituents and total FVFs.

Numerical Results for Non-jammed Preforms

Table 2 list are the input parameters that were kept unaltered for non-jammed preforms.

Figure 18 shows the effect of weave factor and z-yarn linear density on x-, y-, z- and total FVFs of preform. The data of the graphs in the figure were generated from Eqs. (15), (17), (25), and (28). Increase in z-yarn linear density caused decrease in x- and y-yarn FVFs and increased z-yarn and total FVFs. The increase in z-yarn and total FVFs as a result of z-yarn linear density increase is greater for low weave factors due to more through-thickness component of z-yarn in low weave factor structures. Weave factor has no effect on x- and y-yarn FVFs. Increase in weave factor caused a decrease in z-yarn and total FVFs for a given z-yarn linear density due to decrease in the number of through-thickness part of z-yarn and longer floats on the surfaces of preforms.

The effects of weave factor and z-yarn linear density on preform thickness and areal density are shown Fig. 19. The data of the figure were generated from Eqs. (3) and (32). Weave factor has no influence on preform thickness. However, increasing in z-yarn linear density caused an increase in preform thickness. Increase in z-yarn linear density caused increase in preform areal density and the increase is more pronounced for low weave factor due to increase in the number of through-thickness

Table 2 Constant parameters used for non-jammed structures

ρ_{vk}	$\rho_{\ell x} = \rho_{\ell y}$	Φ_k	AR_k	n_y	y-yarn/layer: z-yarn	p_x	p_y
2.5 g/cm³	2,000 g/km	0.75	2	7	1:1	0.46 cm	0.46 cm

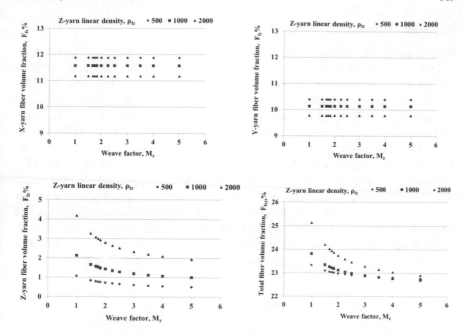

Fig. 18 Effect of weave factor and z-yarn linear density on constituents and total FVFs for non-jammed structure

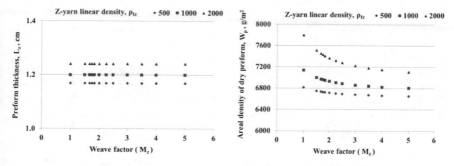

Fig. 19 Effect of weave factor and z-yarn linear density on preform thickness and areal density

components of z-yarn. Increase in weave factor decreased areal density of the preform as a result of decrease in the number of through-thickness components of z-yarn.

It is worth noting from Figs. 18 and 19 that increase in the weave factor from 1 to 3 caused pronounced decrease in z-yarn FVF, total FVF, and areal density of the preform. Further increase in weave factor caused slight decrease in z-yarn FVF, total FVF, and areal density of the preform. These properties tend to level off with weave factor values higher than 4 for all levels of z-yarn linear densities.

Numerical Results for Jammed Structures

Table 3 shows the input parameters that were kept constant for jammed preforms.

Figure 20 depicts the constituent yarns' and total FVFs as a function of weave and z-yarn linear density for jammed preforms using Eqs. (16), (18), (26), and (29). Increase in z-yarn linear density caused reduction in x- and y-yarn FVFs and increase in z-yarn FVF. Reduction in x- and y-yarn FVFs is greater than the increase in z-yarn FVF as a result of z-yarn linear density increase and as a result the total FVF reduced with increase in z-yarn linear density. The increase in weave factor increased x-yarn FVF due to increase in number of x-yarns under weave floats due to reduction in the average x-yarn spacing. The weave factor has no effect on y-yarn FVF since there is no interlacing between z-yarn and y-yarn. As weave factor gets higher, the z-yarn FVF gets lower due to decrease in the number of through-thickness components of z-yarn. Unlike the non-jammed preforms case, weave factor caused slightly increase of the total FVF for jammed preforms due to increase in x-yarn FVF with weave factor that compensated for the reduction of z-yarn FVF with weave factor.

Table 3 Constant parameters used for jammed structures

ρ_{vk}	$\rho_{\ell x} = \rho_{\ell y}$	Φ_k	AR_k	n_y	y-yarn/layer: z-yarn
2.5 g/cm^3	2000 tex	0.75	2	7	1:1

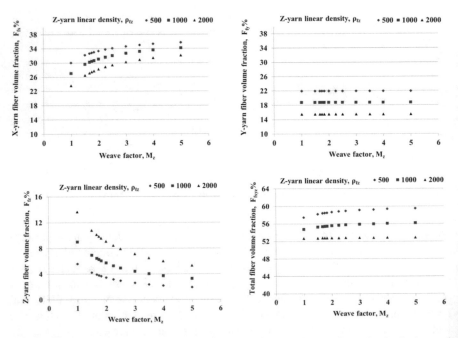

Fig. 20 Constituents' and total FVFs in terms of weave factor and z-yarn linear density for jammed Preforms

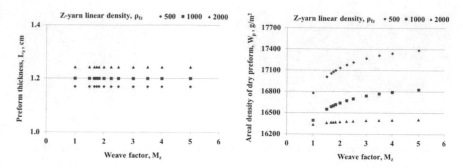

Fig. 21 Preform thickness and areal density in terms of weave factor and z-yarn linear density for jammed structures

Figure 21 shows preform thickness and areal density as a function of weave and z-yarn linear density for jammed preforms using Eqs. (3) and (33). Areal density results for jammed structures are opposite to non-jammed structures. For a given weave factor, z-yarn linear density reduces preform areal density and this reduction is more significant for high weave factors. In contrast to non-jammed preforms, weave factor increases preform areal density of the jammed preforms.

7 Experimental Verification of the Geometric Modelling of 3D Orthogonal Preforms

To verify the model, experimental measurements of geometric parameters of 3D orthogonal woven preforms and their composites from flat glass yarns were conducted. Details of the experimental steps measurements of the geometric parameters are described elsewhere [3]. The parameters measured include: (1) preform thickness, (2) x-, y-, and z-yarns linear density, (3) x-, y-, z-fibers density, (4) preform areal density, (5) mass/m^2 of preform constituents (x-, y-, and z-yarns), (6) total and constituents fiber volume fraction, (7) composite thickness, (8) x-, y-, and z-yarns' width, thickness and aspect ratio, (9) x-, y-, and z-yarns' packing factors, and (10) angle of through the thickness z-yarn component. The parameters (7)–(10) were measured from micrographs (Figs. 22 and 23) of x- and y-yarn cross sections.

The experimental measurements and model prediction of composites and preform geometric parameters of two y-yarn layers (three x-yarns layers) in plain weave are reported in Tables 4, 5, 6 and 7. With the exception of z-yarn angle θ (maximum angle of the z-yarn axis to plane of preform) and fiber volume fraction, the agreement between measured and predicted values of geometrical parameters is fairly excellent. The generalized model can be used to design perform and their composites to meet performance requirements for specific applications.

The x-, y-, and z-yarns' width, thickness and aspect ratio, and x-, y-, and z-yarns' packing factors are functions of preform parameters (yarn size, yarn spacing,

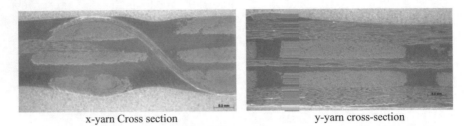

| x-yarn Cross section | y-yarn cross-section |

Fig. 22 Micrographs of x- and y-yarn cross sections of 2 layers plain weave composite [3]

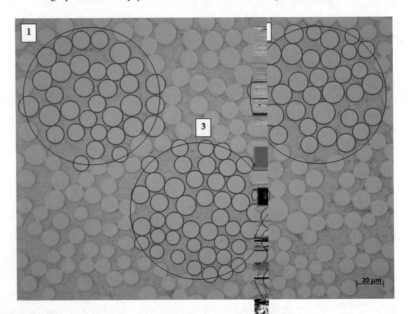

Fig. 23 Enlarged cross section of x-yarn with selected three regions to determine average yarn packing factor [3]

Table 4 Experimental and predicted thickness

	Thickness (mm)
Model	2.44
Dry preform	2.48
Composite	2.62

Table 5 Experimental and predicted constituent yarn mass and preform areal density

	X-yarn mass g/m²	Y-yarn mass g/m²	Z-yarn mass g/m²	Total mass g/m²
Model	1,085	1,078	7	2252.26
Dry preform	1,142 (1.4%)	1,082		2295.62

Table 6 Experimental and predicted total and constituents yarn fiber volume fraction

	X-yarn FVF (%)	Y-yarn FVF (%)	Z-yarn FVF (%)	Total FVF (%)
Model	17.80	17.08	1.46	36.34
Dry preform	18.45	16.86	1.16	36.47
Composite	17.44	15.94	1.09	34.47

Table 7 Experimental and predicted z-yarn angle

	Z-yarn angle, degree
Model	63
Composite	42

number of layers and weave pattern) as its compactness as well as resin infusion processing parameters (vacuum pressure, resin type, resin application method, etc.). With these numerous parameters, it is extremely difficult to predict yarns' aspect ratio and packing factor before composite formation that is the reason behind measuring these parameters from the micrographs of Figs. 22 and 23. To take advantage of the benefits of modelling, further investigations are needed to establish data base of yarns' packing factors in terms of the parameters listed earlier. This work revealed the assumption that z-yarn is completely flexible is not realistic as it can be seen from x-yarn composite cross section of Fig. 22 that affected the deviation between the measured z-yarn angle θ and fiber volume fraction. This assumption may be valid for maximum compaction structures. For non-jammed structures, the z-yarn bending rigidity need to be considered, especially where z-yarn fiber volume fraction is high.

As it can be seen from the numerical results and experimental verification, modelling reveals the potential of fabric architecture parameters of 3D orthogonal woven preforms. For example, the numerical results indicate how to obtain desired x-, y- and z- fiber volume fraction as functions of weave factor and z-yarn linear density. Additionally, for given yarns and weave, the model predicts minimum yarn spacing to achieve maximum construction and hence maximum constituents' and total fiber volume fractions. It is well known fact that fiber volume fraction is a key parameter that decides the performance of final composite. The higher the fiber volume fraction, the higher is the composite performance. Therefore, the model is a powerful tool for designing preforms with predetermined architecture parameters, which influences in-plane and out-of-plane mechanical properties.

8 Modelling Performance of Composite from 3D Orthogonal Woven Preforms

The generalized geometrical models covered in Sect. 6 can be used to develop models to predict mechanical properties of composites from 3D orthogonal preforms manufactured from spun and flat continuous filament yarns since these properties are highly related to the fiber volume fraction, areal density, and amount of through-thickness z-yarn segments.

Midani [20] reported an extensive critical review of models that were developed to predict mechanical performance of composites from 3D orthogonal woven preforms. It was pointed out that some models were mostly based on classical laminate theory and/or 2D woven composites predictive models while others followed computational techniques using finite element analysis [20, 21]. Most researchers, however, focused on modeling the linear elastic region of composite from 3D orthogonal woven preforms with jammed plain woven preforms. Additionally, all previous researchers considered composites from one type of fibers for all x-, y-, and z-yarns and did not address modeling composites from preforms formed from hybrid of different fibers, linear densities, yarn spacing, different yarn cross-section shape, etc. These limitations directed conducting extensive research to develop a generalized models to predict the entire load–extension behavior of composites from 3D orthogonal preforms made from flat filament or spun interwoven with any weave structure and made from any fiber type, including hybrid using finite deformation technique [20, 21]. The finite deformation method was proven useful in predicting the entire uniaxial and bi-axial load-extension of single layer plain and 2 × 2 twill woven fabrics [22–24]. Besides modeling the fabric geometry, this method requires measuring the load-extension of the constituent yarns as input to the model, which provided realistic prediction of the entire load-extension of woven fabrics and not limited to the elastic portion only. To expand the application of the finite deformation technique [25] developed a generalized simple model to predict the entire uniaxial and bi-axial load–extension relationships of any single layer woven fabric from any yarns interlaced to form any weave design. The model also considered woven fabrics from hybrid warp and/or filling yarns and variable thread spacing.

Given this background, Midani et al. [20, 21] conducted research with the objective to develop a generalized simple model to predict the entire load–extension properties of composites from 3D orthogonal woven preforms made of spun or flat yarns, using the finite deformation method utilized in [22–25] and the geometric relationships derived in Sect. 6. Additionally, the tensile properties of the constituent yarns (or fibers) are input parameters to the model. The model predicted the load–extension properties of any composites from 3D orthogonal woven non-jammed and jammed preforms interlaced with any weave architecture. The model also considered preforms of hybrid constituents. The model combined the assumptions and methods followed in [3, 26] [19, 22–25] and Sect. 6. It should be pointed out that the models derived in [20, 22–24] are valid for single layer woven fabric (preforms). The composite load-extension model derived in [20, 21] deals with preforms from x-, y-, z-yarns integrated into 3D structures in a matrix. To deal with such composite structures the rules of blend, which requires the load-extension of resin and x-, y-, z-yarns (or fibers), was followed.

The model was verified experimentally for wide range of composite structures from 3D orthogonal preforms treated with vinylester resin using vacuum assisted resin transfer molding technique, including hybrid structures. It was found that there was a good agreement between the predicted and measured load-extension of composites (Fig. 24). The agreement is closer using the fibers' load-extension (tested at 2.5 cm gauge length) than yarns' load-extension (tested at 25 cm gauge length) due to the resin/fiber interaction. When the preforms are treated with resin to form composites, the yarn inside the composite would reflect fiber properties due much lower effective gauge length due to fiber adhesion with the resin matrix. Inspection of tested samples showed that the failure is due to fiber breakage with no pullout of fibers/yarns, indication of good adhesion between fibers and matrix that verified why the model agrees with fiber properties. The preforms of Fig. 24 were manufactured in three y-yarn layers (four x-yarn layers) using E-glass fibers in y- and z-directions. The x-yarns used were Spectra, Vectran, and Zylon as indicated in the figure.

Petroleum-based synthetic fibers and resins have been dominating the fiber reinforced composite markets for long time due to their high performance that made them established in numerous applications mentioned earlier. Let alone the negative impact of petrol on the environment, petroleum-based synthetic fibers and resins require decades to biodegrade and hence require massive landfill that is becoming a premium. With increasing demand to protect the environment, researchers directed their effort toward conducting research on composite from natural fibers and resins. Extensive coverage of research conducted on composites from natural fibers and resins along with their current and potential applications of composites given in [4, 27]. Unlike synthetic fibers, natural fibers are inherently variable in length, thickness, and mechanical properties. This poses significant challenge to develop composite parts for high performance applications. Significant research and development work is required to account for the natural fiber variability in order to compete and replace synthetic fibers.

Recent research published in [4, 27] focused on experimental and modeling research on composites from 3D orthogonal woven preforms from hemp and flax fibers. There are varieties of hemp and flax fibers with performance similar or better than glass fiber, which is widely used in applications such as automotive, boats, and wind blades. While glass fiber is made from natural materials and inexpensive, it is not biodegradable and possesses high volume density. Reference [27] developed a range of composites from 3D orthogonal preforms from hemp fibers with objective to study their performance in terms of architecture parameters in order to reveal their potential applications and current and future market size. Reference [4] carried out similar research with focus on developing range of composites from 3D orthogonal preforms from flax fiber. Additionally, using the generalized model developed in [21] for predicting load-extension of composites from spun yarns, she verified the model experimentally for composites from flax and hemp. The experimental data of

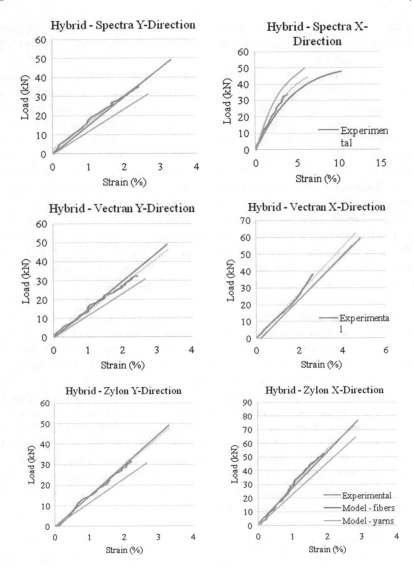

Fig. 24 Model verification of tensile load-extension of hybrid composites from 3D orthogonal preforms of plain weaves in x- and y-axis directions

composites from hemp to verify the model was taken from [27]. The experimental verifications considered the inherent variability found in flax and hemp fibers and yarns. Figure 25 shows the significant variability of flax yarn and fiber load-extension strength.

The variability in yarn load-elongation is much lower than that of fiber. The x-, y-, and z-yarns and resin load-elongation relationships were used as input to the model to predict the composite load-elongation. Example of verifying the model is

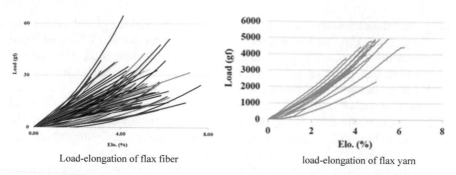

Load-elongation of flax fiber

load-elongation of flax yarn

Fig. 25 Variability of load-elongation of flax fiber and yarn [4]

shown in Fig. 26. To account for variability, load-elongation regression equation was derived from the experimental data along the upper and lower limit of the regression equation. The figure shows the measured and predicted composite from the regression equation along with the upper and lower limits. The experimental (blue curve) is within the upper and lower limit of the predicted load-elongation, which indicates a good agreement between the measured and the modeled data.

Outlook

It has been shown that composites from 3D woven preforms possess higher strength to weight ratio compared to those from 2D stacked woven preforms or unidirectional laminates. This property qualified 3D composites for numerous applications mentioned earlier throughout the chapter. Recently, several market studies have been published such as "Composite Market Size & Forecast Report, 2013–2024" released in July 2018 by Grand View Research [https://www.grandviewresearch.com/ind

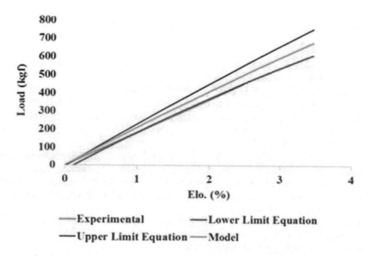

Fig. 26 Predicted and experimental load-elongation of composite from flax fiber [4]

ustry-analysis/composites-market]. The report indicates that the global market size estimated at USD 76,710.5 million in 2017. It is expected to reach 7.7% compound annual growth rate (CAGR) going to 2024. The study include market by end use (automotive, transportation, aerospace, defence, wind energy, electrical electronics, construction, infrastructure, pipes, tanks, marine, and others), by fiber, by resin, by solidification method, and by region. This forecast is based on the increasing demand for harvesting wind energy and lightweight composites for automotive industry. The motivation behind such demand is to produce clean energy and fuel-efficient vehicles to reduce the negative impact of petrol and diesel on the environment.

The market growth requires research and development on such unique materials as more developments take place in improving current fibers and resins, new high performance fibers, improving the uniformity of high performance natural fibers, and the use of hybrid preforms to produce composites parts with performance required for specific end uses. A parallel research in modelling performance of 3D composites is also needed to reduce/eliminate the trials and drive the cost down. Additionally, manufacturing technologies to manufacture net-shape or near net-shape preforms to reduce waste and cost are sorely needed. One such technology is 3D printed fiber reinforced composites. While this technology at its infant stage it is a fast growing technology and it is expected to capture numerous end uses in the field. A critical literature review on 3D printed composites from continuous fibers is published in [28].

This chapter is useful for research students in the majors of mechanical and aerospace engineering, textile engineering and technology, material engineering as well as industry professionals in research and development in these fields.

It is recommended for readers who are interested in structure and formation of 3D fiber-reinforced composites to study listed references and their cited references.

References

1. Bogdanovich, A. E., & Mohamed, M. H. (2009). Three-dimensional reinforcements for composites. *SAMPE Journal, 45*(6), 8–28.
2. Mohamed, M. H., & Zhang, Z. H. (1992). North Carolina State University, Method of forming variable cross-sectional shaped three-dimensional fabrics. U.S. Patent 5,085,252.
3. Ince, M.E. (2013). Performance of composites from 3D orthogonal woven preforms in terms of architecture and sample location during resin infusion. Ph.D. Thesis, North Carolina State University, Raleigh, NC, USA.
4. Eldeeb, H. (2020). Modeling tensile behavior of 3D orthogonal woven composites from high performance natural fibers. Ph.D. Thesis, North Carolina State University, Raleigh, NC, USA.
5. Claunch, E. L. C. (2019). Smart composites: Optic sensors embedded in 3D orthogonal woven fabric for structural health monitoring applications. Ph.D. Thesis, North Carolina State University, Raleigh, NC, USA.
6. Seyam, A. M. (2012). Weaving and weaving preparation at ITMA 2011. *Textile World*, 34–37.
7. Ko, F. K. (2000). 3D textile reinforcements in composite materials. *3-D textile reinforcements in composite materials* (pp. 9–42). CRC Press.
8. Ko, F. K., & Du, G. W. (1997). Processing of textile preforms. In: *Advanced composites manufacturing* (pp. 157–205). Cambridge: John Wiley & Sons, Inc.

9. Brown, D., & Wu, Z. J. (2000). Stiffness properties of 3D orthogonally woven fabric composites. In: *The Proceedings of Advances in Composite Materials & Structures VII; Southampton*, WIT Press, pp. 377–386.
10. Cox, B. N., & Dadkhah, M. S. (1995). The macroscopic elasticity of 3D woven composites. *Journal of Composite Materials, 29*(6), 785–819.
11. Cox, B. N., & Flanagan, G. (1997). *Handbook of analytical methods for textile composites.* Hampton: NASA
12. Gu, P. (1994). Analysis of 3D woven preforms and their composite properties. Ph.D. Thesis, North Carolina State University, Raleigh, NC, USA.
13. Mohamed, M. H., Zhang, Z., & Dickinson, L. (1988). Manufacture of multi-layer woven preforms. In: T.H. Tsiang & R.A. Taylor (Eds.), *The proceedings of the advanced composites and processing technology* (pp. 81–89). Chicago, US.
14. Quinn, J. P., Hill, B. J., & McIlhagger, R. (2001). An integrated design system for the manufacture and analysis of 3D woven preforms. *Composites Part A: Applied Science and Manufacturing, 32*(7), 911–914.
15. Quinn, J., Mcilhagger, R., & McIlhagger, A. T. (2003). A modified system for design and analysis of 3D woven preforms. *Composites Part A: Applied Science and Manufacturing, 34*(6), 503–509.
16. Brown, D., & Wu, Z. J. (2001). Geometric modelling of orthogonal 3D woven textiles. In: *The Proceedings of ICMAC-International Conference for Manufacturing of Advanced Composites*, London, pp. 52–65.
17. Hill, B. J., & McIlhagger, R. (1996). The development and appraisal of 3D interlinked woven structures for textile reinforced composites. *Polymers & Polymer Composites, 4*(8), 535–539.
18. Wu, Z. (2009). Three-dimensional exact modeling of geometric and mechanical properties of woven composites. *Acta Mechanica Solida Sinica, 22*(5), 479–486.
19. Ince, M. E., Seyam, A. M., & Mohamed, M. H. (2014). Generalized Geometric modeling of 3D orthogonal woven preforms: The rectangular cross section case. In: *The Proceeding of the Future Technical Textiles Conference, Istanbul, Turkey.*
20. Midani, M. S. (2016). The influence of weave and structural parameters on the performance of composites from 3D orthogonal woven preforms. Ph.D. Thesis, North Carolina State University, Raleigh, NC, USA.
21. Midani, M., Seyam, A. F., & Pankow, M. (2018). A generalized analytical model for predicting the tensile behavior of 3D orthogonal woven composites using finite deformation approach. *The Journal of the Textile Institute, 109*(11), 1465–1476.
22. Kawabata, S., Niwa, M., & Kawai, H. (1973). The finite-deformation theory of plain-weave fabrics part I: The biaxial-deformation theory. *Journal of the Textile Institute, 64*(1), 21–46.
23. Kawabata, S., Niwa, M., & Kawai, H. (1973). The finite-deformation theory of plain-weave fabrics. Part II: The uniaxial-deformation theory. *Journal of the Textile Institute, 64*(2), 47–61.
24. Kawabata, S., & Niwa, M. (1979). A finite-deformation theory of the 2/2-twill weave under biaxial extension. *The Journal of the Textile Institute, 70*(10), 417–426.
25. Sun, F., Seyam, A. M., & Gupta, B. S. (1997). A generalized model for predicting load-extension properties of woven fabrics. *Textile Research Journal, 67*(12), 866–874.
26. Seyam, A. M., & Ince, M. E. (2013). Generalized geometric modeling of three-dimensional orthogonal woven preforms from spun yarns. *Journal of the Textile Institute, 104*(9), 914–928.
27. Gupta, A. (2019). Composites from natural fibers: hemp fiber reinforced composites from 3D orthogonal woven preforms and their potential applications in the US. Ph.D. Thesis, North Carolina State University, Raleigh, NC, USA.
28. Kabir, S. F., Mathur, K., & Seyam, A. F. M. (2019). A critical review on 3D printed continuous fiber-reinforced composites: History, mechanism, materials and properties. *Composite Structures, 232*, 111476.

Definition and Design of 3D Warp Interlock Fabric

Francois Boussu

Abstract After introducing and explaining the main interest to use 3D warp inter-lock fabrics, a step-by-step methodology will be proposed to design the elementary woven pattern by using a cross-section view with the drawing of the different warp yarns path inside the architecture. Thus, different types of product parameters will be introduced to obtain the final description of the 3D warp interlock fabric. At the end, a general definition of the 3D warp interlock fabric will be given. Different types of 3D warp interlock fabric architectures will be used to illustrate the designing method-ology. Then, to transfer the design onto the weaving loom, a dedicated technical datasheet of the 3D warp interlock fabric will be defined. The different process and product parameters will be integrated and the methodology to define the peg plan with the associated drawing-in plan will be exposed with several examples. Different exercises will be introduced in this chapter with responses provided in the Appendix.

1 Introduction

In the recent years, many researches have highlighted the several advantages to use 3D warp interlock fabrics for many applications. Technical and economic advantages, compared to other textile reinforcements, have been proven in different composite material applications [1–8]. However, this kind of fabric have also some drawbacks due to their specific architecture and production mode [9–12].

According to these research results, also done in our research works [13–18], we have been able to identify the pros and cons of their production process and their architectures which aims to wide diversity of warp and weft yarns arrangements both in plane and through the thickness of the 3D woven structure [19].

Detailed experimental observations on composite material made with 3D fabrics have determined the influence of the textile architecture on the mechanical proper-ties and the related mechanisms of rupture [20, 21]. These structures, due to their

F. Boussu (✉)
ENSAIT, ULR 2461 - GEMTEX - Génie et Matériaux Textiles, University Lille, 59000 Lille, France
e-mail: francois.boussu@ensait.fr

© Springer Nature Switzerland AG 2022
Y. Kyosev and F. Boussu (eds.), *Advanced Weaving Technology*,
https://doi.org/10.1007/978-3-030-91515-5_11

401

specific consolidation type through the thickness, reveal some interesting mechanical properties [22]. 3D warp interlock fabrics improve the resistance to delamination [8, 20, 21, 23, 24] and consequently the impact resistance [25–29]. Thus, the impacted damage zone of 3D warp interlock fabrics tend to be reduced and then seem to be more effective for multi-impact hits [30–32].

However, we have also related in our research works [14, 17] the yarns damage during the weaving process [33] mainly due to yarn-to-yarn abrasion [34] but also by contact with other parts of the weaving loom [35, 10]. This damage effect has been measured and calculated with a loss coefficient [36] all along the production process [37] which have help us to identify the main causes to this yarn degradation [38]. By a result, our weaving loom and more widely our production process have been modified and adapted to minimize the yarns degradation during the weaving stage [39].

Detailed review of advantages and drawbacks of 3D warp interlock fabrics with respect to their mechanical properties have been done in [40] and the different types of composite applications in [41].

However, among all the research works done on 3D warp interlock fabrics, we need to precisely recognize what type of architecture has been used and then point out which process and product parameters of the 3D woven structure are responsible in the resulting effect. Thus, we have provided a general definition of 3D warp interlock fabrics in order to better describe and model these 3D woven architectures.

That's the reason why this chapter will be decomposed into several parts to firstly provide the basic notions and description of the constituents of the 3D warp interlock fabric. Secondly, a general definition will be given using a step-by-step approach to include all the product parameters. Thirdly, based on this general definition, a methodology will be given to precise the peg plan of the elementary woven pattern of the 3D warp interlock fabric. Several exercises for practice will be introduced in the dedicated appendix section of this book.

2   Description of the 3D Warp Interlock Fabric Constituents

From a general point of view, the weaving process creates surfaces, which are obtained by interlacing parallel yarns placed in the direction of the fabric length (warp—axis X) and perpendicular yarns placed in the direction of the fabric width (weft—axis y) [42, 43]. The general definition of a 2D fabric as given in the standard AFNOR NF G 00-001 is: "A woven fabric is a material produced by perpendicular interlace of two sets of yarns. This interlace is achieved during the weaving process on a loom [44–46].

A 3D warp interlock fabric is composed of several layers of 2D fabrics linked through the thickness by a binding warp yarn as represented in Fig. 1.

Several yarns arrangements can be achieved and allow to obtain a wide variety of 3D structures [23]. Contrary to the 2D fabrics that can only vary their woven pattern, 3D warp interlock fabrics reveal more modular options allowing a larger number

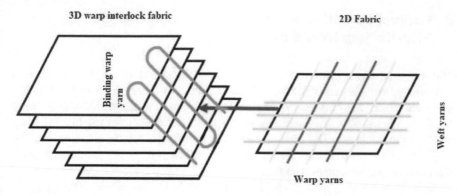

Fig. 1 Geometric view of 3D multi-layers woven structure linked through the thickness by a binding warp yarn

of woven patterns [18]. The schematic representation of a 3D warp interlock fabric (Fig. 2) reveals the main role of each type of yarns inside the woven architecture [47].

1. Surface warp yarns are located on the top and bottom of the woven structure.
2. Weft yarns or fill yarns are perpendicular to warp yarns and inserted at each shed opening on the weaving loom.
3. Binding warp yarns allow to link through the thickness the different layers of the 3D fabric.
4. Stuffer warp yarns are located between fabric layers.

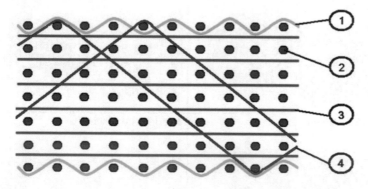

Fig. 2 Schematic representation of a 3D warp interlock fabric by a cross-section view of weft yarns

3   General Definition of a 3D Warp Interlock Fabric_
Step-By-Step Design Process

The proposed classification of Hu [21], related by Ansar et al. [48], involves 4 different types of clusters of 3D warp interlock fabric as represented in Fig. 3. As sum-up by Ha-Minh [16], two geometric parameters are used to distinguish the 3D warp interlock fabrics. The first parameter corresponds to the angle of the binding warp yarns in the thickness. The second parameter corresponds to the depth of binding of these warp yarns in the thickness. By combination of these two parameters, each with two values, we obtain then four clusters of 3D warp interlock fabrics.

This approach of classification [19, 49] allows to include all the binding types; which helps to provide the basis of a general definition of all the 3D warp interlock fabrics.

3.1   General Definition of a 3D Warp Interlock Fabric

Based on the previous research works of Nauman [14] and Cristian [50], several additional parameters introduced in this paper allow to improve the general definition of 3D warp interlock fabric. The synthetic definition of a 3D warp interlock fabric is given in the Eq. (1).

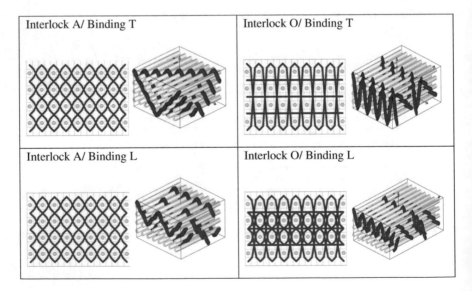

Fig. 3 The 4 clusters of 3D warp interlock fabrics of Hu [21]

$$3D \text{ warp interlock } X1 - X2 \{N\} Y1_k - Y2_k \text{Binding } Wb_k\{B_ki\}$$
$$- \text{Surface Ws } \{Ci\} - \text{Stuffer } \{Si\} \tag{1}$$

where:

- *3D warp interlock* term corresponds to the definition of a 3D warp interlock fabric
- X1 represents the type of angle of binding warp yarn with two values: O (orthogonal) or A (angle)
- X2 corresponds to the type of depth of the binding warp yarn with two values: L (layer to layer) or T (through the thickness)
- $\{N\}$ corresponds to the repetitive sequence of the different number of weft layers for each column of the 3D warp interlock fabric elementary pattern.
- $Y1_k$ is equal to the path of the binding warp yarn of group k
- $Y2_k$ is equal to the depth of the binding warp yarn of group k
- *Binding* term corresponds to the binding warp yarns description
- Wb_k is related to the type of weave diagram visible on the top fabric surface of binding warp yarns of group k
- B_k i contains the numbering of binding warp yarns of group k with inter-ply i position
- *Surface* term corresponds to surface weave warp yarns description
- Ws is related to the type of weave diagram of surface weave warp yarns
- Ci contains the numbering of surface weave warp yarns with inter-ply i position
- *Stuffer* term corresponds to stuffer warp yarns description
- Si represents the numbering of stuffer warp yarns with inter-ply i position.

If the number of weft yarns is different for each column inside the 3D warp interlock fabric architecture, then the repetitive sequence can be defined as: $\{N\} = \{n_{\text{wft}}(j) - n_{\text{wft}}(k)\}$, where the minus operator helps to distinguish the number of each weft yarns column and with $n_{\text{wft}}(j) \neq n_{\text{wft}}(k)$. For a constant number of weft yarns inside each column of the 3D warp interlock fabric architecture, the sequence number $\{N\}$ corresponds to the positive integer value of weft yarns n_{wft}.

B_k i formulae can be written as: $\{bi - \#\}$, with:

- bi corresponds to the different numbering value of the highest cross links of binding warp and weft yarns located at the inter ply i position of the 3D warp interlock fabric following the Wb_k weave diagram.
- minus operator corresponds to an inter-ply position, where the top inter-ply position 0 starts at the top of the 3D warp interlock fabric and from the left side of the formulae
- # symbol corresponds to a lack of highest cross links between binding warp and weft yarns in the inter ply position

Si formulae can be written as: $\{\# - si\}$, with:

- # symbol corresponds to a lack of stuffer warp yarns in the inter ply position

- Minus operator corresponds to an inter-ply position, where the top inter-ply position 0 starts at the top of the 3D warp interlock fabric and from the left side of the formulae
- si corresponds to the different numbering value of stuffer warp yarns located in the inter ply i position of the 3D warp interlock fabric.

More details on this general definition can be found in [51].

3.2 Step-By-Step Methodology to Define a 3D Warp Interlock Fabric

The methodology to define a 3D warp interlock fabric is based on the several following steps.

Introduction of a Three Axes Referential

The three axes referential is represented in Fig. 4:

- The first axis is located in the direction of warp yarns and provides their numbering as: 1, 2, 3, …, X,
- The second axis is located in the direction of weft yarns and provides their numbering as: a, b, c, …, z,
- The third axis provides the number of layers of the woven structure, starting from the top to the bottom of the 3D warp interlock fabric. Its numbering starts is: 1,

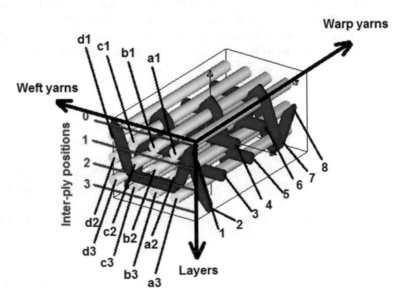

Fig. 4 Introduction of a three axes referential on the 3D warp interlock fabric

2, 3, …, P. Between each layers, an inter-ply can be defined, starting up to the top of the fabric and ending down to the bottom of the fabric. Its numbering is: 0, 1, 2, …, P.

Introduction of the Number of Layers, the Binding Path and Depth of Binding Warp Yarns

The introduction of the following parameters as: the number of layers, the binding path and the binding depth, allows to follow the evolution of the binding warp yarns inside the woven structure [50, 52]:

- The number of layers ($n_{\_wft}$) of a 3D warp interlock fabric represents the number of superposed weft yarns.
- The binding path (x) corresponds to the distance between the two cross-linking points of the binding warp yarn located at the upper inter-ply location. The binding path is supposed to be constant for all the binding warp yarns.
- The binding depth (y) represents the number of necessary layers to link the binding warp yarn inside the woven structure of the 3D warp interlock fabric, which value can be: $y \in [2, …, n_{\_wft}]$.

For instance, the 3D warp interlock fabric represented in Fig. 5 is defined by a number of layers $n_{\_wft} = 3$, a binding path value x = 2 and a binding depth value y = 2.

The 3D warp interlock fabric represented in Fig. 6 is defined by a number of layers $n_{\_wft} = 3$, a binding path value x = 1 and a binding depth value y = 3.

Introduction of Binding Warp Yarns Parameters

As mentioned in the Eq. (1), two different parameters help to describe the binding warp yarns: their woven pattern and their numbering sequence. For instance, let's consider the following 3D warp interlock fabric: O-L 3 3-2 (Fig. 7).

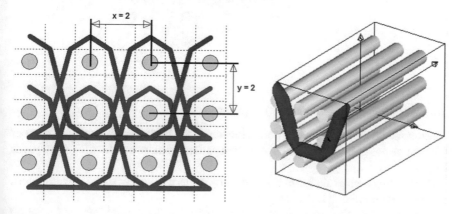

Fig. 5 Representation of 3D warp interlock fabric with 3 layers, binding path at 2 and binding depth at 2 [50]

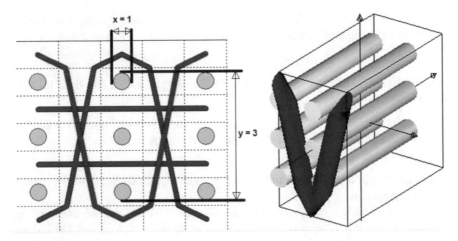

Fig. 6 Representation of 3D warp interlock fabric with 3 layers, binding path at 1 and binding depth at 3 [50]

Fig. 7 Representation of 3D warp interlock fabric: O-L 3 3-2 with numbering of binding warp yarns

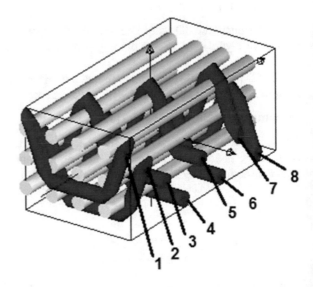

The Woven Pattern of the Binding Warp Yarns

Considering all the cross-linking points located at the first inter-ply position (number 0), it can be possible to recognize the woven pattern of these odd binding warp yarns (Fig. 8).

By isolating each evolution of binding warp yarns, the correponding woven pattern is: Twill 4 weft effect left shift (Fig. 9).

Considering all the cross-linking points located at the second inter-ply position (number 1), it can be possible to recognize the woven pattern of these even binding

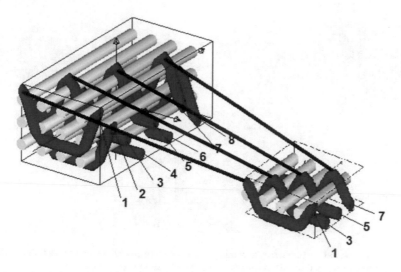

Fig. 8 Recognition of woven pattern of binding warp yarns

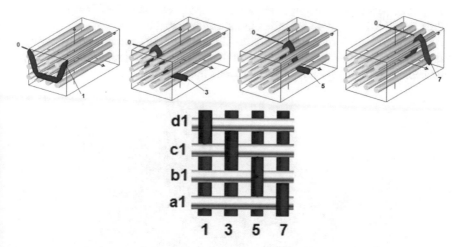

Fig. 9 (Up) Evolutions of each odd binding warp yarns of the 3D warp interlock fabric O-L 3 3-2—(down) corresponding woven pattern of the odd binding warp yarns

warp yarns. By isolating each evolution of binding warp yarns and virutally remove the first layer of weft yarns, the correponding woven pattern is also: Twill 4 weft effect left shift (Fig. 10).

Considering the woven pattern of all the binding warp yarns as: {Twill 4 weft effect left shift}, it can be added to the definition of the 3D warp interlock fabric as: O-L 3 3-2 {Twill 4 weft effect left shift}.

Numbering Sequence of Binding Warp Yarns

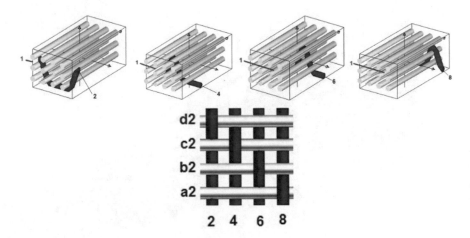

Fig. 10 (Up) Evolutions of each even binding warp yarns of the 3D warp interlock fabric O-L 3 3-2—(down) corresponding woven pattern of the even binding warp yarns

First, we introduce the inter-ply numbering on the representation of 3D warp interlock fabric O-L 3 3-2 {Twill 4 weft effect left shift} in Fig. 11.

By isolating each binding warp yarn (Figs. 9 and 10), it has been possible to record the position of each of their cross-linking points according to the inter-ply numbering and then registered their locations into Table 1.

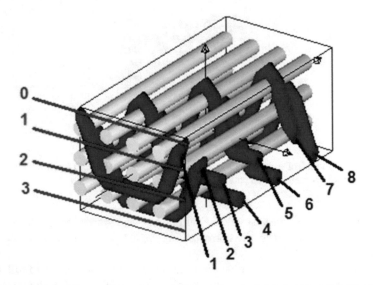

Fig. 11 Numbering of inter-ply locations of 3D warp interlock fabric O-L 3 3-2 {Twill 4 weft effect left shift}

Table 1 Inter-ply position of crosslinking points of each binding warp yarns

Numbering of binding warp yarns	1	2	3	4	5	6	7	8
Location of cross-linking points with their inter-ply numbers	0	1	0	1	0	1	0	1

The numbering sequence of the binding warp yarns of the 3D warp interlock fabric O-L 3 3-2 {Twill 4 weft effect left shift} can then be written as: {1 3 5 7 – 2 4 6 8 – # – #}, where;

- The minus operator corresponds to a layer of the 3D warp interlock fabric,
- The # sign means that at the position of the inter-ply the set of binding warp yarns is empty,
- The odd binding warp yarns numbers: 1 3 5 7 have their cross-linking points located at the inter-ply position 0,
- The even binding warp yarns numbers: 2 4 6 8 have their cross-linking points located at the inter-ply position 1.

Introduction of Surface Warp Yarns Parameters

As mentioned in the Eq. (1), two different parameters help to describe the surface warp yarns: their woven pattern and their numbering sequence. For instance, let considers the following 3D warp interlock fabric: O-L 3 3-2 {Twill 4 weft effect left shift} {1 4 7 10 – 2 5 8 11 – # – #} represented in Fig. 12.

The Woven Pattern of the Surface Warp Yarns

By isolating the surface warp yarns, to check their evolutions inside the 3D warp interlock fabric O-L 3 3-2 {Twill 4 weft effect left shift} {1 4 7 10 – 2 5 8 11 – # – #}, and represented in Fig. 13 (left); the woven pattern of the two pairs of surface warp yarns (3 6 and 9 12) is a plain weave (Fig. 13 (right)) at different inter-ply positions of their crosslinking points.

Numbering Sequence of the Surface Warp Yarns

By isolating each surface warp yarn (Fig. 13), it has been possible to record the position of each of their cross-linking points according to the inter-ply numbering and then registered their locations into Table 2.

The numbering sequence of the surface warp yarns of the 3D warp interlock fabric O-L 3 3-2 {Twill 4 weft effect left shift} {1 4 7 10 – 2 5 8 11 – # – #} can then be written as: {3 6 – # – 9 12 – #}, where;

- The minus operator corresponds to a layer of the 3D warp interlock fabric,
- The # sign means that at the position of the inter-ply the set of surface warp yarns is empty,
- The surface warp yarns numbers: 3 6 have their cross-linking points located at the inter-ply position 0,
- The surface warp yarns numbers: 9 12 have their cross-linking points located at the inter-ply position 2.

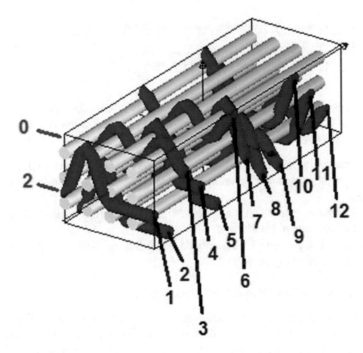

Fig. 12 Numbering of surface warp yarns of the 3D warp interlock fabric O-L 3 3-2 {Twill 4 weft effect left shift} {1 4 7 10 − 2 5 8 11 − # − #}

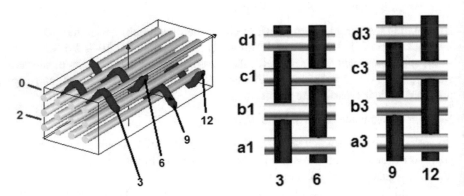

Fig. 13 (Left) Evolution of surface warp yarns of 3D warp interlock fabric O-L 3 3-2 {Twill 4 weft effect left shift} {1 4 7 10 − 2 5 8 11 − # − #} − (right) identification of the plain weave pattern for surface warp yarns

Table 2 Inter-ply position of crosslinking points of each surface warp yarns

Numbering of surface warp yarns	3	6	9	12
Location of cross-linking points with their inter-ply numbers	0	0	2	2

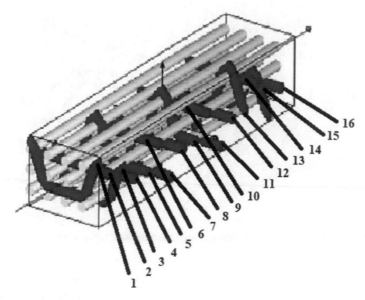

Fig. 14 Numbering of warp yarns of the 3D warp interlock fabric O-L 3 3-2 {Twill 4 weft effect left shift} {1 5 9 13 – 3 7 11 15 – # – #}

Introduction of Stuffer Warp Yarns

As mentioned in the Eq. (1), only one parameter helps to describe the stuffer warp yarns: their numbering sequence. For instance, let considers the following 3D warp interlock fabric: O-L 3 3-2 {Twill 4 weft effect left shift} {1 5 9 13 – 3 7 11 15 – # – #} represented in Fig. 14.

Numbering Sequence of Stuffer Warp Yarns

First, we introduce the inter-ply numbering on the representation of 3D warp interlock fabric O-L 3 3-2 {Twill 4 weft effect left shift} {1 5 9 13 – 3 7 11 15 – # – #} in (Fig. 15).

By isolating each stuffer warp yarn (Fig. 15), it has been possible to record the position of each of their cross-linking points according to the inter-ply numbering and then registered their locations into Table 3.

The numbering sequence of the stuffer warp yarns of the 3D warp interlock fabric O-L 3 3-2 {Twill 4 weft effect left shift} {1 5 9 13 – 3 7 11 15 – # – #} can then be written as: {# – 2 6 10 14 – 4 8 12 16 – #}, where;

- The minus operator corresponds to a layer of the 3D warp interlock fabric,
- The # sign means that at the position of the inter-ply the set of stuffer warp yarns is empty,
- The stuffer warp yarns numbers: 2 6 10 14 are located at the inter-ply position 1,
- The stuffer warp yarns numbers: 4 8 12 16 are located at the inter-ply position 2.

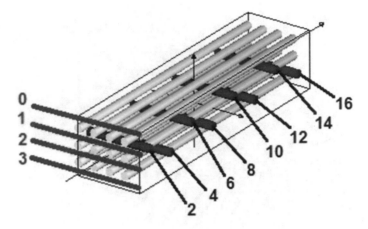

Fig. 15 Numbering of stuffer warp yarns according to their inter-ply locations

Table 3 Inter-ply position of crosslinking points of each stuffer warp yarns

Numbering of stuffer warp yarns	2	4	6	8	10	12	14	16
Location of cross-linking points with their inter-ply numbers	1	2	1	2	1	2	1	2

Application to the General Definition to a 3D Warp Interlock Fabrics Including Binding, Surface and Stuffer Warp Yarns

Let's consider the following 3D warp interlock fabric: O-L 3 3-2 Binding {Twill 4 weft effect left shift} {1 6 11 16 − 4 9 14 19 − # − #} − Surface {Plain weave}{3 13 − # − 8 18 − #} − Stuffer {# − 2 7 12 17 − 5 10 15 20 − #}, represented in Fig. 16.

By isolating all the different types of warp yarns, it can be possible to represent the binding warp yarns (Fig. 17—left), the surface warp yarns (Fig. 17—middle) and the stuffer warp yarns (Fig. 17—right).

Each evolution of warp yarns of the 3D warp interlock fabric: O-L 3 3-2 Binding {Twill 4 weft effect left shift} {1 6 11 16 − 4 9 14 19 − # − #} − Surface {Plain weave}{3 13 − # − 8 18 − #} − Stuffer {# − 2 7 12 17 − 5 10 15 20 − #} can be represented individually in Fig. 18.

Several exercises on the representation and recognition of 3D warp interlock fabrics are given in the appendices.

4   Construction of the General Weave Pattern

The general weave pattern of a 3D warp interlock fabric represents the evolution of all the warp yarns with respect to each weft yarns. The design approach lies on the identification of cross-linking points per layer of the 3D warp interlock fabric, which helps first to define the partial weave pattern scheme. Then, by introducing

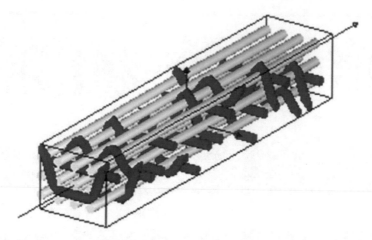

Fig. 16 Representation of the 3D warp interlock fabric: O-L 3 3-2 Binding {Twill 4 weft effect left shift} {1 6 11 16 – 4 9 14 19 – # – #} – Surface {Plain weave}{3 13 – # – 8 18 – #} – Stuffer {# – 2 7 12 17 – 5 10 15 20 – #}

Binding {Twill 4 weft effect left shift}{1 6 11 16 – 4 9 14 19 – # – #}	Surface {Plain weave}{3 13 - # - 8 18 - #}	Stuffer {# - 2 7 12 17 – 5 10 15 20 - #}

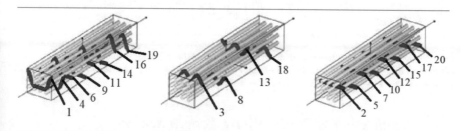

Fig. 17 Distinction between the different warp yarns types of the 3D warp interlock fabric: O-L 3 3-2 Binding {Twill 4 weft effect left shift} {1 6 11 16 – 4 9 14 19 – # – #} – Surface {Plain weave}{3 13 – # -8 18 – #} – Stuffer {# – 2 7 12 17 – 5 10 15 20 – #}

the insertion order of the weft yarns during the weaving process, the weft formulae combined with the partial weave pattern scheme allow to identify the general weave pattern of the 3D warp interlock fabric. To link this general weave pattern to the weaving process, especially for dobby weaving machine, a last step involves to adapt the peg plan and the drawing-in plan to the general weave pattern of the 3D warp interlock fabric.

Let's consider the general definition of the 3D warp interlock fabric O-T 6 1-6 Binding {Plain weave} {1 2 – # – # – # – # – # – #} – Stuffer {# – 3 – 4 – 5 – 6 – 7 – #}. The elementary pattern of the 3D warp interlock fabric is represented in Fig. 19 with the numbering of warp and weft yarns.

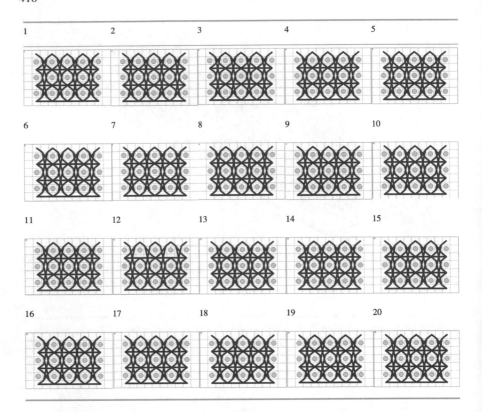

Fig. 18 Scheme of the evolution of each warp yarns inside the 3D warp interlock fabric: O-L 3 3-2 Binding {Twill 4 weft effect left shift} {1 6 11 16 – 4 9 14 19 – # – #} – Surface {Plain weave} {3 13 – # –8 18 - #} – Stuffer {# – 2 7 12 17 – 5 10 15 20 – #}

4.1 Identification of Partial Weave Pattern Scheme

By isolating each warp yarns of the elementary woven pattern, it could be possible to define the evolution per layer of the 3D warp interlock fabric of their cross-linking points with the weft yarns. For instance, if we consider the evolution of the binding warp yarn number 1 (in red colour in the Fig. 19), we could determine its position with respect to the first layer of the 3D warp interlock fabric, with weft yarns a1 and b1. Thus, the binding warp yarn 1 is above to the weft yarn a1 (UP) and below to weft yarn b1 (DOWN). The geometric modelling of this woven pattern is represented in Fig. 20.

Considering the second layer of the 3D warp interlock fabric, the binding warp yarn 1 is above to the weft yarn a2 (UP) and below to weft yarn b2 (DOWN). It can be seen from the Fig. 19 that the evolution of the binding warp yarn number 1 will always be above to all the a1 to a6 weft yarns and below to all the b1 to b6 weft yarns.

Fig. 19 Elementary woven pattern of the 3D warp interlock fabric O-T 6 1-6 Binding {Plain weave} {1 2 − # − # − # − # − # − #} − Stuffer {# − 3 − 4 − 5 − 6 − 7 − #}

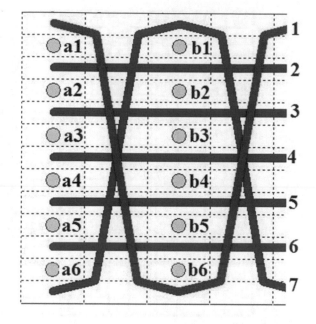

Fig. 20 Geometric model of the position of the binding warp yarn 1 with the first layer of the 3D warp interlock fabric

Then, by applying the same methodology to all the warp yarns with respect to their numberings, the partial weave pattern scheme of the 3D warp interlock fabric O-T 6 1-6 Binding {Plain weave} {1 2 − # − # − # − # − # − #} − Stuffer {# − 3 − 4 − 5 − 6 − 7 − #} is represented in Fig. 21.

4.2 Introduction of the Weft Formulae

The weft formulae corresponds to the order of weft yarns insertion. For the 3D warp interlock fabric and due to the shed opening constraints on the weaving loom, two main types are available. The first, named "inversed N insertion" aims at insert the weft yarns per column and the second, named "U insertion" insert weft yarns from the top to bottom and then from the bottom to the top, as represented in Fig. 22.

If we choose the "N inversed insertion" type, the weft yarns ranking is: a1, a2, a3, a4, a5, a6, b1, b2, b3, b4, b5, b6. Then, using the partial weave pattern scheme of the 3D warp interlock fabric O-T 6 1-6 Binding {Plain weave} {1 2 − # − # − # −

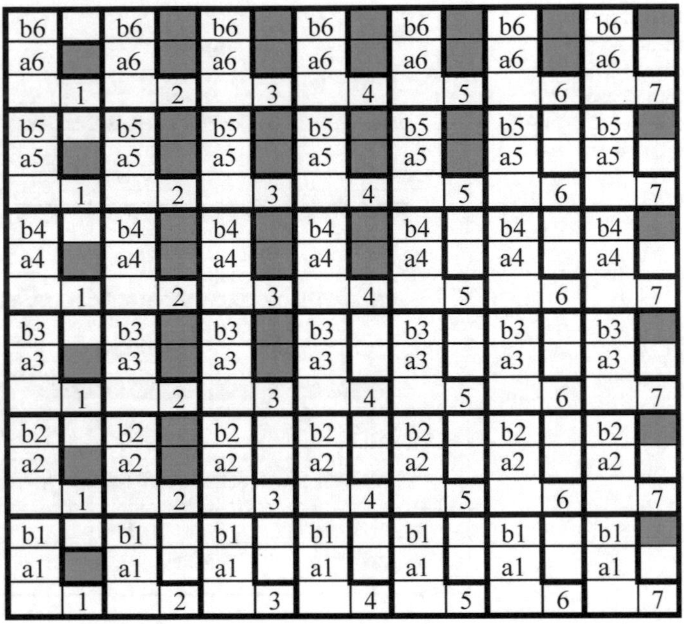

b6		b6	▓	b6		b6	▓	b6		b6	▓	b6	
a6	▓	a6		a6	▓	a6		a6	▓	a6		a6	
	1		2		3		4		5		6		7
b5		b5	▓	b5		b5	▓	b5		b5		b5	
a5	▓	a5		a5	▓	a5		a5	▓	a5		a5	
	1		2		3		4		5		6		7
b4		b4		b4		b4		b4		b4		b4	
a4	▓	a4	▓	a4	▓	a4	▓	a4		a4		a4	
	1		2		3		4		5		6		7
b3		b3		b3		b3		b3		b3		b3	▓
a3	▓	a3		a3	▓		a3		a3		a3		a3
	1		2		3		4		5		6		7
b2		b2	▓	b2		b2		b2		b2		b2	▓
a2	▓	a2		a2		a2		a2		a2		a2	
	1		2		3		4		5		6		7
b1		b1		b1		b1		b1		b1		b1		▓
a1	▓	a1		a1		a1		a1		a1		a1	
	1		2		3		4		5		6		7

Fig. 21 Partial weave pattern scheme of the 3D warp interlock fabric O-T 6 1-6 Binding {Plain weave} {1 2 − # − # − # − # − # − #} − Stuffer {# − 3 − 4 − 5 − 6 − 7 − #}

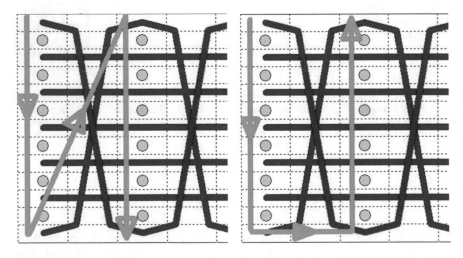

Fig. 22 Weft yarns insertion types: (left) the "N inversed insertion" and (right) the "U insertion"

— # — #} — Stuffer {# — 3 — 4 — 5 — 6 — 7 — #} combined with the "N inversed insertion" type, the general weave pattern is represented in Fig. 23.

If we choose the "U insertion" type, the weft yarns ranking is: a1, a2, a3, a4, a5, a6, b6, b5, b4, b3, b2, b1. Then, using the partial weave pattern scheme of the 3D warp interlock fabric O-T 6 1-6 Binding {Plain weave} {1 2 — # — # — # — # — # — #} — Stuffer {# — 3 — 4 — 5 — 6 — 7 — #} combined with the "U insertion" type, the general weave pattern is represented in Fig. 24.

Fig. 23 General weave pattern of the 3D warp interlock fabric O-T 6 1-6 Binding {Plain weave} {1 2 — # — # — # — # — # — #} — Stuffer {# — 3 — 4 — 5 — 6 — 7 — #} combined with the "N inversed insertio" type

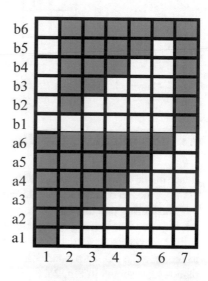

Fig. 24 Gencral weave pattern of the 3D warp interlock fabric O-T 6 1-6 Binding {Plain weave} {1 2 — # — # — # — # — # — #} — Stuffer {# — 3 — 4 — 5 — 6 — 7 — #} combined with the "U insertion" type

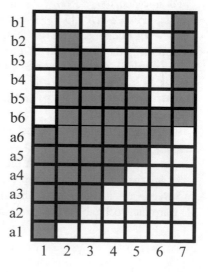

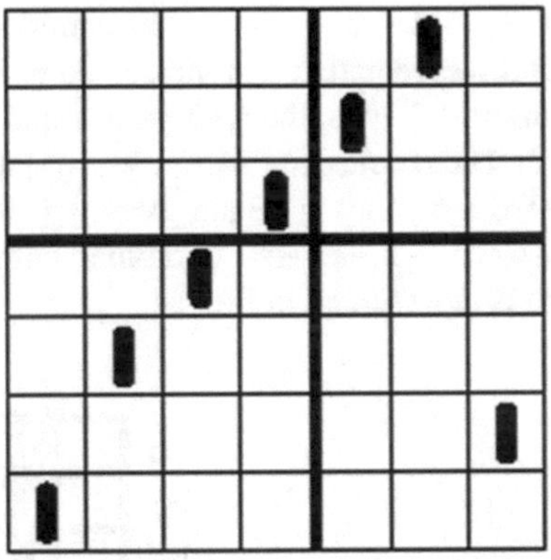

Fig. 25 Drawing-in plan for the 3D warp interlock fabric O-T 6 1-6 Binding {Plain weave} {1 2 − # − # − # − # − # − #} − Stuffer {# − 3 − 4 − 5 − 6 − 7 − #}

4.3 Identification of the Peg Plan and Drawing-In Plan

Based on the dobby design of 2D fabrics, we need to adapt the peg plan and drawing-in plan in order to obtain the general weave pattern of the 3D warp interlock fabric.

Considering that the dobby loom is equipped with the sufficient number of shafts, it is then possible to merge on the two first shafts the binding warp yarns and locate the stuffer warp yarns onto the last 5 remaining shafts. Then, only 7 shafts are necessary to produce the 3D warp interlock fabric on a dobby loom. One possible drawing-in plan is represented in Fig. 25.

For instance, the general weave pattern of the 3D warp interlock fabric O-T 6 1-6 Binding {Plain weave} {1 2 − # − # − # − # − # − #} − Stuffer {# − 3 − 4 − 5 − 6 − 7 − #}, combined with the "N inversed insertion " type represented in Fig. 22 and following the drawing-in plan represented in Fig. 25, is represented in Fig. 26.

The resulted peg plan of the 3D warp interlock fabric O-T 6 1-6 Binding {Plain weave} {1 2 − # − # − # − # − # − #} − Stuffer {# − 3 − 4 − 5 − 6 − 7 − #} with the "N inversed insertion" weft formulae and the given drawing-in plan is represented in Fig. 27.

By the same, the general weave pattern of the 3D warp interlock fabric O-T 6 1-6 Binding {Plain weave} {1 2 − # − # − # − # − # − #} − Stuffer {# − 3 − 4 − 5 − 6 − 7 − #}, combined with the "U insertion" type represented in Fig. 22 and following the drawing-in plan represented in Fig. 25, is represented in Fig. 28.

Several exercises of construction of peg plans with their associated drawing-in of 3D warp interlock fabrics are given in the appendices.

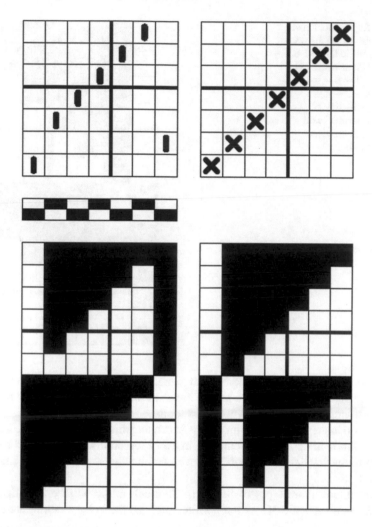

Fig. 26 Representation of general weave pattern of the 3D warp interlock fabric O-T 6 1-6 Binding {Plain weave} {1 2 − # − # − # − # − # − #} − Stuffer {# − 3 − 4 − 5 − 6 − 7 − #} with the "N inversed insertion" weft formulae and the given drawing-in plan

5 Conclusion

The 3D warp interlock fabric can be defined and designed to be produce on existing dooby weaving machine. The definition of the 3D warp interlock fabric is based on a step-by-step procedure which helps to describe the different types of warp yarns (binding, surface and stuffer) as well as their evolutions and locations inside the 3D woven structure. Thanks to this definition, it could be possible to clearly define the 3D warp interlock fabric and find the woven pattern of each warp yarns and inversely.

Fig. 27 Representation of
the peg plan of the 3D warp
interlock fabric O-T 6 1-6
Binding {Plain weave} {1 2
− # − # − # − # − # − #} −
Stuffer {# − 3 − 4 − 5 − 6
− 7 − #} with the "N
inversed insertion" weft
formulae and the given
drawing-in plan

By the same, the design of the 3D warp interlock fabric allows to determine the
suited peg plan to the drawing-in plan. Thanks to the obtained peg plan, it could be
possible to transfer it on a dobby weaving loom.

The main interest of this definition consist in using this type of coding the 3D warp
interlock fabric to describe it accurately in every research works and then be more
efficient to understand the type of 3D woven architectures researchers or engineers
have been used.

In the future, more types of 3D warp interlock fabric will be investigated to
introduce more specific 3D woven architectures as: the hollow fabric, the auxetic
fabric.

Recommendations

All the woven patterns (2D and 3D views) have been prepared using the software:
Wisetex © [53].

All the general weave patterns of the 3D warp interlock fabrics have been design
with the software: DBweave © [54].

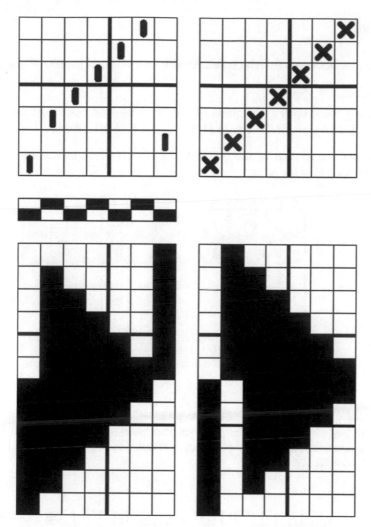

Fig. 28 Representation of general weave pattern of the 3D warp interlock fabric O-T 6 1-6 Binding {Plain weave} {1 2 − # − # − # − # − # − #} − Stuffer {# − 3 − 4 − 5 − 6 − 7 − #} with the "U insertion" weft formulae and the given drawing-in plan

Appendix

Exercises on 3D Warp Interlock Fabric with and Without Stuffer Warp Yarns

Problem 1
Considering the following definition of the 3D warp interlock fabric O-L 3 4-2 Binding {Satin 5 weft effect shift 2} {1 3 5 7 9 – 2 4 6 8 10 – # – #},

1. Provide the scheme of the 3D warp interlock fabric architecture with the numbering of warp yarns.

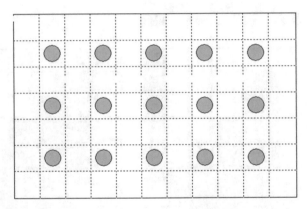

2. Considering a straight drawing-in plan and "N inversed insertion" weft formulae, provide the suited peg plan and general weave pattern for the 3D warp interlock fabric O-L 3 4-2 Binding {Satin 5 weft effect shift 2} {1 3 5 7 9 – 2 4 6 8 10 – # – #}.

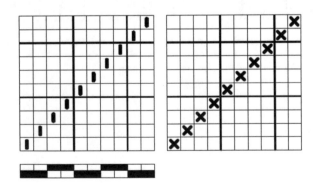

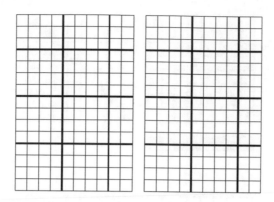

Problem 2

Considering the following definition of the 3D warp interlock fabric A-T 4 7-4 binding {Twill 8 weft effect left shift}{1 2 3 4 5 6 7 8 − # − # − # − #},

1. Provide the scheme of the 3D warp interlock fabric architecture with the numbering of warp yarns.

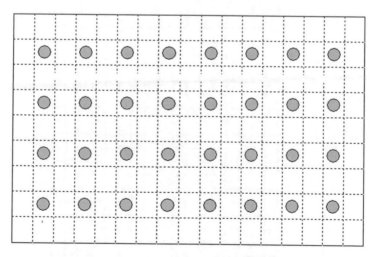

2. Considering a straight drawing-in plan and "N inversed insertion" weft formulae, provide the suited peg plan and general weave pattern for the 3D warp interlock fabric A-T 4 7-4 binding {Twill 8 weft effect left shift}{1 2 3 4 5 6 7 8 − # − # − # − #}.

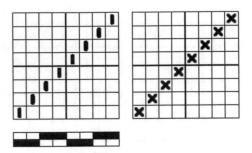

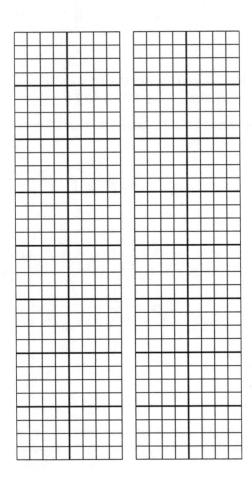

Problem 3

Considering the following definition of the 3D warp interlock fabric A-L 3 3-2 Binding {Twill 4 weft effect left shift} {1 3 5 7 − 2 4 6 8 − # − #},

1. Provide the scheme of the 3D warp interlock fabric architecture with the numbering of warp yarns.

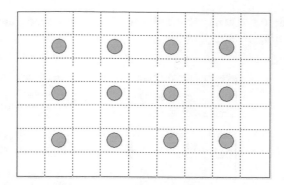

2. Considering a straight drawing-in plan and "N inversed insertion" weft formulae, provide the suited peg plan and general weave pattern for the 3D warp interlock fabric A-L 3 3-2 Binding {Twill 4 weft effect left shift} {1 3 5 7 − 2 4 6 8 − # − #}.

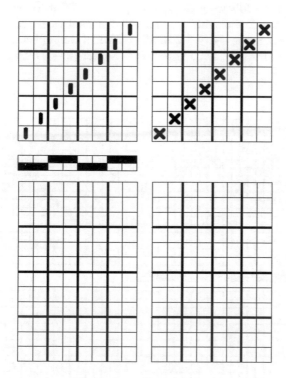

Problem 4

Considering the following definition of the 3D warp interlock fabric O-L 4 3-2 Binding {Twill 4 weft effect left shift} {1 7 13 19 − 3 9 15 21 − 5 11 17 23 − # − #} − Stuffer {# − 2 8 14 20 − 4 10 16 22 − 6 12 18 24 − #},

1. Provide the scheme of the 3D warp interlock fabric architecture with the numbering of warp yarns.

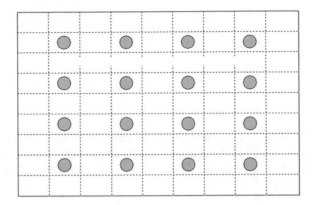

2. Considering the following drawing-in plan and "N inversed insertion" weft formulae, provide the suited peg plan and general weave pattern for the 3D warp interlock fabric O-L 4 3-2 Binding {Twill 4 weft effect left shift} {1 7 13 19 − 3 9 15 21 − 5 11 17 23 − # − #} − Stuffer {# − 2 8 14 20 − 4 10 16 22 − 6 12 18 24 − #}.

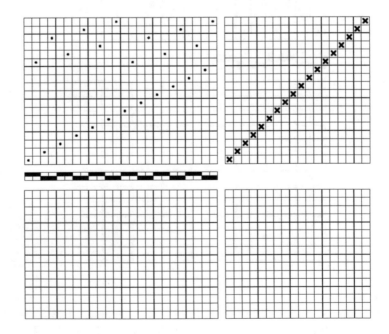

Problem 5

Considering the following definition of the 3D warp interlock fabric O-T 4 5-4 Binding {Twill 6 pattern 1UP5DO – 3UP1DO2UP right shift} {1 3 5 7 9 11 13 15 17 19 21 23 – # – # – # – #} – Stuffer {# – 2 8 14 20 – 4 10 16 22 – 6 12 18 24 – #},

1. Provide the scheme of the 3D warp interlock fabric architecture with the numbering of warp yarns. The binding weave pattern {Twill 6 pattern 1UP5DO – 3UP1DO2UP right shift} is represented below and it also called the Twill 6 "jaw" effect.

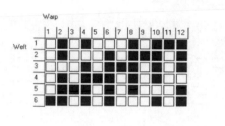
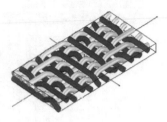

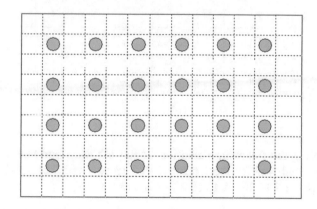

2. Considering the following drawing-in plan and "N inversed insertion" weft formulae, provide the suited peg plan and general weave pattern for the 3D warp interlock fabric O-T 4 5-4 Binding {Twill 6 pattern 1UP5DO – 3UP1DO2UP right shift} {1 3 5 7 9 11 13 15 17 19 21 23 – # – # – # – #} – Stuffer {# – 2 8 14 20 – 4 10 16 22 – 6 12 18 24 – #}.

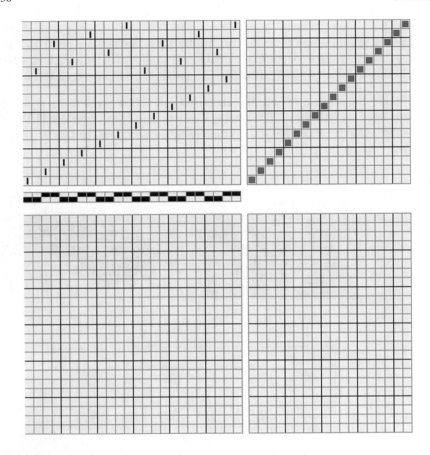

Problem 6

Considering the following definition of the 3D warp interlock fabric A-L 4 3-2 Binding {Twill 4 weft effect left shift} {1 7 13 19 – 3 9 15 21 – 5 11 17 23 – # – #} – Stuffer {# – 2 8 14 20 – 4 10 16 22 – 6 12 18 24 – #},

1. Provide the scheme of the 3D warp interlock fabric architecture with the numbering of warp yarns.

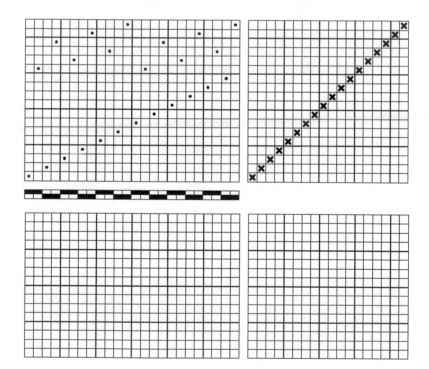

2. Considering the following drawing-in plan and "N inversed insertion" weft formulae, provide the suited peg plan and general weave pattern for the 3D warp interlock fabric A-L 4 3-2 Binding {Twill 4 weft effect left shift} {1 7 13 19 − 3 9 15 21 − 5 11 17 23 − # − #} − Stuffer {# − 2 8 14 20 − 4 10 16 22 − 6 12 18 24 − #}.

Problem 7

Considering the following definition of the 3D warp interlock fabric A-T 4 7-4 Binding {Twill 8 weft effect left shift} {1 3 5 7 9 11 13 15 – # – # – # – #} – Stuffer {# – 2 10 – 4 8 12 16 – 6 14 – #},

1. Provide the scheme of the 3D warp interlock fabric architecture with the numbering of warp yarns.

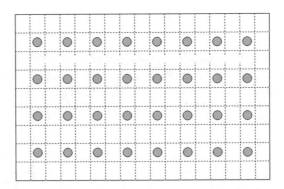

2. Considering the following drawing-in plan and "N inversed insertion" weft formulae, provide the suited peg plan and general weave pattern for the 3D warp interlock fabric A-T 4 7-4 Binding {Twill 8 weft effect left shift} {1 3 5 7 9 11 13 15 – # – # – # – #} – Stuffer {# – 2 10 – 4 8 12 16 – 6 14 – #}.

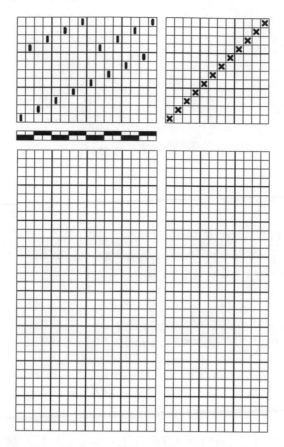

Solutions on 3D Warp Interlock Fabric Without and with Stuffer Warp Yarns

Solution 1

Considering the following definition of the 3D warp interlock fabric O-L 3 4-2 Binding {Satin 5 weft effect shift 2} {1 3 5 7 9 – 2 4 6 8 10 – # – #}.

1. The scheme of the 3D warp interlock fabric architecture with the numbering of warp yarns is:

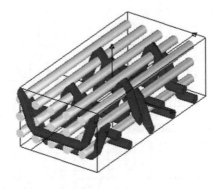

1	2	3	4	5
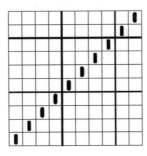				

6	7	8	9	10

2. The general weave pattern for the 3D warp interlock fabric O-L 3 4-2 Binding {Satin 5 weft effect shift 2} {1 3 5 7 9 − 2 4 6 8 10 − # − #} is:

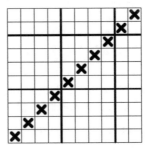

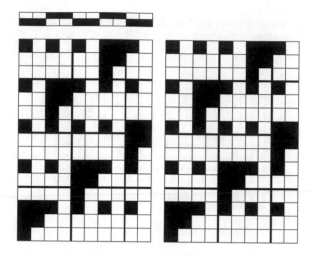

Solution 2

Considering the following definition of the 3D warp interlock fabric A-T 4 7-4 binding {Twill 8 weft effect left shift}{1 2 3 4 5 6 7 8 − # − # − # − #},

1. The scheme of the 3D warp interlock fabric architecture with the numbering of warp yarns is:

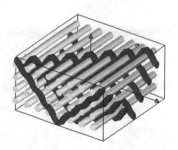

1	2	3	4
5	6	7	8

2.   The general weave pattern for the 3D warp interlock fabric A-T 4 7-4 binding
     {Twill 8 weft effect left shift}{1 2 3 4 5 6 7 8 − # − # − # − #} is:

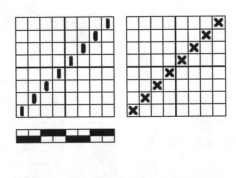

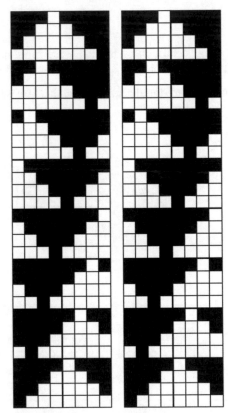

Solution 3

Considering the following definition of the 3D warp interlock fabric A-L 3 3-2
Binding {Twill 4 weft effect left shift} {1 3 5 7 − 2 4 6 8 − # − #},

1. The scheme of the 3D warp interlock fabric architecture with the numbering of warp yarns is:

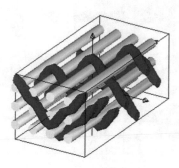

1	2	3	4

5	6	7	8

2. The general weave pattern for the 3D warp interlock fabric A-L 3 3-2 Binding {Twill 4 weft effect left shift} {1 3 5 7 − 2 4 6 8 − # − #} is:

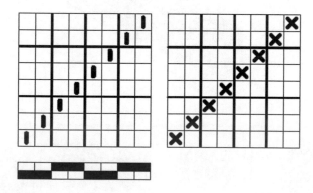

Solution 4

Considering the following definition of the 3D warp interlock fabric O-L 4 3-2 Binding {Twill 4 weft effect left shift} {1 7 13 19 – 3 9 15 21 – 5 11 17 23 – # – #} – Stuffer {# – 2 8 14 20 – 4 10 16 22 – 6 12 18 24 – #},

1. The scheme of the 3D warp interlock fabric architecture with the numbering of warp yarns is:

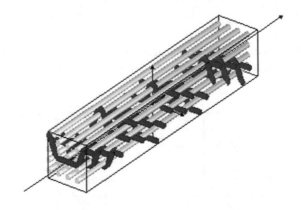

(continued)

(continued)

9	10	11	12
13	14	15	16
17	18	19	20
21	22	23	24

2. The general weave pattern for the 3D warp interlock fabric O-L 4 3-2 Binding {Twill 4 weft effect left shift} {1 7 13 19 − 3 9 15 21 − 5 11 17 23 − # − #} − Stuffer {# − 2 8 14 20 − 4 10 16 22 − 6 12 18 24 − #} is:

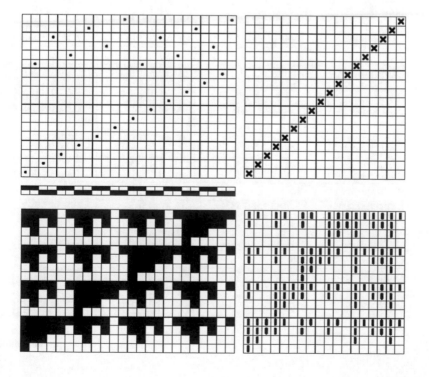

Solution 5

Considering the following definition of the 3D warp interlock fabric O-T 4 5-4 Binding {Twill 6 pattern 1UP5DO – 3UP1DO2UP right shift} {1 3 5 7 9 11 13 15 17 19 21 23 – # – # – # – #} – Stuffer {# – 2 8 14 20 – 4 10 16 22 – 6 12 18 24 – #},

1. The scheme of the 3D warp interlock fabric architecture with the numbering of warp yarns is:

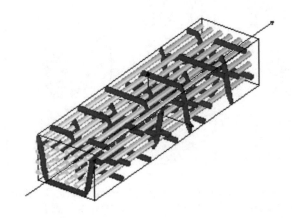

1	2	3	4
5	6	7	8
9	10	11	12
13	14	15	16
17	18	19	20
21	22	23	24

(continued)

(continued)

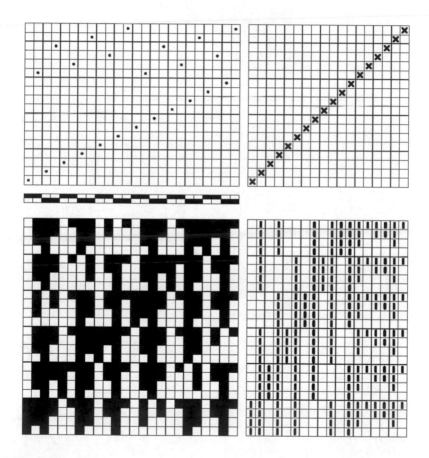

2. The general weave pattern for the 3D warp interlock fabric O-T 4 5-4 Binding {Twill 6 pattern 1UP5DO − 3UP1DO2UP right shift} {1 3 5 7 9 11 13 15 17 19 21 23 − # − # − # − #} − Stuffer {# − 2 8 14 20 − 4 10 16 22 − 6 12 18 24 − #} is:

Solution 6

Considering the following definition of the 3D warp interlock fabric A-L 4 3-2 Binding {Twill 4 weft effect left shift} {1 7 13 19 − 3 9 15 21 − 5 11 17 23 − # − #} − Stuffer {# − 2 8 14 20 − 4 10 16 22 − 6 12 18 24 − #},

1. The scheme of the 3D warp interlock fabric architecture with the numbering of
    warp yarns is:

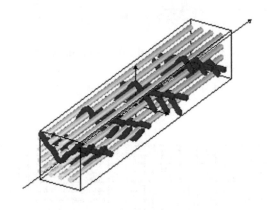

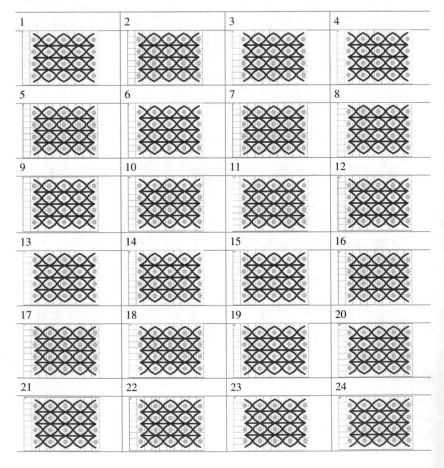

2. The general weave pattern for the 3D warp interlock fabric A-L 4 3-2 Binding {Twill 4 weft effect left shift} {1 7 13 19 − 3 9 15 21 − 5 11 17 23 − # − #} − Stuffer {# − 2 8 14 20 − 4 10 16 22 − 6 12 18 24 − #} is:

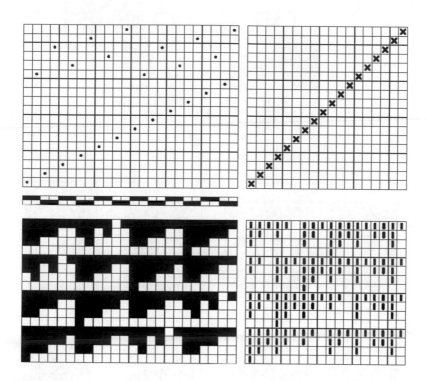

Solution 7

Considering the following definition of the 3D warp interlock fabric A-T 4 7-4 Binding {Twill 8 weft effect left shift} {1 3 5 7 9 11 13 15 − # − # − # − #} − Stuffer {# − 2 10 − 4 8 12 16 − 6 14 − #},

1. The scheme of the 3D warp interlock fabric architecture with the numbering of warp yarns is:

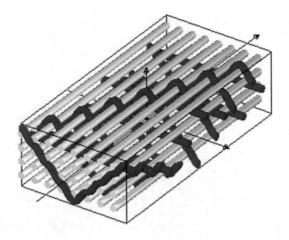

1	2	3	4

5	6	7	8

9	10	11	12

13	14	15	16

2. The general weave pattern for the 3D warp interlock fabric A-T 4 7-4 Binding {Twill 8 weft effect left shift} {1 3 5 7 9 11 13 15 − # − # − # − #} − Stuffer {# − 2 10 − 4 8 12 16 − 6 14 − #} is:

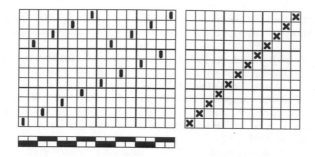

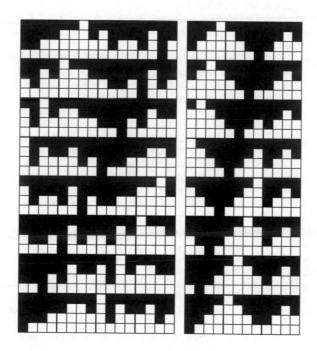

References

1. Boussu, F., & Legrand, X. (2008). Technical and economical performances of 3D warp interlock structures. In: *SAMPE 08, Long Beach, CA, USA, May 18–22, 2008*.
2. Chen, X., & Yang, D. (2010). Use of 3D angle-interlock woven fabric for seamless female body armor: Part 1: Ballistic evaluation. *Textile Research Journal, 80*(15), 1581–1588.
3. Chen, X., Lo, W., Tayyar, A., & Day, R. (2002). Mouldability of angle-interlock woven fabrics for technical applications. *Textile Research Journal, 72*(3), 195–200.
4. Lomov, S., Bogdanovich, A., Ivanov, D., Mungalov, D., Karathan, M., & Verpoest, I. (2009). A comparative study of tensile properties of non crimp 3D orthogonal weave and multi layer plain weave e-glass composites. Part1: Materials, methods and principal results. *Composites Part A: Applied Science and Manufacturing, 40*(8), 1134–1143.

5. Grogan, J., Tekalur, S., Shukla, A., Bogdanovich, A., & Coffelt, R. (2007). Ballistic resistance of 2D and 3D woven sandwich composites. *Journal of Sandwich Structures and Materials, 9,* 283–302.
6. Mohamed, M., Bogdanovich, A., Dickinson, L., Singletary, J., & Lienhart, R. (2001). A new generation of 3D woven fabric preforms and composites. *Sampe Journal, 37*(3).
7. Ko, F. (1999). 3-D textile reinforcements in composite materials. In A. Miravete (Ed.), *3-D textile reinforcements in composite materials* (pp. 9–40). Woodhead Publishing Limited.
8. Mouritz, A., Bannister, M., Falzon, P., & Leong, K. (1999). Review of applications for advanced three-dimensional fibre textile composites. *Composites Part A: Applied Science and Manufacturing, 30,* 1445–1461.
9. Boussu, F., Cristian, I., & Legrand, X. (2009). General specification of warp interlock structure: Application for carbon fiber multi-layer fabrics. In: *Avantex techtextil symposium, Frankfurt, Germany, from 16/06/2009 to 17/06/2009.*
10. Lee, L., Rudov-Clark, S., Mouritz, A., Bannister, M., & Herszberg, I. (2002). Effect of weaving damage on the tensile properties of three-dimensional woven composites. *Composite Structures, 57,* 405–413.
11. Callus, P., Mouritz, A., Bannister, M., & Leong, K. (1999). Tensile properties and failure mechanisms of 3D woven GRP composites. *Composites Part A, 30,* 1277–1287.
12. Cox, B., Dadkhah, M., & Morris, W. (1996). On the tensile failure of 3D woven composites. *Composite Part A Applied Sciences Manufacturing, 27,* 447–458.
13. Nauman, S. (2008). *Modélisation Géométrique de tissu 3D Interlock.* France: Roubaix.
14. Nauman, S. (2011). Geometrical modelling and characterization of 3D warp interlock composites and their on-line structural health monitoring using flexible textile sensors. Lille.
15. Lapeyronnie, P. (2010). Mise en œuvre et comportement mécanique de composites organiques renforcés de structures 3D interlocks. Douai, France.
16. Ha-Minh, C. (2011). Comportement mécanique des matériaux tissés soumis à un impact balistique: Approches expérimentale, numérique et analytique. Lille, France.
17. Lefebvre, M. (2011). Résistance à l'impact balistique de matériaux composites à renforts Interlocks tissés. Application au blindage de véhicules. Valenciennes.
18. Provost, B. (2013). Étude et Réalisation d'une solution à renfort tissé interlock pour la protection balistique de véhicule. Valenciennes.
19. Yi, H., & Ding, X. (2004). Conventional approach on manufacturing 3D woven preforms used for composites. *Journal of Industrial Textiles, 34–39.*
20. Sheng, S., & Hoa, S. (2003). Modelling of 3D angle interlock woven fabric composites. *Journal of Thermoplastic Composite Materials, 16*(1), 45–59.
21. Hu, J. (2008). *3D fibrous assemblies, properties applications and modelling of three dimensional textile structure* (vol. 74). Woodhead Publishing.
22. Cristian, I., Boussu, F., & Nauman, S. (2010).Interesting parameters of 3D warp interlock fabrics influencing the mechanical properties of the final composite structures. In: *10th World Textile Conference, Vilnius, Lithuania, 21–23 June, 2010.*
23. Tong, L., Mouritz, A., & Bannister, M. (2002). *3D fibre reinforced composite materials* (p. 241). Elsevier Applied Science.
24. Nauman, S., Boussu, F., Cristian, I., & Koncar, V. (2009). "mpact of 3D woven structure onto the high performance yarn properties. In: *Second Conference on Intelligent Textiles and Mass Customisation, Textile Composites Workshop, Casablanca, Morocco, November 2009 the 12–14th.*
25. Lapeyronnie, P., Le Grognec, P., Binetruy, C., & Boussu, F. (2010). Angle-interlock reinforcements: Weaving and the mechanical properties of composites. *JEC Composites, 158,* 58–59.
26. Tan, P., Tong, L., & Steven, G. (1999). Micromechanics models for mechanical and thermomechanical properties of 3D through-the-thickness angle interlock woven composites. *Composites Part A, 30,* 637–648.
27. Tsai, K., Chiu, C., & Wu, T. (2000). Fatigue behaviour of 3D multi-layer angle interlock woven composite plates. *Composites Science and Technology, 60,* 241–248.

28. Baucom, J. N., & Zikry, M. (2003). Evolution of failure mechanisms in 2D and 3D woven composite systems under quasi-static perforation. *Journal of Composite Materials, 37*(18), 1651–1674.

29. Tanzawa, Y., Watanabe, N., & Ishikawa, T. (1999). Interlaminar fracture toughness of 3-D orthogonal interlocked fabric composites. *Composites Science and Technology, 59*(8), 1261–1270.

30. Padaki, N., Alagirusamy, R., Deopura, B., & Fangueiro, R. (2010). Influence of preform inter-lacement on the low velocity impact behavior of multilayer textile composites. *Journal of Industrial Textiles, 40*(2), 171–185.

31. Tung, P., & Jayaraman, S. (1991). Three dimensional multilayer woven preforms for composites. In: *High-tech fibrous materials* (pp. 53–80). Washington, DC: ACS Publisher.

32. Coman, F., Herszberg, L., & Bannister, M. (1996). Design and analysis of 3D woven preforms for composite structures. *Science and Engineering of Composite Materials, 5*(2), 83–96.

33. Boussu, F., Cristian, I., & Nauman, S. (2009). Technical performance of yarns inside a 3D woven fabric. In: *ITC International Conference on Latest Advancements in High Tech Textiles and Textile-based Materials, Gent, Belgium, 23–25 September 2009.*

34. Cox, B., & Dadkhah, M. (1995). The macroscopic elasticity of 3D woven composites. *Journal of Composite Materials, 29*, 785–819.

35. Lee, B., Leong, K., & Herszberg, I. (2001). The effect of weaving on the tensile properties of carbon fibre tows and woven composites. *Journal of Reinforced Plastics and Composites, 20*, 652–670.

36. Cristian, I., Nauman, S., Boussu, F., & Koncar, V. (2012). A study of strength transfer from tow to textile composite using different reinforcement architectures. *Applied Composite Materials, 19*(3–4), 427–442.

37. Rudov-Clark, S., Mouritz, A., Lee, L., & Bannister, M. (2003). Fibre damage in the manufacture of advanced three-dimensional woven composites. *Composites Part A, 34*, 963–970.

38. Boussu, F., Cristian, I., Nauman, S., Lapeyronnie, P., & Binetruy, C. (2009). Effect of 3D-weave architecture on strength transfer from tow to textile composite. In: *2nd World conference on 3D Fabrics and their applications, Greenville, South Carolina, USA, 6–7 avril 2009.*

39. Boussu, F., Dufour, C., Veyet, F., & Lefebvre, M. (2015). Weaving processes for composites manufacture. In P. Boisse (Ed.), *Advances in composites manufacturing and processes—Part 1* (pp. 55–78). Woodhead Publishing.

40. Boussu, F. (2016). Les tissus 3D interlocks chaines Définition et applications. In: I. Arser (Éd.), *Editions Universitaires Européennes* (p. 265).

41. Boussu, F., Provost, B., Lefebvre, M., & Coutellier, D.(2019). New textile composite solutions for armouring of vehicles. *Advances in Materials Science and Engineering, 2019*(7938720), 14.

42. Dantzer, D., & de Prat, D. (1949). Les tissus - I - tissus classiques, Librairie Polytechnique Ch. Béranger éd., Paris.

43. Dantzer, J., & de Prat, D. (1949). Les tissus - II - tissus spéciaux, Librairie Polytechnique Ch. Béranger éd., Paris.

44. AFNOR. (1988). Dictionnaire des termes normalisés NF G 00-001.

45. Adanur, S. (2001). *Handbook of weaving.* CRC Press.

46. Lord, P., & Mohamed, M. (1999). *Weaving, conversion of yarn to fabric.* W. publishing.

47. El Hage, C. (2006). Modélisation et comportement élastique endommageable de matériaux composites à renforts tridimensionnels. Compiègne.

48. Ansar, M., Xinwei, W., & Chouwei, Z. (2011). Modeling strategies of 3D woven composites: A review. *Composite Structures, 93*(8), 1947–1963.

49. Ding, X., & Yi, H. (2001). Parametric representation of 3D woven structure. In: *6th Asian Textile Conference (CD Version), Hong Kong, China, 2001.*

50. Cristian, I. (2009). Optimisation des caractéristiques mécaniques des structures tissées multi-couches interlock. Roubaix, France.

51. Boussu, F., Cristian, I., & Nauman, S. (2015). General definition of 3D warp interlock fabric architecture. *Composites Part B, 81*, 171–188.

52. Nauman, S., & Cristian, I. (2014). Geometrical modelling of orthogonal/layer-to-layer woven interlock carbon reinforcement. *The Journal of The Textile Institute*, 1–11. https://doi.org/10.1080/00405000.2014.937560.
53. WISETEX SUITE. https://www.mtm.kuleuven.be/Onderzoek/Composites/software/wisetex. Accessed 14 Aug 2019.
54. DB-WEAVE. https://www.brunoldsoftware.ch/dbw.html. Accessed 14 Aug 2019.

Design and Manufacture of 3D Woven Textiles

Xiaogang Chen

Abstract Various manufacturing processes are available to produce 3D fabric structures by interlacing warp and weft yarns. The manufacturing route is determined by the scale and complexity of the textile geometry and the intended end-use of the 3D textiles. This chapter attempts to make a comprehensive overview on fabrication methods that can be used for making 3D textile woven textiles and give details on the design and manufacture of the four types of 3D woven fabrics. Among the many different definitions of 3D woven fabrics, one common understanding is that 3D fabrics must have substantial dimension in the thickness direction formed by layers of fabrics or yarns. This chapter is written in such a way that it can be used as a textbook chapter for university students studying textile engineering and relevant disciplines.

1 Introduction to 3D Woven Textiles

1.1 What are 3D Woven Textiles

Weaving is a technology unique in interlacing different linear materials from perpendicular directions in many different ways to form integrated, usually flexible, structures. For a long time, it was only necessary to incorporate materials from the warp and weft directions to form a sheet material i.e. a fabric. Fabrics in two-dimensional (2D) sheet form have many properties such as being drapeable, flexible, warmth-keeping, and suitably strong, and all of these make them suitable to be used as materials for clothing and other domestic end-uses. When high performance fibres (such as glass, carbon, and aramid) are used for constructing such 2D fabrics, the woven fabrics

X. Chen (✉)
Department of Materials, School of Natural Sciences, University of Manchester, Sackville Street, Manchester M13 9PL, UK
e-mail: xiaogang.chen@manchester.ac.uk

© Springer Nature Switzerland AG 2022
Y. Kyosev and F. Boussu (eds.), *Advanced Weaving Technology*,
https://doi.org/10.1007/978-3-030-91515-5_12

find many technical applications including textile composite reinforcements for the aerospace industry [1] and body armours for personal protection for the military and the police [2].

The weaving principle is also capable of making structures that have substantial dimension in the thickness direction formed by layers of fabrics or yarns, generally termed as the 3D fabrics. Many of these structures can be made on the conventional weaving machines with no or little modification [3]. The immediate advantages of fabrics with a notable thickness dimension include the structural integrity of the woven structure, the satisfaction of the geometric shapes, and volumes that are required for many end-use applications [4]. In addition, the weaving technology is capable of leaving both crimped and straight yarns in the fabrics to suit the applications [5]. Currently, both the conventional and specially made weaving machines or devices are used in the making of various 3D fabrics, mainly for the composite industry. There have been successful attempts in developing new weaving devices particularly for making 3D woven fabrics [6, 7]. These new technologies arrange warp yarns in a 3D shape and allow weft yarns to be inserted at different levels in one or two directions. The advantages of fabrics made by using such technologies are many. It is fair to say that it is easier than the conventional weaving method to make net-shaped preforms without the need of huge trimming after weaving, therefore reducing the waste of materials. Another obvious advantage is that the weaving devices can make 3D fabrics which are much thicker than the conventional technology. By comparison, the conventional weaving technology can also make various types of 3D fabrics but will be confined in thickness especially when making solid panels. The conventional weaving technology is also popular in making 3D fabrics maybe because of two reasons. The first is that some of the 3D fabrics targeted to make are not always beyond the capability of the conventional weaving machines. The other may be because of the broad base of conventional weaving machines that are readily available for 3D fabric production. The immediate advantages of fabrics with a notable thickness dimension include the structural integrity of the woven structure, the satisfaction of the geometric shapes, and volumes that are required for many end-use applications [4]. In addition, the weaving technology is capable of leaving both crimped and straight yarns in the fabrics to suit the applications [5].

1.2   Classifications of 3D Woven Textiles

The comparison and classification of textile preforms is an arduous task due to the many forms of 3D fabrics available. Besides, the textile preforms can be classified according to different criteria, such as the yarn orientation, manufacturing process, and the geometric features of the textile preforms. [8] listed the hierarchy of the fibre assemblies into four levels i.e., the fibres, yarns, 2D fabrics and 3D fabrics, as shown in Table 1.

Due to the engineering requirements, fibre orientation, structural integrity, and fibre volume fraction of textile preforms are of paramount importance in constructing

Table 1 Fibre architecture for composites [8]

Level	Reinforcement system	Textile construction	Fibre length	Fibre orientation	Fibre entanglement
I	Discrete	Chopped fibre	Discontinuous	Uncontrolled	None
II	Linear	Filament yarn	Continuous	Linear	None
III	Laminar	Simple fabric	Continuous	Planar	Planar
IV	Integrated	Advanced fabric	Continuous	3D	3D

textile assemblies. [9] took into account the methods of manufacture, including the dimension of the structure, fabric architecture, yarn dimensions and directions within the preform. These can be seen in Table 2.

Khokar [10] reported on a type of 3D woven fabrics that require shedding in yarns in the fabric length direction and in the through-the-thickness direction, which permits the weft yarns to be inserted in two directions into the fabric. Based on that, he classified the woven fabrics which are as shown in Table 3. The key criterion for 2D and 3D woven fabrics according to this classification is shedding. Soden and Hill [11] classified the woven fabrics by the weaving process, which comprises of the warp binder, warp interlinked, in-plane interlacing yarns and stuffing warp yarns to produce an integrated structure.

It has been proved that 3D woven fabrics can be manufactured on the conventional weaving machines and on specially made weaving machines/devices. Regardless the types of machines used, the weaving technology is capable of constructing 3D fabrics with many different geometrical types. Reference [12] studied the configurations and geometries of the 3D woven fabrics and classified 3D woven fabrics into four different categories, i.e. the solid, hollow, shell and nodal, which are listed in Table 4.

2   Weaving Technologies for 3D Woven Textiles

3D textile structures have been used in various industries such as aerospace for some time. 3D fabrics, usually with through thickness direction linkages, are important forms of reinforcements and preforms for textile based composite materials. In relation to woven 3D fabrics, both the conventional weaving machines and purpose-made weaving devices are used.

Table 2 Fukuta and Aoki's classification system [9]

Dimension		Axis				
		Non-axial	Mono-axial	Bi-axial	Tri-axial	Multi-axial
1D			Roving Yarn			
2D		Chopped strand mat	Pre-impregnation	Plane Weave	Triaxial weave	Multiaxial weave knit
3D	Linear element		3-D braid	Multi-ply weave	Triaxial 3D weave	Multiaxial 3D weave
	Plane element		Laminate type	H or I beam	Honey comb type	

Table 3 Classification system by Khokar [10]

	Loom	Fabric dimension	Architectural structure	Yarn Interlacement
1	Conventional 2D loom	2D fabrication	Orthogonal	Warp and weft interlacing in one plane, horizontal
2	Conventional 2D loom	2.5D fabrication	Conventionally woven	Interlacing of warp, warp pile and weft pile in 2 mutually perpendicular planes
3	Conventional 2D loom	3D fabrication	Orthogonal with through-the-thickness, also known as multilayer	Interlacing orthogonal warp and weft with through the thickness
4	Conventional 2D loom	3D fabrication	Non-woven	Non-interlacing warp, weft and through the thickness
5	Specialised 3D loom	3D fabrication	Orthogonal	Three sets of orthogonal yarns, vertical, and horizontal shed obtaining a structure of columns and rows

Table 4 3D textile structures and weave architecture [12]

Structure	Architecture	Shape
Solid	Multilayer, orthogonal, angle interlock	Compound structure, with regular or tapered geometry
Hollow	Multilayer	Uneven surfaces, even surfaces, and tunnels on different levels in multi-directions
Shell	Single layer, multilayer	Spherical shells, and open box shells
Nodal	Multilayer, orthogonal, angle interlock	Tubular nodes and solid nodes

2.1 Conventional Looms for 3D Weaving

The conventional weaving technology involves 5 motions of the loom, i.e., shedding, weft insertion, beat-up, take-up and let off, which are list in Table 5. The weaving principle is capable of making structures that have substantial dimension in the thickness direction formed by layers of fabrics or yarns, generally termed as the 3D fabrics. Many of these structures can be made on the conventional weaving machines with no or little modification [3]. The conventional looms support 3D weaving from the following aspects. The powerful shedding methods such as dobby and jacquard are able to handle from tens to thousands individual warp yarn ends. The flexibility in warp supply systems permits warp ends to be sent from multiple weaver's beams or

Table 5 Loom motions and their functions

Loom motions	Functions
Shedding	Separate the warp sheet into upper and lower layers to form a shed, in time for a new pick of weft yarn to be inserted into it to form fabric. The shed formation decides the fabric structure
Weft insertion	Introduce a full length of new pick of the right choice into the shed. There are different methods for weft insertion
Beat-up	Push the newly inserted pick into the cloth fell (the meeting point of the fabric and the warp sheet) to form fabric. The movement of the reed carries out the beat-up
Take-up	Draw away from cloth fell the fabric formed at such a rate that the fabric will have the predetermined weft density. This is also to give space for the following pick to be woven into the fabric
Let-off	Supply the correct amount of warp length from the weaver's beam for weaving. A constant warp tension level must also be kept in order for the weaving process to continue

even creels to cater for independent yarn consumption and yarn tension. In addition, the conventional looms allow very large (virtually infinite) weft repeat in a fabric that is vital for 3D fabric formation.

2.2 Purpose-Made Weaving Devices for 3D Weaving

Many attempts have been successfully made in developing new weaving devices aiming for making 3D woven fabrics. Figure 1a shows the 3D weaving principle developed by Mohamed [6] and (b) the novel 3D woven structure created by Khokar [7].

These new technologies arrange warp yarn in a 3D formation and allow weft yarns to be inserted at different levels in one or two directions. The advantages of fabrics made by using such technologies are many. An obvious advantage is that

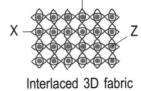

(a) 3D weaving principle [Curtesy Mohamed (2008)]   (b) 3D woven structure [Curtesy Khokar (2008)]

Fig. 1 Purpose-made 3D weaving technology and 3D structure

the weaving devices can make 3D fabrics much thicker than that made with the conventional technology. It is also fair to say that it is easier than the conventional weaving method to make net-shaped 3D shapes without the need of huge trimming after weaving, therefore maintaining the structural integrity and reducing material waste. By comparison, the conventional weaving technology can make various types of 3D fabrics but will be confined in thickness especially when making solid panels. The conventional weaving technology remains to be popular in making 3D fabrics maybe because of two reasons. The first is that some of the 3D fabrics needed are not always beyond the capability of the conventional weaving machines. The other may be because of the vast numbers of conventional weaving machines that are readily available for 3D fabric production.

3   Design and Manufacture of 3D Woven Fabrics

It has been explained that the 3D woven fabrics can be classified into solid, hollow, shell, and nodal, each having their structural characteristics.

3.1   3D Solid Woven Fabrics

3D solid woven fabrics are manufactured through incorporating and manipulating yarns in the length, width and the through-the-thickness directions. The employment of the through-the-thickness yarn within different types of solid fabrics differs greatly, relating to mechanical and other properties. For instance, orthogonal structures are a 3D structure containing straight yarns in the three principal directions, providing a stiffer and stronger 3D woven fabric against tensile loading. Within the multilayer structure the through-the-thickness yarns can be arranged to stitch with any other layers, leaving much room for structural manipulation.

3.1.1   Orthogonal Woven Fabrics

Orthogonal woven fabrics are featured the straight yarns in the warp, weft and through-the-thickness directions. The thickness of the orthogonal structure is defined by the number of layers of the warp (n) or weft yarn (m), with the layers of weft yarn being one more than that for the warp yarn, i.e.

$$m = n + 1 \tag{1}$$

Usually, the binding yarns follow a weave pattern, which is called the binding weave. Choice of binding weave determines the density of yarns that go in the through-he-thickness-direction. An orthogonal structure makes use of one binding weave is termed as an ordinary orthogonal structure. An orthogonal 3D fabric is an enhanced on if it uses two opposite binding weaves. In such a case, there will be two binding yarns with opposite movements running between two stacks of straight warp yarns. Obviously, the vertical links in the thickness direction will be doubled in an enhanced orthogonal fabric. Suppose that the warp and weft repeats of the binding weave are the same, and are expressed by r, then the warp repeat (R_e) and weft repeat (R_p) of the resulting orthogonal fabric weave can be written as following.

For ordinary orthogonal fabrics,

$$R_e = r(n + 1) \tag{2}$$

$$R_p = rm \tag{3}$$

and for enhanced orthogonal fabrics

$$R_e = r(n + 2) \tag{4}$$

$$R_p = rm \tag{5}$$

It is important to create the weave diagram in order to produce an orthogonal fabric. Take an ordinary orthogonal fabric for example, which contains two layers of straight warp yarn, and uses 2/1 Z twill as the binding weave.

The repeat of the binding weave is 3×3, therefore $r = 3$. Because there are 2 layers of straight warp yarn, we have $n = 2$, and $m = 3$. Using Eqs. (2) and (3), we know that

$$R_e = r(n + 1) = 3(2 + 1) = 9$$
$$R_p = rm = 3 \times 3 = 9$$

Therefore the weave diagram for the specified ordinary orthogonal fabric has 9 warp ends and 9 weft yarns. In order to create the weave diagram, it is useful to draw the fabric cross-section along the warp direction. In this case, the cross-section is shown in Fig. 2.

In Fig. 2, S_1 and S_2 are the two layers of the straight warp yarns, B_1, B_2 and B_3 are the three binding warp ends. The ovals in (a) represents the three layers of straight weft yarns, in this case three per layer. The weft yarns should be numbered indicating the sequence for weft insertion. The weave marks can be derived from this cross-sectional view. For B_1, it goes over weft yarns 1–6, and goes under 7–9, therefore the marks are given to weft yarns 1–6 and blanks are left for weft yarns 7–9, as shown in (b). For straight warp yarn S_1, it travels over weft yarns 2, 3, 5, 6, 8, and 9, and six marks are duly given in the column of S_1. Similarly, three marks

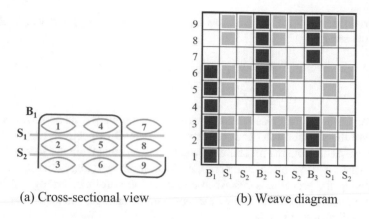

B_1
S_1
S_2

(a) Cross-sectional view   (b) Weave diagram

Fig. 2 Weave creation of ordinary orthogonal fabric

for S_2 are assigned along the S_2 column in (b). In a similar way, weave marks for B_2 and B_3 can be created. Figure 2b is the complete weave for the example orthogonal structure.

For enhanced orthogonal structures, two binding weaves, which are opposite to each other, will be used as a method to increase the amount of vertical straight yarns in the structure. There will be two binding warp ends between any two adjacent stacks of straight want yarns, travelling in opposite directions. The creation of weave diagrams is demonstrated through the example below, whose simulated 3D image is shown in Fig. 3.

Specification of an enhanced orthogonal structure:

Thickness—3 warp layers

Binding weave—3/2/2/1 Z.

Type—enhanced.

Fig. 3 An enhanced
orthogonal fabric

Based on the specification, we know that $n = 3$ hence $m = 4$, and $r = 8$. According to Eqs. (4) and (5),

$$R_e = r(n + 1) = 8(3 + 2) = 40$$
$$R_p = rm = 8 \times 4 = 32$$

Figure 4a shows the cross-sectional view along the warp direction, where S_1, S_2 and S_3 are the straight warp ends for layers 1, 2 and 3 respectively; B_1 is the first binding warp end from the first binding weave and B_1' that from the second binding weave. All 32 weft yarns are numbered, indicating the weft insertion sequence.

Following the procedures described earlier for the ordinary orthogonal fabrics, the weave diagram for the enhanced example can be created. The completed weave diagram is depicted in Fig. 5b.

In any orthogonal 3D fabric, yarns are basically running straight in the warp, weft and the through-the-thickness directions. The through-the-thickness yarns, or the binding warp yarns, travel vertically between any two weft yarn layers. The most common situation is when the binding yarns travel between the top and bottom weft yarn layers. When the through-the-thickness yarns bind weft yarns at any other levels, tapered and figured solid preforms can be created based on the orthogonal principle. Some varieties of the orthogonal structures are indicated in Fig. 5.

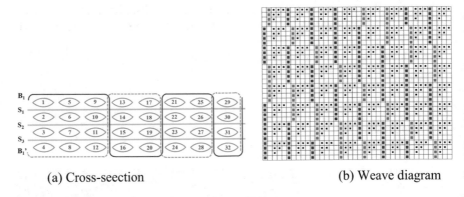

(a) Cross-seection   (b) Weave diagram

Fig. 4 The cross-section and weave diagram of the enhanced orthogonal fabric

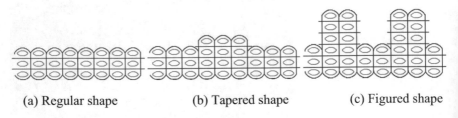

(a) Regular shape   (b) Tapered shape   (c) Figured shape

Fig. 5 Orthogonal woven structures with variations

It is also possible that the orthogonal structures are made using some weft yarns as the binder [13].

3.1.2 Angle-Interlock Woven Fabrics

Original angle-interlock fabrics contain a set of straight weft yarns and a set of warp yarns that weave with the weft in a diagonal direction in the thickness. In most cases, the warp yarns bind diagonally from the top to the bottom and it is also possible to bind the layers of weft yarns not for the full depth [14]. There could be straight warp ends added into the angle-interlock fabrics. Figure 6 shows the cross-sections of angle-interlock fabrics of 7 layers of weft yarns with the warp binding the structure in different ways. It has been reported that when wadding warp yarns are incorporated into the angle-interlock fabrics, the tensile modulus and tensile strength in the warp direction are sharply increased [15].

It has been reported that the angle interlock fabrics have low shear rigidity [16]. In an angle-interlock fabric, such as the one shown in Fig. 6a, there is only on set of warp yarn weaving with the 7 layers of weft yarns, and the warp yarns only interlace with the weft yarns at the top and bottom layers. As such, the frictional resistance exerted when shearing the angle-interlock fabric is lower compared to other forms of 3D woven fabrics that use the same amount of materials. Low shear rigidity leads to good mouldability of the angle-interlock fabrics and this has been found beneficial in forming wrinkle-free doubly curved surfaces for applications such as helmet shells and in female body armour [17–20]. When wadding warp yarns are incorporated structure, it has been reported [15] that the tensile modulus and strength in the warp direction are sharply increased.

Changes in the interlocking depth in this type of fabrics lead to the creation of tapered 3D textiles.

The thickness of an angle-interlock fabric is measured by the number of straight weft yarns layers (n) involved in the fabric. Upon the specification of the straight weft yarn layer, warp and weft repeats (R_e and R_p) can be calculated as follows.

$$R_e = n + 1 \qquad\qquad (6)$$

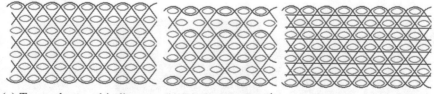

(a) Top-to-bottom binding   (b) Binding to the 5th layer (c) Warp wadded fabric

Fig. 6 Cross-sectional views of angle-interlock fabrics (along the warp direction) with 7 layers of weft yarns

$$R_p = n(n + 1) \tag{7}$$

Similar to the weave creation for orthogonal fabrics, it is beneficial to draw the cross-section along the warp direction before hand. The weft yarns can be arranged in either aligned or staggered formation, with the latter leading to a more stable and more solid 3D fabric. Figure 7 shows the cross-sectional view, the 3D image, and the weave diagram of an angle-interlock fabric with 4 weft yarns, with the staggered weft yarn formation. The number of weft yarns per layer is $(n + 1)$, and in this case there are 5 weft yarns per layer. The weft yarns are numbered to indicate the weft insertion sequence, and the 5 warp ends are designated using labels B_1 to B_5. From the cross-section shown in Fig. 7a, it can be seen that warp end B_1 travels over weft yarns 1–4, 6, 8, 16, and 18–20, and under all the rest. The interlacement of B_1 with all weft yarns is represented in the weave diagram in Fig. 7c. The interlacement of other warp ends can be generated similarly.

In the same piece of 2D angle interlock fabric, the number of weft yarn layers can be reduced or added to, leading to tapered and figured 3D angle-interlock fabrics.

It is noticeable that in an angle-interlock all warp ends are crimped but the weft yarns are straight, which means very different mechanical and possibly other properties. If necessary, wadding warp end can be added to the angle-interlock fabric. Usually, wadding warp ends can be added between any two adjacent straight weft layers, and one between any pair of neighbouring interlacing warp ends. Take the 4-weft-layer angle-interlock fabric for example to demonstrate the wadding yarn addition. The warp and weft repeats of the wadded angle-interlock fabric will be the same and can be expressed below.

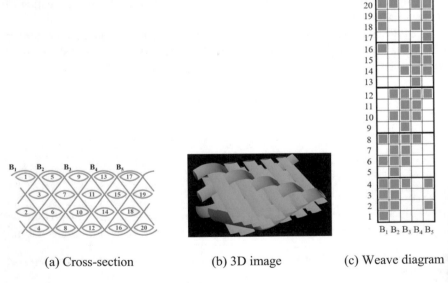

(a) Cross-section   (b) 3D image   (c) Weave diagram

Fig. 7 Weave creation of an angle-interlock fabric

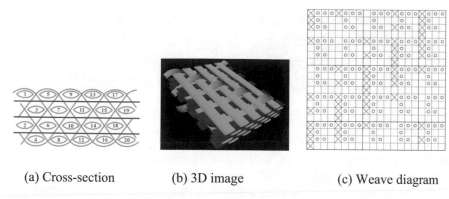

(a) Cross-section   (b) 3D image   (c) Weave diagram

Fig. 8 The angle-interlock fabric with wadding yarns

$$R_e = R_p = n(n+1) \tag{8}$$

Figure 8 displays the cross section, the 3D image, and the weave of the wadded angle-interlock fabric with 4 layers of weft yarns.

3.1.3 Multilayer Woven Architecture

The distinctive feature of the multilayer fabrics is that they have clear definition of fabric layers through the thickness. Each layer is composed of a set of warp ends and a set of weft yarns. The layers are connected together through weaving by either the existing yarns (self stitching) or external sets of yarns (central stitching) [21], 22. A stitch can be introduced from any layer to any other layer, but the stitch introduction needs to satisfy, conventionally, 2 conditions, which are (i) the stitching points should be properly hidden from the fabric layer it stitches to, and (ii) the introduction of a stitch should not alter the weave pattern of the fabric layer from which the stitch is introduced. These basically ensure fabric appearance and will have influences on the fabric performances.

Due to the structural characteristics, all warp and weft yarns in a multilayer fabric are crimped, which leads to relatively low initial modulus under tensile loading along the warp and weft directions. Straight wadding yarns can be added between any two fabric-layers in either warp or weft or both directions, and the additions of wadding yarns will result in higher modulus of the fabric [23]. Figure 9 shows 4-layer fabric with self stitching.

Specification of Multilayer Woven Fabrics

To structurally define a multilayer fabric, it is necessary to specify the following: (i) the number of layers, (ii) weave for each layer, (iii) yarn ratios among the different

Fig. 9 Cross-sectional view of a stitched 4-layer fabric (along the warp direction)

layers, (iv) stitching type, i.e. self stitching or central stitching, and (v) the stitching plan. Upon the specification, the following need to be carried out in order to create the weave diagram and the weaving instruction. A multilayer fabric is specified as follows as an example.

Number of fabric layers: 2

Weaves for each layer: 1st layer—4/1 Z; 2nd layer—1/4 Z

Yarn ratio among layers: warp—1:1; weft—1:1

Stitch type: self stitching

Stitching plan: Stitching once per yarn per repeat.

Weave Repeat Calculation

The weave assigned to each fabric layer may be different, therefore the weave repeat for each weave can be different. Assume that a multilayer fabric has n fabric layers and that the original repeat for the ith layer is r_i ($i = 1, 2, ..., n$). In order to achieve structural continuity in the 3D multilayer fabric, all individual weaves should repeat integral times exactly within the multilayer fabric repeat. Let r' be the readjusted repeat for all layers, which is the lowest common multiple (*lcm*) of the original repeats of weaves for all layers, expressed as

$$r = lcm(r_i)(i = 1, 2, \ldots, n) \tag{9}$$

When the yarn ratio is 1:1: ......:1, the repeat of the multilayer fabric is calculated as follows,

$$R_e = R_p = nr \tag{10}$$

In the case the yarn ratio is not 1:1: ...:1, new repeats for each fabric layer will have to be worked out first according the yarn ratio before calculating the multilayer fabric weave repeat.

Based on the 2-layer fabric specification given earlier where $n = 2$, it is known that the warp and weft repeats for the first and second layers are $r_1 = 5$ and $r_2 = 5$

Fig. 10 Layer indications
and weave superimposition

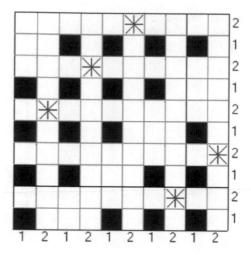

respectively. The lowest common multiple leads to $r = 5$. Because the yarn ratios are 1:1, the warp and weft repeat of the multilayer fabric will be

$$R_e = R_p = n \times r = 10$$

Superimposition of Weaves of All Layers

At this stage, all weaves assigned to individual fabric layer are put together according to the specification. After the overall warp and weft repeats have been worked out, weave design paper should be used to for recording the weaves for the layers. The yarn ratios for warp and weft will be used to mark the layer each warp and weft yarn is for. In Fig. 10, the warp and weft yarns for the two fabric layers are clearly indexed for the lay of fabrics they are for, with "1" for the 1st layer, and "2" for the second.

At the end of this stage, the superimposition of the weaves does not represent the 3D multilayer fabric, as the fabric layers have not been separated yet.

Layer Separation

After inserting weaves for different fabric layers into the weave diagram area, the weave diagram does not yet represent the multilayer fabric structure. In order to achieve the weave representing the multilayer fabric, the layers must be separated by adding some extra marks into the diagram area. The principle used for separating the fabric layers are described below.

When inserting a weft yarn for the ith layer of the fabric,

(i)   all warp ends for layers 1 to $(i - 1)$ will have to be lifted indicated by added marks;

Fig. 11 The 2-layer fabric weave with lifters indicated by "/"

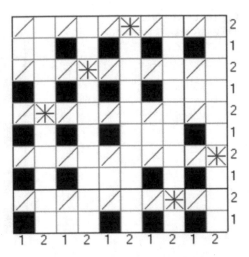

(ii)   for layer i, the separation of the warp yarn sheet is determined only by the weave assigned to that layer; and

(iii)   all warp ends for layers $(i + 1)$ to n will have to remain not lifted, and no extra marks are added.

For the 2-layer fabric example, when inserting weft yarns for the 1st fabric layer, the weave for the first layer alone determines which warp end to be lifted, whilst all warp ends intended for the second fabric layer will have to remain not lifted and there is no need to add any mark to the weave diagram. When a weft yarn is inserted for the second layer, all warp ends for the first fabric layer will have to be raised, meaning extra weave marks will have to be added, while warp yarns for the second layer fabric forms the shed according to the weave only. The added warp-up marks in the weave diagram are termed as lifters. Figure 11 illustrates the 2-layer fabric weave after lifters have been added to separate the two layers.

Stitching

Despite the fact that multiple individual fabric layers can be produced, the multilayer weaving technology is mainly used to make multilayer fabrics with inter layer linkages, or stitching, which are facilitated during the fabric weaving process. A stitching can be introduced using two different methods, i.e. the central-stitching method and the self-stitching method. The former requires a new set of warp or weft yarns that weaves between layers to connect layers together, whereas the latter makes use of the existing warp and/or weft yarns to perform the inter-layer linkage. A selected yarn may go up or down for stitching, but would need to make reference to two considerations:

(i)   The introduction of stitching should not affect the weave features of the fabric layer, from which the stitching is introduced; and

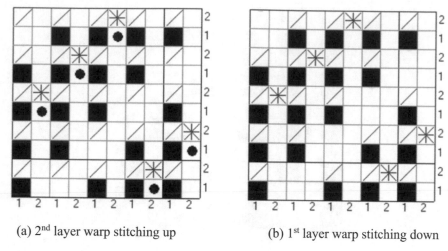

(a) 2nd layer warp stitching up   (b) 1st layer warp stitching down

Fig. 12 Single stitching between two layers

(ii)   At the targeting layer for stitching, the stitching points should be properly concealed underneath the yarn floats, not to affect the existing weave feature of this layer of the fabric.

In relation to the two considerations, the selection of weaves for different layers should be compatible in that top layers should offer longer up-floats with the bottom layers having longer down-floats of yarns. When introducing stitching from a layer, upper floats should be used for up-stitching and lower floats should be used for down-stitching. A stitching point (usually single) should be placed between two adjacent yarn floats in the targeting fabric layer. Figure 12 shows the weave diagrams of the 2-layer fabric with (a) the second layer warp ends are arranged for upper stitching, where dots are used to reflect the up-stitching of the second layer warp ends; and (b) the first layer warp stitching down, where selected lifters have been removed to indicated the down stitching of the first layer warp ends. In each case, the introduction of stitching is one directional, and such forms of stitching are called single stitching. If stitching is dual directional, then it is called double stitching. Figure 13 displays the 2-layer weave diagram with double stitching.

A Four-Layer Fabric Example

This section demonstrates the creation of weave diagram of a 4-layer fabric. The four weaves specified for layers 1 to 4 are 4/1 Z, 3/2 Z, 2/3 Z, and 4/1 Z. The yarn ratios for both warp and weft are 1:1:1:1. Accordingly, the warp and weft repeats of the multilayer fabric are both 20. Figure 14 is the weave diagram of this 4-layer fabric with second layer warp yarn stitching to layers 1 and 4, third layer warp yarn stitching to layers 1 and 4.

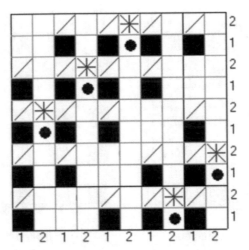

Fig. 13 Weave diagram of the 2-layer fabric with double stitching

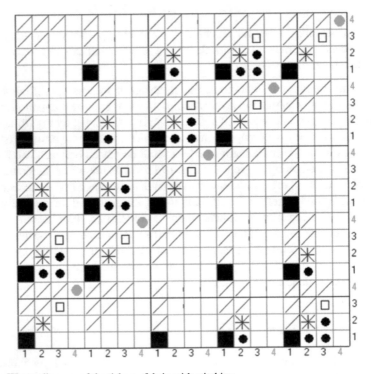

Fig. 14 Weave diagram of the 4-layer fabric with stitching

4 3D Hollow Woven Preforms

3D hollow woven preforms can be generally divided into two types. One is with flat top and bottom surfaces and the other with uneven surfaces [24, 25]. Both can be manufactured based on the conventional weaving technology.

4.1 Hollow Fabrics with Flat Surfaces

This type of hollow fabrics incorporates three or more layers of fabrics. In the case where three layers of fabrics are used, the layer of fabric connecting the top and bottom layers will be woven with longer length than the top and bottom layers. The length of the middle layer is determined by the thickness of the fabric and the config-uration of the fabric cross-section. The cross-sectional shape can be trapezoidal, triangular, or rectangular [26], as shown in Fig. 5. When this type of fabrics is made from yarns with suitably rigid yarns, such as glass, the structure is self opening when resin is wicked into the fabric.

The hollow fabrics can be made in single level or multiple levels. Diesselback and Stahl [27] produced a triple-level flat surfaced structure with a triangular inner core. It was reported that due to their 'expanding' nature these fabrics are air-inflated once taken off the loom.

Rheaume [28] invented a 3D hollow structure with flat surfaces comprising of two independent levels of plain woven fabric, connected by vertical sections of fabrics. This is a special case of hollow fabric where the whole structure is foldable. The structure was woven flat and then opened up when taken off the loom. During weaving, each fabric section was woven using a designated shuttle, enabling the structure to be woven in a flattened 2D configuration.

Chen and Wang [25] modelled the mathematically hollow fabrics with flat surfaces and created an algorithm for such fabrics to be designed and manufactured. Based on the specification of the cross-sectional geometry and the weave type for each layer, the algorithm generates the overall weave for the hollow structure that can be used directly for manufacture.

A similar type of hollow fabric is the woven version of the "spacer fabric", which is composed of two parallel layers of fabrics connected by vertical yarns/fibres. The spacer fabrics are mainly made using the knitting technology and their air perme-ability and thermal conductivity are much dependent on the fabric parameters and fibres used [29]. When such fabrics are made by weaving, the face-to-face velvet weaving technology is used, such as the Van de Wiele looms [30]. In this principle, the two substrate layers of fabrics will be made simultaneously while the pile warp yarn travels between the layers to serve as the spacer yarn.

4.2   Hollow Fabrics with Uneven Surfaces

Takenaka and Eiji [31] fabricated a 3D hollow fabric based on the multilayer principle, where the adjacent layers of fabrics are combined and separated at arranged intervals. The combining and separating of adjacent layers allow the whole fabric to be opened up after weaving to become a 3D structure with cellular shaped cells in the cross-section. Because of the opening of the structures, the top and bottom surfaces of this type of hollow structure will not be flat. Chen et al. [24] studied the mathematical modelling of this structure and an algorithm was created for the computerised design and manufacture of this type of hollow fabric.

In a more systematic study [24, 32, 33], the 3D hollow fabrics with uneven surfaces were defined based on the number of fabric layers involved and the lengths of the two types of cell walls. A free cell wall of a hexagonal cell is one that is formed by one layer of fabric, whereas a bonded cell wall is a wall that is created by combining two adjacent fabric layers. Any hollow fabric of this type can be denoted by the general expression $xL(y + z)P$, where x is the number of layers of fabrics involved in making the hollow fabric, y is the length of the bonded walls and z is the length of the free walls, both measured in number of picks. In this notation, L stands for layers of fabrics and P for picks respectively. When the bonded walls have the same length as the free walls, the notation reduces to $xLyP$. As an example, the $8L3P$ hollow fabric is one that involves 8 fabric layers and is with regular shaped hexagonal cells in the cross-section, whose free wall and bonded wall both involve 3 picks.

5   3D Shell Woven Fabrics

3D shell woven fabrics refer to fabrics that form doubly curved shell structures where fibre continuity is maintained. Such fabrics may assume spherical shapes and may be in the form of cubic shapes. Such shell shaped fabrics can be made using different techniques, such as weaving with discrete take-up, weave combination, moulding, and the origami method.

5.1   Weaving with Discrete Take-Up

Busgen [34] invented a method for the creation of 3D shell fabrics by direct weaving. Modifications were carried out to the conventional loom on the take-up and let-off mechanisms. The conventional one-piece take-up roller was replaced by one made up by many discs which are electronically controlled to perform individual take-up movement. The discrete take-up of the fabric results in double curvatures in the fabric. Weft density of the fabric can be reduced at the top of the curvature and some extra weft yarn can be added to ensure uniform material density in the fabric.

Modifications to the let-off mechanism were also made in such a way that each yarn is controlled independently. Chen and Tayyar [35] employed an easy-to-use add-on device to the conventional loom for making 3D shell fabrics. They used a profiled take-up roller to take up the fabric for the creation of the dome shape. Changes in the let-off were made to accommodate the take-up discrepancies as in the previous case. When the add-on device is removed, the loom resumes to its original set-up for making ordinary fabrics.

5.2   Use of Combined Weaves

In making everyday woven fabrics, the aim is to create flat fabrics with even material distribution. When weaves with different float lengths are used side by side in a fabric, cares need to be taken that the tension difference caused by different weaves must be evened out in order not to affect the flatness of the fabric. However, in some cases the effect of an uneven fabric is desired, and one notable example is the honeycomb weave, which is a combination of short-float weaves (typically the plain weave) and long-float weaves. When short-float weaves are used more intersections between the warp and weft yarns will take place than in the fabrics with long-float weaves. This leads to the situation where the short-float fabric section takes up more areal space than the long-float fabric section. In another word, the fabric section made from short-float weave will tend to expand whereas the fabric section made from long-float weave will tend to shrink. This is especially the case when the fabric is removed from the loom and is free of tension. This creates the bubbled effect of the fabric. This same principle was used to create 3D shell fabrics [35]. Concentric rings were planned and gradient change of weaves was employed in the areas between two adjacent rings.

5.3   Shell Fabrics by Moulding

Due to the extensibility of yarns and fabrics and the allowance of shear, most flat woven fabrics can be moulded into doubly curved surfaces to some extent. One example is that fabrics are moulded into 3D shapes as lining material for the inside of car doors. However, the extent that a woven fabric can be moulded is quite limited mainly due to the friction between the warp and weft yarns at the cross-over points and space limitation brought about by the fabric structure [20]. It is reported [16] that the 3D angle interlock fabrics have notably lower resistance against shear than other 2D and 3D fabrics under the same circumstances. This finding has helped the development of helmet shells and female body armour.

5.4   Use of Origami Principle for Box Shells

Shell shapes may also be in the form of an open box. For this type of 3D shell structures it is expected that the fibres will be continuous throughout the box shell. Reference [36] worked on such shell structures and proposed that the origami (paper folding) principle can be used to fold the box shell as the first step. For a given box shape which is to be woven, there are usually different ways to get it folded in such a way that is suitable for weaving. The optimal solution for folding should be the one that gives the best fibre orientation and hence the most appropriate performance and the one that can be woven most conveniently.

Once an open box has been folded flat, regions can be classified according to the structural features of the folded geometry.

6   3D Nodal Woven Fabrics

A 3D nodal fabric refers to a fabric that facilitates a network formed by different tubular or solid members joining together. Figure 14 illustrates one type of the nodal fabrics [37]. In such a case, each member is a tube whose wall may be of a single layer fabric or any form of the 3D solid fabrics [38]. The fabric is woven flat and when taken off the loom the fabric is pulled into shape.

Many have worked on the design, manufacture and application of the 3D nodal fabrics for different end-uses. Reference [39] developed woven nodal structures from the plain weave using thermal reactive fibres which enable the tubular members to have adaptable dimensions. Day et al. [40] create a multitude of possibilities of nodal structures based on the use of a jacquard loom. The woven flanges are required to have an area for defining and joining the two cloths together by stitching to obtain the tubular structures. Reference [19] described that tubular members can be formed within a 3D fabric and the relationships of these tubular members can be one above the other or intersectional.

Taylor [38] worked on the design and creation of nodal structures and developed a procedure for the design of generic nodal fabrics. The design process involves the nodal structure specification, structure flattening, pattern segmentation, weave assignment, and weave combination. One important issue in manufacturing nodal fabrics is the smooth intersection of two tubes with or without the same diameter. The algorithm led to a solution that ensures high quality intersection between tubes. Based on this achievement, Smith [37] computerised the algorithms and produced software to assist the design and manufacture of 3D nodal fabrics.

7 3D Woven Architectures from Specially-Made Devices

The conventional weaving machine is basically development for making 2D fabrics, primarily for domestic applications. As it has been demonstrated, it can be used for making various types of 3D fabrics, but would run into problems when the thickness of the 3D fabric becomes substantial. In addition, the conventional weaving machine allows the weft yarns to be inserted only in one direction. This would pose limitation to the creation of 3D fabrics. In parallel to the use of conventional weaving machines, different forms of weaving devices have been developed for making 3D fabrics.

King [41] developed a 3D textile assembly using a purpose-made weaving device from static yarns oriented in the Z direction. The X yarns are inserted first then beaten into place, and then the Y yarns are inserted and beaten. This is repeated to produce a compact structure until the desired height is achieved, producing a 3D rectangular cross-sectional configuration. There is no shedding involved in this technology and the final product is an orthogonal structure which demands no interlacements. Figure 18 illustrate the principle of this non-interlacing weaving device.

Mohamed and Zhang [42] patented a method for creating variable cross-sectional shaped 3D fabrics based on the orthogonal structure. Apart from the binding warp yarns, the straight warp yarns are made into several layers and the weft yarns at different levels are inserted into the fabric simultaneously from one side or from both sides of the fabric. If inserted from both sides, the weft yarns may be inserted simultaneously or alternately from each side of the warp. The binding warp yarns bind the structure together due to shedding. Depending on shedding, the 3D fabric could be orthogonal, angle interlock, warp interlock, or combination of all three in the same piece. Figure 19 shows the principle of the weaving machine and some of the products [6].

Based on the belief that weaving must involve yarns interlacing with one another, Khokar [10] developed a weaving device that is capable of shedding in two directions. The linear-linear dual-directional shedding operation alternately displaces the grid-like arranged warp yarns Z to enable creation of sheds in the fabric-thickness and fabric-width directions. Consequently, two mutually perpendicular sets of corresponding weft yarns Y and X can be inserted into the created sheds. This enables the warp yarns to interlace with the vertical and horizontal weft yarns to create an interlaced woven 3D fabric. Varieties of 3D fabric structures can be made using this method. Such made 3D fabrics offer ultimate structural integrity even when the fabric is cut or damaged.

8 Conclusions

As a matter of fact, the textile technology is the only approach that can organise unidirectional materials into 2D and 3D assemblies with desired shape, dimension and to some extent material orientation. 3D fabrics have been widely discussed to be used as preforms for advanced composites because of their many advantages, including the structural integrity, lightweight, and high performance. This paper has reviewed different types of 3D woven fabrics and routes for manufacturing them.

In the discussion of 3D fabric structures, this paper reviewed the 3D woven fabrics in the categories of 3D solid, 3D hollow, 3D shell, and 3D nodal fabrics. The 3D solid fabrics, which can be made using different weave geometries, not only can be used as reinforcements to board shaped and tapered composites, they can also be used as constituent sections in other 3D fabrics such as 3D nodal and shell structures. The 3D hollow fabrics lead to lightweight composites that are good for energy absorption and impact protection. The 3D shell fabrics can be used to reinforcing composites that have doubly curved shapes with structural integrity. The origami method is used to produce box shapes fabrics that can be further refined for more complex components. 3D nodal structures are particularly useful for making truss structures from textile composites.

3D woven fabrics are basically made through two routes nowadays, one by using the conventional weaving technology, another by using specially developed weaving devices or heavily modified conventional looms. The paper highlighted two types of specially engineered weaving devices, one enabling simultaneous multi-level weft insertion and the other employing dual-shedding. Whilst these are capable of making 3D woven fabric with considerable sizes, the availability of such devices is much limited in comparison to the prevalence of the conventional weaving machines.

Acknowledgements Appreciations go to Ms. Yajie Gao, a final year Ph.D. student waiting for her viva, who kindly helped in creating some of the illustrations used in this Chapter.

References

1. Long, A. (Ed.). (2005). *Design and manufacture of textile composites*. Woodhead Publishings Ltd.
2. Chen, X., & Sun, D. (2009). Textile materials in personal protection equipment for the police and military personnel. *Journal of Xi'an Polytechnic University, 23*(2), 67–74.
3. Chen, X., & Tayyar, A. E. (2003). Engineering, manufacture and measurement of 3D domed woven fabrics. *Textile Research Journal, 73*, 375–380.
4. Chen, X., & Hearle, J. W. S. (2008). Developments in design, manufacture and use of 3D woven fabrics. In: *TEXCOMP9 (International Conference on Textile Composites), University of Delaware, USA*.
5. Ko, F. K. (1999). 3D textile reinforcements in composite materials. In: A. Mirvete (Ed.), *3D textile reinforcements in composite materials*. Cambridge, England: Woodhead Publishing Ltd.
6. Mohamed, M. H. (2008). Recent advances in 3D weaving. In: *Proceedings to The 1st World Conferences in 3D Fabrics and Their Applications, Manchester, UK*.
7. Khokar, N. (2008). Second-generation woven profiled 3D fabrics from 3D-weaving. In: *Proceedings to The 1st World Conferences in 3D Fabrics and Their Applications, Manchester, UK*.
8. Scardino, F. L. (1989). Introduction to textile structures. In: T.W. Chou, & F.K. Ko (Eds.), *Textile structural composites*. Covina, CA, USA: Elsevier.

9. Fukuta, K., & Aoki, E. (1986). 3D fabrics for structural composites. In: *Proceedings to the 15th Textile Research Symposium, Philadelphia, PA, USA.*
10. Khokar, N. (2001). 3D-weaving: Theory and practice. *Journal of Texttile Institution, 92,* Part 1(2), 193–207.
11. Solden, J. A., & Hill, B. J. (1998). Conventional weaving of shaped preforms for engineering composites. *Composite Part A: Applied Science and Manufacturing, 29*(7), 757–762.
12. Chen, X. (2007). Technical aspect: 3D woven architectures. In: *NWTexNet 2007 Conference, Blackburn, UK.*
13. Banister, M., & Herszberg, I. (1995). Mechanical performance and modelling of 3D woven composites. In: *Proceedings to the 4th International Conference on Automated Composites (ICAC95), Nottingham, UK.*
14. Chen, X., & Potiyaraj, P. (1999). CAD/CAM of the orthogonal and angle-interlock woven structures for industrial applications. *Textile Research Journal, 69*(9), 648–655.
15. Han, Z. (1999). Production and tensile properties of 3D woven structures as composite reinforcements. MPhil *Thesis,* Department of Textiles, University of Manchester Institute of Science and Technology, UK.
16. Chen, X., Knox, R. T., McKenna, D. F., & Mather, R. R. (1992). Relationship between layer linkage and mechanical properties of 3D woven textile structures. In: *Proceedings to the International Symposium on Textile and Composite Materials for High Functions, Tampere, Finland.*
17. Chen, X., & Yang, D. (2010a). Use of 3D angle-interlock woven fabric for seamless female body armour, Part I: Ballistic evaluation. *Textile Research Journal. 80*(15), 1581–1588. https://doi.org/10.1177/0040517510363187.
18. Chen, X., & Yang, D. (2010b). Use of 3D angle-interlock woven fabric for seamless female body armour, Part II: Mathematical modelling. *Textile Research Journal. 80*(15), 1589–1601. https://doi.org/10.1177/0040517510363188
19. Chen, X., & Zhang, H. (2006). Woven textile structures, Patent No. GB2404669.
20. Roedel, C., & Chen, X. (2007). Innovation and analysis of police riot helmets with continuous textile reinforcement for improved protection. *Journal of Information and Computing Science, 2*(2), 127–136.
21. Grosicki, Z. (1977). *Watson's advanced textile design: Compound woven structures* (4th ed.). Newnes-Butterworths.
22. Newton, N., Georgallides, C., & Ansell, P. (1996). A geometrical model for a two-layer woven composite reinforcement fabric. *Composite Science and Technology, 56,* 329–337.
23. Georner, D. (1989). *Woven structure and design—Part : compound structures.* Leeds, UK: British Textile Technology Group.
24. Chen, X., Ma, Y., & Zhang, H. (2004). CAD/CAM for cellular woven structures. *Journal of the Textile Institute, 95*(1–6), 229–241.
25. Chen, X., & Wang, H. (2006). Modelling and computer aided design of 3D hollow woven fabrics. *Journal of the Textile Institute, 97*(1), 79–87.
26. Koppelman, E., & Edward, A. R. (1963). Woven panel and method of making same, US Patent No. 3090406.
27. Diesselback, D., & Stahl, D. (1983). Dimensionally stable composite material and process for manufacture thereof, Patent No. US4389447.
28. Rheaume, J. A. (1970). Three-dimensional woven fabric, Patent No. US3538957.
29. Yip, J., & Ng, S.-P. (2008). Study of three-dimensional spacer fabrics: Physical and mechanical properties. *Journal of Materials Processing Technology, 206,* 359–364.
30. Van de Wiele website. (2010). http://www.vandewiele.com/src/.
31. Takenaka, K., & Eiji, S. (1988). Woven fabric having multilayer structure and composite material comprising the woven fabric, Patent No. US5021283.
32. Chen, X., Sun, Y., & Gong, X. (2008). Design, manufacture, and experimental analysis of 3D honeycomb textile composites, Part I: Design and manufacture. *Textile Research Journal, 78,* 771–781.

33. Tan, X., & Chen, X. (2005). Parameters affecting energy absorption and deformation in textile composite cellular structures. *Materials & Design, 26*, 424–438.
34. Busgen, A. (1999). Woven fabric having a bulging zone and method and apparatus of forming same, Patent No. US6000442.
35. Chen, X., & Tayyar, A. E. (2003). Engineering, manufacturing, and measuring 3D domed woven fabrics. *Textile Research Journal, 73*(5), 735–380.
36. Chen, X., & Tsai, L.-J. (2009). Weaving of complex composite preforms of 3D shapes based on the origami principles. In: *Proceedings to the 3rd Annual Conference of Northwest Composites Centre, Manchester, UK*.
37. Smith, M. A. (2009). CAD/CAM and geometric modelling algorithms for 3D woven milti-layer nodal textile structures. Ph.D. Thesis, School of Materials, The University of Manchester, UK.
38. Taylor, L. W. (2007). Design and manufacture of 3D nodal structures for advanced textile composites. PhD Thesis, School of Materials, The University of Manchester, UK.
39. Lowe, F. J. (1987). Articles comprising shaped woven fabrics, Patent No. 4668545.
40. Day, G. F., Robinson, F., & Williams, D. J. Composite articles, Patent No. US4923724.
41. King, R. W. (1976). Apparatus for fabricating three-dimensional fabric material, Patent No. US3955602.
42. Mohamed, M. H., & Zhang, Z.-H. (1992). Method of forming variable cross-sectional shaped three-dimensional fabrics, Patent No.US5085252.

3D Weaving Process Adapted for Natural and High Performance Fibres

Frédérick Veyet, Francois Boussu, and Nicolas Dumont

Abstract The 3D warp interlock weaving process, derived from the 2D weaving loom, can face to many problems during the production. Based on our previous research works using several looms and different types of yarns, with variations on the raw material and linear density, the 3D warp interlock weaving could led to yarn-to-yarn and yarn to loom parts friction due to high warp yarns density and thickness values as well as difficulties on the yarns location into the 3D woven structure. Some yarn breaks, jams and structure distortion will be major issues to fix to avoid any weaving rupture during production. Moreover, the yarn degradations and the inaccuracy of the structural geometry will have an impact on the mechanical properties of the 3D woven fabric. This chapter will provide tips and advises to any weaver or designer to improve the specific weaving process of 3D warp interlock woven fabrics. From loom adaptation to yarn's placement, during the preparation step on the weaving machine, and according to the various nature of yarns (natural or high performance synthetic or inorganic yarns), these advices shown in this chapter can appear as a solution to overcome technical barriers.

1 Introduction

The weaving of 3D warp interlock fabric is mainly based on the weaving technique of 2D fabrics [1] (Fig. 1). The elements of the weaving machine such as frames, heddles and the insertion of the weft thread through the shed are identical. The major difference lies in the distribution of the warp yarns, which are divided into three types: binding, surface and stuffer warp yarns [2, 3].

In contrast to 2D weaving, 3D warp interlock weaving requires several warp yarn feeding systems, which preferably correspond to the number of different evolutions of these yarns into the 3D woven structure (Fig. 2). The weaving principle is then

F. Veyet (✉) · F. Boussu · N. Dumont
ENSAIT, ULR 2461 - GEMTEX - Génie et Matériaux Textiles, University Lille, 59000 Lille, France
e-mail: frederick.veyet@ensait.fr

© Springer Nature Switzerland AG 2022
Y. Kyosev and F. Boussu (eds.), *Advanced Weaving Technology*,
https://doi.org/10.1007/978-3-030-91515-5_13

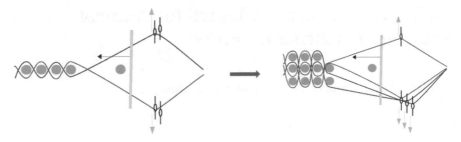

Fig. 1 Comparison between the 2D weaving (left) and the 3D warp interlock weaving (right) [4]

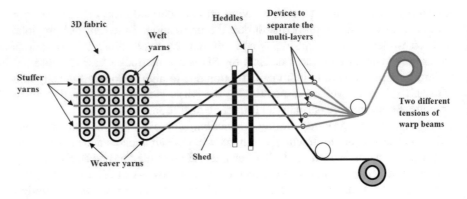

Fig. 2 Management of the different warp yarns onto the 3D weaving loom [6]

similar, we can recover the motion of the frames and the shed opening which allow the insertion of the weft yarn in the different layers of the 3D woven structure [5].

Based on these main motions to produce a woven structure and according to our research work [7–13], we have adapted our production tools to the weaving of 3D warp interlocks fabrics, in order to avoid problems of "weave-ability" or non-conformity of the 3D woven geometry [4].

Companies such as Shape 3 [14] and recently Dornier [15] have also developed and adapted machines to the weaving of 3D warp interlocks, including simultaneous insertion of weft yarns into the different warp sheet sheds, thus increasing production speeds.

However, the increase of the number of layers into the 3D woven structure to reach a high thickness (more than 4 layers for instance) may face to many issues during the weaving and could be ended after few inserted weft yarns. Moreover, depending on the physical and mechanical behaviour of the yarn used for the weaving such as a low elongation and high stiffness as well as an untwisted multi-filaments structure, the 3D warp interlock weaving process could be very difficult and laborious.

Thus, this chapter will be focus on the different approaches to reach a successful weaving process of 3D warp interlock fabric. The first approach is dedicated to the fibre degradation and the way to avoid or reduce it. The second approach is focused on the warp yarns density and its consequence during the 3D warp interlock weaving to obtain thick structure. The third approach concerns the geometry of the 3D woven structure and how to optimize its yarns' placement.

2 Fibre Degradation

Fibre degradation is one of the main problems occurring during the 3D warp interlock weaving. Higher value of warp yarns density can lead to higher fibre degradation and then makes harder the weaving process.

Warp yarns are generally more affected then weft yarns by fibre degradation, which can be explained by higher contact surface with loom's parts. Generally, technical yarns have stronger tensile resistance and modulus all along the fibre directions rather than in transverse direction, that implies low rigidity values to bending and compression [16]. A damaged multi-filaments yarn by friction reveals lower cohesion and cut filaments implies more hairiness at the surface of the yarn.

The applied bent on crossover points between warp and weft yarns has been analysed and measured by the use of a stiffness loss criterion in the study of Cox et al. [17]. It has been related that at a crossover point between a warp and weft yarns, a change of inclination is occurring and a resulted bending is applied to yarns, more intense for the weft yarn than the warp yarns having an initial tension load given by the warp beam. Thus, the fibre degradation due to this bending, correlated with the warp yarns density, may induce some stress concentration areas. By the same, in the study of Cox [18] dedicated to tensile tests on the same 3D warp interlock fabrics, it has been revealed three different deformation phases: elastic, hardening and rupture. The hardening phase, detailed by John et al. [19], highlights the longitudinal and transverse degradations on yarns promoting the rupture initiation of the composite structure.

In the study of Lee et al. [20], during the weaving process of 3D warp interlock fabrics, a loss of 12% of carbon yarns tenacity has been measured due to the fibre degradation. At each steps of the weaving process, yarn-to-yarn contacts and yarns in contact with moving parts are eroded by friction. Moreover in Lee et al. [21], during the weaving of an orthogonal 3D warp interlock fabric, fibre degradation of yarns has been characterized by tensile tests on yarns and a decrease of 50% of the tensile strength and 7% of the tensile modulus have been measured. During the Jacquard weaving process of E-glass yarns to produce 3D warp interlock fabrics, losses of mechanical properties are due to fibre degradation by yarn-to-yarn friction and loss of the sizing agents [22].

In the study of Abu Obaid et al. [23], the woven structure geometry of a 3D warp interlock fabric has influenced the fibre degradation, up to 35% of tenacity loss for S2-glass yarns. By the same, in the study of Archer et al. [24], different measurements of carbon yarns tenacity have led to a final fibre degradation value of 9,3%; where 2,8% is directly due to the woven structure geometry.

According to our research works done on 3 different architectures of 3D warp interlock fabrics made with 6 K carbon yarns of IM7 type (223 Tex) as: Interlock A-T 5-5-5, Interlock A-L 3-3-5 and Interlock A-L 2-2-5 [25–27], a major loss of the 6K carbon yarns tenacity [28, 29] has been measured thanks to load transfer criterion, defined by Cristian et al. [30]. In the study of Nauman et al. [31], it has been revealed that the shed opening and dobby frames motion have led to a tenacity loss of 42% for 3D warp interlock fabrics made with carbon yarns.

In the research works of Lefebvre [10], at each step of the 3D warp interlock weaving, different measurements of tensile strength have been done on several para-aramid yarns Kevlar-29 (3300 dTex) [32], inserted in different 3D woven architectures [33], and produced on our manual weft insertion weaving loom [34]. The maximum loss values have been measured, respectively 20% for the tensile modulus, 12% for the elongation at break and 10% for the yarn tenacity. Abrasion of yarns has decreased the presence of sizing agents, which have reduced the transverse shear resistance.

In the research work of Tourlonias et al. [35], friction interactions that occurred between warp yarns have been experimentally simulated on a dedicated bench. Among the several process parameters tested, it has been revealed that the coefficient of friction decreases with the increase of normal load applied on warp yarns.

In the PhD report of Decrette [36], the fibre degradation occurs between 45 and 85% of the several observed events that can disturb the weaving process during full production. Additionally, in the study of McHugh [37], it has been revealed that several parameters of the shed opening have an influence on the fibre degradation of warp yarns during the full weaving production.

Based on our studies of the weaving of 3D warp interlock fabrics with carbon or HDPE yarns (Fig. 3), due to fibre degradation the warp yarns tends to interpenetrate between each other and after few cycle of weaving, an un-desired web of broken filaments appears during the shedding. This web avoids any clear shed to allow a safe insertion of weft yarn and inverse the up or down lift position of trapped warp

Fig. 3 Consequence of fibre degradation on the shed opening of 3D warp interlock fabrics made with carbon yarns (left) and HDPE yarns (right)

yarns inside this web. This lead to yarn breaks and blocks any shed opening during the weaving process.

In the study of Decrette et al. [38], it has been also revealed a clear effect of shedding on warp yarns degradation thanks to the quantification of damages. An optimized configuration of shedding parameters can be found to reduce the fibre deterioration involved in high-density multilayer woven fabrics.

Considering natural yarns as hemp or flax fibres, the rough surface of these yarns will increase the phenomena of friction. According to the study of Corbin et al. [39], it leads most of the time to a faster degradation of the yarn and consequently to the breaking of yarns.

However, according to our experiments done on several 3D warp interlock fabric productions, it can be possible to reduce or avoid the fibre degradation of yarns done by friction. Let considers this quick overview of the different adaptations that can be made on the loom, especially for the production of 3D woven structures.

2.1 Use Specific and Adapted Material

The motion of the yarn done by a mechanical part of the loom will create abrasion on fibre everywhere there is a contact between yarns and loom parts, as: yarn guide, heddle eyes and reed's dent. Weaving accessories providers chose specific raw material with low friction coefficient property to produce their devices, as: ceramics, glass and hard steel. These accessories can also be designed with a specific shape, particularly adapted for flat tow of multi-filament yarns with high linear density value (see heddle for heavy tow carbon yarn in Fig. 4).

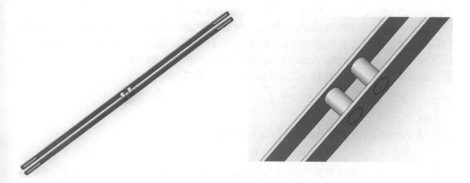

Fig. 4 Double flat steel heddles

2.2   Reduce the Loom Motions

By reducing the motion of warp yarns during the shed opening to the minimum needed to produce the woven architecture, a lower tension can be applied and consequently can reduce the friction between warp yarns and loom parts. It can be introduced, just at the design step, by avoiding as less as possible the crossing of warp yarns among each other. It can be adjusted by the suited choice of the order of weft yarns insertion and find the suited numbering of weft yarns. During the production step, Jacquard or dobby mechanisms have to select and lift the needed heddle or frame to allow the shed opening. However, some mechanisms bring systematically all the heddles or frames in the closed shed fabric plan in order to make the selection, which create un-useful and additional movement responsible of higher yarn friction. According to the study of Decrette et al. [38] on an Unival 100 Jacquard machine, a multi-shed and multi-filling loom allows, in one motion, to produce one weft row, which minimize the friction among yarns and offer higher productivity.

2.3   Avoid Any Contact During the Weft Insertion

During the weft insertion, the minimum of contact between the two different warp yarns plans resulting from the shed opening and the filling device will consequently reduce the yarn degradation. For instance, water-jet or air-jet insertion loom devices feed the shed with weft yarns without any contact the warp yarns. For conventional loom with mechanical insertion devices, the contact between the warp yarns of the shed and the inserted weft yarn is completely avoided. For instance, on a narrow weaving loom as represented in Fig. 5, the bent needle allows a complete insertion of two weft yarns length during one full shed opening cycle.

   To comply with this technical issue for larger fabric width, we have designed a specific shuttle device, represented in Fig. 6, to ensure both the accurate insertion, without any contact with warp yarns, and placement of the weft yarn direction and also store a sufficient length of weft yarn onto the shuttle bobbin.

2.4   Decrease the Yarn Friction Coefficient

Yarn coating or sizing will give a smooth surface to the yarn and reduce its friction coefficient. Add a few twist to the yarn will bring cohesion between fibres or filaments. The yarn will increase its tenacity and then reduce its breakage by defibrillation with contact with other yarns during the weaving process. Using a melt spinning machine, it could be possible to sheath the multi-filaments yarn during its production and then obtain a thermoplastic cover with low friction coefficient value.

Fig. 5 Needle for narrow fabric weaving loom

Fig. 6 Shuttle insertion device inspired by narrow fabric weaving loom [40]

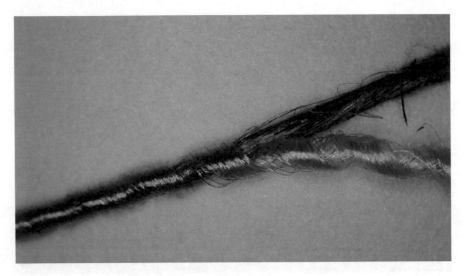

Fig. 7 Hemp yarn covered by PA12 multi-filaments yarn

Using a wrap yarn spinning machine, it can be also possible to twist around the multi-filament yarns and then obtain a yarn covering with lower friction coefficient. For instance, the covering can be a thermoplastic polymer PA12 used as a multi-filament yarn and wraps around the hemp yarn to be used for a thermo-compression process of composite application (Fig. 7) [41].

According to the research works of Vilfayeau [12] and Trifigny [42], we have adapted an industrial weaving machine to produce 3D warp interlock fabrics made with two different types of E-glass roving (untwisted multi-filaments yarn), such as: 300 Tex from Owens Corning ® [43] and 900 Tex from NEG ® [44]. However, we have previously observed that the application of a twist to multi-filament para-aramid yarn provides additional cohesion to the yarn and prevents the formation of significant fibrils during weaving [33], as well as on flax fibres roving [45]. These fibrils prevent a correct shed opening during weft insertion, making this operation difficult and very degrading for the warp yarns, until they break.

It has been observed that adding a slight twist to an E-glass multi-filaments yarn, between 10 and 30 twists/meter, can improve its mechanical properties [46] and especially the contact friction with another yarn.

2.5 Manage the Loom Parts Motion

High tenacity yarns as: carbon, E-glass or aramid, have very low elongation properties. High speed lifting motions applied to the warp yarns during the shed opening led to high longitudinal stress value. These high values of tension applied to warp yarns can play a significant role in the fibre degradation by the increase of friction

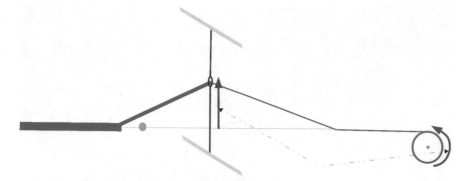

Fig. 8 Tension control following the frame motion

between yarns or hard steel loom parts. To solve this issue, a coordination between the shaft lift and the warp beam could manage the warp yarns tension to the lowest value and to keep these tension values during the multi-cycle of the shed opening the more regularly as possible (Fig. 8).

The first adaptation carried out on the industrial weaving machine, rapier weaving loom from DORNIER© type (HTVS-S/4) [47] equipped with 24 frames at a maximum width of 140 cm, has consisted in selecting the suited drawing-in scheme to minimize the yarns contact according to their evolutions in the 3D woven structure, the type of shed (oblique or rectilinear) and the reed beating type (opened or closed shed).

Indeed, the greater the number of selected frames of the weaving machine, to achieve the necessary number of warp yarn evolutions of the 3D warp interlock fabric, the greater the stroke of the frames is increased to keep the same value of shed angle, which increases the dynamic forces exerted on the transverse section of the warp yarns. In our study [48], we have optimised the use of only 13 needed frames on the weaving machine to make the three types of 3D warp interlock fabrics by allocating the first 8 frames to the evolution of the binding warp yarns (from frame 1–4 for the 1st evolution and from frame 5–8 for the 2nd evolution, opposite to the 1st evolution), then frames 9–11 for the 3 layers of stuffer warp yarns, and finally frames 12 and 13 allocated to the selection of the leno device to select the warp yarns and block the fabric selvedge. The leno warp yarns are mounted on independent bobbins, are not made of glass fibres and do not interfere with the binding and stuffer warp yarns (Fig. 9).

Two types of shed opening can be obtained: a rectilinear shed and an oblique shed (Fig. 10). The rectilinear shed is characterized by a constant stroke of the warp yarns introduced into the heddles and the oblique shed by a variable stroke [1].

In the case of the 3D warp interlock weaving, the tension applied on the warp yarns is important, the opening angles are slightly reduced and the shed is not very deformable to let the weft yarns pass through. It is therefore preferable to use the oblique shed, which guarantees a unique and constant shed angle whatever the evolution of the frames.

Fig. 9 Allocations of dobby shafts to produce different 3D warp interlock fabrics [48]

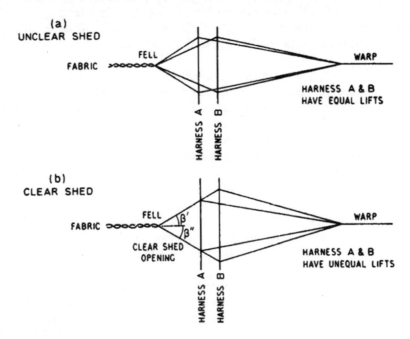

Fig. 10 Schemes for the opening of rectilinear shed (**a**) and oblique shed (**b**) [49]

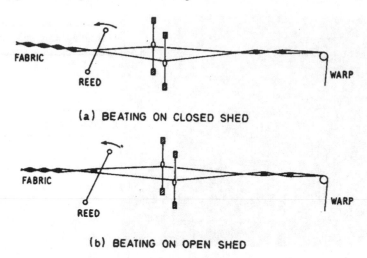

(a) BEATING ON CLOSED SHED

(b) BEATING ON OPEN SHED

Fig. 11 Schemes for the reed beating type in closed shed (**a**) and in open shed (**b**) [1]

Following the insertion of the weft yarn into the shed, its packing at the fabric forming line by the dynamic motion of the reed can be synchronized with the warp yarns frames, by adjustment, and in two configurations; either in open shed or in closed shed (Fig. 11) [1]. In the open shed, the packing takes place just before the crossing of the warp layers by the movement of the frames, and in the closed shed just after.

During the weaving of the 3D warp interlock fabrics [1], the placement of the weft yarns at the different layers in the thickness of the 3D woven structure was carried out with less friction and less fibre degradation by selection of an open shed than in closed shed.

The second adaptation achieved on the industrial rapier weaving machine [2], has consisted in positioning the stuffer warp yarns on one beam and the binding warp yarns another beam, thus making it possible to manage the significant differences in consumption of the warp yarns within the 3D interlock warp woven structure (Fig. 12).

Finally, during the 3D warp interlock weaving, we have positioned a bar (aluminium cylinder) at the fabric forming line, which becomes the equivalent of a full-width templet, in order to restrict the vertical oscillation of the already formed fabric, due to the tension phenomena taking place during the 3D warp interlock weaving (Fig. 13).

On the basis of all these adaptations carried out on the DORNIER© industrial weaving machine (HTVS-S/4) [2], done within the research works of Vilfayeau [12] and Florimond [50], three types of 3D warp interlock architectures combined with two E-glass yarns linear densities (300 and 900 Tex), have been produced on the same loom with the same process parameters. The physical charactheristics of these 3 different type of 3D warp interlock fabrics are precised in Table 1.

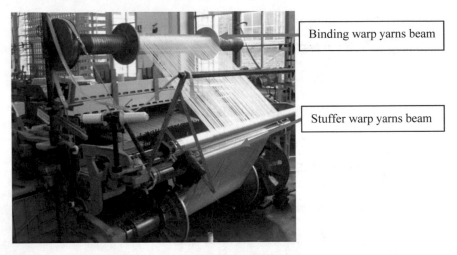

Binding warp yarns beam

Stuffer warp yarns beam

Fig. 12 Storage of binding warp yarns and stuffer warp yarns onto two separate beams [48]

Fig. 13 Location and use of a "full-width fabric templet" (cylinder rod) at the fabric forming line [48]

Observations of the woven surface of the 3D warp interlock fabrics, for the different architectures combined with the two linear densities of E-glass yarns, as well as local 3D observations of fabric samples through tomographic images, allowed us to validate the regularity and quality of the production (Figs. 14, 15 and 16).

Table 1 Areal weight and thickness of the three different 3D warp interlock fabrics [48]

Linear density (Tex)	Interlock O-T 1-4-4 with stuffer warp yarns (plain weave diagram)		Interlock O-T 2-4-4 with stuffer warp yarns (twill 2-2 weave diagram)		Interlock A-T 7-4-4 with stuffer warp yarns (satin 8 weave diagram)	
	300	900	300	900	300	900
Areal weight (g/m^2)	1732	2722	1762	3060	1657	3090
Thickness (mm)	1.5	3.1	1.7	4	2.4	4.3

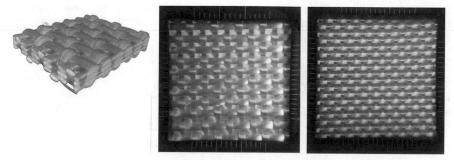

Fig. 14 Different views of the 3D warp interlock fabric O-T 1-4-4 with stuffer warp yarns (plain weave diagram) with a 3D tomographic picture (left)—a top view with E-glass yarn 900 Tex (middle)—a top view with E-glass yarn 300 Tex (right) [48]

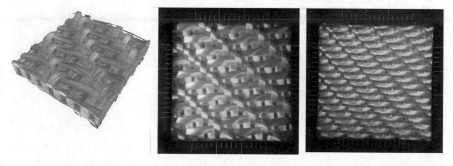

Fig. 15 Different views of the 3D warp interlock fabric O-T 2-4-4 with stuffer warp yarns (twill 2-2 weave diagram) with a 3D tomographic picture (left)—a top view with E-glass yarn 900 Tex (middle)—a top view with E-glass yarn 300 Tex (right) [48]

3 Density of Warp Yarns and Thickness

A high-density of warp yarns is one of the main factors that will increase the friction phenomenon and leads to significant damage on warp tows. In the study of Bessette et al. [51], it has been revealed that significant differences have been observed on the warp yarns tension values during the different shifting load of the shed opening.

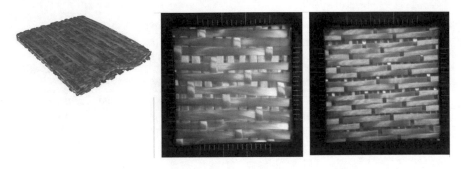

Fig. 16 Different views of the 3D warp interlock fabric A-T 7-4-4 with stuffer warp yarns (satin 8 weave diagram) with a 3D tomographic picture (left)—a top view with E-glass yarn 900 Tex (middle)—a top view with E-glass yarn 300 Tex (right) [48]

Even before any yarn degradation, the high density of warp yarns leads to contact surface value so high that all the yarns will be stuck together. This problem of space appears at the different steps of the weaving process, as on: warp beam, lease rods (Fig. 17), heddles and reed (Fig. 18). Indeed, warp yarns coming from warp beams or bobbins from a creel have to be merged from a 2D plan into a thick several layers

Fig. 17 Yarns jam at the lease rods

Fig. 18 Yarns jam at the reed

of yarns to produce the 3D fabrics. Therefore, the needed space to side by side yarns is much larger than the current space on the loom.

To reduce the impact of a high-density value of warp yarns, several solutions can be proposed, as: to separate the warp yarns into several groups, to avoid some loom element contact and to locate yarns at their closest locations they will have into the final product.

3.1 Manage the Warp Yarns Storage

Warp yarns can be stored both on bobbins for individual selection and tension load or on a beam for a common selection and tension load. Considering the warp yarns density value into the final 3D fabrics multiply by the fabric width value, the total number of warp yarns can be calculated. Thus, depending on the available number of bobbins slots, it can be possible to use a creel rather than a warp beam. However, when the number of warp yarns overcomes the creel capacity then it becomes necessary to use several warp beams. These beams will be achieved according to the different type of warp yarns consumption and evolutions inside the 3D woven structure to ensure a constant warp yarns tension. The number of these beams will be optimized to both minimize the density of warp yarns inside each warp beams according the linear density of the warp yarn, by reducing the contact surface, and to reduce the storage space of warp yarns, as it can be seen on this creel of warp beams of our prototype loom (Fig. 19).

Fig. 19 Creel of warp beams optimized for the 3D warp interlock weaving of para-aramid yarns [33]

3.2 Use of Yarns Condenser

Lease rods are very helpful to recover the numbering type of warp yarns (even or odds), and then useful after a yarn break to relocate its correct position into the warp yarns web. However, in 3D warp interlock weaving, several layers of yarn are merging into a unique plan and then it becomes critical to keep a separation system of even or odds warp yarns due to the higher density.

Then, one solution used on our prototype loom, consists in guiding all the warp yarns through condenser boards, as represented in Figs. 20 and 21. This solution brings advantages. Material of condenser board provides low friction coefficient value with other materials. These condenser boards allow to reach high-density value of warp yarns thanks to their holes positioning (staggered rows are also possible). Warp yarns are also guided into the 3D fabric thickness to ensure a better placement inside the 3D woven structure. Upper layers are located at the top, lower layers are located at the bottom of this guided warp yarns web, and then yarn-to-yarn contacts

Fig. 20 Jacquard holes board used to guide warp yarns (left) coming from the bobbins or warp beams creel and (right) guided in the thickness direction

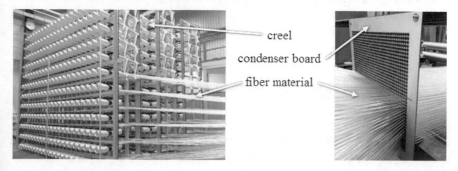

Fig. 21 Creel set-up with 1512 positions and condenser board for the 3D weaving machine at DORNIER© Composite Systems Technology Center [15]

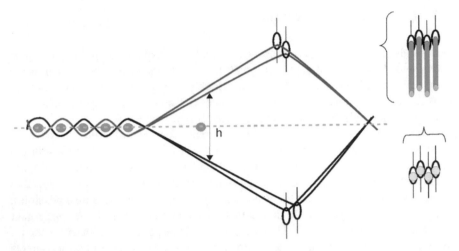

Fig. 22 Shift set up of the frame and its influence on the shed

are fully avoided. Lease rods function have been replaced by the condenser board as each yarn is inserted into its own and unique hole.

3.3 Adapt the Inclination of Warp Yarns

Warp yarns can be located into different plans during the shed forming stage by varying the position and the stroke of the heddles. Nevertheless, it has to be balanced with the height of the shed. The warp yarns placement in several plans will create a non-straight shed and then reduce its opening, as represented in Fig. 22.

To increase the density of the weft yarns after their insertions, a slight inclination of the fabric plan during the beating of the reed can be produced by the modification of the position and stroke of the heddles during the shed forming, as represented in Fig. 23.

3.4 Select the Adapted Drawing-In Plan

A satin draft for the drawing-in scheme, compared to a straight draft drawing-in scheme represented in Fig. 24, will allow to reduce the friction between yarns and heddles in case of high density of warp yarns. Indeed, due to the applied tension on warp yarns, the trajectory of a single warp yarn coming from the first heddle-eye will be in higher contact surface with other heddle-eyes for the straight draft than the satin draft of the drawing-in schemes.

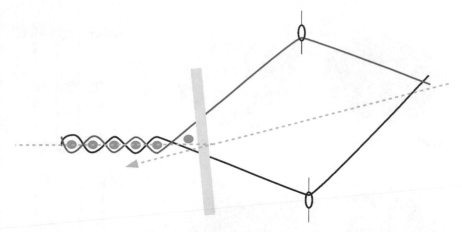

Fig. 23 Inclination of the fabric plan to increase the weft yarns density

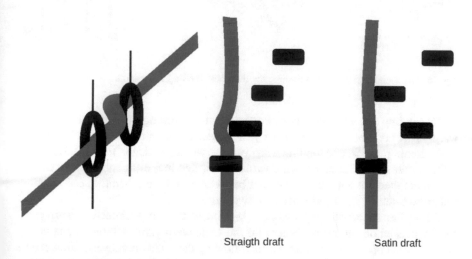

Straigth draft   Satin draft

Fig. 24 Different drafts of drawing-in schemes

3.5 Adapt the Weaving Reed Stitching

To produce 3D warp interlock fabrics, the stitching of the weaving reed is significant to both reach high-density of warp yarns value and minimize the yarn-to-yarn friction inside each dent of the weaving reed. This optimization is based on the positioning of each warp yarns coming from one warp column into one reed dent if they do not cross each other. Thanks to this reed stitching strategy, yarn-to-yarn friction is minimized as the selected yarns are fully prepared to be located in their future positions inside the 3D woven structure.

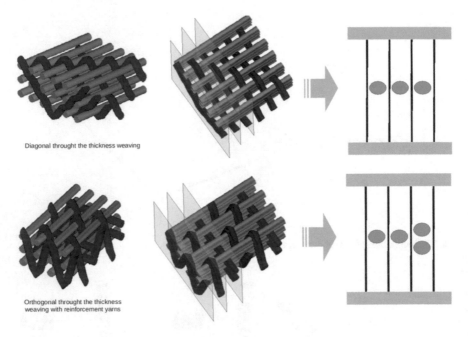

Fig. 25 Reed stitching adapted to through the thickness binding warp yarns

Taking into account the diagonal/orthogonal and through-the-thickness binding warp yarns architectures, each warp yarn physically creates its own column. The reed stitching will be one binding warp yarn between two dents (Fig. 25). However, if the stuffer or reinforcement warp yarns are located into different layers of the 3D woven architecture without any contact between each other, then the reed stitching can be two stuffer warp yarns between two dents.

Taking into account the layer-to-layer binding warp yarns architecture, warp yarns can be stacked one above the other into the same warp yarns column. Without any yarn-to-yarn contact, all the warp yarns following the same binding evolution can be merged into the same column of warp yarns and be stitched between two weaving reed dents (Fig. 26).

4 Integrity of the 3D Woven Geometry

During the 3D warp interlock weaving, depending on several product and process parameters the expected 3D woven architecture can be respected or not. It has been revealed in our studies [52, 6] that the modification of the final geometry led to different mechanical behaviours. Among all these parameters, a focus is given on the process parameters, as the take-up roller and the fabric selvedge, and the product parameters, as the warp yarns tension and the weft insertion order.

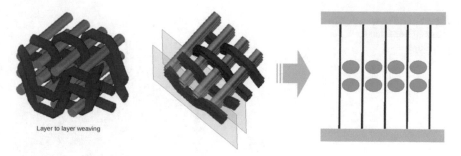

Fig. 26 Reed stitching adapted to layer-to-layer binding warp yarns

4.1 Adapt the Take-Up Roller

During the weaving of thick structure, when the stacking of fabric layers is significant, the weaving process will deform the structure geometry. For a thin 2D woven structure, the take-up of the woven fabric is achieved by pressure on fabric surface done by several rollers. For a thick 3D woven fabric, yarns without any direct contact with the cylinder have the possibility to slip from the woven structure, as represented in Fig. 27.

In addition, different warp yarn consumptions will appear with the curvature value of the cylinder. The warp yarn of the upper layer will have a longer length than the warp yarn of the lower layer in contact with the roll. A thick 3D warp interlock fabric

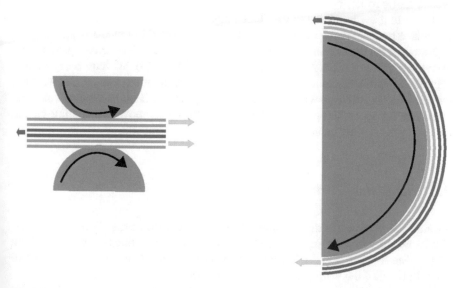

Fig. 27 Yarn slippage on the take-up rollers

Fig. 28 Difference of yarn
consumptions due to
curvature

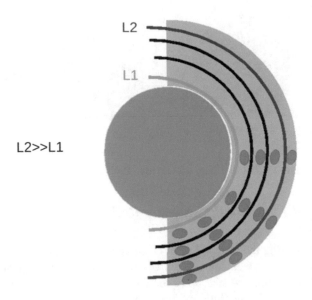

L2

L1

L2>>L1

winded on a cylinder will exhibit a significant difference of final curvature values
(Fig. 28).

Twin Rolls can be placed in opposition to cancel the difference of yarn consumption, as represented in Fig. 29. It can compensate the local deformation of the 3D
woven structure done onto the first fabric roll and then inverse the phenomena on the
second fabric roll.

To avoid these problems of yarn consumption, the cylindrical take-up roller can
be replace by a flat and horizontal take-up, as represented in Fig. 30.

Based on the horizontal position of warp yarns, we have developed a prototyping
weaving machine dedicated to 3D warp interlock fabric [53]. We have designed and
manufactured this machine in order to allow the motion of the loom while weaving
instead of a fix machine with horizontal take-up device (Fig. 31). Layers of warp
yarns are set up on the prototype loom, then the loom moves depending on the
required weft density value to produce the 3D woven structure.

4.2 Adapt the Fabric Selvedge

Several leno yarns and a tight plain weave diagram constitute the classical selvedge.
Until now, there is no existing device to create a selvedge for thick 3D warp interlock
fabric. One solution is to use a shuttle loom. The continuity of the yarn in the return
trip will blocked the warp yarn at each fabric edges.

However, with a shuttle insertion device, warp yarns located at the fabric edge
have to be selected by the weft yarn during the return trip of the shuttle to ensure a
complete and full interlacement of warp and weft yarns. It has already been observed

Fig. 29 Opposite rolls to
cancel the yarn consumption
gap

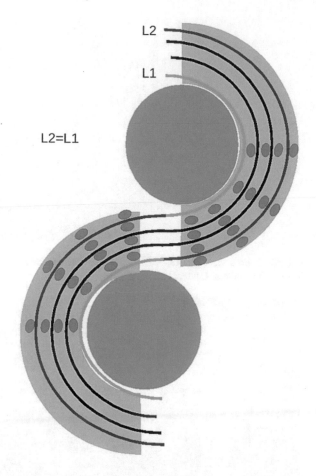

for layer-to-layer binding warp yarns structure. To avoid this phenomenon, the shuttle
path has to be inverted or a warp yarn has to be added on each fabric edges to ensure
the full interlacement between weft yarn and multi-layers of warp yarns (Fig. 32).

4.3 Manage the Warp Yarns Tension

Depending on the woven architecture, the consumption in length of warp yarns in a 3D
warp interlock fabric can be significant. For the through-the-thickness binding warp
yarn architecture, the yarn consumption in length is directly linked to the thickness
value of the 3D fabric. According to the warp yarns tension, an adapted value helps
to better consume the binding warp yarn length, as represented in Fig. 33 (right), and
then allow to obtain a more orthogonal woven architecture, compared to a higher

creel

Jacquard machine

weaving machine

horizontal take up

Fig. 30 Overall view of DORNIER© 3D weaving machine set-up at DORNIER© Composite Systems Technology Center [15]

Fig. 31 Overall view of our prototype weaving loom at GEMTEX laboratory at ENSAIT [53]

Fig. 32 Shuttle path adapted
to layer-to-layer binding
warp yarn structure

value that rather contributes to deform the final geometry, as represented in Fig. 33 (left).

The first and most common idea to control the warp yarn tension is to use a creel of bobbins for individual selection. This solution allows the tuning of each warp yarns independently from each other. However, 3D warp interlock woven structures with high density or numerous numbers of layers often need a significant number of warp yarns, depending on the fabric width value. So, using a creel of bobbins for numerous independent warp yarns will require a lot of space and will consume a significant length of warp yarn due to the distance between the farer located bobbins of the weaving loom. One solution to save space and length consumption lies in the use of a creel of beams to facilitate the control of the warp yarns tension. To manage the warp yarns tension quite independently, each warp beam can be allocated to each type of warp yarns evolution according to the 3D woven architecture, as represented in Fig. 19.

4.4 Use the Dedicated Weft Yarns Insertion Order

In the research study of Nauman et al. [7], it has been revealed that the weft yarns insertion order can influence the final geometry of the 3D woven structure. The "U"

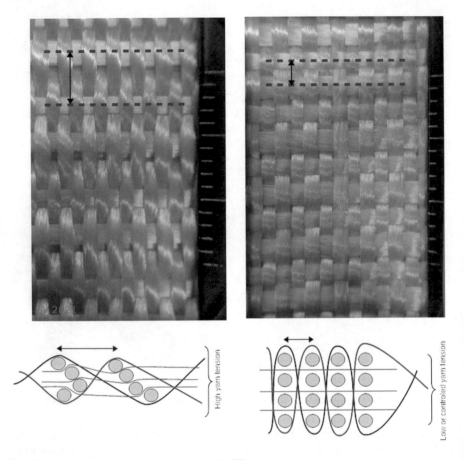

Fig. 33 Impact of the warp yarns tension on the 3D woven geometry

shape of weft yarns insertion order leads to more compact 3D woven structure than an "inverted N" shape, as represented in Fig. 34. Consequently, the "U" shape of weft yarns insertion tends to minimize the warp yarns motion during the lifting process of the shed forming from one weft yarns column to another.

5 Conclusion

The aim of this chapter aims at providing some advices to facilitate the weaving of thick fabrics like the 3D warp interlock structure. It has been revealed that the main factors to improve the weaving of natural or technical yarns are the decrease of the yarn friction and the reduction of the warp yarns contact surface. Separation and protection of warp yarns, adapted placements and an accurate control of warp yarn tension are one of the main solutions to these previous issues. Some loom adaptation

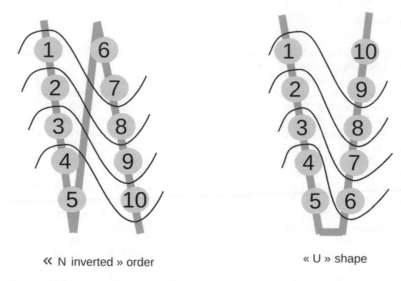

Fig. 34 "N inverted" and "U" shape of weft yarns insertion order

and dedicated set up are able to improve the behaviour of the yarn motion during the weaving. All these solutions contribute to less yarn and fibres degradation as well as the respect of the final geometry of the 3D woven structure, which directly improve its resulted mechanical properties. Moreover, thick 3D warp interlock fabrics can be produced more easily on the adapted 3D weaving loom.

New opportunities for 3D warp interlock woven fabric can be done thanks to recent and upcoming developments both on loom and adapted devices for 3D weaving. The horizontal take-up device substituting the cylindrical take-up roller is now available on a Dornier© loom [15]. The controlled shed opening, achieved by an Unival Staubli© machine (https://www.staubli.com/en/textile/textile-machinery-solutions/jacquard-weaving/unival-100/ [36]), allows to have several shed forming positions with multi-insertion of weft yarns, which contributes to the improvement of the yarn placement into the woven structure. A tuned closed shed on selvedge from an Unival Staubli© machine [38] may improve the yarn placement into the woven structure to obtain adapted fabric selvedge for thick 3D fabrics. By the same, an adapted selvedge device linked to a multi-shuttle insertion loom from Mageba© can also help to produce complex shapes of 3D woven fabrics [54]. Specifics creel and warp yarns tension device are still in development to decrease their cost and reduce yarn waste. Complementary to the weaving loom development, loom parts are also increasingly adapted to the different yarns shape and linear density as a flat roll heddle from Derix© [55] or a specific steel for weaving reed adapted for carbon yarns by Groz-Beckert© [56].

References

1. Lord, P. R., Mohamed, M. H. (Eds.). (1982). *Weav* (2nd edn, p. vi). Woodhead Publishing. https://doi.org/10.1016/B978-1-85573-483-8.50004-2.
2. Mouritz, A. P., Bannister, M. K., Falzon, P. J., & Leong, K. H. (1999). Review of applications for advanced three-dimensional fibre textile composites. *Composites. Part A, Applied Science and Manufacturing, 30*, 1445–1461. https://doi.org/10.1016/S1359-835X(99)00034-2
3. Rudov-Clark, S., Mouritz, A. P., Lee, L., & Bannister, M. K. (2003). Fibre damage in the manufacture of advanced three-dimensional woven composites. *Composites. Part A, Applied Science and Manufacturing, 34*, 963–970. https://doi.org/10.1016/S1359-835X(03)00213-6
4. Hu, J. (2008). 3-D fibrous assemblies 1st edition properties. In: *Applications and modelling of three-dimensional textile structures* (1st edn). Woodhead Publishing.
5. Bhatnagar, A. (2006). *Lightweight ballistic composites: Military and law-enforcement applications.* https://doi.org/10.1533/9781845691554.
6. Boussu, F., Dufour, C., Veyet, F.,& Lefebvre, M. (2015). *Weaving processes for composites manufacture.* https://doi.org/10.1016/B978-1-78242-307-2.00003-8.
7. Nauman, S. (2011). *Geometrical modelling and characterization of 3D warp interlock composites and their on-line structural health monitoring using flexible textile sensors.* University of Lille 1. http://www.theses.fr/2011LIL10010.
8. Lapeyronnie, P. (2010). *Mise en oeuvre et comportement mécanique de composites organiques renforcés de structures 3D interlocks.* Univ Lille 1. https://www.theses.fr/en/2010LIL10126.
9. Ha-Minh, C. (2011). *Comportement mécanique des matériaux tissés soumis à un impact balistique : approches expérimentale, numérique et analytique.* Lille 1. https://www.theses.fr/en/2011LIL10184.
10. Lefebvre, M. (2011). *Résistance à l'impact balistique de matériaux composites à renforts Interlocks tissés. Application au blindage de véhicules.* 'Université de Valenciennes. https://www.theses.fr/en/2011VALE0030.
11. Provost, B. (2013). *Etude et évaluation d'une solution composite à renfort tissé interlock pour la protection balistique de véhicule.* University of Valenciennes. https://www.theses.fr/en/2013VALE0003.
12. Vilfayeau, J. (2014). *Modélisation numérique du procédé de tissage des renforts fibreux pour matériaux composites.* INSA, Lyon. https://www.theses.fr/en/2014ISAL0026.
13. Boussu, F. (2016). *Les tissus 3D interlocks chaines Définition et applications.* Editions U.
14. Shape 3. Innovative Textiltechnik GmbH n.d. http://www.shape3.com/Frameset_Shape3.htm. Accessed 12 Aug 2021.
15. Klingele, J. (2016). Cost-efficient and flexible production of high quality 3D woven fabrics for composite applications. In: F. Boussu, & X. Chen (Eds), *7th World Conference on 3D Fabrics and Their Applications, Roubaix, France*, pp. 105–14. ISBN 978-2-9557912-0-2.
16. Gasser, A., Boisse, P., & Hanklar, S. (2000). Mechanical behaviour of dry fabric reinforcements. 3D simulations versus biaxial tests. *Computational Material Science, 17*, 7–20. https://doi.org/10.1016/S0927-0256(99)00086-5.
17. Cox, B. N., & Dadkhah, M. S. (1995). The macroscopic elasticity of 3D woven composites. *Journal of Composite Materials, 29*, 785–819. https://doi.org/10.1177/002199839502900606
18. Cox, B. N., Dadkhah, M. S., & Morris, W. L. (1996). On the tensile failure of 3D woven composites. *Composites. Part A, Applied Science and Manufacturing, 27*, 447–458. https://doi.org/10.1016/1359-835X(95)00053-5
19. John, S., Herszberg, I., & Coman, F. (2001). Longitudinal and transverse damage taxonomy in woven composite components. *Composites. Part B, Engineering, 32*, 659–668. https://doi.org/10.1016/S1359-8368(01)00047-6
20. Lee, B., Leong, K. H., & Herszberg, I. (2001). Effect of weaving on the tensile properties of carbon fibre tows and woven composites. *Journal of Reinforced Plastics and Composites, 20*, 652–670. https://doi.org/10.1177/073168401772679011

21. Lee, L., Rudov-Clark, S., Mouritz, A. P., Bannister, M. K., & Herszberg, I. (2002). Effect of weaving damage on the tensile properties of three-dimensional woven composites. *Composite Structures, 57*, 405–413. https://doi.org/10.1016/S0263-8223(02)00108-3

22. Rudov-Clark, S. (2007). *Experimental investigation of the tensile properties and failure mechanisms of three-dimensional woven composites.* RMIT University.

23. Abu Obaid, A., Andersen, S. A., & Gillespie, Jr. J. W. (2021). Effects of the weaving process on Tensile strength distribution of S2 glass fiber tows extracted from different 3D woven fabrics n.d.:2. https://www.ccm.udel.edu/wp-content/uploads/2014/09/MS_Abu-Obaid_Effects_2009.pdf. Accessed 12 Aug 2021.

24. Archer, E., Buchanan, S., McIlhagger, A. T., & Quinn, J. P. (2010). The effect of 3D weaving and consolidation on carbon fiber tows, fabrics, and composites. *Journal of Reinforced Plastics and Composites, 29*, 3162–3170. https://doi.org/10.1177/0731684410371405

25. Boussu, F., Legrand, X., Nauman, S., & Koncar, V. (n.d.). Geometric modelling of 3D angle interlock fabrics. In: *Autex 08, World Textile Conference, Biella, Italy.*

26. Nauman, S., Boussu, F., Legrand, X., & Koncar, V. (n.d.). Geometrical modelling of 3D textile composite application to warp interlock carbon fabrics. In: *13th European Conference on Composite Materials, Stockholm, Sweden.*

27. Nauman, S., Cristian, I., Boussu, F., Legrand, X., & Koncar, V. (n.d.). Weaving of 3D interlock layer to layer carbon-glass reinforcement on a conventional loom. In: *2nd International Conference on Textile Cloth, Lahore, Pakistan.*

28. Boussu, F., Cristian, I., & Nauman, S. (n.d.). Technical performance of yarns inside a 3D woven fabric. In: *ITC International Conference on Latest Advanced High Tech Textile Materials, Gent, Belgium.*

29. Boussu, F., Cristian, I., Nauman, S., Lapeyronnie, P., & Binetruy, C. (n.d.). Effect of 3D-weave architecture on strength transfer from tow to textile composite. In: *2nd World Conference on 3D Fabrics and Their Applications, Greenville, South Carolina, USA.*

30. Cristian, I., Nauman, S., Boussu, F., & Koncar, V. (2012). A study of strength transfer from tow to textile composite using different reinforcement architectures. *Applied Composite Materials, 19*. https://doi.org/10.1007/s10443-011-9215-x.

31. Nauman, S., Boussu, F., Cristian, I., & Koncar, V. (n.d.). Impact of 3D woven structure onto the high performance yarn properties. In: *Second Conference on Intell. Text. Mass Cust. Text. Compos. Work., Casablanca, Morocco.*

32. Lefebvre, M., Boussu, F., & Coutellier, D. (n.d.). Degradation measurement of fibrous reinforcement inside composite material. In: *ICCE-19 19th Annual International Conference on Composite or Nano Engineering, Shanghai, China.*

33. Lefebvre, M., Boussu, F., & Coutellier, D. (2013). Influence of high-performance yarns degradation inside three-dimensional warp interlock fabric. *Journal of Industrial Textiles, 42.* https://doi.org/10.1177/1528083712444298.

34. Begus, V. (2006). *Nouvelle solution textile pour composite anti-balistique.* ENSAIT-GEMTEX.

35. Tourlonias, M., & Bueno, M.-A. (2016). Experimental simulation of friction and wear of carbon yarns during the weaving process. *Composites. Part A, Applied Science and Manufacturing, 80*, 228–236. https://doi.org/10.1016/j.compositesa.2015.07.024

36. Decrette, M. (2014). *Tissage Jacquard : étude de paramètres et optimisation du tissage 3D haute densité.* University of Mulhouse. https://www.theses.fr/en/2014MULH7952.

37. McHugh, C. M. (2009). The use of recent developments in conventional weaving & shedding technology to create 3D one-piece woven carbon preforms. *Society for the Advancement of Materials and Process Engineering, 5.*

38. Decrette, M., Osselin, J.-F., & Drean, J.-Y. (2019). Motorized Jacquard technology for multilayer weaving damages study and reduction: Shed profile and close shed profile. *Journal of Engineered Fibers and Fabrics, 14*, 1558925019833025. https://doi.org/10.1177/155892501 9833025

39. Corbin, A.-C., Boussu, F., Ferreira, M., & Soulat, D. (2020). Influence of 3D warp interlock fabrics parameters made with flax rovings on their final mechanical behaviour. *Journal of Industrial Textiles, 49.* https://doi.org/10.1177/1528083718808790.

40. Veyet, F., & Boussu, F. (n.d.). *Conception et fabrication d'un prototype de machine à tisser automatique capable de réaliser des préformes carbones.* Le Havre, France.
41. Corbin, A.-C., Labanieh, A.-R., Ferreira, M., & Soulat, D. (2019). Amélioration de la tissabilité des préformes pour applications composites par l'utilisation de rovings guipés chanvre/PA12. 21ème Journées Natl. sur les Compos., Bordeaux, Talence, France.
42. Trifigny, N. (2013). *Mesure in-situ et connaissance des phénomènes mécaniques au sein d'une structure tissée multicouches.* Univertity of Lille1. https://www.theses.fr/en/2013LIL10152.
43. Corning, O. (n.d.). *SE1200 Versatility for broad applications.* https://www.owenscorning.com/en-us/composites/product/se1200.
44. Nippon Electric Glass. Hybon™ 2001 n.d. https://www.neg.co.jp/en/assets/file/product/fiber/e-roving/e-roving_list/Hybon_2001.pdf.
45. Goutianos, S., Peijs, T., Nystrom, B., & Skrifvars, M. (2006). Development of flax fibre based textile reinforcements for composite applications. *Applied Composite Materials, 13,* 199–215. https://doi.org/10.1007/s10443-006-9010-2
46. Rajwin, A., Giridev, V. R., & Renukadevi, M. (n.d.). Effect of yarn twist on mechanical properties of glass fibre reinforced composite rods. *Indian Journal of Fibre and Textile Research, 37,* 343–346. http://www.nopr.niscair.res.in/bitstream/123456789/15220/1/IJFTR%2037%284%29%20343-346.pdf.
47. Dornier. (n.d.). Rapier Weaving machine. https://www.lindauerdornier.com/en/weaving-machines/.
48. Boussu, F. (2014). Compréhension des paramètres de produit et de procédé de fabrication des tissus 3D interlocks chaines.Applications en tant que renfort fibreux de solutions de protection à l'impact. l'Université de VALENCIENNES ET DU HAINAUTCAMBRESIS.
49. Lord, P. R., & Mohamed, M. H. (1982). 11—Shedding and beating. In: P. R. Lord & M. H. Mohamed (Eds.), *Weav* (2nd edn, pp. 191–218). Woodhead Publishing. https://doi.org/10.1533/9781845692902.191.
50. Florimond, C. (2013). Contributions à la modélisation mécanique du comportement de mèches de renforts tissés à l'aide d'un schéma éléments finis implicite. INSA Lyon. https://www.theses.fr/en/2013ISAL0136.
51. Bessette, C., Decrette, M., Tourlonias, M., Osselin, J.-F., Charleux, F., Coupé, D., et al. (2019). In-situ measurement of tension and contact forces for weaving process monitoring: Application to 3D interlock. *Composites. Part A, Applied Science and Manufacturing, 126,.* https://doi.org/10.1016/j.compositesa.2019.105604
52. Boussu, F., Veyet, F., & Lefebvre, M. (2010). Recent advances in textile composite for impact protection. *World Journal of Engineering,* 53–66.
53. Boussu, F., Veyet, F., & Dumont, N. (n.d.). *Développement d'une machine prototype de tissage pour la réalisation d'échantillons textile 2,5D.* Lille, France.
54. Mageba. (n.d.). *Multi-shuttle loom.* https://www.mageba.com/de/textilmaschinen/webmaschine.html.
55. Derix. (n.d.). *Weaving loom accessories.* https://www.derix.biz/en/.
56. Groz-Beckert. (n.d.). *Weaving reed for carbon yarns.* https://www.groz-beckert.com/.

Multi-axis Weaving

Ahmad Rashed Labanieh and Xavier Legrand

Abstract The conventional weaving technology restrains the orientation of the in-plane yarns to two orthogonal angles only (0° and 90°) relative to the fabric main axis. The multi-axis weaving technologies are developed to overcome this limitation of the conventional bi-dimensional and three-dimensional woven fabrics. The principle of the multi-axis weaving technologies, which counts on obtains in-plan yarns oriented in $\pm\theta°$ with respect to the fabric main axis, is explained with demonstrating examples. The geometry of the multi-axis woven fabric is analyzed with highlighting the requirements associated to the yarn pattern and orientation. The technical restrictions of the used weaving mechanisms to produce a multi-axis woven fabric are identified. Further, the innovated techniques in this field are explored and the architecture of the produced structure is analyzed. Certain technical innovations promote the development to attain a multi-axis weaving machine with reasonable production speed. That aids to manufacture the woven preform with a controlled in-plan yarn orientation to response to some specific technical applications.

1 Introduction

The mechanical properties and performance of the fiber reinforced composite material depend essentially on the fiber orientation within the structure relative to the applied loads directions. So the fiber orientations and placements in the textile preform have to be considered at the preform design stage in order to the respond correctly to the application conditions of the final composite part. For particular application conditions, specific fiber orientation is required. That can't be necessarily obtainable by the conventional textile manufacturing processes such as weaving, braiding, knitting and stitching. Thus new preform fabrication techniques for each particular application were innovated by modifying and combining the conventional textile manufacturing processes.

A. R. Labanieh (✉) · X. Legrand
ENSAIT, ULR 2461 - GEMTEX - Génie et Matériaux Textiles, University Lille, 59000 Lille, France
e-mail: ahmad.labanieh@ensait.fr

© Springer Nature Switzerland AG 2022
Y. Kyosev and F. Boussu (eds.), *Advanced Weaving Technology*,
https://doi.org/10.1007/978-3-030-91515-5_14

For a specific application in which quasi-isotropic in-plane tensile properties and good shear strength and stiffness are required for a thick plate of composite part with a good inter-laminar resistance and through thickness properties, the conventional 2D weaving process is not adequate to produce the adapted preform. The weaving process produces single-layered structure with only two in-plan orthogonal yarn groups [1]. The properties of this obtained structure are far away from respond to the specified requirement. Other 3D manufacturing process such as 3D warp interlock permits to form multi-layered structure, in which the difference in-plane layers are combined by a through-the-thickness binding yarn and all layers are composed of only two in-plan orthogonal yarn groups like the 2D woven structure [2]. The obtained 3D woven structure has a good inter-laminar strength thank to the presence of binding yarns, but the lack of in-plane yarns in a direction other the two orthogonal main direction leads to low shear and out-of-axis properties [3, 4]. Laminate and Non-crimp-fabric techniques allow obtain structures with a specific in-plane properties thank to the possibility of control the in-plan yarn layer orientation [5]. These multi-layered structures can be reinforced in the through-thickness direction by stitching process using a reinforcement yarn to bring the required inter-laminar strength [6]. However, some defects linked to the stitching needle insertion and stitch yarn, such as in-plane yarn degradation, in-plan yarn deflection around the stitch and the formation of resin pocket, limit the use of stitching technique. As a conclusion, all of these presented techniques don't respond properly to the pronounced requirement [7, 8]. Therefor the multi-axis weaving process is created based on the 3D weaving process to form a multi-layered structure, in which the reinforcement yarn is inserted during weaving process, with the possibility to control in-plane yarn orientation to obtain bias yarns ($\pm\theta°$) in addition to the main orthogonal yarn groups; warp and weft, oriented in 0 and 90°.

This chapter will introduce the definition of the textile structure relative to the yarn orientation. Then the multi-axis weaving technology requirements and the constituent yarn movement exigencies will be detailed. Further the basic geometry features of the bias yarn layers will be introduced to describe the bias yarn layer architecture. Finally the main innovated multi-axis techniques are presented.

2   Definition

The textile products can be manufactured in different architectures with variant geometrical and functional features. Its deformability and its performance depend mainly on the material nature and type, orientation and arrangement of the constituting fibers. Further the mechanical behavior of the fiber-reinforced composite materials, lightweight with high stiffness and strength structural material, depends on the fiber orientation within the matrix domain and its proportion in each material direction. Therefor the orientation of the fibers is one of the basic characterizing parameters of textile structures and it is an occupation for textile and composite designers. The textiles products can be classified according to the fiber orientation by indication the number of the material axis along with the fibers are oriented within the structure. However, the fibers can be randomly oriented within the structure such as non-woven web.

In addition to the fiber orientation, the textile products are categorized into three groups according to the number of the non-neglected dimensions of the structure; one, two and three-dimensions. That helps to determinate the adequate manufacturing technology.

The Mono-axial structure consists of parallel fibers aligned with the main product axis (the longitudinal axis) such as a roving and a unidirectional lamina, Fig. 1. Otherwise, these two products are in a different category in term of the structural dimensions. The roving is a linear (1D) structure produced in very long length and neglected width and thickness, Fig. 1a. However, the unidirectional lamina is thin planar (2D) structure. Its thickness is neglected against the two considerable membrane dimensions (length and width), Fig. 1b.

The 2D woven fabric is an example of bi-axial structure built of two sets of orthogonal yarns denoted warp and weft, oriented in 0° (y-axis in Fig. 2) and 90° (x-axis in Fig. 2) and interlaced in a specific weave pattern. Multilayer of the in-plane yarn sets (warp and weft) can be combined together during the weaving process via a group of the warp yarns passing through the thickness and interlaced with weft yarns of the different layers, Fig. 3. The obtained multilayer woven fabric is classified as a three-dimensional structure when it contains a considerable number of layers with a non-negligible thickness. This type of woven structure is known as 3D warp interlock fabric [2]. When the binder yarns pass through all the fabric thickness with considerable fabric thickness as illustrated in Fig. 3 the 3D woven fabric is considered tri-axial structure comprising fiber along the three axes (x, y and z).

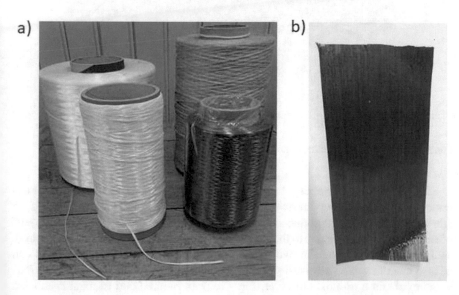

Fig. 1 Mono-axial structure: Roving **a** and unidirectional prepreg of carbon fiber **b**

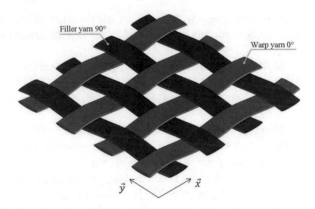

Fig. 2 Schematic of 2D woven fabric with plain weave pattern

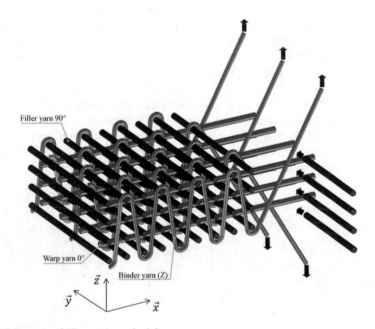

Fig. 3 Schematic of 3D weaving principle

The bi-axial structure can be produced via braiding technology with non-orthogonal directions for the two constituting yarn sets unlike in the 2D woven fabric. Bi-axial braid is produced with two sets of yarns oriented in $+\theta°$ and $-\theta°$ other than $0°$ and $90°$ relative to the braid main axis (axis $\vec{y}$ in the presented schema), as illustrated in Fig. 4a. The two yarn sets are named $\pm$ bias yarn. The braid can be also manufactured with axial yarns in the $0°$ direction in addition to the two bias yarns to obtain a tri-axial structure, Fig. 4b. Both bi-axial and tri-axial braids are produced in tubular form which can be opened into a flatten fabric.

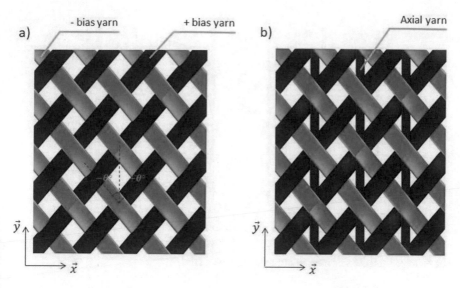

Fig. 4 Schematic of bi-axial braid **a** and tri-axial braid **b**, $\theta°$ is the braiding angle

The multi-axis 2D woven fabric is characterized with four distinct orientations (or more than three) for the constituting in-plane yarns; warp, weft, +bias and -bias aligned with the 0° (y), 90° (x), +θ° and −θ° axis respectively as illustrated in Fig. 5. Similar to the 3D warp interlock fabric, multilayer of the four in-plane yarn sets combined together by a binder yarn (Z) forms a multiaxial 3D woven structure. Figure 6 represents a schematic of the multi-axis 3D woven fabric.

Fig. 5. 2D multiaxis woven fabric

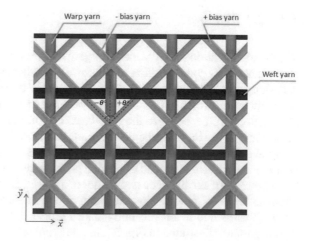

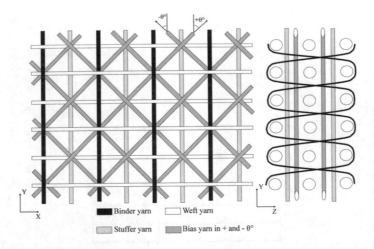

Fig. 6 Representative scheme of multi-axis 3D woven fabric

3   Principle of Multi-axis Weaving

The 2D and 3D multi-axis fabrics are produced via an innovated technique derived from the conventional weaving process. These techniques give the possibility to insert the bias yarns in $\pm\theta°$ directions during the weaving process. Before describing the multi-axis weaving technology and the required adjustment to the weaving loom, the principle of the 2D and 3D weaving technology are presented.

3.1   2D Weaving

The 2D woven fabric is used widely for garment, furniture and 2D-curved composite parts thanks to its good handling and deformability behavior. That is attributed to the cohesion brought by the interlacement between the two constituting yarn sets: warp and weft, as illustrated in Fig. 2. On the weaving loom, the warp yarns are extended parallel to each other between the warp beam and the fabric wound on the cloth beam, Figs. 7 and 8. On the warp beam the warp yarns are juxtaposed and spread over a width equal to the fabric width. The warp yarns are straightened between the two beams and each one is passed through a hole located in the center of a heddle then passed through a slot between two blades of a closed reed, Figs. 7 and 8. The heddles alternate between two positions (up and down) at each weft insertion to separate the warp yarns into two layers and create a proper space, termed shed, to insert the weft apparatus, illustrated in Fig. 7. The weft yarn is inserted across the fabric width perpendicularly to the warp yarns. Then the reed moves forward towards the fabric with closing the reed to pack the inserted weft yarn to the fabric and starting a new weaving cycle.

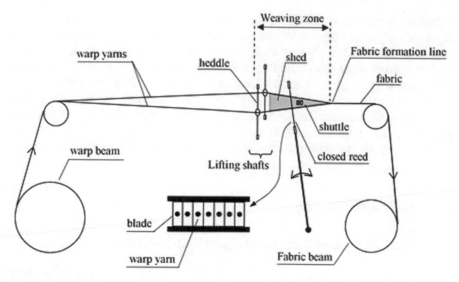

Fig. 7 Schematic of the conventional weaving loom—side view

Fig. 8 Simplified schematic of a 2D weaving loom—top view

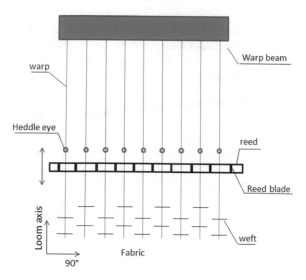

3.2 Multilayer Weaving

The demand of thick preform increases considerably specially for aeronautic, transport and protection applications. This type of structure is produced usually by overlapping several layers of 2D fabric or unidirectional lamina without a fiber reinforcement combining the layer in the through thickness direction. In order to improve the delamination strength and impact resistance of this stack, the multilayer weaving technique is developed. This technique enables to combine a number of in-plane layers of warp and weft yarns by a group of the warp yarn passing through the thickness of the fabric and interlaced with the weft yarns of the different layers. This fabric is composed of three sets of yarns; stuffer (non-interlaced warp), weft and binder (warp passing through the thickness), and it is defined as 3D warp interlock fabric [2]. This type of woven fabric can be produced on the standard 2D weaving loom, but with several adjustments relative to the warp feeding system (multiple beams) and packing operation for the heavy fabric. For the very thick fabric with high number of layers, a special machine are constructed with multiple weft insertion system.

3.3 Multi-axis Weaving

3.3.1 How Do the Bias Yarns Obtain the ±θ° Orientation?

The multi-axis woven fabric contains + and − bias yarns in addition to the in-plane warp and weft yarns as explained previously and shown through Fig. 5. To obtain this bias orientation inside the fabric, the bias yarns have a movement and path in the weaving zone differs from that of the warp yarn. At the beginning of each weaving cycle, the bias yarns are aligned with the loom axis parallel to the warp yarns after the last packing operation performed by the reed, Fig. 9a. Then, with the preparation for the new weft insertion and creation of the shed, the bias yarns are displaced transversally (along with the weft yarn axis) contrary to the warp yarns which are

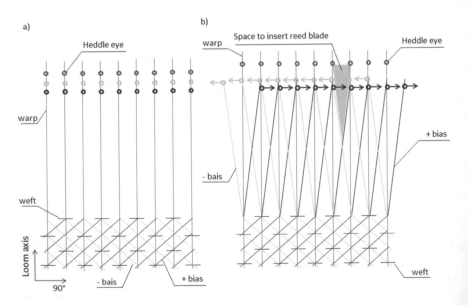

Fig. 9 Principle of the multiaxis weaving

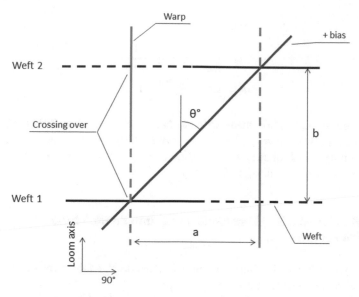

Fig. 10 Geometry of the bias yarn orientation inside the woven fabric, solid line represents the top position while the dashed line represents lower position

kept parallel to the loom axis, Fig. 9b. The +bias yarns displace transversally in the direction of the weft axis-90° while the −bias yarns displace transversally in the opposite direction, Fig. 9b. The bias yarns in the weaving zone are moved by a specified step equal to an integer number of the inter-warp yarn distance. For example the bias yarn can be moved to the next warp yarn position for each weft insertion or to the second or third yarn position next to it. This step depends on the desired angle for the bias yarns inside the fabric. To define this angle and to consider the other weaving parameter influencing this orientation, the geometry of the bias yarn inside the fabric is schematized on Fig. 10. The bias yarns are trapped between the interlaced warp and weft yarns at the crossing over the area. Thus the orientation of the bias yarns is a result of the ratio between the distances (a) and (b). These distances correspond to the distance between two crossing points with bias yarn along weft and warp yarn respectively. With considering the weaving parameters, the distance (a) represents the inverse of the number of warp yarns per centimeter (nwa) times a factor (p) equal to the size of the displacement step (number of warp yarns to which the bias yarn is displaced). The distance (b) represents the inverse of the number of weft yarns per centimeter (nwe) times a factor (t) equal to the number of weft insertion done for one displacement of the bias yarn. Thus the orientation ($\theta°$) of the bias yarn inside a woven fabric is defined by the Eq. (1):

$$\tan \theta = \frac{a}{b}$$

$$a = \frac{1}{nwa}p; \ b = \frac{1}{nwe}t \tag{1}$$

$$\theta = \tan^{-1}\left(\frac{p}{t}\frac{nwe}{nwa}\right)$$

For example, in a fabric made with similar number of yarns in both directions warp and weft (nwa = nwe) and the bias yarns are displaced for one warp yarn (next warp position) (p = 1) at each weft insertion (t = 1), the bias yarns become oriented within the fabric along an angle ($\theta = 45°$).

3.3.2 How Are the Bias Yarns Maintained into Position Inside the Fabric?

The bias yarns are maintained into position inside the fabric by means of two techniques prisoning and interlacing.

In the 2D multiaxis fabric, the bias yarns are prisoned at the crossing area between the warp and weft yarns which are interlaced by moving selectively the warp yarns up and down, as in the woven fabric Fig. 5. Thus, by this technique the bias yarns are not interlaced with the warp and/or weft yarns and the opposite bias yarns are crossed in the crossing area of the warp yarns and the weft yarns, Fig. 11a. In the other technique the weft yarns are interlaced between warp and bias yarns. Here all warp yarns are raised in the space between the crossed bias yarns then the weft yarn is inserted, Fig. 11b. Thus the bias yarns don't pass by the crossing area of the warp

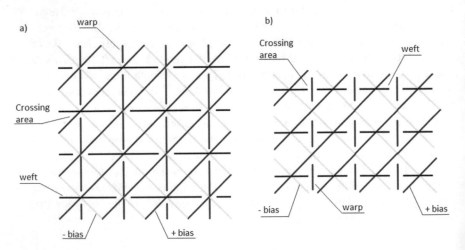

Fig. 11 Representative scheme of multi-axis 2D woven fabric in top view showing two patterns of interlacement between warp and weft to maintain the bias yarns; **a** by prisoning and **b** by interlacement

yarns and the weft yarns and the opposite bias yarns are crossed in the space between the warp and weft yarns. However, this pattern type results in a more open structure (big space between the yarn = low cover factor).

In the same manner the bias yarn layers are maintained into position within the 3D woven fabric. The binder yarns are passed thought the thickness of the fabric in the space between the crossed bias yarns then the weft yarn is inserted in the created shed between the binder yarn and fabric surface. Two opposite groups of binder yarns can be used to produce a symmetric fabric; one pass from top to bottom and the other from bottom to top face of the fabric.

4   Geometry and Architecture of Bias Yarn Layer

Different position control apparatus and techniques are used to give $\pm\theta°$ orientation to the bias yarns within the woven fabric. That results in a variation in geometry and architecture of the formed bias layers and their placement among other fabric layers (warp and weft). Therefor it is necessary to define the characterizing geometric and architectural features of the bias yarn layer to classify these techniques, identify the required adjustment to the weaving loom and assess the mechanical performance of the produced fabric.

By exploring the developed techniques and the architecture of the produced multi-axis fabrics in the published patents and research papers, the following features are defined in a previous work by the authors [9]. They describe the architecture and the geometry of the bias yarn layers in both 2D and 3D woven multi-axis fabric:

– Number bias yarns layers through the thickness of the fabric. In the 2D fabric 1 layer or 2 associated opposite layers ($\pm\theta°$) are inserted. While in the 3D structure more than one layer in same orientation can be inserted. For example, on the 3D weaving loom built by the author [9] two $+\theta°$ layers and two $-\theta°$ layers are inserted.
– Position of bias yarns layers through the thickness of the produced preform. This feature concerns only the 3D fabric. Some position control techniques of the bias yarns in the weaving zone can be placed only on the outer surfaces of the fabric. Thus the formed bias yarn layers are on the top or/and bottom faces of the fabric. They can't be inserted in another order in the thought-the-thickness of the fabric. That limits the possibility front of the preform designer to improve the delamination resistance of the composite by controlling the layer orientation order. Another technique permits to integrate the bias layer inside the 3D structure not only on the outer fabric surfaces.
– Number of associated bias yarns layers formed by the bias yarn position control device. Some techniques form a single independent bias layer, whereas other techniques insert a pair of associated opposite + and $-\theta°$ bias layers.
– Pattern of bias yarns path across the preform width. Among of the developed devices, two patterns are revealed; zig-zag and edge-to-edge pattern.

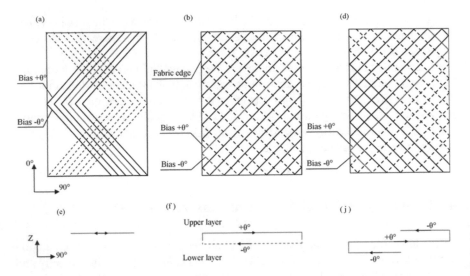

Fig. 12 Architecture pattern of the bias yarn layers; zig-zag pattern **a**, edge-to-edge pattern with uniform and non-uniform pair of opposite bias yarn layers **b** and **d** respectively, motion path of the bias yarn positioning guide across the fabric width under each pattern **e, f, j**. The solid line represents the upper layer while the dashed line belongs to the lower layer

- By zig-zag pattern one layer can be produced in which the bias yarns bundle sweep the fabric width, Fig. 12a. The yarns alternates their orientation ± to $-/ + \theta°$ once the yarns reach the fabric edge, the path of the motion of the positioning guide of the bias yarns is plotted in Fig. 12e. The width of the bias layer in less than the fabric width.
- By edge-to-edge pattern each individual bias yarn sweep across the fabric width from an edge to another, Fig. 12b and d. When the bias yarn reaches the fabric edge is folded and oriented in the opposite bias orientation.

- Uniformity of the bias yarn layer. The orientation of the bias yarn is switched to the opposite direction when it reaches the fabric edge or when the positioning guide in the zig-zag pattern reaches its transversal motion limit. When the switched yarn is kept in the same bias layer, non-uniform bias yarn layer is formed. While if the switched yarn is folded to the next associated bias layer in opposite orientation, pair of uniform layer of bias yarn is formed. In this manner, all the bias yarns in the same layer have the same orientation $+$or $-\theta°$. However, in the non-uniform layer the same yarns are kept in the layer (are not folded to other associated layer) but their orientation is not the same overall the fabric length.

5 Multiaxis Weaving Machine

As demonstrated previously, obtaining the $\pm\theta°$ orientation to the bias yarns inside the fabric requires a transversal movement for these yarns in the weaving zone contrary to the warp yarns which are kept straight. This transversal movement is carried out by means of special mechanisms integrated to the weaving loom with adjusting the other weaving mechanisms or by constructing a completely new weaving machine with adapted weaving operation mechanisms. These completely new machines are specially constructed for the 3D multiaxial fabrics or for a multiaxial profile structure with specific cross section shape. The 2D multiaxias fabric is produced by adjusting 2D weaving loom. The brought modifications to the weaving operations depend essentially on the concept of the used bias yarn position control technique and architecture of the formed bias yarn layers.

The reader is requested to discover the main weaving operations done in the conventional loom demonstrated in another book chapter to better understand the multiaxis weaving process.

5.1 Shedding and Packing Operation

The main adjusted operation mechanisms on the weaving loom are that put obstacle blocking the transversal movement of the bias yarns in the weaving zone such as shedding and packing operations. On the weaving loom the shedding operation is achieved via heddles through its central eye warp yarn is passed, as shown in Fig. 7. Each heddle is attached from its two extremities to dobby or jacquard system control the movement of the warp yarns up and down to create a shed. In case of feeding the bias yarns behind the heddles, this system has to be modified and moved out of the weaving zone at each weaving cycle (initiation of transversal stepwise movement of bias yarns). Another obstacle is related to the packing operation that uses a closed reed in which the warp yarns are passed through the spaces created between the reed blades. The warp yarns are enclosed inside the reed and it can't be moved out of the weaving zone. Further the bias yarns can't be inserted in the reed at the end of the weaving cycle after their transversal displacement in the weaving zone to be packed to the fabric. Thus, for all multiaxis technique only an open reed can be used for packing operation. It is moved out at the beginning of each weaving cycle then inserted again to pack the weft and bias yarn to the fabric at the end of the weaving cycle.

5.2 Bias Yarn Feeding System

In addition to the necessary modifications brought to the shedding and packing operations on the weaving loom the feeding system of the bias yarns has to be adjusted in accordance with the path and consumption of the yarns inside the fabric.

The warp yarns are wound juxtaposed to each other on the warp beam and they are fed parallel to the loom axis toward the fabric, as illustrated in Figs. 7 and 8. This feeding system can be used for the bias yarns incorporated in zig-zag pattern, Fig. 12a, and edge-to-edge non-uniform pattern, Fig. 12d. But a tension control system has to be used to maintain the yarn tension constant since the length of the yarn path on the machine is not the same and varies during weaving.

The edge-to-edge uniform pattern requires rotating movement of the guide unit of the bias yarns within a closed path. Thus, using this feeding system for bias yarn from fixed bobbins or beams induces entanglement of the yarns behind the guide device. Therefore, the rotary feeding system has to be used for the bias yarns to ensure the continuous production.

5.3 Bias Yarns Position Control Mechanism

Different techniques have been invented to achieve the transversal movement of the bias yarns in the weaving zone resulting in variant architecture and geometry to the formed bias yarn layers as detailed previously. Some techniques are presented herein in 2D and 3D multiaxis woven fabric.

5.3.1 2D Multiaxis Weaving Machine

Split Reed and Jacquard Shedding Mechanism

Mood modified the 2D weaving loom to produce four layers multiaxis woven fabric (warp, filler, + and −bias) by means of a split reed and jacquard shedding mechanism. Instead of a closed reed on the weaving loom a split reed into two open reeds is used. The upper one is able to move transversally into the weaving zone while the lower one is fixed. The bias yarns are distributed on heddles attached to a selective jacquard system (one yarn is passed through the eye of one heddle). The selected bias yarns to be displaced transversally are moved up by the jacquard system and put in the upper reed. Whereas the rest of the bias yarns, which are kept in position, are moving down towards the fixed lower reed. The distribution of the bias yarns into two layers (upper and lower) by jacquard system permits also to create the required shed to insert the weft yarn. The bias yarns are prisoned between the interlaced warp and weft yarns at the crossing area (Fig. 13).

Fig. 13 Multiaxis four layers woven fabric **a**, one case of distribution the bias yarns by jacquard mechanism into the upper movable reed and lower fixed reed **b**, [10]

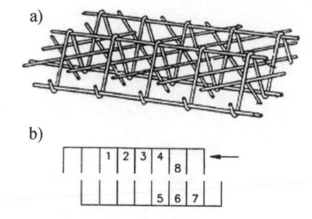

a)

b)

Multiweave Concept

Four interlaced yarn sets; warp, weft, + and −bias, compose a 2D woven fabric manufactured on a modified weaving loom. The bias yarns are fed from beams then pass via a tension compensation device, as shown in Figs. 3 and 4 of [11], towards the fabric. The bias yarns are displaced in a stepwise transversal movement to form a pair of opposite uniform layers. During this movement the controller of warp yarns and the open reed are moved out of the weaving zone. All warp yarns are raised in the space created between the crossed +bias and −bias yarns forming with the bias yarn layers a shed through it the weft yarn is inserted, as illustrated in [24]. However, the open fabric structure is obtained with low fiber content as shown in Fig. 11.

Controlled Bobbin Carrier

Braider carriers are used by Nayfeh et al. [12] to feed the bias yarns on a weaving loom adjusted to produce woven profile with a specific cross section shape. A carrier holds only one bobbin feed the yarn in a manner to maintain its tension constant with the help of a tension control system provided with each carrier. The carriers are mobile following a specific path leading to insert the bias yarn in the desired orientation and forming the fabric in the predetermined shape. A schematic of this technical concept is presented in Fig. 14.

5.3.2 Multilayer Multiaxis Weaving Machine

Lappet Apparatus

Lappet-weaving apparatus has been developed by Ruzand and Guenot [13] by modifying standard weaving loom to produce a multilayer multiaxi woven fabric. Hooks

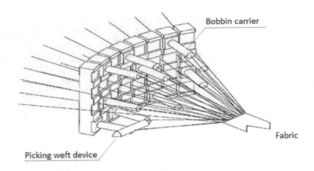

Fig. 14 Schematic of multiaxis weaving machine developed by Nayfeh [12]

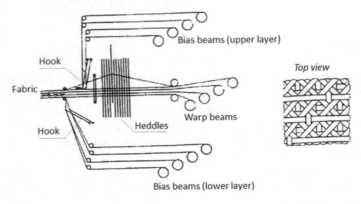

Fig. 15 Schematic of the modified lappet multiaxis weaving by Ruzand and Guenot [13] with produced architecture

(one for each bias yarn) are disposed on a segmented transversal bar called lappet bar. The lappet bar has two movements; vertical movement to enable the weft yarn insertion and beating up operation by the reed and horizontal transversal movement to give the required orientation to the bias yarns. The lappet apparatus and the feeding unit are placed upper and lower the warp beams and heddles to avoid crossing the bias yarns with heddles. Thus the bias yarn layers can be only inserted on the top and bottom surface of the fabric. The lappet bar width is higher than the fabric width. So the lappet with the bias yarn when it reaches the fabric edge it is folded to the next layer in the opposite direction. Thus a pair of edge-to-edge uniform bias yarn layer is formed (Fig. 15).

Tube Rapier

Bilisik and Mohamed [14] developed a tube rapier weaving method to produce multi-axis three-dimensional fabric. Both warp and bias yarns pass through tube rapiers which are arranged in a matrix corresponding to a cross section shape of demanded preform. Weft and binder yarns are inserted by needles passing between the tube rapiers of bias and warp yarns. That reduces the damage incurred by insertion of the needles to warp and bias yarns met in other technologies. Tube rapiers of bias yarns displaces only transversally with respect to the loom main axis [14]. The yarn orientation is inversed when it reaches the fabric edge but it is still in the same layer resulting in zig-zag non-uniform bias layer. On this loom, the bias yarns layer could be placed in any layer in through the thickness of the fabric, contrary to lappet apparatus, allowing producing a 3D multiaxis woven fabric. The warp and bias yarns are fed from fixed bobbins (one yarn from a bobbin). That enables individual control of yarn tension during the process since they have different path in the fabric.

Screw Shafts

Anahara et al. [15] used a screw shafts to displace the bias yarns transversally in the weaving zone across the entire width of the fabric forming non-uniform edge-to-edge layer of bias yarn, Fig. 16. The screws can be placed in any layers through the thickness of the fabric as in tube rapier technique. The different layers of weft, bias and warp yarn are combined by binder yarns passing by means of a rapier through all the fabric thickness, Fig. 16. The warp, binder and bias yarns are fed from independent beams as shown in Fig. 16 since the path and consumption of each type is different. The bias yarns are fed from a traditional beam as warp yarns, Fig. 16, and they are pulled out of the beam parallel to each other.

Tube Carrier

Bilisik and Mohamed [14, 16, 17] developed another weaving method using a tube carrier in order to get pair of opposing uniform edge to edge bias yarns layer. Each bias yarn is fed from individual bobbin then it passes through one tube. The tubes move transversally inside a box and when one tube reaches the fabric edge, it is shifted up or down to the next opposite bias yarns layer so it changes its orientation from $(-\theta/+\theta)$ to $(+\theta/-\theta)$ Fig. 17. Thereby, the tubes are in a circular movement requiring circular moving for feeding bobbins. The weft yarns and the binder yarns are inserted in the same manner as a tube rapier weaving method by means of rapier needles.

Guide Blocks

Yasui et al. [18] developed 3D multiaxis weaving machine using movable guide blocks to displace the bias yarns in the weaving zone instead of screw shafts, Fig. 18. The guide blocks are actuated to move in closed cycle crossover the fabric width, Fig. 18b, forming a pair of opposite edge-to-edge uniform bias yarns layers. In

a)

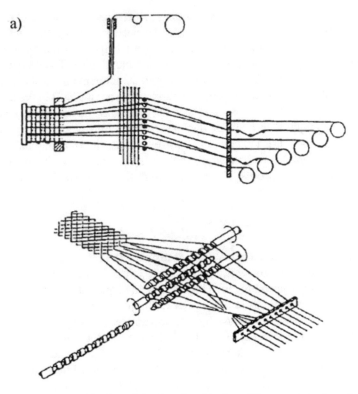

Fig. 16 a Schematic of a side view of developed multiaxis multilayer weaving loom by Anahara et al. [15] using screw shaft and rapier needle, **b** Isometric view of produced architecture, **c** top view of the bias yarns path forming an edge-to-edge nun-uniform layer

accordance with this cycle motion of the bias guide the bias yarn is fed from individual bobbins mounted on rotary support to ensure continuous feeding without intertwining the bias yarns. The rapier needles are also used to insert binder yarns in the through thickness direction to fabricate orthogonal multilayer fabric.

Uchida et al. [19] used as well the guide blocks to control the position of bias yarns in the weaving zone. Different method, Fig. 19, is used to actuate the guide blocks where special means are used to displace guide blocks sideways (transversally) and shift them up or down to fold the bias yarns once they reach the fabric edge so alternating its orientation from $(-\theta/+\theta)$ to $(+\theta/-\theta)$ but bias yarn still in the same layer. As a result, a pair of edge to edge non-uniform bias yarns layer is formed. Thereby, there is no need for circularly moving of bias yarns supply mechanism and the bias are fed from the beam. In order to control the variation of their tension during the weaving process, a tension device is developed. A pair of chain conveyors is used at the interval of fabric width to retention woven fabric in order to stabilize its width after weaving zone and to prevent high shrinkage caused by folded bias yarns.

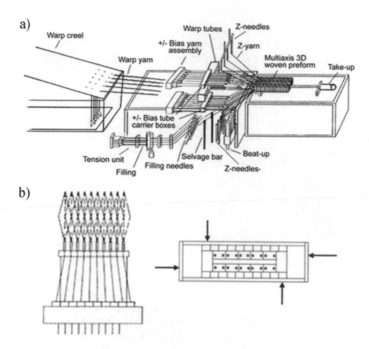

Fig. 17 Tube carrier weaving method [17]; **a** scheme of developed multiaxis 3D weaving machine, **b** representation of circular moving of each assembly of tube carrier for one bias yarns layer

Author et al. [20] constructed a 3D weaving machine using the guide block apparatus principle with the possibility of incorporate a warp yarn layer separating the two successive opposite layers of bias yarns, Fig. 20. That improves the delamination resistance of the produced composite and delayed the damage initiation into it. On this machine each bias yarn is fed from a bobbin with individual tension control system. The bobbin carriers are mounted on rotating chain ensuring the continuous feeding for the bias yarn without entangling the yarns. This unit can be repeated on the machine to stack a desired number of the bias yarn layer in any order through the fabric thickness.

5.4 Polar Multilayer Multiaxis Weaving

Polar multiaxis 3D weaving loom, developed by Yasui et al. [21], capable of fabricate a fabric of 24 layers of circumferential weft, bias yarns and axial warp yarns, Fig. 21. These layers are combined by a radial binder yarn passing through the different layers. The concept of this machine is based on exchanging 144 bobbins between two rotating and axially translating disks. However, the large shed size formed by exchanging the bobbins causes fibre damages.

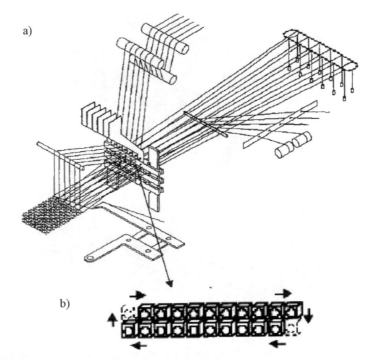

Fig. 18 Scheme of multiaxis weaving machine developed by Yasui et al. [18] using guide blocks (**b**) to control the position of bias yarns with a rotating motion for feeding mechanism

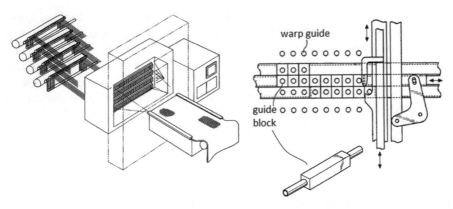

Fig. 19 Scheme of multiaxis 3D weaving machine developed by Uchida et al. [19] using guide blocks to control the position of bias yarns

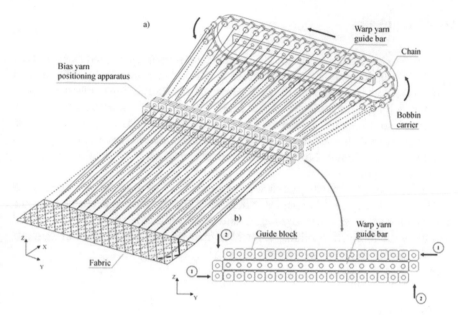

Fig. 20 Schematic of guide block apparatus developed by authors to control the bias yarn orientation with an individual rotating feeding system

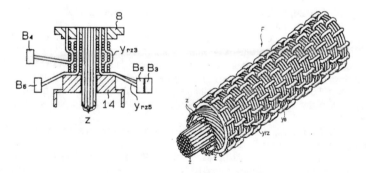

Fig. 21 Polar multiaxis weaving machine developed by Yasui et al. [21], **b** architecture of fabricated woven preform

Bilisik [22] developed a multiaxis 3D circular weaving loom based on the 3D braiding principle. The tubular fabric is composed of axial warp yarns, circumferential weft yarns and ± bias yarns fed by braider bobbins and placed only on the outside and inside surfaces of the tube. Also here radial yarns are used to combine the different layers. It is inserted by means of special carrier units passing through the layers radial thickness. This machine can produce thick and thin cylinders and the presence of bias yarns improves its torsion properties, as reported by Bilisik [23].

6 Conclusion

The multi-axis weaving technique is an innovative technology derived from conventional 2D weaving and developed at three structure dimension scales; 2D, 3D and polar. It is a novel branch of textile preform manufacturing aiming to obtain a fibrous structure with more than two in-plane fiber orientations. That permits to improve in-plane shear properties and reduce the structure sensibility to out-off axis loads as a comparison with 2D and 3D woven fabric. Further, it permits to obtain a multi-axis structure by one manufacturing step that reduces production costs as a comparison with stitching process.

The different technical apparatus have been developed in the last decades as detailed previously in this chapter, but these solutions have not reached yet the commercialized stage especially for 3D and polar multi-axis fabric. The sequential motion of the bias yarn's guiding element in the weaving zone is the main obstacle towards profitable production speed.

References

1. Behera, B. K., & Hari, B. K. (2010). *Woven textile structure: Theory and application*. Woodhead Pub.
2. Boussu, F., Cristian, I., & Nauman, S. (2015). General definition of 3D warp interlock fabric architecture. *Composite Part-B, 81*, 171–188.
3. Guenon, V. A., Chou, T. W., & Gillespie J. W. (1989). Toughness properties of a three-dimensional carbon-epoxy composite. *Journal of Material Science, 24*, 4168–4175.
4. Agrawal, S., Singh, K. K., & Sarkar, P. (2013). Impact damage on fibre-reinforced polymer matrix composite - A review. *Journal of Composite Material, 48*(3), 317–332.
5. Kaw, A.K. (2006). *Mechanics of composite material*. United States of America.
6. Tong, L., Mouritz, A. P., & Bannister M. K. (2002). *3D fibre reinforced composites*. Elsevier Science Ltd.
7. Mouritz, A. P., & Cox, B. N. (2000). A mechanistic approach to the properties of stitched laminates. *Composite Part A, 31*(1), 1–27.
8. Mouritz, A. P., & Cox B. N. (2010). A mechanistic interpretation of the comparative in-plane mechanical properties of 3D woven, stitched and pinned composites. *Composite Part A Applied Science Manufacture, 41*(6), 709–728.
9. Labanieh, A. R., Legrand, X., Koncar, V., & Soulat D. (2016). Development in the multiaxis 3D weaving technology. *Textile Research Journal, 86*(7), 1869–1884.
10. Mood, G. I. (1996). Multi-axial yarn structure and weaving method, United States patent: 5-540-260.
11. Lima, M., Fangueiro, R., Costa, A., Rosiepen, C., & Rocha V. Multiweave: New weaving technology of a multiaxial structure for technical application. In *8th Brazilian Conference on Design Research and Development* (pp. 2564–2573). Brasil, 2008.
12. Nayfeh, S. A., Rohrs, J. D., Rifni, O., Akamphon, S., Diaz, M., & Warman, E. C. (2006). Bias weaving machine, United State Patent: 7-077-167.
13. Ruzand J. M., & Guenot, G. (1994). Multiaxial three-dimensional fabric and process for its manufacture, International patent WO: 94/20658.
14. Bilisk, A. K., & Mohamed, M. H. (1994). Multiaxial 3D weaving machine and properties of multiaxial 3D woven carbon/epoxy composites. *Proceeding of American society of composite*, 868–882.

15. Anahara, M., Yasui Y., & Omori H. (1991). Three-dimensional textile and method of producing the three dimensional textile, European patent: 0-426-878-A1.
16. Bilisk A. K., & Mohamed M. H. (2010). Multiaxis three-dimensional (3D) flat woven preform-tube carrier weaving. *Textile Research Journal, 80*(8), 696–711.
17. Mohamed, M. H., & Bilisik, A. K. Multi-layer three-dimensional fabric and method for producing, U.S. Patent, 5465760.
18. Yasui, Y., Anahara, M., Hori, F., & Mita Y. (1993). Production of three-dimensional woven fabric, Japanese Patent Publication: 05-106-140.
19. Uchida, H., Yamamoto, T., Takashima, H., Yamamoto, Tet., Nishlyama, S., & Shinya, M. (1999). Three-dimensional weaving machine, United States patent 6-003-563.
20. Duchamp, B. (2016). *Contribution à l'élaboration de préforms textiles pour le renforcement de réservoir souple.* Lille University.
21. Yasui, Y., Anahara, M., & Omori, H. (1992). Three dimensional fabric and method for making the same. *United States Patent, 5091246,* 25.
22. Bilisik A. K. (2000). Multiaxial three dimensional (3D) woven carbon preform, United States patent: 6-129-122.
23. Bilisik, K. (2012). Multiaxis three dimensional weaving for composite: A review. *Textile Research Journal, 82*(7), 724–743.
24. Lima M., Fangueiro R., Costa, A., Rosiepen, C., & Rocha V. (2009) Multiweave _a prototype weaving machine for multiaxial technical fabrics. *Indian Journal of Fibre and Textile Research, 34,* 59–63.

Weaving in 3D Shape and Volume

Xavier Legrand

Abstract Traditional textile technologies such as weaving have been designed to produce fabric rollers. The resulting products are flat structures (2D). In addition, it should be remembered that the mechanical properties are given by the main directions of the fibres of the structure. In other words, the directions in which there is no fiber, the mechanical properties are low. This is why there is a need to propose new technologies for the structuring of so-called 3D textiles. This chapter presents two of these new technologies. Also, in this chapter, we try, first of all, to define what we call 3D, for this purpose, a subclass of 2D, named 2D+. In a second time, a particular technology will be presented: 3D weaving technology for tetrahedral shell. In order to do this, the 3D structuring technology called BWS for "Braindring - Weaving System" will be detailed.

Keywords Example chapter · Advanced weaving book · Springer

1 Introduction

The classical woven textile product is a flat stuff as illustrated into some of the previous chapters. Into this chapter, two technologies to produce 3D shape and 3D volume woven products are introduced.

2 Definition

Because we need to define the objective of this chapter, it is necessary, first of all, to propose a classification of textile structures in relation to their number of dimensions (1D, 2D, 3D…). [1–6].

X. Legrand (✉)
ENSAIT, ULR 2461 - GEMTEX - Génie et Matériaux Textiles, University Lille, 59000 Lille, France
e-mail: xavier.legrand@ensait.fr

© Springer Nature Switzerland AG 2022
Y. Kyosev and F. Boussu (eds.), *Advanced Weaving Technology*,
https://doi.org/10.1007/978-3-030-91515-5_15

Thus, it is widely agreed that a fibre, yarn or tow, because their length is very much greater (several dozen or even a hundred times) than the other dimensions, are considered to be one dimension (1D). Similarly, because the thickness of a fabric is several times smaller than its length or width, then, it is very generally considered that a woven structure is a 2-dimensional (2D) textile structure. In this chapter, we are interested in so-called 3D woven structures. So, three classes are traditionally agreed: class 1D, class 2D and class 3D.

A special case is the interlock woven structures, widely presented and detailed in one of the previous chapters. The objective of interlock technology is to bond several layers of fabric together in the thickness. The thickness of such woven structures is then more important. It therefore seems that these structures do not belong to class 2D. However, the thickness remains small compared to the other two dimensions of the interlock woven structure. This is why, in this chapter, we will not classify interlock structures in the 3D class. It seems obvious that we need to create an intermediate class (between class 2D and class 3D), let's call it, class 2D+. This class could be defined as "This class 2D+ includes textile structures whose thickness is "less negligible" in relation to their length and width".

Let's study the case of the 3D class, now. In this class we distinguish, as indicated by the title of the chapter, the 3D structures shell or shape and the 3D structures volume.

Again, the difference between the two is the importance of thickness. In fact, in the case of structures in the 3D volume subclass, it is more difficult or even impossible to define the notion of thickness. Symmetrically, as its name suggests, woven structures belonging to the 3D shape subclass have a thickness considered negligible compared to their other dimensions. This classification of textile structures can be summarized in Fig. 1.

3　Principle of 3D Shape at 0°/90°

The object that is taken as an example of woven textile structure geometry in the 3D shape subclass in this chapter is a tetrahedron, geometry illustrated in Fig. 2.

3.1　3D Shape Obtained by Stamping

The first idea to make such a structure is to stamp a flat fabric (2D). Thus, from a blank—a ply cut from a fabric—it is possible to obtain a product with the desired 3D shape by cold or hot forming. The blank is clamped between a punch and a die to obtain a product with a non-developable shape.

The Fig. 3 illustrate a flat fabric which has been stamped by a tetrahedron stamp (Fig. 3a) into a tetrahedron 3D shape (Fig. 3b).

		randomly	mono axial	bi axial	tri axial	multi axial
1D			fibre, tow, yarn			
2D		non woven	uni directional	Woven fabric, knit, bi axial braid	tri axial braid (flat braid, over braid)	NCF stitch bonded, multi axial woven fabric
2D+		thick non woven		woven interlock		multi axial multi layer interlock
3D	surface	shaped non woven		tetrahedron ply	over-braiding, preforming, stamping of braid or/and multi axial	
				stamping		
	volume			assemblies by tufting, stitching, nailing/z-pinning	cross of shaped	BWS, cross of stiffeners (integral 3D)

Fig. 1 Classification of textile structures

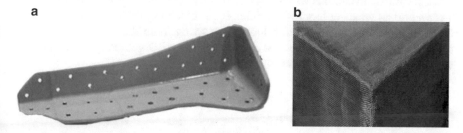

Fig. 2 Tetrahedron 3D shape, a- in aluminium, b- ply at 0°/90° in carbon tows

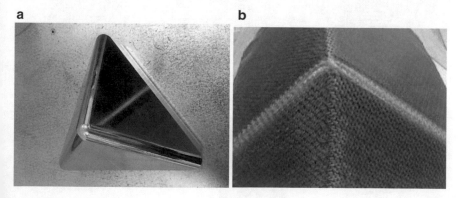

Fig. 3 Stamped tetrahedron 3D shape

3.2 3D Shape Obtained 2D Weaving

Based on the knowledge of traditional weaving techniques, the first solution is to make the tetrahedron geometry by tubular weaving. This technique, indeed, ensures continuity at the weaving edge by using the weaving shuttle.

In the first step, from warp yarns, the weaving shuttle is inserted according weaving pattern. So, a double layer fabric is produced (Fig. 4a). In the second step, in order to set an angle of 45°, step by step, some of warp yarns are leave when the shuttle is patterning (Fig. 4b). In the third step, the exterior warp yarns are cut (Fig. 4c). Then, the part is open to form a tetrahedron geometric ply (Fig. 4d, e, f).

The Fig. 5 presents the obtained product by tubular weaving. On the Fig. 5a, which is a view of the inside of the tetrahedron ply, it can be observed that warp yarn are orthogonal to weft yarns on each face. Continuity of yarn are preserved between faces. But, on shown on Fig. 5b, because warp yarn have been cut, a edge of free yarn exist on a face. This side is composed partly by the upper layer and by the lower layer of the woven part.

This discontinuity is not acceptable for the part in question. The conclusion of this first test with a conventional weaving loom, based on pocket weaving, is therefore that a conventional weaving loom is not suitable for obtaining such a woven geometry. In other words, a classic weaving loom, a weaving loom that has been designed to produce 2D geometries, does not allow the desired 3D geometry to be obtained without discontinuity in the produced part.

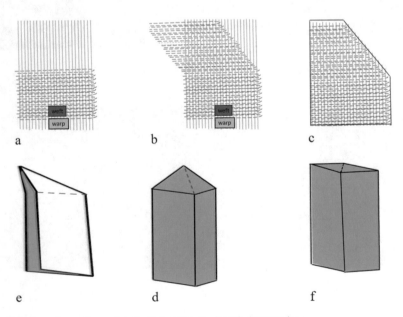

Fig. 4 Process to produce a tetrahedron geometry by tubular weaving

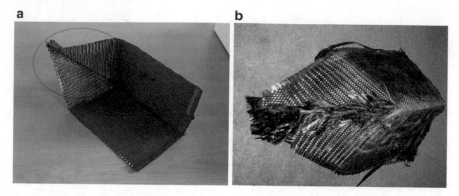

Fig. 5 Obtained product by tubular weaving

3.3 3D Shape Obtained 3D Weaving !

From the previous conclusion, it was necessary to rethink the weaving technology. The idea is no longer to weave in 2D, but to weave directly in 3D. The desired ply, as shown in Fig. 6, must include three continuous yarn networks (green yarns, blue yarn AND red yarns).

Indeed, as presented in [7–9], the innovation to produce such a geometry responds to the double constraint: Continuity of fiber and Control of fiber orientation.

As the orientations of fibres must be well assured (0°/90°) and combined with complete continuity, a new device to realize the surface 3D weaving part has been developed. The corner fitting 3D ply with the fibre direction of 0°/90° can be manufactured in the three steps illustrated on Fig. 7.

First step: as for traditional loom, the first step for manufacturing fabric concerns the fibres beaming and piquing in the direction X (cf Fig. 7a).

Second step: this step is still a traditional weaving loom process. The fibres in the direction Y are weaved with the ones in the direction X to build the side 1 (cf. Fig. 7b).

Fig. 6 Yarn orientations on the desired 3D ply

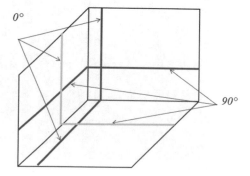

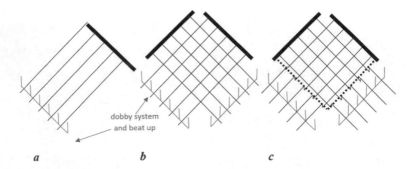

a   *b*   *c*

Fig. 7 CF ply in 3 steps

Third step: the free fibres in the directions X and Y become the Z fibres. Then, XY fibres are weaved with Z fibres as shown in Fig. 7c. The insertion of XY fibres is non-rectilinear but in the path of "L". In this way the two faces (the sides 2 and 3) can be constructed together.

The desired orientations of fiber are set by controlling the tow tension during the whole structuring (interlacing) process, which is a traditional solution used in textile industry. The X and Y fibers produced in the steps 1 and 2 are in tension controlled. In order to guarantee the same tension force on each tow, a spring is employed for each loop of carbon tow during the step 3. All the X direction springs are connected together by a transverse bar self-controlled on tension, to assure global tension applied on X tow. Each "L" weft (inserted tow during the step 3) is also tensioned after its insertion between X and Y tows. Therefore the tow is maintained its right place and the desired orientations could be assured.

The 3D corner fitting ply with the fibre direction of 0°/90° is shown in Fig. 8a. The 3D corner fitting ply with the fibre direction of −45°/45° is performed in the same way. It is shown in Fig. 8b.

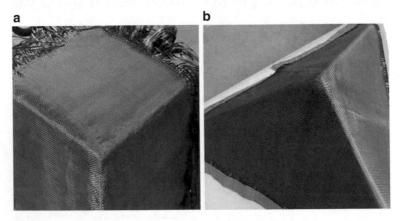

Fig. 8 Obtained CF plies

There are no winkles observed in these final 3D surface structures. The fibres in warp and weft directions stay perfectly perpendicular on the three faces.

This innovative technology opens up new possibilities for realization. This technique may indeed be easily extended to other geometries requiring "3D surface shape weaving. The main limitation of this innovation is that it leads to stratification preforms made of plies not bound together.

The obtained preform is a 3D ply. Its thickness is the same then a classical woven fabric using the same contexture. But, it is not flat mike a classical woven fabric. The global shape is tetrahedron. So, it is 3D woven fabric with a "classical" thickness!

The obtained product (and its technique) is classified in the surface 3D, our new sub group of group of 3D textiles structures.

4 Principle of 3D Volume

The 3D weaving technology propose by Khokar [10] is a very clever idea. Called Noobing by its brilliant inventor, this one consists in the insertion of threads in the plane perpendicular to a matrix of warp threads as illustrated in the paper [11]. It can be considered that this beautiful technology is a 3D dobby weaving. The technology presented hereafter can then be considered in a 3D Jacquard weaving.

Indeed, in order to structure stiffeners with variable section profiles, an innovative technology called BWS (Braiding Weaving System) is proposed [12, 13].

Drawing on existing textile technologies and analyzing the needs for reinforcements, an innovation has been developed. Thus, BWS technology is a hybridization of Weaving and Braiding. As shown in Fig. 9a and b, the BWS technology switches the tow (1D) to the preform of stiffener (3D volume). So, using this technology several cross section stiffeners have been produced (cf. Fig. 9c).

The originality of BWS consist in the hybridization of two existing textile technologies: braiding and weaving. It means that weft yarn are inserted following predefine angles into the braided structure. Each yarn is independent and used to create

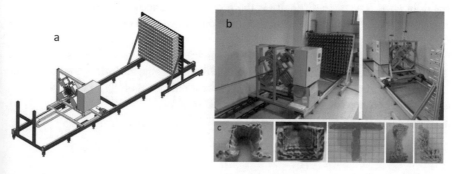

Fig. 9 BWS **a** CAD, **b** photos, **c** cross section of stiffeners

complex shaped structure, with tailored interlacing and non-regular architecture. The principle is based on previous patent [11] and described in [12]. The two sets of yarns used are warp (mainly in the production direction) and weft (orthogonal to this direction). Each warp yarn is tension controlled and guided by the specific case. All the guides together in row and column form the production matrix. It allows vertical and horizontal displacement of guides. Those displacements are totally automated and offer wide range of interlacing, for example dual orthogonal shedding for 3D weaving or four-step braiding. After this displacement, the warp set is rearranged and can be stabilized or improved by the insertion of weft yarn in transverse direction. This insertion is also tailored with the control of the insertion space and the insertion angle with a complete rotation device. The row and column cases matrix mobility is a combination between 3D braiding and 3D weaving which brings new interlacing possibilities for preforms. Each case can move within the matrix to reach any position. Case displacement and insertion will interlace tows together by interference. Also, to design 3D structure, because it is important to predict interlacement and its result on global textile architecture, consequently this device is associated to a numerical tool able to anticipate position of yarns during the process. This software [13] is also interfaced with software as TexGen [14] in order to visualize preform' architectures.

The obtained structure is not a braid either a woven fabric. It is a mix of both. The yarn are superposed/interlaced into a 3D pattern according to orientations draw by needing mechanical properties. So, the structure is 3D but it is not surface 3D as the previous one. It is volume 3D. Besides, this technic allows us to structure tows as a Volume 3D with surface 3D parts…it is possible also to get from it either a developable or an undevelopable structure.

5 Conclusion

3D structuring textile technologies are simply extensions of existing robust textile technologies for 2D products. Therefore, it is easy to define the right technology for the right need. The main difficulty is the problem of congestion caused by the large number of threads, leading to degradation of the threads and reduction of the density of the fibrous networks.

This brief presentation of some 3D weaving technologies, both surface and volume, shows the versatility of textile technologies. Thus, according to the specifications of the final part, it is possible to find the textile technology likely to structure in 3D the fibers according to the desired orientations.

References

1. Tong, L., Mouritz, A. P., & Bannister M. K. (2002). *3D fibre reinforced composites*. Elsevier Science Ltd.
2. Behera, B. K., & Hari, B. K. (2010). *Woven textile structure: Theory and application*. Woodhead Pub.
3. Guenon, V. A., Chou, T. W., & Gillespie, J. W. (1989). Toughness properties of a three-dimensional carbon-epoxy composite. *Journal of Material Science, 24*, 4168–4175.
4. Kaw, A. K. (2006). *Mechanics of composite material*. Taylor & Francis Group.
5. Mouritz, A. P., & Cox, B. N. (2000). A mechanistic approach to the properties of stitched laminates. *Composite Part A, 31*(1), 1–27.
6. Mouritz, A. P., & Cox, B. N. (2010). A mechanistic interpretation of the comparative in-plane mechanical properties of 3D woven, stitched and pinned composites. *Composite Part A Applied Science Manufacture, 41*(6), 709–728.
7. Legrand, X., Piana, M., Tsarvarishki Georgi, Charles, J., Blot, P., & Guittard, D. Airbus France, Three-dimensional surface weave, WO2008049877 (A1), 2008-05-02.
8. Legrand, X., Tsarvarishki Georgi, Charles, J., & Blot, P. Airbus France, System for weaving a continuous angle, WO2008049883 (A1), 2008-05-02.
9. Legrand, X., Boussu, F., Blot, P., & Guitard, D. (2009). A new technique of weaving 3D surface: Application to carbon/epoxy corner fitting plies, *International Journal of Material Forming, 2*(1 Supplement), 185–187.
10. Khokar, N. *Weaving 3D fabrics for reinforcing advanced composite materials,* Melliand International, Issue 1/2018.
11. Khokar, N. (2002). Noobing: A Nonwoven 3D fabric-forming process explained. *Journal of the Textile Institute, 93*, 52–74.
12. Legrand, X. ENSAIT, Matrix for shaping elongated textile elements for manufacturing a three-dimensional textile part and method for manufacturing such a three-dimensional textile part, EP2489768, 2012-08-02.
13. Risicato, J. V., Legrand, X., Soulat, D., & Koncar, V. (2014). Innovative geometrical pre-mesh modelling strategy for 3D fibre preform manufacturing. *Journal of Industrial Textiles, 44*(3), 447–462.
14. Long, C., & Brown, L. P. (2011). Composite reinforcements for optimum performance: Modelling the geometry of textile reinforcements for composites: TexGen, edited by P Boisse, Woodhead Publishing Ltd, ISBN: 978-1-84569-965-9. http://www.woodheadpublishing.com/en/book.aspx?bookID=223.

Various Applications of Woven Structures

The Weaving of Optical Fiber for Cancer Treatment

Cédric Cochrane and Vladan Koncar

Abstract In this chapter, novel light-emitting fabrics (LEF) realized by the weaving process are presented. These LEFs may be used in various applications such as light flexible panels for signalization or advertising, flexible displays for car dashboards, dynamic light-emitting flags, fashion, etc. However, one of the most promising and important applications covers the medical area. It focuses on a new generation of flexible woven light sources dedicated to a skin cancerous state in its initial phase called Actinic Keratoses (AK). A medical treatment process supposed to eradicate AKs is well known and rather efficient, it is called Photo Dynamic Therapy (PDT). A successful PDT requires a specific photosensitizer, oxygen, and light of a specific wavelength and power. Today Photodynamic Therapy (PDT) is administrated on patients with Light Emitting Diode (LED) Panels. These panels deliver a non-uniform light distribution on the complex shapes of human body parts. It is because human anatomy is not a flat surface (head vertex, hand, shoulder, etc.). For efficient photodynamic therapy (PDT), a light-emitting fabric (LEF) was woven from plastic optical fibers (POF) aiming at the treatment of dermatologic diseases called Actinic Keratosis (AK). Plastic optical fibers (POF) (Toray, PGR-FB250) have been woven in textiles to create macro-bendings and thus emit out the injected light directly to the skin. The light intensity and light-emitting homogeneity of the LEF have been improved thanks to Doehlert Experimental Design. During the treatment with PDT, the photosensitizers were activated in the cancerous cells. These cells may be visualized, as they show a characteristic fluorescent under UV light, which is called Fluorescence Diagnosis (FD).

Keywords Light Emitting Fabric (LEF) · Plastic Optical Fiber (POF) · Photodynamic Therapy (PDT) · Weaving · Fluorescence Diagnosis (FD)

C. Cochrane (✉) · V. Koncar
ENSAIT, ULR 2461 - GEMTEX - Génie et Matériaux Textiles, University Lille, 59000 Lille, France
e-mail: cedric.cochrane@ensait.fr

V. Koncar
e-mail: vladan.koncar@ensait.fr

© Springer Nature Switzerland AG 2022
Y. Kyosev and F. Boussu (eds.), *Advanced Weaving Technology*,
https://doi.org/10.1007/978-3-030-91515-5_16

1   Introduction

1.1   AK and PDT

Actinic Keratosis (AK) is a precancerous condition due to chronic UV light exposure that may develop into non-melanoma skin cancer [1, 2]. Thus, the treatment of AK is highly recommended. The AK lesions are characterized by red, scaly, and crusty plaques or papules [3, 4]. This skin disease mainly affects fair-skinned individuals (face, bold head, forehead, etc.) [5].

There are many treatment options for AK. Cryosurgery, curettage, and Photodynamic Therapy (PDT) are the most common treatments. Cryosurgery is an operation to destroy the tissue by using freezing temperatures performed with liquid nitrogen or carbon dioxide. This method is efficient on the thinner lesions but less successful on the thick lesions and may result in scarring [6]. Curettage is used to scrape larger, hypertrophic lesions with a curette. The drawbacks of this technique are the necessitate of local anesthesia and the scars [7]. Moreover, in the case of AK fields that may develop in some cases on the large area of the face or the vertex, it is not possible to remove all the skin covered by the AKs.

Photodynamic therapy (PDT) is a noninvasive method, particularly used to treat pre-cancerous or cancerous lesions with the combination of a photosensitizer and an appropriate light.

PDT has an increasing use to treat the AK, as it is efficient as other techniques are given before but also has excellent cosmetic results. PDT is repeatable and does not kill the healthy cells (selective cell-killing) [8, 9]. This treatment is also suitable for uncancerous lesions as psoriatic, acne vulgaris, pre-cancerous lesions as AK and Bowen, and cancerous lesions as basal cell carcinoma (BCC) and squamous cell carcinoma (SCC) [9, 10].

The PDT leads to selective destruction of the cancerous cells by activated photosensitizing agents, methyl aminolevulinate (MAL) in Europe and 5-aminolevulinic acid (ALA) in the United States. The photosensitizers (PS) are activated with an appropriate light, which is red light (630 nm) in Europe and blue light (450 nm) in the United States in the presence of oxygen [11]. The activation of the PS generates singlet oxygens ($_1O^2$) which causes chemical reactions inside the cancerous cells as they are rich with molecular oxygen [12]. This is so-called "selective cells killing".

In Europe, PDT is performed by using a drug photosensitizer Methyl AminoLevulinate (MAL by Metvix, Galderma) on cancerous lesions and a 3-h interval to start enlightenment [13]. The entire treated area is illuminated by a red light source (narrow spectrum around 630 nm) [14]. The activation of Metvix required a dose of 37 J/cm^2 [15, 16]. The light dose is determined by such factors as the size of the light field, the distance between the lamp and the surface of the skin, and the illumination duration. Therefore, it is not possible to treat numerous patients per day since the treatment of a single patient takes about 5 h. It is possible to reduce the light exposition time by increasing the light dose, but the pain rate will also rise.

Today, PDT is administrated by LED (light-emitting diode) rigid panels [17]. This is an effective method without side effects with good cosmetic results. However, LEDs do not emit the same light dose over the entire treatment area and do not adapt to the complex shapes of the body [13, 18]. Another disadvantage of this method is pain due to the dose of pure light that effective treatment with LED panels [19, 20]. Indeed, if the light output was less, we could limit the pain caused by the PDT.

PDT needs to evolve despite its benefits, to make this treatment more effective and less painful. The market innovates and proposes inventions that allow surmounting the inconveniences of topical PDT with LED panels.

1.2 Light Diffuser from Optical Fibers

Another possibility is to use light diffusers based on optical fibers. This approach allows to move the source away from the treated area and to have a more flexible and lighter diffuser. Optical fibers are used to transmit information minimizing losses. To transform an optical fiber into a light emitter, losses could be caused. In telecommunications, the losses are well known and shown schematically in Fig. 1.

From optical fibers originally designed and utilized for telecommunications, there are 2 methods to put out the light laterally: (i) use of a process for degraded cladding; A mechanical process consists of creating scratches and micro-holes on the surface of the POF, more precisely in their cladding (sandblasting or toothed roll) [21]; Or a chemical process is focused on applying a solvent which degrades the outside of the optical fiber (its cladding) and passes light [22], or (ii) the "geometrical" method consisting of creating macro-bends (bindings in macroscopic size). By this technique, the total internal reflection of light rays is not satisfied and leakages are created [23, 24].

Another way is to directly produce POF with some defaults or inclusions to have self-diffusing fiber [25].

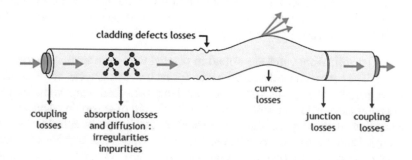

Fig. 1 Schematic description of possible propagation losses

Fig. 2 Light power
distribution along with the
light-emitting optical fiber

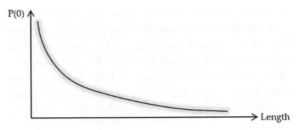

To produce panels for PDT treatment, these methods must be able to produce surfaces to reach from 1D diffusers to 2D. Thus the textile process and more particularly weaving process are suitable. Weaving is a method of textile production that interlaces warp and weft yarns to form a fabric (Fig. 4).

The quantity of the emitted light decreases with the distance to the light source as seen in Fig. 2. The light transmission loss in an optical fiber is defined by the following equation according to scientific literature [26, 27]:

$$\alpha = \frac{10}{L} \log\left[\frac{P(0)}{P(1)}\right] \tag{1}$$

where α represents the light transmission loss (dB/km), the length of the optical fiber (km), P(0) represents optical input power, P(1) represents optical output power. Thus, if defaults are constant along with fiber, the amount of light emitted decreases along with the fiber away from the source. (Fig. 2). One of the keys will be to try to compensate for this decrease: non-constant defaults, sources on both sides, etc.

2   2D Weaving Machines for OF

On the conventional 2D weaving loom shown in Fig. 3, the fabric is made by interlacing two sets of yarns (warp and filler), which are perpendicular, with specific weave pattern by a series of weaving operations in the following order:

1.   Warp left-off: Warp yarns are wound in parallel with even tension on the warp beam. The beam is turned a specific step for each weaving cycle, corresponding to demanded filler yarns count per unit length, to feed warp yarns into the weaving zone under controlled tension with the aid of tensioning devices.
2.   Shedding: each warp yarn passes through a heddle eye (one heddle for one warp). These heddles are attached to lift shafts. Warp yarns that have the same state in weave pattern (that means the same sequence of interlacement up/down with filler yarns) their heddles are attached to the same shafts. For each weaving cycle, a part of the lift shaft moves up while the others move down to separate the warp yarns into two layers creating appropriate space between them across the

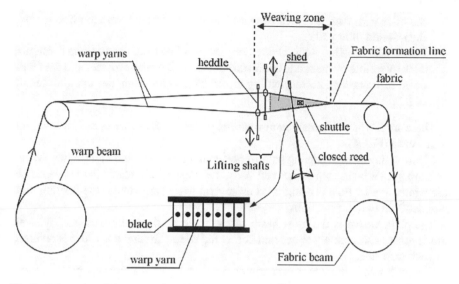

Fig. 3 Schematics of the conventional 2D weaving loom in 2D

Fig. 4 Fundamental weaves: plain weave, twill 3–1, satin 4, respectively

loom, called shed, Fig. 8. The shed allows the insertion of filler yarn across loom width. This proper space is essentially required to avoid crossing and rubbing warp yarns while inserting the filler yarn. The distribution of the warp yarns on lift shafts and the sequence of lifting the shafts (up/down) across weaving carry out the demanded weave pattern for fabric.

3. Filler insertion: For each weaving cycle, a filler yarn is inserted through created shed and it passes across the loom width. Different apparatuses are used to carry filler yarn such as shuttle, rapier, projectile, and air-jet.

4. Beating: the filler yarn, after being inserted, is packed tightly to the fabric by using a reed which is closed combs through the gaps created between its blades the warp yarns are passed. And the number of warp yarns in one space in addition to the count of blades per unit length defining the count of warp yarns per unit length of fabric on a loom. The reed has an alternative movement, it moves back away from the fabric formation line to allow creating the shed and insertion of

the filler yarn then it moves forward toward the fabric formation line to pack the inserted filler yarn.

5.   Fabric take-up: The new packed filler yarn is pulled from the weaving formation zone by rolling up the fabric on the fabric beam. The take-up roller and the fabric beam turn one step corresponding to the filler yarns count per one-unit length of fabric.

There are 3 types of fundamental weaving patterns; plain weave, twill weave, and satin weave (Fig. 4).

Figure 4 shows the repeating pattern presentations of fundamental weaves. When the warp yarn is on the top of the weft yarn, it is presented in black; in the other case, it is white. A weft float is designated as several warp yarns under the floating weft yarn between two intersections.

The plain weave is the most basic of the three fundamental weaves. The weft thread passes successively above and below the warp yarn and this order is reversed for each weft line.

The twill weave has a harness number equal to 1, and the crossing points form a diagonal. It consists of floats longer than plain weave.

The crossing points of the satin weave are defined with the harness number, and it is higher than 1. Satin weave has longer floats compared to other fundamental weaves.

Thus the choice of the pattern will influence the density of the textile in optical fibers. Weaving has a direct influence on the amount of light emitted.

3   LEF Preparation and Connections

The dimensions of LEF prototypes have been fixed to 5×20 cm^2, containing 184 optical fibers of a diameter of 0.25 mm. To measure the emission of light from the LEF, it is first necessary to connect this textile to a laser source. All the optical fibers are first gathered in an extensible silicone tube (Fig. 5a), using sleeve pliers ($\emptyset$ 1.2 $\rightarrow$ 11.5 mm), to create beams that will be connected to laser sources. This step is used to facilitate the re-entry of the fibers into stainless-steel connectors (Fig. 5b).

The outside diameter of the stainless-steel connector imposed by the laser sources' dimensions is fixed for connection to the laser source of the same diameter ($\emptyset$ 6 mm). The internal diameter of this connector allows containing 92 optical fibers with minimum space between them. This implies the tests to be carried out on the half-widths of the fabrics (approximately 2.5 cm width of the LEF sample) corresponding to a quantity of 92 optical fibers. The technical drawing of the connectors used is given in Fig. 6.

The connectors require to be used with resin to hold the optical fibers in the connector. This facilitates the polishing of the connector to obtain a smooth finish. The smooth finish is essential so that the fibers lose as little incident light as possible in rear reflection.

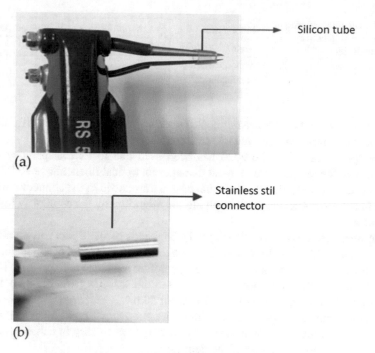

Fig. 5 **a** and **b** Insertion of optical fibers in a stainless-steel connector

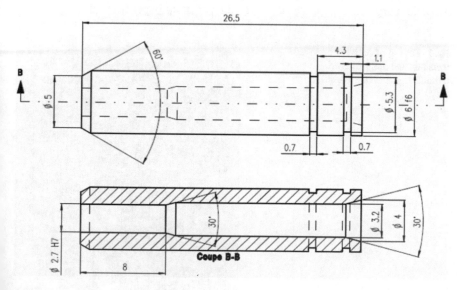

Fig. 6 Stainless-steel connectors sketch (mm unit)

Table 1 The mixture ratio of the epoxy resin Araldite 2020, Huntsman

Mixing ratio	Weight proportion	Volume proportion
Araldite 2020 A	100	100
Araldite 2020 B (hardener)	30	35

A mixture of epoxy resin (Araldite 2020, Huntsman) is prepared with a hardener mixture ratio (Table 1) indicated on the technical sheet.

The epoxy resin Araldite 2020 has been used due to its transparency and its hardening at the temperature to avoid damages on optical fibers above 70 °C.

The ends of optical fibers' beams have been introduced in the connector and kept 5 min in the resin Araldite 2020 implying the migration by the capillarity inside the connector.

The fibers are then suspended (Fig. 7). The ends of the fibers can be placed in an oven at around 50 °C for 6 h to speed up the polymerization of the resin. Epoxy resins are thermosetting, their hardening is therefore related to temperature. The time required for the minimum shear strength of Araldite 2020 is given in Fig. 8. Finally, the resin optical fibers were cured at 20 °C for 30 h.

When the curing is complete, the fibers are cut using a razor blade 2 mm from the end of the connector. Then, manual polishing is necessary to achieve a smooth finish and thus reduce the loss of incident light.

Fiber optic beam polishing is accomplished by drawing an "8" and holding the connector perpendicular to the surface of the polishing paper (Fig. 9).

Fig. 7 Drying of optical fibers inserted and resinated in stainless-steel connectors

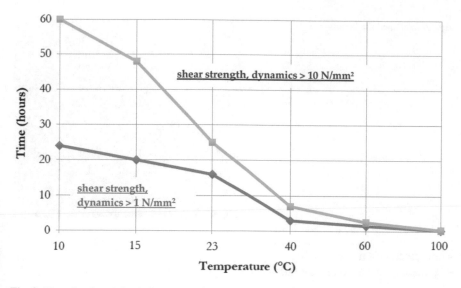

Fig. 8 Time for the minimal shear strength

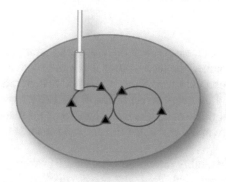

Fig. 9. 8-shaped fiber optic polishing

Hard epoxies and PMMA optical fibers are easily polished using silicon carbide polishing paper (20, 30 μm, etc.). We started with a rough manual dry polishing (30 μm), then a wet manual polishing (9, 3 then 0.3 μm) and finally a manual polishing with Acryglas paste (3–5 min) (see Fig. 10). Acryglas is a paste used to remove scratches and scratches from plastics such as Plexiglas.

The time required to prepare a single sample is approximately 3.5 days. The main steps that make up this preparation are:

- fiber counting and their insertion in a connector,
- resin the fibers inserted in the connector,

Fig. 10 Polished OF beams

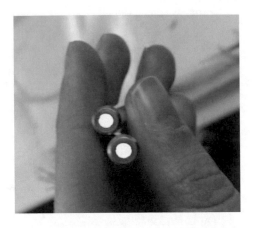

- polymerization
- polishing of the connector ends.

4 Measurement and Evaluation Technique for Emitted Light Quality

To finalize the light fabric measurement device, the light emitted fabric must first be connected to a laser source. The laser source is portable containing two 635 nm red diodes (Fig. 11). The output power of the box is 400 mW in total, or about 200 mW injected into a strand of 92 fibers on each side of the LEF.

A power meter (Ophir, Nova II) was used to measure the distribution of light over the length of the fabric (mW / cm^2). A 1 cm × 1 cm sensor head is connected to the power meter.

To perform the light quality measurements, the LEFs were placed on a flat white surface in a room without light. First, the power (mW/cm^2) of the laser outputs is

Fig. 11 Portable laser source

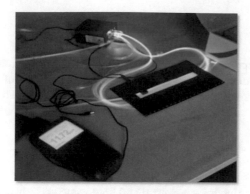

Fig. 12 Light power measurement (mW/cm²)

measured. Then, the remaining light power at the connector outputs is measured when the laser source is connected on one side of the fabric. After recording these values, the laser sources are connected to the connectors at both ends to measure the power per cm² generated by the LEF laterally and the uniformity of the light emitted is analyzed (see Fig. 12).

The power is measured for each square centimeter over the entire length (21 cm) of the LEF and on two lines as illustrated in Fig. 13. A power meter (Ophir, Nova II) was used to measure the distribution of light.

42 values are collected by measuring the power of a line in each sample of the LEF. It is important to quantify these values in terms of light efficiency and uniformity so that samples can be compared more easily.

Two formulas are used to describe each luminous textile. These formulas are inspired by the arithmetic mean used to know the characteristic of data dispersion.

The value which represents the luminous intensity of the luminous textile (P) is obtained by calculating the average of the power values in length (only 18 values are taken into account, the values located 1.5 cm from the ends are not added because the radii of curvature are different due to poor support of the fabric at the ends).

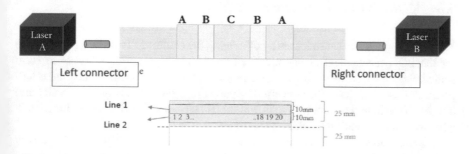

Fig. 13 Light power measurement

$$P \; experimental = \overline{x} = \frac{\sum_{i=1}^{n} x_i}{n} \qquad (2)$$

The value that represents the heterogeneity of the light distribution of the LEF (H) is obtained by calculating the sum of the quadratic deviations of the power values divided by the square of the light intensity (P) of the light fabric (only 18 values are taken into account, values located 1.5 cm from the ends are not added).

$$H \; experimetal = \frac{\sum_{i=1}^{n} (x_i - \overline{x})^2}{(P \; experimental)^2} \qquad (3)$$

A luminous fabric having a powerful and homogeneous light emission has a large P-value and a small H value. Thus, the criterion for optimizing luminous textiles consists of maximizing P and minimizing H.

5 Light-Emitting Fabric Adapted to a POF Treatment Using Materials' Modifications

5.1 Materials

Optical fibers are used to transmit information minimizing losses. To transform an optical fiber into a light emitter, losses could be caused. The easiest way is to work with PMMA (poly(methyl methacrylate)) fibers because of their good refractive index (1.49) and their ease of supply because they are used in telecommunications. For fibers with less attenuation, it is possible to use polystyrene (PS), and for fibers more suitable for high temperatures, polycarbonate (PC). Commonly PMMA POF is available with a diameter of 0.25, 0.50, 0.75 mm, etc.

5.2 Methods and Associate Results

The treatment aims to cause losses by damaging the cladding. Two ways are possible for the POF, mechanical or chemical action. Mechanical action consists to form some holes on the surface with abrasive medium, toothed roller, sandblasting, micro perforator, etc. [28, 29]. The side emitted power of light depends on the density and the size of the hole. The first works on this subject have been carried out by Bernasson who used micro-sanding machine [21].

The main advantage of this process is the possibility to modify the hole density along with a fiber or along with fabric. Thus it is possible to compensate for the depletion of light or localize the light emission. On another side, if craters are too larges or too numerous, a very luminous spot could appear and interfere with the medical application (poor homogeneity) [30, 31].

Other methods of abrading the surface with a medium (paper, toothed wheel) have been developed. They aim to produce fibers with repetitive spots like LEDs on fiber [29].

These light spots are not required in the context of PDT treatment but by fine sanding and control, the spot effect can be reduced and medical applications exist. The most well-known application is the treatment of neonatal jaundice (icterus) where a flexible panel containing light-emitting optical fiber is placed in the contact with the baby's skin [32, 33]. Inside the panel, optical fibers are not always woven, the light (instantaneous) intensity is weak, and the baby moves.

The chemical method allows the cladding of the optical fiber to be damaged by spraying a solvent in the form of an aerosol to create surface defects [34]. By modifying the process conditions (temperature and concentration of the solvent) it is possible to change the quantity and the quality of damage to the cladding of the optical fiber. The formulas and mixtures of solvents used are patented and confidential, and specific to each company. Among the common and known solvents which can damage POFs, we can give acetone as an example.

Optical fiber could be woven before or after treatment, it depends on the desired effect and process control. If the light emission has to be located only at one place on the fabric, it is easier to treat after weaving. But if a fully illuminating fabric is a need, using pre-treated fibers is a good solution. From this fabric is it also possible to hide some areas with an opaque coating. In the case of treatment on fabric (solvent, sanding…), it must be ensured that the other materials can withstand the processing.

Jacquard weaving allows more complex structure and effect [35] than conventional weaving, for example, to hide some area with opaque yarn [30] or to design pixels [22]. But these effects are not used in the medical field.

6 Light-Emitting Fabric Adapted to a POF Treatment Using Structural Modifications

6.1 Materials

For OFs, the core of the fiber has a refractive index greater than the cladding's and therefore the light is confined thus completely reflected (Snell's Law, Eq. 4). When a macro-bend is formed with an OF, light enters into the OF with an angle larger than the critical angle (Eq. 5) undergoing multiple reflections.

$$\sin \alpha . n_{air} = \sin \theta_1 . n_{core} \tag{4}$$

$$\theta_{critical} = \arcsin(n_{cladding}/n_{core}) \tag{5}$$

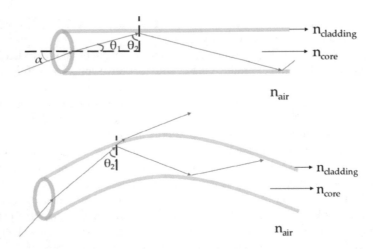

Fig. 14 Multiple reflections in a bent optical fiber

As a consequence, when the fiber is bent, the light rays outside of the bent section will be emitted; the others will continue to meet internal reflection as seen in Fig. 14.

The majority of works use polymer fibers and more precisely PMMA (poly(methyl methacrylate)) core enveloped by a cladding made of fluorinated polymers because of its good curvature tolerance and their flexibility more important than glass fibers.

6.2   *Methods and Associate Results*

The goal is to insert the POF into a textile structure is to cause the bending and allow and control the emission of light from the surface of the textile. Embroidery, knitting, and weaving can allow this control of the bending of the OF. Figure 15 shows the cross-section of a plain weave and macro bending of yarn. For these three textiles technologies, it is important to choose the structure and the other yarns of the structure which will maintain the curvatures of the OFs.

To weave the optical fibers, the rapier loom is mostly used because they do not support the spooling to be placed in the shuttles. Industrial Dornier weaving machines can be used, for instance by adapting it, especially at the insertion speed. Jacquard weaving machine could be used to create a more complex pattern and/or effect [36].

About the general concept, in 1989 Parker developed a light panel made up of several layers of woven polymer optical fibers coated with a diffusing layer [37].

POF bending is obtained by weaving but light emission is not well control and the addition of diffusing layer allows to homogenize emission of light by the surface. This multilayer structure is rigid.

In order to maximize the comfort of the PDT and remove the disadvantages of multilayer structures, Inserm (Institut National de la Santé et de la Recherche

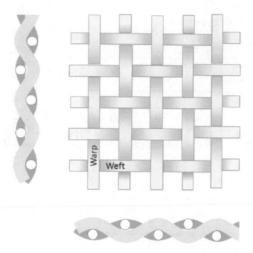

Fig. 15 Schematics of a plain weave

Médicale, France) and ENSAIT (École Nationale Supérieure des Arts et Industries Textiles, France) have developed LEF composed only of PMMA optical fibers (POF) and polyester yarns microfibers [38, 39].

From fundamental weaves (Fig. 4 and Table 2) different weaving patterns were woven to measure their attenuation along woven. As previously written, the loss of the light in an OF (or for a woven), depends on the radius of the bending curvature, the number of bending points (density of bending), and the wavelength of the signal. The number of bending is related to the number of crossing points and the longer of float to the bending radius. Other parameters have an effect like the diameter of the warp yarn, their rigidity, their tension…

Oguz et al. have prepared LEF samples from fundamental weaves with 3 different warp tensions during weaving. Each fabric is characterized by its theoretical density of bending (/cm) and "up and down yarn ratio" and evaluated in terms of total emitted light intensity (mW) [40]. The results are presented in Table 2.

From experiments summarized in Table 2, profiles of light emission are plotted for fabrics woven with different warp yarns tensions (Fig. 16 (a) 40 g/yarn, (b) 63.5 g/yarn and (c) 80 g/yarn). Light intensity is expressed in mW/cm²/W because it is normalized by the light power of the source. This expression makes it possible to visualize the efficiency of fabrics.

These results show that only a satin and unbalanced twill allow good total light intensity distribution, even for an intermediate tension for the satin 6. This pattern could be gathered according to their theoretical density of bending (/cm) and up and down yarn ratio. Thus acceptable area for effective total light intensity is visible in Fig. 17.

With these simple weave patterns, even if the fabric is light on both sides it is difficult to obtain homogeneous power, there is often a lack of light in the middle. To

Table 2 Fundamental weaving patterns

		Samples									
Name		Plain Weave	Satin 4	Satin 6	Satin 8	Twill 1/3	Twill 1/5	Twill 1/7	Twill 2/2	Twill 3/3	Twill 4/4
Weaving patterns											
The theoretical density of bending (/cm)		10.1	5.1	3.3	2.5	5.1	3.3	2.5	5.1	3.3	2.5
Up and down yarn ratio		1/1	1/3	1/5	1/7	1/3	1/5	1/7	2/2	3/3	4/4
Total light intensity (mW)	40.0 g/yarn	0.8	16.6	14.3	86.2	89.3	10.6	0.0	0.9	9.7	1.0
	63.5 g/yarn	0.0	206.3	109.6	31.2	11.6	6.3	7.8	0.0	9.8	6.4
	80.0 g/yarn	2.8	214.9	186.9	113.4	0.0	75.7	200.7	0.0	118.1	7.5

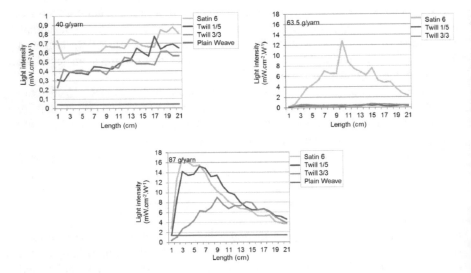

Fig. 16 Light emission profile of fabric woven with different yarns tension

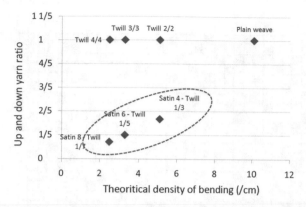

Fig. 17 Classification of fundamental weaves according to the theoretical density of bending (/cm) and up and down yarn ratio. Circled weaves are acceptable in terms of total light intensity

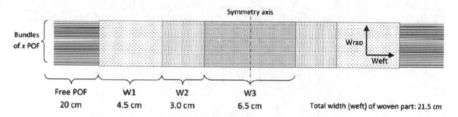

Fig. 18 Structure and dimension of textile light diffuser composed of different weave areas

get around this difficulty, it is possible to design a specific wave pattern consisting of a mix of fundamental weaves (Fig. 18).

The full weaving pattern of the structure is given in Fig. 19. It is composed of a mix of satin 4 and 8 in area W1 and W3 and satin 6 in area W2. The textile light diffuser pattern is done so that a bundle of x POF, through the fabric in the weft direction, are woven in satin 4 (W1 area) then in satin 6 (W2 area) then in satin 8 (W3 area) then in satin 6 (W2 area) and then in satin 4 (W1 area). Symmetrically, the following bundle of x POF is woven in satin 8 (W1 area) then in satin 6 (W2 area) then in satin 4 (W3 area) then in satin 6 (W2 area), and then in satin 8 (W1 area). This alternated design is used to compensate, in addition to light decreases, the various relative shrinking of satin weave thus the fabric stays flat.

For this complex weaving pattern, the warp yarns tensions are a significant parameter for fundamental weaves. To optimize the warp yarns' tension during weaving, Oguz et al. used the design of experiments based on Doehlert table [38]. This work tested tensions between 40 and 70 g/yarns on the W1, W2, and W3 weave area (Fig. 19) and made it possible to maximize, at the same time, the emitted light intensity and the light homogeneity. Figure 20 shows equations of the quadratic additives models for light intensity (Fig. 20 up) and the light homogeneity (Fig. 20 down) and the graph of adequacy between the model and the measurements.

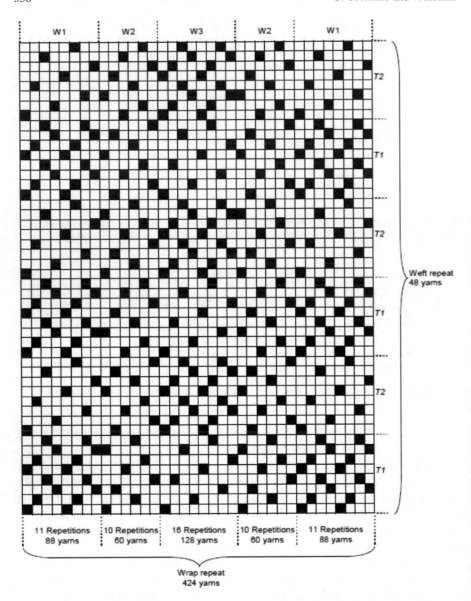

Fig. 19 Full weaving pattern of textile light diffuser composed of a mix of satin 4, 6 and 8

From both models, response surfaces are plotted and allow visualizing the compromise to be made to obtain good light homogeneity and good light intensity.

For medical application, this compromise is found for the adjustment: W1 = −1, W2 = −1 and W3 = −0.6 corresponding to 40, 40 and 52 g/yarn respectively (black cross on Fig. 21. From this theoretical setting, a sample was woven and compared to a non-optimized sample. Pictures of the two light-emitting fabrics (before and after

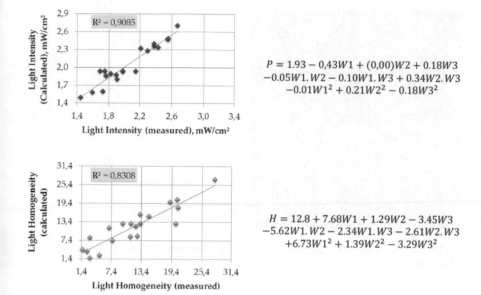

$$P = 1.93 - 0,43W1 + (0,00)W2 + 0.18W3$$
$$-0.05W1.W2 - 0.10W1.W3 + 0.34W2.W3$$
$$-0.01W1^2 + 0.21W2^2 - 0.18W3^2$$

$$H = 12.8 + 7.68W1 + 1.29W2 - 3.45W3$$
$$-5.62W1.W2 - 2.34W1.W3 - 2.61W2.W3$$
$$+6.73W1^2 + 1.39W2^2 - 3.29W3^2$$

Fig. 20 Equations of the quadratic models and graph of adequacy between the model and the measurements for light intensity (Fig. 20 up) and light homogeneity (Fig. 20 down)

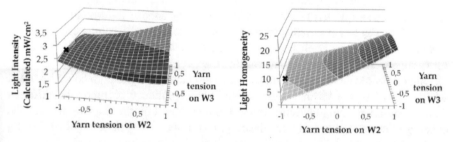

Fig. 21 Surface response from model and position of best compromise between light intensity (maximization) and light homogeneity (minimization) [38]

optimization) are shown in Fig. 22. In addition, the two light intensity profile was plotted in the graphic (Fig. 22, right) and show clearly the qualitative gap between samples.

7 Conclusion

The work presented in this chapter is focused on the design and manufacturing, using the weaving technique, of the light-emitting fabrics (LEF) able to provide a suffi-cient light fluence to 3D complex shapes of the human body suffering from Actinic

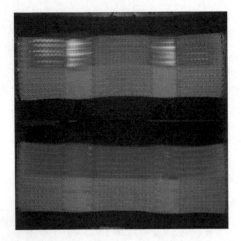
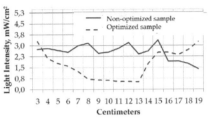

Fig. 22 Light-emitting fabric before (up) and after (down) optimization and light-emitting profile of both textile (right)

Keratoses. The weaving technique has been optimized and the design plan has also been used to find the best possible warp yarns tensions to optimize the compromise between the emitted light power and homogeneity over the whole surface. The property of light leakage laterally from optical fibers when bent above the threshold angle has been exploited to obtain mild and homogeneous light-emitting flexible fabrics. The light is inserted into those fabrics through OF beams embedded in metallic connectors and polished to minimize the reflective losses. The light sources for Europe were Diode Lasers that emitted the red light at 630 nm.

Other possible development supposed to give interesting results could be based on the use of light-emitting highly flexible OFs produced currently by different companies. Therefore, our future work will be focused on the improvement of the existing solution together with the development of the new generation of LEF using light-emitting flexible OFs and the comparison of their optical properties.

Finally, it is important to stress that the clinical trials have successfully have been realized with the developed solutions showing their interest from the medical point of view.

References

1. Zalaudek, I., Giacomel, J., Schmid, K., Bondino, S., Rosendahl, C., Cavicchini, S., Tourlaki, A., Gasparini, S., Bourne, P., Keir, J., et al. (2012). Dermatoscopy of facial actinic keratosis, intraepidermal carcinoma, and invasive squamous cell carcinoma: A progression model. *Journal of the American Academy of Dermatology, 66*, 589–597. https://doi.org/10.1016/j.jaad.2011.02.011
2. Krouse, R. S., Alberts, D. S., Prasad, A. R., Yozwiak, M., Bartels, H. G., Liu, Y., & Bartels, P. H. (2009). Progression of skin lesions from normal skin to squamous cell carcinoma. *Analytical*

and *Quantitative Cytology and Histology, 31*, 17–25.

3. Anwar, J., Wrone, D. A., Kimyai-Asadi, A., & Alam, M. (2004). The development of actinic keratosis into invasive squamous cell carcinoma: Evidence and evolving classification schemes. *Clinics in Dermatology, 22*, 189–196. https://doi.org/10.1016/j.clindermatol.2003.12.006

4. Moy, R. L. (2000). Clinical presentation of actinic keratoses and squamous cell carcinoma. *Journal of the American Academy of Dermatology, 42*, S8–S10. https://doi.org/10.1067/mjd.2000.103343

5. Smits, T., & Moor, A. C. E. (2009). New aspects in photodynamic therapy of actinic keratoses. *Journal of Photochemistry and Photobiology, B: Biology, 96*, 159–169. https://doi.org/10.1016/j.jphotobiol.2009.06.003

6. Ceilley, R. I., & Jorizzo, J. L. (2013). Current issues in the management of actinic keratosis. *Journal of the American Academy of Dermatology, 68*, S28–S38. https://doi.org/10.1016/j.jaad.2012.09.051

7. Sheridan, A. T., & Dawber, R. P. (2000). Curettage, electrosurgery and skin cancer. *Australasian Journal of Dermatology, 41*, 19–30. https://doi.org/10.1046/j.1440-0960.2000.00383.x

8. Kawczyk-Krupka, A., Bugaj, A. M., Latos, W., Zaremba, K., Wawrzyniec, K., & Sieroń, A. (2015). Photodynamic therapy in colorectal cancer treatment: The state of the art in clinical trials. *Photodiagnosis and Photodynamic Therapy, 12*, 545–553. https://doi.org/10.1016/j.pdpdt.2015.04.004

9. Lee, Y., & Baron, E. (2011). Photodynamic therapy: Current evidence and applications in dermatology. *Seminars in Cutaneous Medicine and Surgery, 30*, 199–209. https://doi.org/10.1016/j.sder.2011.08.001

10. Svanberg, K., & Bendsoe, N. (2013). Photodynamic therapy for human malignancies with superficial and interstitial illumination; Woodhead Publishing Limited. ISBN 9780857097545.

11. Buggiani, G., Troiano, M., Rossi, R., & Lotti, T. (2008). Photodynamic therapy: Off-label and alternative use in dermatological practice. *Photodiagnosis and Photodynamic Therapy, 5*, 134–138. https://doi.org/10.1016/j.pdpdt.2008.03.001

12. Dinehart, S. M. (2000). The treatment of actinic keratoses. *Journal of the American Academy of Dermatology, 42*, S25–S28. https://doi.org/10.1067/mjd.2000.103338

13. Kalka, K., Merk, H., & Mukhtar, H. (2000). Photodynamic therapy in dermatology. *Journal of the American Academy of Dermatology, 42*, 389–413. https://doi.org/10.1016/S0190-9622(00)90209-3

14. Allison, R. R., Sibata, C. H., Downie, G. H., & Cuenca, R. E. (2006). A clinical review of PDT for cutaneous malignancies. *Photodiagnosis and Photodynamic Therapy, 3*, 214–226. https://doi.org/10.1016/j.pdpdt.2006.05.002

15. Salvio, A. G., Ramirez, D. P., de Oliveira, E. R., Inada, N. M., Kurachi, C., & Bagnato, V. S. (2015). Evaluation of pain during large area photodynamic therapy in patients with widespread actinic keratosis of upper limbs. *Photodiagnosis and Photodynamic Therapy, 12*, 326–327. https://doi.org/10.1016/j.pdpdt.2015.07.013

16. Gholam, P., Denk, K., Sehr, T., Enk, A., & Hartmann, M. (2010). Factors influencing pain intensity during topical photodynamic therapy of complete cosmetic units for actinic keratoses. *Journal of the American Academy of Dermatology, 63*, 213–218. https://doi.org/10.1016/j.jaad.2009.08.062

17. Kuonen, F., & Gaide, O. (2014). Nouvelle lumière sur la thérapie photodynamique cutanée. *Revue Médicale Suisse, 10*, 754–759.

18. Khan, T., Unternahrer, M., Buchholz, J., Kaserhotz, B., Selm, B., Rothmaier, M., & Walt, H. (2006). Performance of a contact textile-based light diffuser for photodynamic therapy. *Photodiagnosis and Photodynamic Therapy, 3*, 51–60. https://doi.org/10.1016/S1572-1000(05)00182-1

19. Warren, C. B., Karai, L. J., Vidimos, A., & Maytin, E. V. (2009). Pain associated with aminolevulinic acid-photodynamic therapy of skin disease. *Journal of the American Academy of Dermatology, 61*, 1033–1043. https://doi.org/10.1016/j.jaad.2009.03.048

20. Attili, S. K., Lesar, A., McNeill, A., Camacho-Lopez, M., Moseley, H., Ibbotson, S., Samuel, I. D. W., & Ferguson, J. (2009). An open pilot study of ambulatory photodynamic therapy

using a wearable low-irradiance organic light-emitting diode light source in the treatment of nonmelanoma skin cancer. *British Journal of Dermatology, 161*, 170–173. https://doi.org/10. 1111/j.1365-2133.2009.09096.x

21. Bernasson, A. (1994). Fibre optique a eclairage lateral multi-ponctuel. European Patent.
22. Deflin, E., & Koncar, V., & Weill, A. Bright optical fiber fabric: A new flexible display. Text. Asia 2002.
23. Harlin, A., Myllymäki, H., & Grahn, K. (2002). Polymeric optical fibres and future prospects in textile integration. *Autex Research Journal, 2*.
24. Selm, B., Gurel, E. A., Rothmaier, M., Rossi, R. M., Scherer, L. J. (2010). Polymeric optical fiber fabrics for illumination and sensorial applications in textiles. Journal of Intelligent Material Systems and Structures, *21*, 1061–1071, https://doi.org/10.1177/1045389X10377676.
25. Luxeri, L. Lighting lateral optical fiber. Available online: https://www.luxeri-int.com/lateral-lighting-optical-fiber-c2x30101026. last accessed 2021/09/01
26. Spigulis, J., Pfafrods, D., Stafeckis, M., & Jelinska-Platace, W. (1997). The glowing optical fibre designs and parameters.
27. Sasaki, I., Nishida, K., Morimoto, M., & Yamamoto, T. (1991). Light-transmitting fiber.
28. Koncar, V. (2005). Optical fiber fabric displays. *Optics & Photonics News, 16*, 40–44.
29. Harlin, A., Makinen, M., Vuorivirta, A., & Mäkinen, M. (2003). Development of polymeric optical fibre fabrics as illumination elements and textile displays. *Autex Research Journal, 3*, 1–8.
30. Midlightsun Midlightsun products. Available online: https://www.midlightsun.com/. last accessed 2021/09/01.
31. Lumigram Lumigram products. Available online: https://lumigram.com/en/. last accessed 2021/09/01.
32. BiliBlanket BiliBlanket, Jaundice treatment Available online: https://biliblanketbaby.com/. last accessed 2021/09/01.
33. NeoMedLight NeoMedLight, Jaundice treatment Available online: https://www.neomedlight. com/bilicocoon/.
34. Bernasson, A., & Peuvergne, H. (1998). Optical fiber with multiple point lateral illumination. US Patent.
35. Berzowska, J., & Skorobogatiy, M. Karma Chameleon : Bragg Fiber Jacquard-Woven photonic textiles. In *Proceedings of the TEI 2010*; Cambridge, Massachusetts, USA, 2010; pp. 297–298.
36. Wang, J., Huang, B., & Yang, B. (2013). Effect of weave structure on the side-emitting properties of polymer optical fiber jacquard fabrics. *Textile Research Journal, 83*, 1170–1180. https://doi. org/10.1177/0040517512471751
37. Parker, J. R. (1989). Fiber optic light emitting panel and method of making same. US Patent.
38. Oguz, Y., Cochrane, C., Koncar, V., & Mordon, S. R. (2016). Doehlert experimental design applied to optimization of light emitting textile structures. *Optical Fiber Technology, 30*, 38–47. https://doi.org/10.1016/j.yofte.2016.01.012
39. Cochrane, C., Mordon, S. R., Lesage, J. C., & Koncar, V. (2013). New design of textile light diffusers for photodynamic therapy. *Materials Science and Engineering C, 33*, 1170–1175. https://doi.org/10.1016/j.msec.2012.12.007
40. Oguz, Y. Conception et réalisation de structures textiles lumineuses pour le traitement et le monitoring de la thérapie photo dynamique; Thèse de Doctorat, 4 avril 2017, Université de Lille (France), 2017.

3D Weaving for Composites

M. Amirul "Amir" Islam

Abstract This chapter discusses woven three-dimensional (3D) fabrics and structures, such as, contour structures, polar structures, load bearing joints, including T, tapered T, X, cruciform, Pi, tapered Pi, intersecting Pi, single and double blade joints, I-beam, H, and thick panels. Their applications are discussed in this chapter. Various 3D weaves are also discussed as are the different manufacturing technologies used to produce 3D structures.

Keywords 3D woven fabrics and structures · 3D weaves · 3D weaving technologies · Near net shape woven preforms · Woven textile composites · Polar and contour weaving · Woven TPS

1 Introduction: 3D Woven Fabrics and Structures

Three-dimensional (3D) fabrics and textile structures have been used in various industries, such as, aerospace, NASA, medical, automotive, armored vehicles, marine, boat and shipbuilding, construction, wind energy, electronic textiles (E-textiles) and smart textiles wearable, and other industrial applications. The aerospace and other industries have been using three-dimensional (3D) structures for some time. Putting several plies of fabrics on top of one another, cutting them into shapes, and stitching them together were the original methods of forming 3D structures. These techniques are labor intensive, time consuming, and expensive. Moreover, the fibers are damaged due to stitching and the joints are weak because no continuous fibers run through the intersections. Current 3D structures are generally created by weaving, using a specially built loom, braiding, knitting, filament winding and pultrusion, amongst other methods. In this chapter only a few of those 3D structures created using a loom, loom mechanisms and specially built machines are discussed.

The term 'yarn' is commonly used to describe both warp and fill (weft) in a regular woven fabric. In a 3D woven composite structure it is preferable to use the fiber tows

M. Amirul "Amir" Islam (✉)
Bally Ribbon Mills, Bally, Pennsylvania 19503, USA
e-mail: amirislam@ballyribbon.com

© Springer Nature Switzerland AG 2022
Y. Kyosev and F. Boussu (eds.), *Advanced Weaving Technology*,
https://doi.org/10.1007/978-3-030-91515-5_17

without any twist. The term 'fiber tows' is used to refer to the fiber throughout this chapter. The warp yarns or fiber tows are referred to as '0°' or 'X direction fibers' or as 'machine direction fibers. Warp yarns or fiber tows numbers are expressed in ends per inch. That means, '0°' or 'X direction fibers' or as 'machine direction fibers', or warp ends indicate the same fiber in vertical direction. The weft yarns or fibers are called 'fill fibers. The fill yarns or fiber tows are referred to as 'Y direction fibers' or '90° fibers. Fill yarns or fiber tows numbers are expressed in picks per inch. That means, the weft, fill fibers, 'Y direction fibers', '90° fibers', and picks indicate the same fiber in horizontal direction. The yarns or fiber tows through-the-thickness are referred as 'Z direction fibers' in this chapter. Layers are referred to as 'ply' in the composite industries.

1.1 Definition of 3D Fabrics and Structures

There are various definitions of 3D fabrics and structures because researchers and scientists differ in their opinions [1–4]. In this chapter any fabrics which take the shape of three dimensions are considered to be 3D fabrics. Woven structures with substantial thicknesses, thick structures, orthogonal structures or, which have been made using standard loom or special loom mechanisms and any woven three-dimensional shaped structures are considered to be 3D woven structures in this chapter.

1.2 3D Fabrics

3D fabrics may be classified as:

- Single–layer fabrics with an overall 3D shape
- Multi-layers flat fabrics
- Contour fabrics
- Spacer fabrics
- Shell fabrics
- Tubular fabrics (straight, bifurcate, tapered, or flared).

1.3 3D Woven Structures

3D woven structures may be classified as:

- Solid panel (layer to layer, angle interlock, orthogonal, and combination weaves)
- Load bearing joints, such as, T, Tapered T, X, Cruciform, double cruciform, Pi, Tapered Pi, Intersecting Pi, I-beam, single and double blade joints
- Fillets (noodle)

- Polar structures/spiral/arched structures
- E-textile structures
- Hollow structures
- Spacer preforms, such as, truss core, double truss core
- Complex structure.

1.4 Designing of 3D Woven Fabrics and Structures

The followings are the prerequisites to designing 3D woven fabrics and structures:

- Knowledge of fibers, and fiber properties
- Knowledge of 3D weaves
- Warp Preparations
- Knowledge of various looms
- Full understanding of loom mechanisms
- Minimize the fiber damage during weaving

  Various fibers and their properties are not discussed here.

2   3D Weaves

The weaves used to manufacture woven 3D fabrics and structures are considered to be 3D weaves. There are many types of 3D weaves and their selection is determined according to the end use and performance criteria of the 3D preforms. 3D weaves are generally based on the three basic weaves (plain, twill and satin) and are adapted using modifications, alterations, or combinations of these.

2.1   Layer to Layer Interlock Weaves Either by Warp or Filling

A typical layer to layer interlock weave based on a plain weave is shown in Fig. 1 [2, 8, 9] and another example based on a twill weave connection is shown in Fig. 2 [2, 8, 9]. There are various twills and connections between layers could be followed accordingly. Both figures show four layers joined together, but any number of layers can be joined together using the same technique. In this example the dark circles denote the warp ends and the curved lines indicate the filling path. The warp layers are joined by filling. The layers could however be joined by the warp ends, in which case the dark circles would denote the picks and the curved line would indicate the warp ends. The decision as to whether the layers should be joined either by the warp end or filling is based on the performance requirements of the 3D preforms.

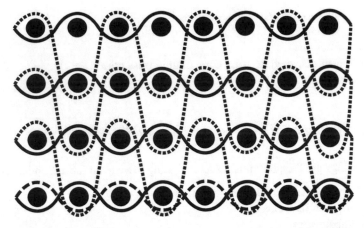

Fig. 1 References [2, 8, 9]: Typical layer to layer (ply to ply) interlock weave based on a plain weave

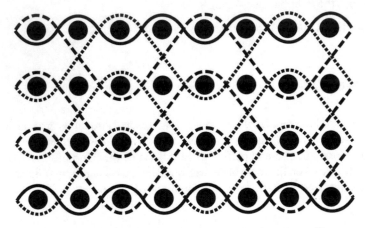

Fig. 2 References [2, 8, 9]: Typical layer to layer interlock weave based on twill weave connection

2.2 Layer to Layer Interlock Weave with Z Fibers

A typical layer to layer interlock weave with Z fibers is shown in Fig. 3 [2, 8, 9]. Z fibers are added to the layer to layer interlock weave. The dark circles denote the warp ends and the curved lines show the filling path. Z fibers run through the fabric thickness in the warp direction. In the example figure, Z fibers are perpendicular to the warp, but Z fibers can also be inserted at various angles, such as, 30°, 45° and 60°.

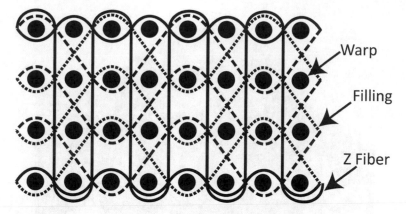

Fig. 3 References [2, 8, 9]: Layer to layer interlock weave with Z fibers

2.3 Layer to Layer Interlock Weave Based on Satin Weave

Figure 4 [2, 8, 9] shows a typical layer to layer interlock weave based on satin weave. The curved line shows warp and the circles denote the fill. There are 5 warp layers and 6 fill layers, which are joined by the warp ends. The layers of 3D fabric as shown in Fig. 5 [5] are joined together using this method.

Figure 5 [5] shows a 3D fabric made from carbon fibers. This fabric was made up of six layers. Each layer was connected with the next by a layer-to-layer interlock weave based on satin weave. The fabric was drapeable and formed three dimensions without wrinkles.

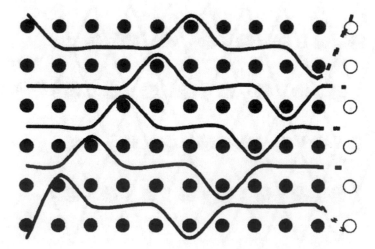

Fig. 4 References [2, 8, 9]: Layer to layer interlock weave based on satin weave

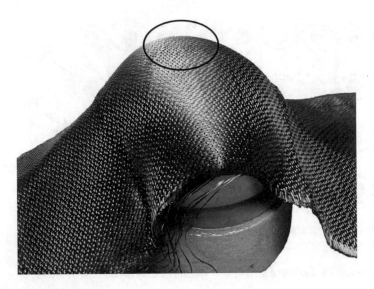

Fig. 5 Reference [5]: 3D fabric

2.4 Layer to Layer Angle Interlock Weave Either by Warp or Filling

A typical layer to layer angle interlock weave is shown in Fig. 6 [2, 8, 9]. In this type of weave, either the warp or the filling runs through the fabric thickness at an angle.

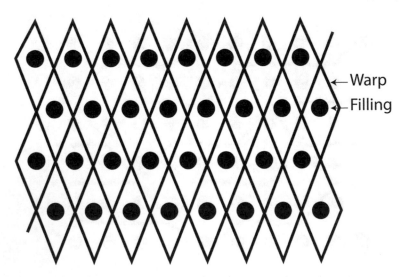

Fig. 6 References [2, 8, 9]: Layer to layer angle interlock weave

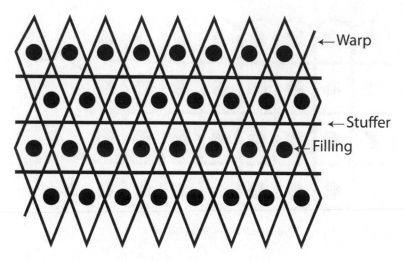

Fig. 7 References [2, 8, 9]: Layer to layer angle interlock with stuffer threads

The joining of layers, either by the warp ends or the filling, is selected based on the performance requirements of the end product.

2.5 Layer to Layer Angle Interlock with Stuffer Threads

Stuffer threads can be added to the layer to layer angle interlock structure to increase the thickness and tensile strength. A typical angle warp interlock with stuffer threads is shown in Fig. 7 [2, 8, 9].

2.6 Orthogonal Weave

In this type of structure, the warp and the filling run straight without crimping and the Z fiber binds all the layers. In a regular woven fabric, there are crimps in the warp and the fill due to their interlacement. Such crimps are undesirable for maximum composite mechanical properties. The use of a Z fiber throughout the fabric thickness effectively eliminates delamination as a failure mode. An orthogonal weave with Z fibers is shown in Fig. 8 [2, 8, 9]. Z fibers run through the thickness in the warp direction. This type of structure is also called non-crimp fabric (NCF).

The composite industry prefers this type of orthogonal weave. The benefits of orthogonal weave architecture are that the fibers in all three directions run straight which makes the orthogonal structure rigid. Higher fiber volume can be achieved for this type of structure due to absence of fiber crimp in all three directions. However,

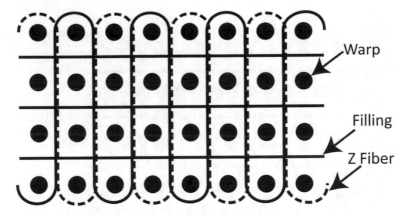

Fig. 8 References [2, 8, 9]: Orthogonal weave architecture

the disadvantage of this type of weave is that, if the Z fibers break, there is nothing to hold all the layers together.

2.7 Different Numbers of Layers in Various Locations for Complex Shapes

Different numbers of layers are used in various locations to achieve various thicknesses in this type of structure. A typical 4-layer weave architecture is shown in Fig. 9 [2, 8, 9], in which 4 layers have been reduced to one layer. The circle shows the warp fibers and the line indicates the filling path. The same technique is used to make a tapered structure. The gradual tapering in thickness is achieved by terminating one

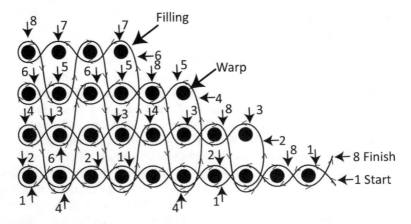

Fig. 9 References [2, 8, 9]: Weave architecture for tapered or complex shapes

layer at a time. The base of Pi as shown in Fig. 28 [5] has 8 layers and each layer was dropped from 8 layers to a single layer.

2.8   Combination Weaves for Complex Shapes

A combination weave or modified weave can be used in order to create complex structures and to achieve specific properties. A combination weave made up of an orthogonal weave and a plain weave is shown in Fig. 10 [2, 8, 9]. The top and bottom layers are plain weave and the layers in-between are orthogonal weave. There are two straight warp layers between the top and bottom plain weaves and Z fibers bind all the layers. Layer to layer interlock weave is more flexible, but it has less tensile strength due to interlacement and more fiber damage during weaving. An orthogonal weave is more rigid, but has a higher tensile strength due to the lack of interlacement and lower fiber damage during weaving. Layer to layer interlock weave based on plain, twill or satin and orthogonal weaves can be combined in many ways. The sequence can be varied, for example, x number of layers joined by layer to layer, and then y number of layers orthogonal, or x layer, then y layer and so on. The optimum combination can be selected based on the performance requirements of the end structure.

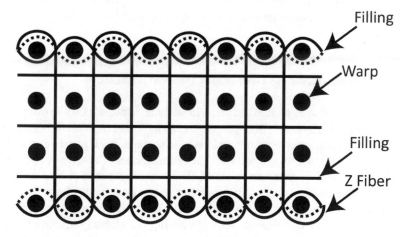

Fig. 10   References [2, 8, 9]: Combination weave architecture

2.9 Properties of 3D Weaves

The selection of 3D weaves is very important. The selection of weave architecture depends on the end use of 3D fabrics and 3D structures, and their required performance criteria.

2.9.1 3D Weave Based on Satin Weave

If 3-D fabrics have to conform to a three-dimensional shape, then the layers have to be joined using satin weave as shown in Fig. 4. Multi-layer satin weave fabrics are drapeable and easily conform to the shape. A multi-layer thick panel can be woven by joining layers using satin weave.

2.9.2 3D Weaves Based on Plain Weave

Multi-layer panels or 3D structures can be woven using a layer to layer interlock weave based on the plain weave as shown in Fig. 1. The warp end or fill are used to join the adjacent layers. If the joining threads are broken, then only a portion of one layer will fall apart, not the whole structure. This is one of the greatest benefits, but it means a lower tensile strength because the fibers are crimped by their interlacement. The likelihood of fiber damage during weaving is increased due to this high number of interlacements.

2.9.3 3D Weaves Based on Orthogonal Weave Architecture

In orthogonal weave architecture the fibers in all three directions (XYZ) run straight, as shown in Fig. 8. Maximum fiber properties can therefore be achieved due to non-crimping. Orthogonal weave architecture is used to make thick panels or structures as shown in Figs. 43, 44, 45 and 46. The maximum number of warp fibers (X fibers), picks (Y fibers), and Z fibers can be inserted because of the lack of interlacements. That means, that a higher fiber volume percentage can be achieved from orthogonal weave architecture, however, the disadvantage is that the whole structure falls apart if the Z fibers break. Fiber damage during weaving is less likely because there are no interlacements.

2.9.4 3D Weaves Based on Combination Weave

Often, a combination weave is selected, based on the end product's performance criteria. An orthogonal weave is used to make thick panels and to achieve maximum performance properties. A combination weave made up of layer to layer interlock weave and orthogonal weave would be selected to achieve maximum performance properties without the layers falling apart.

3 Contour Fabrics

A contour fabric is shown in Fig. 11 [5]. This fabric, made of carbon fibers and woven on an Iwer (rapier) loom, was made into a composite for a jet engine. Contour weaving technology was used for the circumferential frame applications. Contour fabrics take the shape of the contours they are based on, meaning that they fit well, not only on cylindrical surfaces with different diameters, but also on flanges and complex curvatures. Regular fabrics do not fit so well on the contour surfaces and tend to wrinkle. When infused with resin, contour fabrics also make good composite parts. When used in this way it is important that the warp and fill fibers align close to 0°, 90° without much angle variation.

Contour Weaving: It is based on the same principles as standard weaving except that the take up mechanism is modified to accommodate the different lengths of the warp fibers based on the configuration of the contour fabric. The take up rate of the warp fibers remains the same along the whole width when weaving a regular fabric, which is why they will not conform to curvatures without cutting or darting them. When weaving a contour fabric, the length of every single warp fiber is calculated based on the specified configuration and the take up is adjusted accordingly to achieve the right fiber length and orientation. Contour fabrics can therefore, easily conform to curvatures without wrinkling.

Contour weaving technology can be used for stiffener and aircraft engine fan blade containment. Aramid fibers are usually used for the latter. The benefits of contour weaving technology are as follows.

- reduced manufacturing costs eliminating manual cutting and darting
- wrinkles are eliminated
- warp and fill are aligned close to 0°, 90° without much angle variation

Fig. 11 Reference [5]: Contour fabric

- continuous and consistent parts are produced
- both convex and concave profiles can be created
- Structures can be made in small scale without a full-scale configuration

Various processes, such as, RTM (resin transfer molding), VARTM (vacuum assisted resin transfer molding) and RFI (resin film infusion) can be used to make composite parts.

4 Polar Structures/Spiral/Arched Structures

Polar/Spiral weaving:

One 3D weaving technique, which forms a two-dimensional (2D) flat fabric into a three dimensional shape is called polar weaving. A typical polar woven material is shown in Fig. 12 [5]. The solid model of a typical polar structure is shown in Fig. 13. The polar weaving process modifies the typical 0°, 90° fiber orientation of a woven product and produces a circular woven material, with the fibers orientated in the circumferential (hoop) and radial directions. The material produced on a loom, modified for polar weaving, produces a continuous woven cloth with a set inner and outer diameter. When producing polar woven material, the polar woven material will spiral on top of itself like a spring, and make a thick wall cylinder. The thickness of the cylinder will depend on the thickness of each polar woven layer and the number of layers making the cylinder.

Since the fill fiber is orientated in the radial direction, the material has a gradient from inside diameter (ID) to outside diameter (OD) and the fabric areal weight and material thickness will decrease from inner diameter to outer diameter. It can be compensated by adding broken/short picks to the outer diameter. 3D multi-layers Silicon Carbide woven polar material is shown in Fig. 14 [5] with a very small ID (inside diameter) and a large OD (outside diameter). The take up system is modified by designing a set of take up rollers based on ID and OD. As an example, ID: 4″ and OD: 10″ and then the width of polar woven fabric would be half of (OD-ID), that is, 3″. The roller has to be designed such a way so that the roller diameters at narrow side is 4″and 10″ at wider side with a distance of 3″ or a proportion of it. Several variations of polar weaving have been developed to make more complex shapes or preforms. Figure 15 [5] shows a curved T structure. One variation incorporates traditional weaving with polar weaving, which produces a single woven material with a section of traditional woven and another section with a polar weave creating a cylinder with a flanged end. The polar/spiral woven material change the typical 0°,90° woven materials into a circular structure with hoop and radial fiber alignment. The polar/ woven materials allow the woven materials to include sections of typical 0°,90° woven and polar woven (hoop/radial) material into and single profile which rotates around a single axis of rotation. Figure 16 [5] shows a polar/typical 0°,90° woven structure.

Fig. 12 Reference [5]:
Typical polar structure

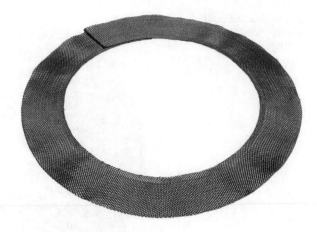

Fig. 13 Solid model of a
typical polar structure

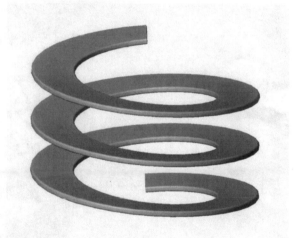

A modified polar/spiral weave technology was used to weave an aircraft window
frame. A composite window frame is shown in Fig. 17 [5]. The window frame
weaving technique was developed [5] to overcome the limitation of contour weaving.
The window frame technique will allow weaving flanged "L" preforms. The upright
section of the L preform will always remain 0°,90° but the flanged section of the L
preform changes from polar to 0°,90° and back to polar.

Fig. 14 Reference [5]:
Silicon carbide polar
structure

Fig. 15 Reference [5]:
Curved T structure

5 Load Bearing Joints

5.1 T Structure

Figure 18 [5] shows a T structure made from carbon fibers. Each flap consists of four
layers and all four layers are connected by layer to layer interlock weave. IM7-12K

Fig. 16 Reference [5]: Polar/ typical 0°,90° woven structure

Fig. 17 Reference [5]: Composite window frame

Fig. 18 T structure

carbon fibers were used. A tapered T was woven with tapering all three sides. The tapering was created by dropping one layer at a time. A narrow fabric shuttle loom was used to weave the T structure.

This T structure was woven in a narrow fabric 2D loom in a folded form as shown in Fig. 19.

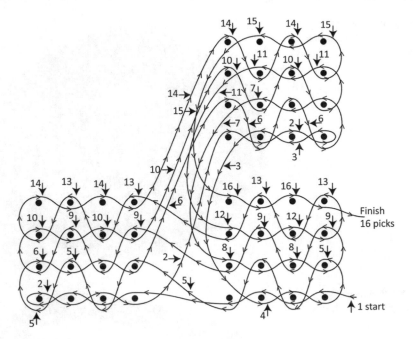

Fig. 19 Reference [6]: Typical filling path of T structure

Weave Architecture of T Structure

In this chapter, the weave architecture means the schematic view of a complete weave pattern. It shows the filling path in relation to warp. The structure means the shape of the woven material.

The typical weave architecture of a four-layer T structure created using layer to layer interlock weave based on plain weave [2, 8, 9] is shown in Fig. 19 [6]. The warp fibers of both sides of T are shown spaced apart for clarity. The black circle shows the warp fiber and the curved line denotes the filling. The four layers are joined together by a layer to layer interlock weave based on plain weave and the upright flap is joined with the base flap by filling.

An integral multiple T stiffener is shown in Fig. 20. This T stiffener was woven in flat form as one piece. 4 layers interlock weave was used in each section.

Figures 21 [5] and 22 [5] show a single blade joint and double blade joint. In both cases IM7-12K carbon fibers were used and the layers were connected using a layer to layer interlock weave based on the plain weave. A narrow fabric shuttle loom was used to weave both structures. Both structures were woven in flat forms in a narrow fabric 2D shuttle loom.

A cruciform structure is shown in Fig. 23 [5]. The 100% fill fiber of each flap goes through the intersection. Each flap consists of four layers and again they are connected by layer to layer interlock weave based on a plain weave. The typical construction of each layer is 16 × 16 ends and picks per inch in the case of 6K

Fig. 20 Reference [5]:
Integral multiple T stiffener

Fig. 21 Reference [5]:
Single blade joint structure
without tapering

Fig. 22 Reference [5]:
Double blade joint without
tapering

Fig. 23 Reference [5]:
Cruciform structure

Fig. 24 Reference [5]:
I-beam structure

carbon fibers and 16×8 ends and picks per inch in the case of 6K carbon warp fibers and 12K carbon fill fibers. A double cruciform with all sides tapered was also woven using a modified narrow fabric shuttle loom.

An I-beam is shown in Fig. 24 [5]. This I beam was woven using carbon fibers and carbon rods.

The I- beam can be used as structural elements of aircraft fuselage.

5.2 Fillets (Noodles)

Figure 25 shows a solid model of a fillet and a woven carbon fillet (noodle) is shown in Fig. 26. A bifurcate carbon fillet (noodle) is shown in Fig. 27. These fillets are used at joints to avoid resin rich areas.

Fig. 25 Solid model of a fillet

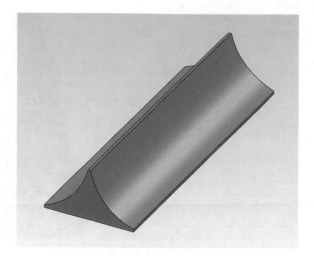

Fig. 26 Reference [5]: Carbon fillet (noodle)

Fig. 27 Reference [5]: Carbon bifurcate fillet (noodle)

5.3 PI Structure

A typical Pi structure with all sides tapered is shown in Fig. 28 [5]. This Pi structure was woven from 6K and 12K IM7 carbon fibers. The base and clevis are 8 layers and upright legs are 4 layers. A layer to layer interlock weave was used. IM7-6K carbon fiber as warp and IM7-12K carbon fiber as fill were used at the base and IM7-12K carbon fibers were used as both warp and fill for the upright legs. The construction

Fig. 28 Reference [5]: Pi
structure

of each layer consisted of 16×9 ends and picks per inch. The gradual tapering was achieved by terminating one layer at a time.

The Pi structure had been woven from carbon, fiberglass, ceramic fiber, silicon carbide, and even from metallic wires at Bally Ribbon Mills. This Pi structure has been patented [6, 7] and it is widely used in the aerospace industry for structural joints. The base, clevis and upright legs of the Pi structure can consist of any combination of layers based on the performance requirements. This Pi structure was woven in a narrow fabric 2D shuttle loom as shown in Fig. 50.

The Pi and double cruciform structures designed by Bally Ribbon Mills were used in the NASA Composite Crew Module (CCM) [10]. One of the significant benefits of 3D woven joints [8, 9] over traditional metallic structures is that they can be easily formed into complex shapes that may be structurally efficient.

Weave Architecture of Pi Structure

The typical filling path for a 4-layer base, 4-layer clevis and two-layer upright legs for a Pi structure without tapering is shown in Fig. 29. The warp fibers of the clevis are shown spaced apart for clarity. The filling path for an 8-layer base, 8-layer clevis and 4-layer upright legs can be shown in the same way. The black circle shows the warp fiber and the curved line denotes the filling. The four layers are joined together by a layer to layer interlock weave based on plain weave [2, 8, 9] and the upright legs are joined with the base by filling.

Figure 30 [6, 7] shows the weave architecture of a tapered Pi preform in which all the sides are tapered. The base and clevis are made up of 8 layers and the upright legs of 4 layers. The gradual tapering was achieved by dropping one layer at a time. All layers are joined together using a layer to layer interlock weave based on a plain weave [2, 8, 9]. The upright legs are joined to the base by the filling. The warp fibers at the clevis are shown spaced apart for clarity. Any combination of layers for a tapered Pi preform can be woven in the same way. The isometric view of the

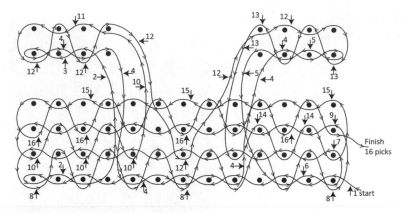

Fig. 29 Reference [6]: Typical filling path of Pi structure

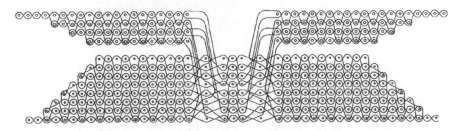

Fig. 30 References [6, 7]: Weave architecture of a tapered Pi preform

tapered Pi [6, 7] is shown in Fig. 31. The tracer fibers, comprising a colored strand and an x-ray opaque, shown using arrows, are used at the beginning of the tapers and

Fig. 31 References [6, 7]: Isometric view of the tapered Pi preform

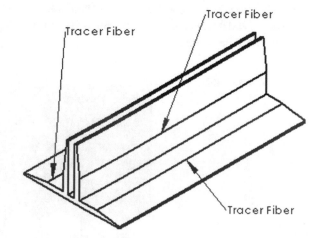

Fig. 32 Reference [5]:
Intersecting Pi structure

identify where the thickness of the legs or base begins to decrease for an accurate dimensional inspection of the preform.

An intersecting Pi structure is shown in Fig. 32 [5]. This intersecting Pi preform was woven as one piece in a modified narrow fabric shuttle loom. Each flap consists of four layers and they are connected by layer to layer interlock weave. IM7-12K carbon fibers were used. All sides of this intersecting Pi were woven edges.

5.4 Hollow Structures

Figure 33 shows a hexagonal structure. Other hollow structures are shown in Fig. 34 (hexagonal and diamond shapes) and Fig. 35 (diamond shape). Many types of hollow structures can be woven.

Fig. 33 Reference [5]:
Hexagonal structure

Fig. 34 Reference [5]:
Multiple cell structure

Fig. 35 Reference [5]:
Diamond shaped cell
structure

5.5 Truss Cores

An interwoven triangular truss core with face sheets is shown in Fig. 36 [5]. A double
interwoven truss core with face sheets is shown in Fig. 37 [5]. The integrally woven
cells were made to reduce the delamination of the composite materials to withstand
multiple impacts.

Fig. 36 Reference [5]:
Triangular truss core

Fig. 37 Reference [5]:
Double truss core

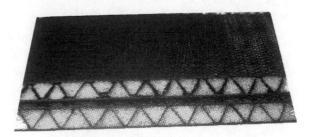

5.6   Complex Structures

An outlet guide vane (OGV) for a jet engine is shown in Fig. 38 [5]. A combination weaves was used and the thickness varied in both width and length directions. Snecma (SN) and Albany Engineered Composites (AEC) have developed woven composites fan blades for jet engines [11]. 3D woven preforms produced by layer to layer interlock pattern are processed into composites by resin transfer molding (RTM). The advantage of composites fan blades includes improved damage tolerance and increased thrust to weight ratio by lowering the mass of the fan.

A woven Sine-Wave structure is shown in Fig. 39 [5]. This sine-wave structure was woven in a narrow fabric 2D shuttle loom in flat form. A complex woven preform is shown in Fig. 40 [5]. Electronics and photonics can be placed into those slots for various applications.

SONACA of Belgium [12] in partnership with Bally Ribbon Mills developed a new concept of structural European launcher cryogenic upper inter-stage composite ring as part of the European Space Agency (ESA) Future Launchers Preparatory

Fig. 38 Reference [5]:
Composite woven engine
vane

Fig. 39 Reference [5]:
Sine-wave structure

Fig. 40 Reference [5]:
Complex woven Preform

Fig. 41 Reference [5]:
Composite ring

Program (FLPP). This challenging program targeted the replacement of an ARIANE traditional aluminum ring by a lighter and affordable multi-functional composite structure. Figure 41 shows a composite ring. This composite ring was woven in a narrow fabric 2D shuttle loom with modified take up system.

5.7   Solid Panels

Bally Ribbon Mills has been working with NASA on their woven thermal protection systems (TPS).

- Heat shield for Extreme Entry Environment Technology (HEEET) Program. HEEET is developing a mass –efficient, dual –layers carbon/phenolic system tailored for planetary probe missions for potential Discovery and New Frontiers class missions.

- 3D-MAT (3D Woven Multifunctional Ablative Technology), structural ablative material for the Orion compression pad. Orion compression pads are the interface between Crew Module (CM) and Service Module (SM) to withstand launch and ascent structural loads, pyro-shock (CM& SM separation event), and earth re-entry (high heating, ablation). 3D –MAT is a structurally robust 3D woven Quartz fibers panel with Cyanate Ester resin to meet the needs of the Orion Exploration Mission compression pad.
- Adaptable Deployable Entry and Placement Technology (ADEPT). The objective of ADEPT is to develop a semi-rigid low –ballistic coefficient aeroshell entry system concept to perform and entry descent landing (EDL) functions for planetary missions. This concept would be used to safely deploy scientific payloads or enable long-term exploration to other planets with their associated cargo needs.

Figure 42 [13] shows components of Orion Multi-purpose crew vehicle. 75 mm thick 3D-MAT Quartz panel is shown in Fig. 43 [5].

Fig. 42 Reference [13]:
Component of orion
multi-purpose crew vehicle

Fig. 43 Reference [5]:
75 mm thick 3D MAT
Quartz panel

Fig. 44 Reference [5]:
50 mm thick 3D Orthogonal
Carbon Panel

Figure 44 [5] shows a 3D woven panel of 50 mm thick made from AS4 12K carbon fibers. Orthogonal weave architecture has been used for this panel. 3D orthogonally woven structures consist of non-interlaced reinforcing fibers that generally oriented perpendicular or nearly perpendicular to each other in the X, Y, and Z directions. The orthogonally woven structures offer superior mechanical properties and corrosion resistance after infusing resin into a composite. The panel as shown in Fig. 44 has 61 layers of warp, 62 layers of fill and Z fibers. The typical fiber volume of the panel is around 55% and X, Y, Z fiber volume fraction are around 40%, 40%, and 20% respectively. The typical Z fiber volume fraction can vary from 3 to 33% depending on the performance requirements of end use.

The panels can be woven using different fibers in each direction (X, Y, and Z). The fiber concentrations in each direction can be varied to achieve desired fiber volume fraction. The different fibers can also be used in the same direction or other directions. The hybrid panel can easily be made. X and Z fibers can be the same fibers and the fill fiber is a different fiber. The different fibers can be used in different layers even in X direction. The layout of fibers in various layers is decided based on the performance requirements of the finished panel. A 50 mm thick Quartz panel is shown in Fig. 45 [5], where Quartz fibers have been used in all three directions.

Fig. 45 Reference [5]:
50 mm thick 3D Orthogonal
Quartz Panel

Fig. 46 Reference [5]:
50 mm thick 3D Orthogonal
Pitch Panel

Figure 46 [5] shows 50 mm thick 3D Orthogonal Pitch Panel. It is difficult to weave a thick panel using all pitch carbon fibers due to poor bending properties of pitch fibers. Pitch fibers are therefore generally served with rayon or polyvinyl alcohol (PVA) fibers, which can withstand the friction and abrasion during weaving. The term 'serving' is used to describe the process of wrapping the core fibers with another fiber. In this case, the pitch fibers were wrapped with rayon. The selection of fibers, fiber volume and fiber volume fraction are determined by the performance requirements of the end product.

5.8 3D Weaving Calculations

When designing 3D textile preforms for the composite industry, often the textile engineer is required to calculate the 3D preform construction from a cured composite specification, supplied by the composite manufacturer or the end user. The composite design engineer typically supplies the textile engineer with the following information about the cured composite:

- Cured Composite Part Thickness (CPT)
- Fiber Volume (V_f)
- Fiber Orientations or Fiber Percentages (%X, %Y, and %Z)
- Fiber Type/Fiber Density (ρ_f)/Fiber Linear Density (yd/lb)

Many times the end user will not have all items specified and it is up to the textile engineer to understand the composite part requirements in order to recommend a suitable preform construction. In particular, the fiber linear density is a variable which needs close review by the textile engineer. For high performance composite applications, the composite design engineer may specify fine denier fibers that exceed the jacquards capacity or worst yet, are physically impossible to weave. If there is a problem with the customer's supplied specification(s), the textile engineer must find a solution, often making compromises between performance and manufacturability.

To find out 3D preform construction for each direction (X, Y, Z directions), the areal weight (AW) is first calculated by the following equation, and then multiplied by the specified fiber volume fraction percentage and fiber yield.

Areal weight (lb/in^2) = Cured part thickness (CPT) × Fiber density (ρ_f)
× fiber volume   (1)

where Areal weight (AW) in lb,

Cured part thickness in inch,

Density (ρ_f) in lb/in^3, and

Fiber volume in %

Fibers required in each direction = Areal weight × fiber volume fraction
× fiber yield × 36   (2)

where Fibers in each direction in inch,

Areal weight in lb,

Fiber volume fraction in %,

Fiber yield in yards/lb.

By using the above equations, the total ends per inch (EPI) and total picks per inch (PPI) and Z fibers are easily calculated. Depending on the weave pattern and fiber crimp the 3D preform construction may need adjusting. The number of warp layers and fill layers depend on the Jacquard harness configuration and other factors.

For example, a 3D woven preform is needed to make a 2.0″ thick composite panel using standard modulus carbon fiber with an epoxy matrix. The composite panel is required to have a fiber volume of 52%. A 12K standard modulus carbon fiber should be used in all three directions with a density of 0.065 lb/in^3 and a yield of 567 yards/lb. An orthogonal weave pattern with 40% of the fiber orientated in the warp (0°/X) direction, 40% of the fiber orientated in the fill/weft (90°) direction, and 20% of the fiber orientated through the thickness (Z) direction would be used. Based on this information the ends per inch, picks per inch, and Z fiber of the 3D orthogonally woven preform could be calculated using Eqs. 1 and 2 with 61 warp layers and 62 fill layers.

Answer:

Areal weight (AW) = 2.00″(thickness) × 0.065(density)
×0.52(fiber volume) = 0.0676 lb

Total ends per inch = 0.0676(AW) × 0.40 (warp fiber fraction)
×567(yield) × 36 = 552

$$\text{Ends/inch/layer} = \frac{\text{Total ends/inch}}{\text{Warp layers}} = \frac{552}{61} = 9.05$$

Total picks per inch (PPI) $= 0.0676(\text{AW}) \times 0.40(\text{fill fiber fraction})$
$\times 567 \text{ (yield)} \times 36 = 552$

$$\text{Picks/inch/layer} = \frac{\text{Total picks/inch}}{\text{Fill layers}} = \frac{552}{62} = 8.90$$

To calculate the total number of Z fibers required per square inch of 3D orthogonal woven preform, the preform thickness must be taken into account. It is assumed that each Z fiber travels through the entire preform thickness.

$$\text{Z fibers per in}^2 = \frac{0.0676(\text{AW}) \times 0.20 \text{ (Z fiber fraction)} \times 567(\text{yield}) \times 36}{2.0 \text{ (thickness)}} = 138$$

6 Applications

Joints entirely made from composites are being used in aerospace applications to alleviate the need for fasteners. Fillets are being used at the joints to avoid resin rich areas. The aerospace industries are also looking into the possibility of composite window frames. Composite truss cores are used in fuselage areas and the composite vane and fan blades are used in the jet engine. Polar woven carbon material is used for carbon/carbon brakes. The automobile industries are using carbon panels and the construction and boat industries are using fiberglass panels. NASA, meanwhile, in partnership with Bally Ribbon Mills, is developing woven 3D structures to use as a thermal protection system (TPS) [14] for their future missions to Mars and beyond.

7 Advantages of Woven 3D Structures

3D woven structures are woven directly into net shapes. The advantages of woven 3D preform are as follows:

- No plying, cutting, and stitching
- Near net shape
- Minimum machining to achieve required shape and size
- Automated process
- Consistent products
- Increased production
- Eliminates delamination, which is the primary damage mechanism in laminated composites
- Reduces cost
- Better performance

The cost of manufacturing preforms is a major portion of the total cost. The benefits of using composites instead of metallic structures are that they are light weight, stronger and corrosion free. The saving in weight is tremendous. Prichard [15] suggests that the use of 3D fabrics and structures could lead to a 30% reduction in aircraft weight and every pound of weight saved could save one million dollars over the life of the aircraft. The saving in weight also reduces the aircraft fuel cost. The greatest advantage of using 3D woven composites and structures is the elimination of delamination.

8 Manufacturing Technologies

8.1 Weaving Preparation

Yarn preparation for regular fabrics has been discussed in other chapters. The scope of yarn preparation for weaving 3D structures is not the same as that for a regular fabric. The following steps are used.

8.1.1 Yarns or Fibers

The preference for a 3D woven composite structure is to use the fiber tows without any twist. Occasionally this is difficult to achieve, in which case, a very small amount of twist is inserted. If the customer does not want any twist, fiber tows with a higher sizing percent are used to facilitate weaving. As an example, Carbon fiber tows usually come with a less than 1% sizing.

Some fibers, such as, pitch carbon and silicon carbide are difficult to weave. Pitch carbon fibers break easily due to their very poor bending rigidity. For this reason, 'difficult to weave' fibers are usually served with polyvinyl alcohol (PVA) yarn or rayon to facilitate the weaving process. A machine called a 'serving machine' is used for this purpose. The serving machine wraps the 'difficult to weave' fibers with rayon or PVA. The rayon or PVA takes the abuse of weaving and the core fibers remain undamaged. Rayon is removed using heat and PVA is removed by dissolving it in warm water after weaving.

8.1.2 Winders

Fiber manufacturing companies usually supply fiber in large packages. Since these fibers are costly, it is necessary and economical to transfer fiber from larger to smaller packages on a suitable winding machine.

8.1.3   Creels

Unlike weaving regular fabrics, warp beams are not usually used for weaving 3D structures and instead, creels are the preferred method of feeding the warp. It is very important to maintain uniform tension on each package. A good creel is a prerequisite for weaving any 3D structures and complex shapes. A reciprocating creel, which takes out the slack during weaving, is used for more complex shapes.

Many thousands of ends are needed to weave a thick wide panel. Creels, which can accommodate such a large number of ends are very expensive and occupy a lot of space. In these cases, warp beams are used for the X or $0°$ fibers and a creel is used for the Z fibers. A combination of beams and creels can be used for tapered and complex structures.

8.2   Weaving Technologies

The following machinery can be used in the production of woven 3D fabrics and structures:

- Conventional 2D loom
- Modified 2D loom
- 3D loom
- Specially built loom or machine

8.2.1   Conventional Loom

Conventional 2D loom mechanisms have been discussed in other chapters. The shedding mechanism is one of the primary motions of a loom. A Jacquard shedding mechanism is used for 3D structures because the individual warp fibers are controlled independently by Jacquard harness cords. A 3D fabric can be woven on a 2D conventional rapier loom, but there is a fabric thickness limitation when using any conventional 2D loom.

Loom Set Up

Proper loom set up plays a significant role in reducing warp breaks during weaving and is therefore key in increasing productivity. Loom set up is even more important when weaving 3D fabrics and structures in order to minimize fiber damage during weaving to achieve maximum mechanical properties. The maximum translation of fiber or yarn properties to fabric for tensile strength [16] is dependent upon:

- The damage suffered by the fiber or yarn during weaving
- The fabric structure such as:

    - Fiber or yarn size (linear density)

- Fiber or yarn absolute diameter depending on fiber density
- Ends and picks per inch
- Warp and filling yarn crimp
- Warp and filling yarn interlacement angle
- Cover factor
- Weave type and number of binding points between warp and fill

Tensile strength realization in the fabric is primarily dependent on the weaving process, meaning that minimal fiber or yarn damage during weaving will improve the fabric tensile strength and weaveability. Research [16] shows that fiber damage during weaving can be minimized in the following ways:

- By setting up the harness on an elliptical path in order to keep the same warp tension
- By adding optimum twist to the fibers and by reducing the friction and abrasion at every contact point of the fiber during weaving
- By reducing beat up force
- By using round eye heddles to reduce the coefficient of friction between the fiber and the heddle
- By reducing the warp tension variation
- By increasing the optimum open-air space in the reed dent

Fiber damage

In Fig. 47, it appears that the damage to the warp increases gradually as the warp passes through the heddles and reed. There is a somewhat steeper increase in warp damage between the heddle and the cloth fell and a very large increase in warp damage between the region immediately in front of the cloth fell and the fabric itself.

The movement of the warp during the weaving process causes various stresses and strains [17, 18]. The let-off and shedding motions, together with the movements of the back-rest roller and of the cloth fell, influence the total warp tension. Shedding

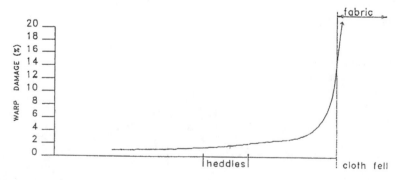

Fig. 47 Reference [16]: The warp damage

is one of the main factors which contribute to the total warp tension during normal loom operation.

Taking into account the mathematical relationship between tension on the two sides of the heddle eye:

$$\frac{T_f}{T_b} = e^{\mu\theta} \tag{3}$$

where e = the base of Napierian Logarithms, μ = the coefficient of friction between warp yarn and heddle eye, θ = the lap angle at the heddle eye in radians, T_b = the warp tension before passing through heddle eye, and T_f = the warp tension after passing through heddle eye.

Since tension = strain × elastic modulus, Eq. (3) can be written in terms of the strain ratio assuming that the strain is entirely due to shedding.

$$e^{\mu\theta} = \frac{W_f \text{ (Warp strain in front shed)}}{W_b \text{ (Warp strain in rear shed)}} \tag{4}$$

Let: P_f = AD/AB in Fig. 48 the normalized distance from the cloth fell to the heddle eye during shed formation, P_b = BD/AB in Fig. 48, the normalized distance from the back rest to the heddle eye during shed formation, H = CD/AB, the normalized shed height, F = AC/AB, the normalized front shed length, α = the front shed opening angle, and β = the rear shed opening angle. Then:

$$P_f = \sqrt{F^2 + H^2} \text{ in terms of F and H} \tag{5}$$

$$P_f = \frac{\sin\beta}{\sin(\alpha + \beta)} \text{ in terms of } \alpha \text{ and } \beta \tag{6}$$

$$P_f = \sqrt{F^2 + (F\tan\alpha)^2} \text{ in terms of F and } \alpha \tag{7}$$

Similarly, for P_b

Fig. 48 Reference [17]: Geometry of the warp yarn path due to shedding

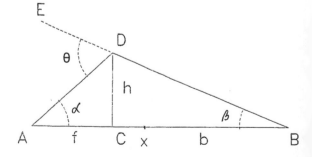

$$P_b = \sqrt{(1 - F)^2 + H^2} \tag{8}$$

$$P_b = \frac{\sin \alpha}{\sin(\alpha + \beta)} \tag{9}$$

$$P_b = \sqrt{(1 - F)^2 + (F \tan \alpha)^2} \tag{10}$$

Let X be the point on the closed shed line AB, which is in contact with the heddle eye at D. $F\mu = AX/AB$, the normalized length of yarn which forms the front shed during shedding. The warp strain increments in the front shed $W_{f\mu}$ and rear shed $W_{b\mu}$ are given by:

$$W_{f\mu} = \frac{P_f - F_\mu}{F_\mu} \tag{11}$$

$$W_{b\mu} = \frac{P_b - (1 - F_\mu)}{1 - F_\mu} \tag{12}$$

From Eq. (4):

$$\frac{P_f - F_\mu}{F_\mu} = e^{\mu\theta} \frac{(P_b - 1 + F_\mu)}{1 - F_\mu} \tag{13}$$

Hence:

$$P_f - P_f F_\mu - F_\mu + F_\mu^2 = e^{\mu\theta} F_\mu (P_b - 1 + F_\mu)$$

Or,

$$F_\mu^2 (e^{\mu\theta} - 1) + F_\mu (e^{\mu\theta} P_b - e^{\mu\theta} + P_f + 1) - P_f = 0$$

Therefore,

$$F_\mu = \frac{-C + \sqrt{C^2 + 4(e^{\mu\theta} - 1)P_f}}{2(e^{\mu\theta} - 1)} \tag{14}$$

where $C = e^{\mu\theta} P_b - e^{\mu\theta} + P_f + 1$.

The warp strain increments using Eq. (11) and (14) is shown in Fig. 49.

If the front shed opening angle and the normalized distance from the cloth fell to the harness are measured or calculated, then the warp strain can be found from the equations and from Fig. 49. This figure is also a guide for how to set up the harness in order to maintain the same warp tension or to minimize the warp tension variation. It is very important to minimize the fiber damage for 3D woven composites.

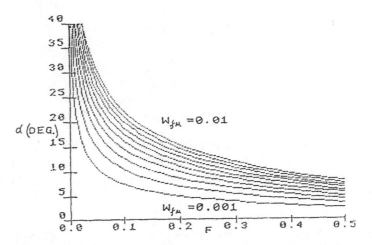

Fig. 49 Dependence of warp strain increment (W$_{F\mu}$) in the front shed on α and F when $\mu = 0.30$ at the heddle eye (Eqs. 11 and 14). Curves of constant strain increment at intervals of 0.001 [17]

For example: in Fig. 48

$$AC = 6.75'', \quad CB = 23.00'', \quad CD = 1.65'' \text{ and}$$

$$AB = 6.75'' + 23.00'' = 29.75'' \text{ then}$$

$$\text{H, the normalized shed height} = \frac{CD}{AB} = \frac{1.65}{29.75} = 0.0555$$

$$\text{F, the normalized front shed length} = \frac{AC}{AB} = \frac{6.75}{29.75} = 0.2269$$

$$\tan \alpha = \frac{CD}{AC} = \frac{1.65}{6.75} = 0.2444$$

α, the front shed opening angle $= \tan^{-1} 0.2444 = 13.7339°$

$$\tan \beta = \frac{CD}{CB} = \frac{1.65}{23.00} = 0.0717$$

β, the rear shed opening angle $= \tan^{-1} 0.0717 = 4.1033°$

θ, the lap angle at the heddle eye $= \alpha + \beta = 13.7339° + 4.1033°$
$= 17.8372° = 0.3113$ radians

P_f, the normalized distance from the cloth fell to the heddle eye during shed formation is given by:

$$P_f = \sqrt{F^2 + H^2} \text{ in terms of F and H}$$
$$= \sqrt{(0.2269)^2 + (0.0555)^2}$$
$$= \sqrt{0.0515 + 0.0031}$$
$$= \sqrt{0.0546}$$
$$= 0.2337$$

$$P_f = \frac{\sin \beta}{\sin(\alpha + \beta)} \text{ in terms of } \alpha \text{ and } \beta$$
$$= \frac{\sin 4.1033}{\sin(13.7339 + 4.1033)}$$
$$= \frac{\sin 4.1033}{\sin 17.8372}$$
$$= \frac{0.0716}{0.3063}$$
$$= 0.2337$$

$$P_f = \sqrt{F^2 + (F \tan \alpha)^2} \text{ in terms of F and } \alpha$$
$$= \sqrt{(0.2269)^2 + (0.2269 \times 0.2444)^2}$$
$$= \sqrt{(0.2269)^2 + (0.0555)^2}$$
$$= \sqrt{0.0515 + 0.0031}$$
$$= \sqrt{0.0546}$$
$$= 0.2337$$

Similarly, P_b, the normalized distance from the back rest to the heddle eye during shed formation is given by:

$$P_b = \sqrt{(1 - F)^2 + H^2} \text{ in terms of F and H}$$
$$= \sqrt{(1 - 0.2269)^2 + (0.0555)^2}$$
$$= \sqrt{(0.7731)^2 + (0.0555)^2}$$
$$= \sqrt{0.5977 + 0.0031}$$
$$= \sqrt{0.6008}$$
$$= 0.7751$$

$$P_b = \frac{\sin\alpha}{\sin(\alpha+\beta)} \text{ in terms of } \alpha \text{ and } \beta$$

$$= \frac{\sin 13.7339}{\sin(13.7339 + 4.1033)}$$

$$= \frac{\sin 13.7339}{\sin 17.8372}$$

$$= \frac{0.2374}{0.3063}$$

$$= 0.7751$$

$$P_b = \sqrt{(1-F)^2 + (F\tan\alpha)^2} \text{ in terms of F and } \alpha$$

$$= \sqrt{(1-0.2269)^2 + (0.2269 \text{ x } 0.2444)^2}$$

$$= \sqrt{(0.7731)^2 + (0.0555)^2}$$

$$= \sqrt{0.5977 + 0.0031}$$

$$= \sqrt{0.6008}$$

$$= 0.7751$$

$$e^{\mu\theta} = e^{0.30 \text{ x } 0.3113} = e^{0.0934} = 1.0979$$

Assuming μ, the coefficient of friction between warp yarn and heddle eye $= 0.30$. From Eq. (14)

$$C = e^{\mu\theta} P_b - e^{\mu\theta} + P_f + 1$$

$$= 1.079 \text{ x } 0.7751 - 1.0979 + 0.2337 + 1$$

$$= 0.8510 - 1.0979 + 1.2337$$

$$= 2.0843 - 1.0979$$

$$= 0.9868$$

$$F_\mu = \frac{-C + \sqrt{C^2 + 4(e^{\mu\theta} - 1)P_f}}{2(e^{\mu\theta} - 1)}$$

$$= \frac{-0.9868 + \sqrt{(0.9868)^2 + 4(1.0979 - 1)\text{x } 0.2337}}{2(1.0979 - 1)}$$

$$= \frac{-0.9868 + \sqrt{0.9738 + 4 \text{ x } 0.0979 \text{ x } 0.2337}}{2 \text{ x } 0.979}$$

$$= \frac{-0.9868 + \sqrt{0.9738 + 0.0915}}{0.1958}$$

$$= \frac{-0.9868 + \sqrt{1.0653}}{0.1958}$$

$$= \frac{-0.9868 + 1.0321}{0.1958}$$

$$= \frac{0.0453}{0.1958}$$

$$= 0.2314$$

$W_{F\mu}$, the warp strain increments in the front shed

$$W_{f\mu} = \frac{P_f - F_\mu}{F_\mu}$$

$$= \frac{0.2337 - 0.2314}{0.2314}$$

$$= \frac{0.0023}{0.2314}$$

$$= 0.0099$$

$W_{b\mu}$, the warp strain increments in the rear shed

$$W_{b\mu} = \frac{P_b - \left(1 - F_\mu\right)}{1 - F_\mu}$$

$$= \frac{0.7751 - (1 - 0.2314)}{1 - 0.2314}$$

$$= \frac{0.7751 - 0.7686}{0.7686}$$

$$= \frac{0.0065}{0.7686}$$

$$= 0.0085$$

- The results have been shown up to 4 decimal points for all calculations.

If the warp strain increments either in front shed or rear shed exceeds the breaking strain of the warp, then the warp ends will break. If the values of AC, CB, and CD in Fig. 48 are known, then the warp strain increments either in front shed or rear shed can easily be calculated following the above example.

The damage to warp can be minimized by reducing the strains caused by shedding motion, whether normal or distorted by traps. Damage can also be minimized by using round heddle eyes and by reducing strain due to friction and abrasion at all contact points. Minimum fiber damage during weaving will improve the fabric's tensile strength and weaveability.

The 3D structure on a conventional 2D loom is woven in a flat form. The first step in weaving of 3D structures on a conventional 2D loom is to figure out, how to flatten

Fig. 50 Reference [5]:
Narrow fabric shuttle loom
weaving Pi structures

[19] the structure? Once it is determined, it is followed by the subsequent set up of the loom and the weave architecture. The filling path, based on the performance requirements of 3D structures, has to be carefully selected. Figure 50 [5] shows weaving of carbon tapered Pi structures in a narrow fabric shuttle loom.

8.2.2 Modified Loom

The 2D conventional loom is modified to weave various 3D woven preforms. What kind of modification(s) should be carried out depends on the nature of structures to be produced. The take up and shedding motions are modified in many ways along with other tools that are also installed. Modified looms are used to produce thick panels, contour materials, polar materials, and complex 3D structures. Unlimited number of 3D structures and complex shapes can be woven on a modified loom with electronic Jacquard. Figure 51 [5] shows a narrow fabric shuttle loom weaving 50 mm thick carbon panel. An orthogonal weave has been used with a dead pick motion so that X, Y, and Z fibers are perpendicular to each other.

Fig. 51 Reference [5]:
Modified 2D loom weaving
50 mm thick carbon panel

Fig. 52 Reference [5]: Rear section of the machine

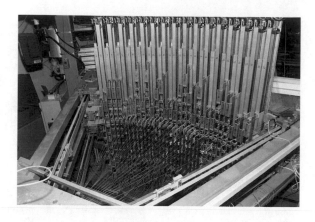

8.2.3 3D Looms

A few companies have developed 3D looms [20–25]. These looms are not discussed here. Quinn et al. [26] shows that, by increasing the amount of Z fibers in 3D orthogonal structures, the damage area after impact was reduced and low crimp, multi-axis, multi-layer woven structures offer improved damage tolerance.

8.2.4 Specially Built Machine

Bally Ribbon Mills (BRM) [5] has developed a patented weaving machine [27–29] where fibers can be inserted in X, Y, Z, +45°, and −45° directions. Various 3D structures, such as, Pi, T, cruciform have been woven. The test shows improved results in Pi structures produced in this machine over woven Pi structure with + 45°/−45° ply (Fig. 52).

Figure 53 [5] shows the front section of the loom, whereas Fig. 52 [5] shows the rear section of the BRM's specially built machine. The X direction (O°) fibers were controlled by an electronic Jacquard harness which formed the shedding. Z fibers were also controlled by an electronic Jacquard harness. The filling insertion was carried out by a specially built shuttle operated by a servo motor. +45° and −45° fibers were also controlled by servo motors. The beat up was carried out by individual reed blade. The take up was controlled by a servo motor. There were multiple vertical columns for bias fibers. There were 9 positions in each vertical column. Eight positions were filled with fiber bobbins and one position was left empty. The bobbins in each vertical column can move vertically (up and down) as well as move horizontally by servo motors. The bias fibers can be placed in any locations up to eight locations between the layers and outside the structure.

Many structures, such as, Pi, T, cruciform can be woven using fibers in five directions.

BRM has built other weaving machines in order to weave various complicated 3D structures including thick panels up to 6 inches thick.

Fig. 53 Reference [5]: Front section of the machine

9 Future Trends and Applications

As 3D woven fabrics, structures and composites become more popular and inexpensive, they will be increasingly used in various industries for a range of new applications, such as, consumer products, sporting goods, boat industries, automotive, construction and bridges, bicycle, alternative energy sector.

Although there are some simulations available for a few 3D structures, the weave design software currently available is limited. As a result, engineers have to spend a considerable amount of time figuring out how to weave the particular structure and coming up with a weave design. As software and the range of simulations improve, the time spent designing the weave structure will significantly decrease and the whole process of manufacturing 3D woven fabrics and structures will become much faster and cost effective.

The properties of 3D preform are dependent upon fiber properties, weave architecture, and fiber damage during weaving. It is very important to minimize fiber damage in woven reinforcements for composites. Progress has been made to predict the properties of 3D structures based on the weave architecture. The loom set up and fiber damage have to be considered in addition to the weave architecture for modeling the properties of the woven composites. Unfortunately, composite engineers lack textile knowledge and textile engineers lack composite knowledge. An improved understanding and knowledge of one another's subject areas would significantly improve the situation. Advanced research and developments in the areas of computerized 3D weave design and simulation would also be beneficial.

Acknowledgements The author would like to express his sincere thanks to Raymond Harries, CEO of Bally Ribbon Mills for allowing him to write this chapter. The author is thankful to Mark Harries, Curt Wilkinson, and Lorraine Hornig for their help with the line drawings and photographs. The author would also like to express his gratitude to his wife Ruksana Rita Islam for her continued support and inspiration.

References

1. Mohamed, M. H. (1990). Three-dimensional textiles. *American Scientist, 78.*
2. Islam, M. A. 3-D woven near net shape structures. In *ASME International Congress & Exhibition, November 14–19.1999*
3. Chen, X., & Hearle, J.W.S. Developments in design, manufacture and use of 3D woven fabrics. In *9th International Conference on Textile Composites, Recent Advances in Textile Composites, University of Delaware, October13–15, 2008.*
4. Khokar, N. Second generation woven profiled 3D fabrics from 3D weaving. In *First World Conference on 3D Fabrics and their Applications, April 10–11, 2008.*
5. Courtesy of Bally Ribbon Mills, Bally, Pennsylvania 19503, USA.
6. Schmidt, R., Bersuch, L., Benson, R., & Islam, A. Three-dimensional weave architecture. US Patent No. 6,712,099 B2, March. 30, 2004.
7. Schmidt et al. Woven preform for structural joints. US Patent # 6, 874,543 B2, Apr. 5, 2005.
8. Islam, M. A. Designing of narrow woven fabrics: 3-D Woven near net shape structures. In *Narrow Fabrics Conference of Clemson University, November 10-11, 1999, Charlotte, North Carolina, USA*
9. Islam, M. A. 3-D woven near net shape and multi-axial structures. In *The Textile Institute 80th World Conference, 16–19 April, 2000, Manchester, UK*
10. NASA Engineering and Safety Center (NESC) Web site on Composite Crew Module.
11. Dambrine et al. Development of 3D woven, resin transfer molded fan blades. In *9th International Conference on Textile Composites, Recent Advances in Textile Composites, University of Delaware*, Newark, Delaware, USA, *October 13–15, 2008.*
12. Fischer, H. et al. Composite ring made of 3D woven preform injected by RTM. In *SAMPE Conference*, Long Beach, California, USA, *May 23–26, 2016.*
13. Feldman et al. Development of an ablative 3D Quartz/Cyanate Ester composite for the Orion Spacecraft Compression Pad. In 249th ACS National Meeting, Denver, Colorado, USA, March 22–26, 2015.
14. NASA Office of Chief Technologist Press Release on "Development of New Game Changing Technologies" November, 2011.
15. Prichard, A.K. *Presentation at First World Conference on 3D Fabrics and their Applications*, Manchester, UK, April 10–11, 2008.
16. Islam, M.A. The damage suffered by the warp yarn during weaving. Ph.D. thesis, The University of Manchester, UK, 1987.
17. Islam M. A. Shed and shedding trap geometry. M.Sc. thesis, The University of Manchester, UK, 1985.
18. Islam, M. A. Designing of narrow woven high temperature fabrics. In *25th International SAMPE Technical Conference, October 26–28 1993, Philadelphia, Pennsylvania, USA*
19. Greenwood, K., Islam, M. A., & Downes, L. Research into narrow fabrics opportunities. In *Melliand Narrow Fabric and Braiding Industry Magazine*, Vol. 29, 1992.
20. Mohamed et al. US Patent # 5, 085, 252 Feb. 4, 1992 and US Patent # 5, 465, 760 Nov. 14, 1995, North Carolina State University, North Carolina, USA.
21. 3Tex, Carry, North Carolina, USA.
22. Buckley, M. A Report on 3D Preform Technologies for Advanced Aerospace Structures, Airbus, UK.
23. Khokar, N. Woven 3D Material. US Patent # 6,338, 367 B1, Jan. 15, 2002.
24. Uchida et al. Three-Dimensional Weaving Machine. US Patent # 6,003,563 Dec. 21, 1999.
25. Wilson et al. SPARC 5 Axis, 3D Woven, Low Crimp Preforms, Bombardier Aerospace, Short Brothers plc, Resin Transfer Molding SAMPE Monograph No. 3, 1999.
26. Quinn et al. 5 axis weaving technology for the next generations of aircraft-mechanical performance of multi-axis weave structures. In *9th International Conference on Textile Composites, Recent Advances in Textile Composites*, University of Delaware, Newark, Delaware, USA ,October 13–15, 2008.

27. Bryn, L., Islam, M. A., Lowery, W. L., Harries, H. D. Three Dimensional Woven Forms with Integral Bias Fibers and Bias Weaving Loom. US Patent # 6, 742, 547 B2, June 1, 2004.
28. Bryn, L., Nayfeh, S. A., Islam, M. A., Lowery, W. L., & Harries, H. D. Loom and Method of Weaving: Three Dimensional Woven Forms with Integral Bias Fibers. US Patent # 6, 892,766 B2, May 17, 2005.
29. Nayfeh, S. A., et al. Bias Weaving Machine. US Patent # 7,077,167 B2, July 18, 2006.

M. Amirul "Amir" Islam is a Commonwealth Research Scholar and a Fellow of The Textile Institute, UK since 1994. He has a B. Sc. Tech., M. Sc., and Ph. D. in Textiles from India, Japan, and England respectively. He has over 40 years hands-on experience in weaving, as well as, designing 3D fabrics and structures, E-textiles and medical textiles. He is internationally recognized as an expert on 3D woven fabrics and structures. Currently he is a Director of Research and Development at Bally Ribbon Mills, USA and has been since 1992. Dr. Islam can be reached at **AmirIslam@ballyribbon.com; Dramirislam1@gmail.com.**

3D Woven Fabrics-A Promising Structure for Women Soft Body Armor Development

Mulat Alubel Abtew, François Boussu, and Pascal Bruniaux

Abstract Personal protective body armor is one of the most important pieces of equipment which helps to guard human beings from different kinds of critical and fatal injuries. Besides, for the last few decades', women's participation has been also significantly increased in various threat exposed tasks such as military, law enforcement offices, personal guards, etc. Even though women revealed different body morphology than men, they were mostly enforced to wear men's body armor with smaller sizes. Such kinds of body armor outfit gave a less ballistic performance, discomfort and affect the wearer's psychology. A dedicated designing technique along with appropriate material is then required to develop women's body armor to accommodate their unique shapes with better ballistic performance and comfort. In this chapter, general information on the different personal ballistic protective equipment (body armor), their categories and different material used will be introduced. The different techniques used for developing soft body armor particularly for women including the cut-an-stitch, folding, molding methods, etc. will be discussed. The different interesting behaviors of three-dimensional (3D) woven fabrics which make it as prominent material for developing women soft body armor panels will be discussed in detail. Finally, some examples of the applications of 3D warp interlock for the developments of women's soft body armor through the molding process will be outlined.

Keywords 3D woven fabrics · Women's soft body armor · Body armor designing techniques · Ballistic material

M. A. Abtew (✉) · F. Boussu · P. Bruniaux
ENSAIT, ULR 2461 - GEMTEX - Génie et Matériaux Textiles, University Lille, 59000 Lille, France
e-mail: mulat-alubel.abtew@ensait.fr

M. A. Abtew
Bahir Dar University, Ethiopian Textile and Fashion Design Institute, Bahir Dar, Ethiopia

© Springer Nature Switzerland AG 2022
Y. Kyosev and F. Boussu (eds.), *Advanced Weaving Technology*,
https://doi.org/10.1007/978-3-030-91515-5_18

1 Introduction

Generally, the protection of the human body from various threats including sharp object and combat projectile dates back to the history of mankind. Previously, protection clothes were made and passed to different kinds of materials including animal skin (leather), wood, stones, copper, steel, etc. For example, the material used from padded hide jackets to a coat with metal scales, and then to chain mail covered by armored plate were commonly used [1]. Such equipment was used by warriors and soldiers in most of the world for many years. Nowadays the ballistic protective garment (personal armor) is also mainly used by peoples working in security, police department and military battlefields to protect themselves from fragments. Even then it has shown great improvement to protect against swords, arrows, military maces, firearms, etc. [2]. Normally such kinds of equipment includes body armor (ballistic vest) to protect the torso, different kinds of helmets to cover the cranium, different visors, glasses and goggles in order to protect the face and eye, other equipment to protect the pelvic and neck parts of the body [1, 3, 4]. Emerging of projectile guns also forces to further develop heavier weight armor for better protection. Different textile and laminated materials made of traditional fibers including linen, cotton, silk, and nylon then introduced to produce such protecting equipment [5–7]. Body armor made of leather on the Grecian shields, different layers of silk in ancient Japan and, armor suits with chain mail were some of the examples. Even further development of sophisticated firearm technology pushes harder the different stakeholders to exploit extensively new materials and designs to accomplish the new-age requirements of configuring lightweight and flexible body armor with adequate energy-absorbing capacity.

The development of novel synthetic fibrous materials with high-strength and high-modulus fibers in the late 1960s was a great success to design soft body armor systems with better resistance. The first synthetic fiber applied in ballistic material was Nylon 6.6 as model M-1951, which a nylon basket-weave flexible pad used to cover the upper chest and shoulders, and the other part made of Doron plates to protect the lower chest. However, the developments of aramid fibers in 1965 and commercialized in 1972 (DuPont) steered the new era. Moreover, modern military operations along technology-driven ammunition drive for developing body armor which provides not only good ballistic resistance but also damage-resistant, flexible, fitted and lightweight [8, 9]. Continuous and immense research and development were carried out to fill such requirements while developing body armor considering various parameters [10–12]. High-performance fibers have been employed to develop different phases of applications such as yarns, fabrics or related textile reinforcement composites to deliver much better performances of soft body armor [13–15]. The ballistic performance of the soft body armor has been also enhanced by different factors including designing technic, panel arrangement, fabric finishing, etc. [16–20].

Moreover, women participation in the field of law enforcement, privet security, and military forces has been also significantly increased across the world [21, 22]. Specifically designed soft body armor for the woman becomes very vital due to their unique morphological shape than men. However, for the last many decades, women officers were fitted with small-sized male-based body armor systems that impose both physiological differences and disproportionate on fitness, comfort, and bad ballistic protection. Later, considering the body shape difference, different designing techniques including molding along with proper material selection were come out to reduce such problems [8, 23–30]. Besides, 3D warp interlock fabrics became a promising fabric structure for developing women body armor [31–33] due to its good mechanical and ballistic performance along with good delamination [34, 35], low shear rigidity and excellent formability behaviors [31, 36].

This chapter will deliver information in order to understand the different kinds of body armor and it's in general, and the different designing techniques and how the 3D woven warp interlock fabrics became a promising fabric structure to design and develop women soft body armor with better ballistic performance along with flexibility and fit in particular.

2 Body Armor and Its Category

Body armor is the common ballistic and sharp weapon protecting clothing items that have been used for the last many decades by military and civilian which subjected to fragments [2]. Several law enforcement agencies have also made it mandatory for their officers to wear body armor while on the field. Such requirements have contributed not only to its huge need but also on progressive improvement of the body armor both on the material and design techniques. Nowadays, a huge variety of body armor is available on the market which comes in a wide range of levels that are individually catered to different types of threats. However, not all body armor offers the same levels of protection from different dangers. Body armor can be classified as covert and overt based on their flexibility, wearing styles and wearing situations. For example, covert body armor is designed as thin as possible to be worn comfortably under standard uniforms or other garments. It is applied where the line of the work requires only a certain level of protection such as security operatives, cops, and close protection officers. Whereas, overt body armor is usually thicker and bulkier design made from harder bulletproof panels and wore over the garments. Such body armor styles are usually used in some kinds of work that put an officer a lot of pressure on the security operative and high-risk including riot control, military operations, war zone journalism, humanitarian missions, and security operations. However, all the mentioned body armor composed of three sections, namely carriers, panel cover and ballistic panels as shown in Fig. 1. The carriers used to support and secure the panels to the wearer's and Cover mainly protects the ballistic materials from the environment, whereas, Panels helps to provide ballistic protection.

Besides, depending on the level of firearm threats, modern body armor worn by police, law enforcement agencies and military categorized into two categories, namely, soft body armor and hard body armor [10]. This subsection will briefly

Fig. 1 The compositions of
typical body armor system

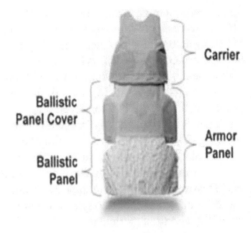

discuss the threat level, materials used and applications of the two types of body
armor.

2.1 Soft Body Armor

Soft body armor is a kind of body armor that can suffice to perform against low to
medium level firearm threats (Level IIA, II and IIIA) which could go up to velocity
500 m/s according to the National Institute of Justice (NIJ) standards. It usually
comprises multiple layers of flexible dry woven fabric arranged and stitched together
in different forms (Fig. 2). The soft body armor offers not only significant contributions to ballistic protection but also tailored to conform very well to the body contour

Fig. 2 Soft body armor and
its ballistic panels

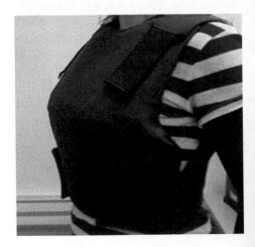

to provide good levels of comfort. However, it is commonly designed to protect from small arms fire (handgun bullets). Unidirectional (UD) fabric is other types of textile fabrics which used to develop an improved and desirable soft body armor systems in terms of higher protection without sacrificing mobility and comfort [37]. Besides the fabric type, all the different multiscale structural hierarchy of soft body armor from fibril to fiber/filament to yarn to fabric to panel determines not only the ballistic impact performances but also can cause a probabilistic penetration response [38].

Various companies including DuPont, Honeywell, and DSM have developed various kinds of high-performance fibers to improve the ballistic performances of the body armor. Para-aramids and ultra-high molecular weight polyethylene (UHMWPE) are commonly used and well-known high-performance fibers to produce soft body armor. High-performance fibers usually possess high modulus, high resistance-to-impact damage, and high breaking strength with great capacity of absorbing the impact energy of high-velocity projectiles [39–41]. Twaron®, Kevlar®, Dyneema®, and Spectra® a registered trademark of Teijin, DuPont, DSM, and Honeywell respectively are among the well-known high-performance fibers which extensively used for flexible body armor due to their desired engineering properties [42–44]. Zylon, Spectra®, M5 Vectran, Technora, and Nextel are also other commonly used high-performance fibers from Toyobo, Allied Signal, Hoechst Calaneses, Teijin, and 3M Ceramic Fiber Products. Moreover, even though it is costly, a very recent material such as PBO (p-phenylene-2, 6-benzobisoxazole) and M5® (polyhydroquinone-diimidazopyridine) are also used to improve the ballistic performance [45]. The mechanical properties and their comparisons are given in Fig. 3 (a and b).

Conventional soft body armor is usually made up of a large number (20–50) and weighing (below 4.5 kg) of woven fabric layers made of yarns of high-performance fibers. Thus, it becomes even heavy, bulky, and inflexible, interfering which affects the mobility and comfort of the wearer. This puts a great challenge for researchers and armor develope to device appropriate methods and materials for achieving lightweight, flexible, and comfortable armor panels, with enhanced impact performance to resist the new firearm technology [47].

2.2 Hard Body Armor

Hard armor is another category of body armor which delicately designed to resist the projectile having a velocity of more than 500 m/s according to the NIJ standard (beyond level IIIA). It is worn at a high risk of attack not only to stop the projectile but also fragments from explosions as well as bullets from handguns. Hard armor usually composed of different materials including metallic plates, ceramic tiles, silicon carbide or boron carbide plates, laminated composites, etc. in conjoined with soft body armor to ensure sufficient protection (Fig. 4). The hard armor plate could weight by itself up to 1.4 to 3.0 kg. Ceramics such as alumina (Al_2O_3 and silicon carbide (SiC)) are commonly used armor plates for hard armor ballistic applications.

Fiber type	Fiber diameter (μ)	Fiber density (g/cm³)	Tensile strength (GPa)	Tensile modulus (GPa)	Elongation at break (%)	Melting point (C°)
Aramid						
Twaron CT, Teijin	12	1.45	3.2	115	2.9	450
Kevlar 29, DuPont	12	1.43	2.9	70	4	530
Kevlar 129, DuPont	12	1.45	3.4	99	3.3	530
High performance						
Polyethylene-HPPE (Dyneema SK60, Toyobo)	3.8	0.97	3.5	110	3.5	150
Ultrahigh strength polyethylene-UHSPE (Spectra 900, Honeywell)	38	0.97	2.59	117	3.6	147

(a)

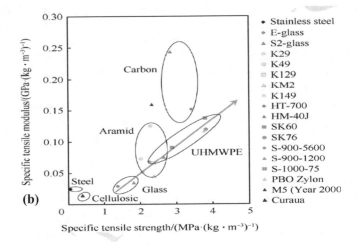

(b)

Fig. 3 **a** The mechanical properties of typical ballistic fibers [46] and **b** the relationship of specific tensile modulus and strength among the fibers [37]

Moreover, the hard armor also composed of ceramics parts combined with composite reinforcement with epoxy resin in order to obtain better flexibility, light-weight and ballistic performance as compared with pure ceramic tiles [48].

Nowadays, fiber-based textile laminated composite systems (woven fabrics impregnated with resin) become the widely used armor panel materials in the market and gets higher attention for researchers mainly due to its reduction of weight by maintaining the impact resistance. However, characterizing the composite material before the application will help to obtain an armor plate with better ballistic resistance and comfort. For example, a small amount of resin (20%) while impregnating the woven fabrics gives better energy absorption by maintaining flexibility than higher resin values due to the resin restricted yarns do not share the load effectively

Fig. 4 Commercial hard ballistic armor for high protection level with its hard steel plates [37]

[49]. Incorporations of vinyl ester resins coated Dyneema® fabrics in the soft body armor (NIJ-level II) could also give a ballistic resistance up to NIJ standard level III (approximately up to 840 m/s). Moreover, various research studies have also intensively investigated and characterized the ballistic impact performances and flexibility properties of the different fiber-based textile composite used as hard body armor.

3 Designing Methods of Women Soft Body Armor

Even though various countries make mandatory to wear body armor by their officers, many personnel's did not feel comfortable to wear the vests while on duty. This is due to the fact that as you increase the ballistic protection capabilities of the body armor, it becomes very heavy and rigid. Unless it is designed with light weighted, comfortable and fitted along with the ballistic resistance performances, wearing a heavy and inflexible vest for long hours could generate excessive heat and bring less mobility [8, 9]. This problem even became worse while developing characterized body armor due to the curvaceous shape of their body [52]. Moreover, as mentioned in the introduction, the participation of women in different dangerous tasks including in the field of law enforcement, privet security, and military forces has been also significantly increased across the world for the last few decades [21, 22]. For example, in 2008, 16.7% of police officers in the EU were women. In the recent year of 2016, among the total of 324 police officers per 100,000 people in the EU, women police officers' percentages were increased to 21%. This means that one in five polices officers is a woman [50]. Like other countries, the entire UK police forces were also male at the start of the 20th century and even their numbers were very small since the 1970s. However, the number of women law enforcement members has been recent increases and make up a significant minority. According to the study, in March 2016, 28.6% of police officers in England and Wales were women. This was an increase from 23.3% in 2007 [51]. Besides, in March 2017, the percentage of female officers even increased to 29.1% in England and Wales, 29% in Scotland and 28.5% in Northern Ireland [51–53]. The women were made up of a 61% majority of non-sworn police staff and 45% of Police Community Support Officers. Moreover, the percentage of women was 5.0% in 1980 and 9.8% in 1995 for the police force in the United States [53, 54]. This number even grew further and recorded 11.2% in 2005 and 11.9% in 2014 [55, 56].

Such increments of women's participation in the law enforcement and police office department make it mandatory to uniquely develop a dedicated designing and manufacturing woman soft body armor based on their unique morphological shape. Such features of body armor optimization specifically for women enable them more protected without conceding fitness and comfort while wearing the body armor. As discussed earlier, designing women's body armor would be quite challenging due to their unique body shape and even affected by various factors including race, geographic locations, etc. However, for the last many decades even though the numbers of women personnel are significant, they were fitted with small-sized male-based body armor systems. Such fitting of male-based armor for women personnel could be not acceptable not only due to the physiological difference but also imposes the disproportionate on fitness, comfort, and worse ballistic protection. Nowadays, after considering the body difference, various research studies have brought solutions to reduce such problems through either designing techniques or designing with proper material applications [24–30, 57]. The woman body armor designing techniques were mainly used to form the curvilinear shape and accommodate the bust area. Even though the different methods had their own drawbacks, designing women's body armor through such methods is far better than the male-based body armor by forming the curvilinear shape and accommodates the bust shape properly and provides better fitness and comfort. This subsection will discuss briefly some of the most common designing techniques for developing women body armor.

3.1   2D Pattern Making Methods

For the last many years, traditional cut-and-sew designing methods are commonly used design methods in the industry to develop women's body armor to form the required domed bust shape for better comfort. This method involves different kinds of dart designs in different forms to accommodate mainly the women's body bust area. So far various research scientists and armor developers have a design, develop and patented the women's soft body armor using cut-and-stitch design approaches. However, designing women's soft body armor with such a design principle faces weakness points at the stitches of the dart seams line against projectile during impact. This could ultimately reduce the ballistic performance of the material. Moreover, the accumulated material at darts areas also cinches the body near-certain positions of the bust and creating uncomfortable when worn by the wearer [19, 58]. Some of the research studies even further studied on the optimizations of ballistic performance and the comforts of women body armor design designed by the cut-and-stitch methods.

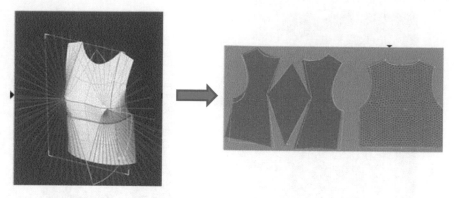

Fig. 5 Developments of women ballistic vest with dart rotation methods [59]

For example, concealable soft body armor was designed using a plurality of layers of ballistic material, where the different layers of the material include a series of folded pleats arranged at selected angles and intervals along with the layer and sewn along their length [8]. Another woman's body armor having dual cups has been also designed for accommodating, protecting and supporting the breasts and protecting the woman torso from trauma [27]. Body armor with contoured front panel, all-fabric, lightweight panel composed of a plurality of superposed layers of protective plies of fabric made of aramid polymer yarns has been also developed. This panel was contoured using overlapping seams joining two side sections to a central section of the panel to the curvature of the bust of a female wearer to convey good ballistic protection and comfort [24]. Ballistic vest with front and the rear ballistic panel which also has elongated side portion to provide ballistic protection for the side of a female wearer's breast conformed to the body of a wearer by a stretchable outer garment located over the ballistic panels has been also developed [29]. One of the studies also optimize the protection zones for tailor-made bullet-proof vest using a virtual body given by the body scanner for more comfort and to accelerate the sizing process through improving the measurement process [59] as shown in Fig. 5.

Here, it was tried to replace the traditional two-dimensional design method by three-dimensional design process using the darts rotation techniques for better fitness based on the different female morphology. Even though the three-dimensional shape has been developed on the two-dimensional flat fabric using different darts and seams, apart from improved fitness by getting precise dart design, location, and quantity because of the 3D design process but it leaves difficulty to get the desired level of ballistic performance. This was mainly the methods still used many darts and seams to create the required shape, which does not overcome the same issue that the traditional method meets. Another approach was also proposed a new geometric definition of the vest by introducing different body parameters (Fig. 6). The 3D virtual body was generated using Body Scanner to makes the tailor-made bullet-proof vest more comfortable and improve the measurement process during sizing. Such a study was mainly focused on the optimization of the different protection zones of the basic

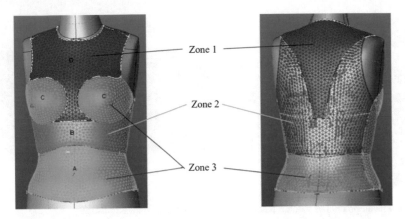

Fig. 6 Designing of a women body armour considering different front and back protection zone [60]

vest by materials distribution based on the vulnerability of the body toward the impact [60].

During development, the material distribution on various parameters that affect the garment cost, ballistic performance, weight, and comforts were considered. Finally, the developed pattern was generated and draped on the attained virtual 3D female dummy using Design Concept and its various protection zone were delimited with the help of various darts positions.

3.2 Folding and Overlapping Methods

Fabric folding and overlapping are barely used designing techniques to develop the women's body armor [8, 61]. For example, the folding technique is mainly used to shape the materials into a three-dimensional form by folding the materials and stitched at the side in a certain shift. This also revealed the limitation of ballistic performance due to material discontinuities and weak stitching area around folded material. One of the women's body armor with multiple layers of penetration-resistant material was designed using a series of V-shaped darts in successive layers to shape and fit the bust of women. The V-shaped material was then folded on itself to form a pleat and is angularly offset from one another in directions so that the added thickness is distributed substantially evenly, thereby avoiding bulges or stiffness and improving the wearing comfort [61]. However, one of the disadvantages of folding techniques is that many folded fabrics at the side with sharp edges create thicker panels while folding which causes itching, less personal mobility, and discomfort.

Another method of developing women's body armor is superposing the layers of ballistic material to develop contoured surfaces to accommodate the frontal women part called overlapping. The methods are also applied overlapping seams in order

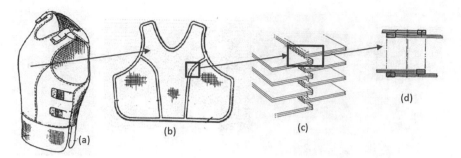

Fig. 7 Women Body armor with overlapping seams, **a** a perspective view of a complete woman body armor **b** a plan view of the woman front body armor panel, **c** exploded perspective view of different ply joined with overlapping **d** a vertical section through the overlapping seams [24]

to join the different superimposed layers. Even though involving overlapping seams could give both the contour surface to accommodate the frontal body and much strength in fabric stitching; however, the small ballistic missiles still can penetrating by severing the loop of threads among the seams [24, 62]. One of the patents tried to develop a light-weight, all-fabric and contoured body armor panel to protect the torso of a woman against small arms missiles using overlapping methods [24]. The frontal panels were contoured to the curvature of the women bust using overlapping seams while joining the two side section to the center section of the panel as shown in Fig. 7. The multiple layers which were superimposed were comprised of ballistic-resistant fabric made of aramid polymer yarns.

3.3 Molding Methods

The cut-and-stitches, folding and overlapping designing techniques provide better fitness and comfort as compared with fitting the women personnel with the unisex outfits. However, the mentioned designing techniques also face a problem of obtaining an accurate surface data for women's breasts due to its ambiguous border-line with the skin surface. Normally, the effectiveness of the women's ballistic protection depends on the adaptiveness of the body armor to the human morphology. The more the ballistic vest resembles the required shape of the body; the better the fitness and efficient protection will be obtained. Therefore, developing women's body armor which properly accommodates the bust area and fitting of the different curvilinear areas with the required dome-shape will improve not only the fit and comfort but also its ballistic protections for different women morphology [63, 64]. Molding is a new designing technique usually creates the required dome-shape through the forming process without applying any kind of cutting and stitching to develop women's soft body armor. Thus, it helps to develop seamless frontal body armor by mimicking the bust area without the need of cut-and-stitch or any other finishing methods. However,

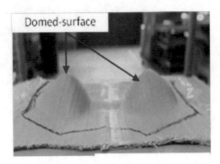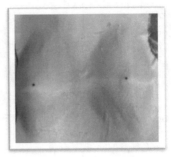

Fig. 8 Female body armor through the molding process [65]

the methods need proper material in order to form the required shape without facing shear deformation, panel thickness variations, and wrinkles on the material and body size limitations.

Various body armor developers and researchers have been used such designing techniques to develop women's soft body armor. For example, Teijin Aramid in collaboration with Triumph International claims the concept of manufacturing the women's body armor through the molding process as shown in Fig. 8. The molding process was carried out using pressure and heat on the 2D woven p-aramid (CT709) fabric in order to create the molded panels [65]. Each molded layers were then kept one over the top of the other to obtain a garment part and, joined by a base seam, e.g., in the middle of the panel. Women's body armor for ballistic protection and increased comfort and ease of movement using a three-dimensional molded multilayer laminated woven structure was also developed with better retaining and conforms of the female torso [26]. Moreover, some researchers have also used a three-dimensional pattern for the frontal panels of women's body armor in order to avoid the problem of the traditional designing method. 3D modeling and 3D shaping in women body armor design using scanning process leads to woman chest separated from the body, parameterized and then adapted to individually in relationship of given customer's morphology. For example, parametric design methods for generating human body models of varying sizes according to different anthropometric measurements in the 3D domain and 3D presentation were used to develop warp-knitted seamless garments [66]. A new design approach has been also proposed for developing female body armor by introducing the different personal parameters of the body inside the new geometric definition of the vest. It uses the real 3D body measurement to optimize the assembly process and projection zone, which later leads for reduction of weight and waste quantity during the cutting operation. Afterwards, it can be possible to create a base virtual dummy with scalable measurements, depending on a certain number of measures in order to recreate the internal surface of the bullet-proof vest [60]. Another three-dimensional seamless women body armor were also designed by researcher combining CAD software for design and seamless knitting technology for production [67] as shown in Fig. 9. This method claimed to eliminate the traditional cut-and-sew method since it has negative effect on the

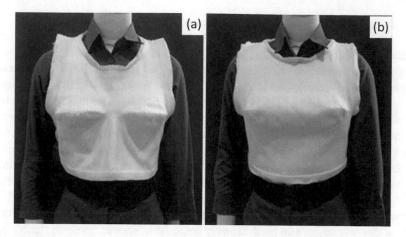

Fig. 9 The developed 3D seamless women body armor vests using knitted fabrics on officer uniform dressed manikin and the different fabrics used **a** Bra-vest design, **b** loose-vest design using combinations of Knitted Kevlar–wool and 100% Kevlar Knitted fabric [67]

protective and comfort performance while accommodating the bust contour. Two different styles of 3D seamless women body armor vests, namely the loose-vest and bra-vest have been developed.

Besides using appropriate clothing design system, developing and using an appropriate material to accommodate the required women body contour without affecting the overall performance is inevitable and make the system sustainable. The traditional two-dimensional flat fabric is not suitable to form the three-dimensional body contours without the involvement of darts. However, 3D woven fabrics, such as angle-interlock woven fabric, has been proved its ballistic performance and capability of shaping deep dome designs without cutting due to its extraordinary mouldability properties [68, 69]. The next section will deal with the 3D warp interlock fabrics and its performance to develop the women's soft body armor.

4   3D Woven Fabrics for Women Soft Body Armor Development

Due to the dynamic ballistic impact process which is usually completed within 50–200 μs, the body armor should be developed with effective and resistive response material structures. As discussed earlier, 2D woven and UD textile fabrics made from high-strength fibers are the most commonly used textile material for body armor development due to their excellent mechanical properties. Their ballistic performances and energy absorptions capabilities based on various influential parameters were also well investigated [70–75]. Besides, apart from the ballistic performance, the soft body armor material should possess sufficiently light in weight and flexible without affecting the final performances of the product in order to afford a relatively normal movement. As discussed earlier in our introduction part, 3D warp interlock fabric structures have been found as a good material for the development of women's body armor due to its specific characteristics. The fabrics have good elastic behavior properties [76], good formability and formability [33] due to their low shear rigidity as compared to other woven fabric structures. So, the fabric could be used in forming techniques to develop the frontal panels of women's soft body armor to resemble the female torso shape by properly accommodating the bust [64, 77]. In response, the designed product could then give both the comfort and enhanced ballistic performance by removing the stitching and other material weakening surface during traditional manufacturing. However, understanding the different formability characteristics of the fabric prior to applying to any kind of technical application i.e. body armor and composite parts are very important. Various research studies have investigated and compared the forming characteristics of 3D warp interlock fabrics. Even though different researchers have defined and classify the 3D warp interlock fabric [78–80], but it was more clearly proposed for the description of 3D warp interlock based on orientation and interlacing angles of the different weft layers by binding yarns [34]. Figure 10 shows the proposed classifications of the 3D warp interlock fabrics.

Based on the study, it is classified into four categories; orthogonal layer to layer (O-L), orthogonal throughout the thickness (O-T), angle interlock layer by layer (A-L) and angle interlock throughout the thickness (A-T). The binding warp yarn was used to interlace either for complete or different weft layers of the structure following a diagonal direction in the A/T and A/L configuration respectively. Whereas, the different weavers warp yarn tried to link the entire thickness or some of the layers

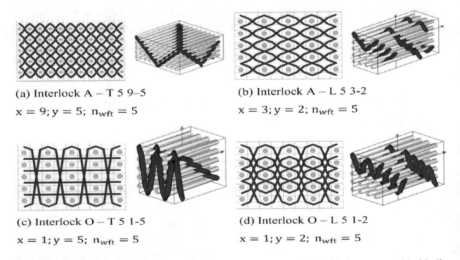

(a) Interlock A − T 5 9–5
$x = 9; y = 5; n_{wft} = 5$

(b) Interlock A − L 5 3-2
$x = 3; y = 2; n_{wft} = 5$

(c) Interlock O − T 5 1-5
$x = 1; y = 5; n_{wft} = 5$

(d) Interlock O − L 5 1-2
$x = 1; y = 2; n_{wft} = 5$

Fig. 10 Representation of the four classes of 3D warp interlock fabrics types with binding parameters and number of layers [34]

following an orthogonal direction for O/T and O/T configuration respectively. Due to this different structure inside the fabrics, the 3D warp interlock fabric provides good mechanical properties and deliver good ballistic performance along with good delamination properties [34, 35, 63] as compared to the traditional (2D and UD) fabrics. This is mainly due to its different structure inside the fabrics that could give various advantages including low shear rigidity and excellent formability behaviors. For example, studies have successfully worked on developing a single piece of the helmet using the 3D warp interlock reinforcement composite by avoiding cutting the fabric [14, 81–84]. The 3D warp interlock structures also exhibited less fabric damage mechanisms during forming processes [36].

4.1 Some Examples of 3D Woven Fabric Applications for Women Body Armor Design

As mentioned in the above section, soft armor products are mainly developed based on cut-and-stitch methods using 2D woven and UD fabrics. Even though various research study has proposed 3D warp interlock fabrics for women body armor due to its good mouldability, only a few were revealed a practical development process using this particular fabric. This subsection will introduce and discuss some of the women's body armor development methods using the molding process on the 3D warp interlock fabrics. Among the different patented women armor design, a molding process on the base functional materials made of multi-layer laminated woven structure to substantially conform the intended women's frontal body morphology was invented [26] as shown Fig. 11. Moreover, the thermoplastic material was applied during the molding process to fuse the fibers within the 3D woven material for better lamination and improved ballistic impact performances. The material forming on the upper and lower panel shape was designed based on the women's frontal shape to provides not only good ballistic resistant but also comfort and ease of movement of

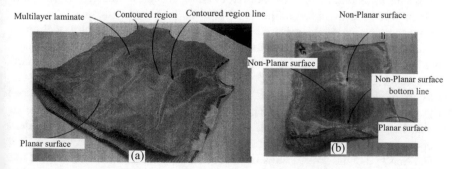

Fig. 11 **a** Multilayer laminate for ballistic protective wear in vest form constructed to conform to the natural curvature of a female torso, and **b** its sectional or component form

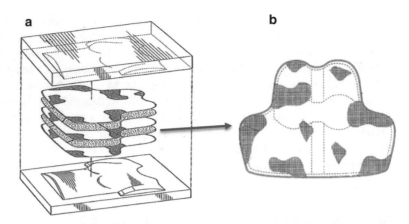

Fig. 12 Molding of 3D woven materials for developing soft women body armor with 3D woven material [85]

the wearer. Such design was claimed for its capability of retaining the molded shape for better comfort and ease of movement when the protective wear is in use.

A contoured, form-fitting and flexible panel through the molding process using three-dimensional woven material was also invented (Fig. 12). The required number of layer fabric is placed jointly and substantially parallel with each other and molded to adapt to the women's torso. To facilitate the torso conforming feature of the product, each layer of the fabric was made of weft and warp yarns with long float weave in one or more directions. The Adhesive resin can be used for the impregnations of yarn in each sheet. As shown in Fig. 12, the multiple woven ballistic fabric layers are placed on the preformed mold between the upper and lower portion to conform to the desired body contour. Then, the fabric sheets with applied adhesive will be compressed between the two portions and subsequently the long floats of the fabric weave help to conform to the required shape of the mold. Even though such methods provide better ease of movement and comfort, however, it also creates different creases and folds due to the inflexibility of fabric which ultimately might cause problems with fit and comfort [85].

For example, one of the research studies presented mathematical modeling to determine the pattern to develop seamless women body armor frontal panel considering both body size and bra size for better ballistic performance in speedy pattern development [58]. The mathematical model was constructed by combining both the bust-cup (model) and no bust-cup areas (block projection) based on the classic coat block pattern by omitting the dart. Then, the bust area was geometrically modeled with the help of Rhinoceros software. According to the study, three different models to represent and accommodate the bust-cup areas of the women were tested.

The first two-dome-shaped model to accommodate the bust area was found not appropriate due to excessive fabric creation while flattening of the domed area during the pattern making the process. The second model was designed to illuminate the excessive fabrics exists in the first model by linking the two doomed shape entirely.

Even such models were impractical for real body armor developments. The last model was likely an appropriate model developed considering the shape of both the frontal mannequins and bust area and dividing it into seven different geometrical shapes. Finally, the mathematical model based on half of the female body armor front panel was computed based on a size 12 standard mannequin and then a simple pattern for the first frontal panel layer was generated and fitted with 3D warp angle-interlock fabric to validate the model through a molding process. Further, a separate pattern for various panel layers was developed considering the relationship between the thickness of the fabric and the pattern block projection in continuation of the previous work [63]. In this specific research work, a 3D woven fabric was molded according to the shape of the frontal women's body armor of the different layers. The deformational stress, while flattening; on the fabrics might not be uniform throughout the front panel and brought some problem with the final ballistic performances of the body armor [86]. Another study also shows the interrelationship of bras (bust sizes D-cup) and the soft body armour to develop a new bra that may contribute to better fit of soft body armour as well as improved comfort for the wearers [87]. Nowadays, in order to solve the problem facing by the traditional pattern making systems, computer-aided design (CAD) systems were involved in the design and developments of soft body armor. Primarly, such systems is used to precisely separate the bust area and non-bust area. For examples one of our study has introduced a 3D design process to generate an adaptive bust (lower (Fig. 13a) and higher (Fig. 13b) breast volume) on the virtual manikin based on its morphological evolution. The adaptive bust was introduced into a 3D design process directly on the bust (Fig. 13c) for the creation of a well fitted 3D bra pattern (Fig. 13d).

The proposed method was found easier to implement and, could generate the 2D bra patterns with satisfactory fit and comfort [88]. This is due to the fact that the developed 3D pattern could perfectly follow the exact body contours than assuming

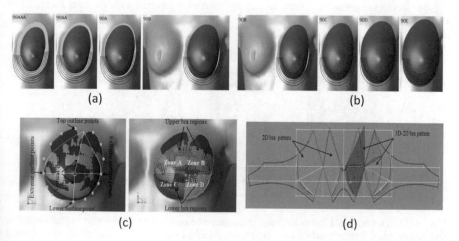

Fig. 13 Application of CAD for developing an adaptive bust on a virtual mannequin and generating well-fitted bra patterns [88]

the measurement as a 2D graphical method did. Based on the developed 3D female adaptive virtual mannequin, an efficient reverse engineering approach (2D-3D-2D) was carried out to develop 2D patterns for women seamless soft body armor front panel. The methods mainly help to attain the required bust volume in the pattern generation process for eliminating the darts design element [64].

The methods were basically involved horizontal and vertical projection lines in front of the already developed 3D virtual female mannequin as shown in Fig. 14 to generate the block pattern. Such a projection system gave an accurate measurement for curvy body couture, and the developed flattened pattern provides a better shape to accommodate the deformation with low distortion throughout the fabric surface while flattening. Considering the generated first layer, a novel method of systematic 3D design approach through the parametrization process was applied to generate block patterns for the successive layer of seamless frontal body armor panels as shown in Fig. 15 [89]. Such methods mainly consider the already developed first-panel layer on the virtual mannequin with grading values of zero.

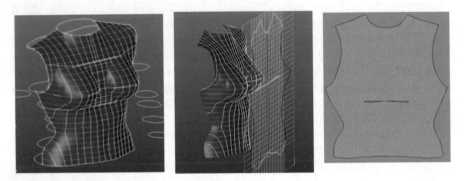

Fig. 14 Reveres engineering pattern developing system for generating the 1st layer for seamless female frontal soft body armor panel [64]

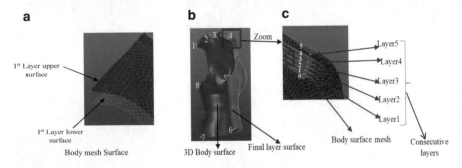

Fig. 15 Developing multi-layers for frontal seamless women soft body armor panel **a** Mesh creation and generation of 1st layer surface mesh **b** developed multi-layer panels with its thickness **c** Zoomed views of corresponding mesh (5 layers) with the thickness of each successive layers (t_1, t_2, t_3 … t_n) [89]

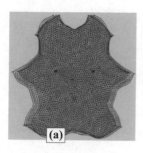
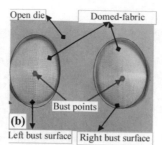
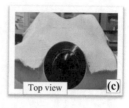

Fig. 16 **a** Pattern block projection **b** Flattened multi-layer pattern and **c** Draping of the 3D warp interlock fabrics to develop seamless women soft body armour panels on the physical mannequin with the right bust volume for fitness evaluation [89]

Later, the pattern for each successive layer of the complete body armor was generated based on the thickness of each layer in the 3D design database with appropriate coordinates as shown in (Fig. 16a). The flattened multi-layer panel's pattern were then applied on the 3D warp interlock fabrics using moulde by customized reversed bust-shaped punching bench and draped on the standard female physical mannequin as shown in Fig. 16c and d. According to the result, the new multi-layer pattern generation approach and its manufacturing system were found easy and for seamless female soft body armor development. Molding property of the textile structure is the capability of a flat textile material to be directly deformed to fit a three-dimensional surface without the formation of different defects including wrinkles, kinks or tears [90]. Besides, such a process could increase both the productivity as well as the mechanical performances of the final product by maintaining the integrity within the material. Various research studies have studied the formability properties of 3D angle interlock fabric structures. One of the research works has also experimentally studied the different molding behaviors of 3D warp interlocks and its counterpart 2D plain p-aramid fabrics made of similar p-aramid yarn using hemispherical punch [91]. The result shows that 2D fabric sample required high punching loads to deform but lower drawing-in values with high recovery as compared to 3D warp interlock fabrics. Unlike the hemispherical punch; the bust-shaped punch molding was also employed to compare the different forming characteristics of 3D warp interlocks and 2D plain p-aramid fabrics made of similar p-aramid yarn [92] as shown in Fig. 17.

The results show that Unlike 3D warp interlock fabric panel, 2D plain weave fabric preform exhibited more wrinkles in its majority surfaces during the molding process as shown in Fig. 17b and c. The domed shapes of panels made of 2D fabrics have also shown a very fast deformational recovery in both dimensions and become flat very quickly. Whereas, the molding recovery of the 3D warp interlock panels in all directions were insignificant for an extended period of time. Such stability behavior after deformation with fewer surface damages makes the 3D warp interlock fabrics a good moldable structure not only in the hemispherical punch but also in non-uniform punches including bust-shapes.

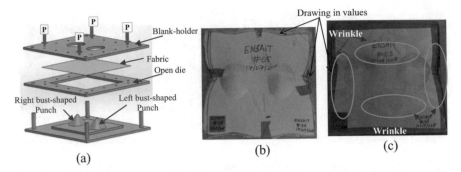

Fig. 17 Deformations of target panel to resemble the women frontal body shape **a** schematic view of adapted manual-based forming bench set up **b** deformed 3D warp interlock target panel and **c** deformed 3D warp interlock target panel

5 Conclusion

The ballistic material in the applications of soft body armor should provide a balance between ballistic impact protection, fitness, flexibility, and comfort. This would be even more complicated in women's body armor systems due to their unique and curvaceous body shapes. Unless the body armor fits the women's body shape properly, it will face not only inadequate coverage for ballistic impact protection but also lacks comfort and less likely to be worn. Fitting of women with small-sized male-based body armor systems could not be acceptable not only disproportionate on fitness, comfort, and worse ballistic protection but also it could impose the physiological difference. So, developing women's soft body armor with proper designing techniques and appropriate material is very important. This requirement has been aggravated even due to an increment of women's participation in the law enforcement and police office department. The traditional design techniques, such as cut-and-sew, fabric folding, and overlapping were tried to solve problems of comfort and fit by accommodating the curvilinear shape. However, the mentioned design techniques were also faced with their own drawbacks. For example, the cut-and-sew method by involving different dart designs to accommodate the bust area brought the weakest point against projectile impact at a dart seam line. Fabric folding methods also had a limitation of ballistic performance due to material discontinuities and weak stitching area around folded material, and many folded fabrics at the side with sharp edges could cause itching, less personal mobility, and discomfort. The overlapping method, even though it uses strong overlapping seams while fabric stitching; however, also faced problems of projectile penetration at the seams.

In order to avoid such problems developing women's body armor frontal panel with better impact performance and properly accommodating the bust area without the need for cutting, stitching, stretch folding or folding is very imperative. Molding is the most appropriate designing technique to avoid the above problem by developing the required 3D shape through dome-formation. While designing women's body armor panel using the molding method, it is possible to develop a seamless frontal front by mimicking the bust area. Due to this, it ultimately gives better comfort, fitness and better ballistic protection than any other design methods. However, unless a proper material used, the methods might also create shear deformation, panel thickness variations, and wrinkles on the material and body size limitations. Nowadays 3D warp interlock fabrics became a promising fabric structure for developing women's body armor due to its good mechanical and ballistic performance along with good delamination, low shear rigidity, and excellent formability behaviors.

Acknowledgements This work has been done as part of the Erasmus Mundus Joint Doctorate Programme SMDTex-sustainable Management and Design for Textile project, which is financially supported by the European Erasmus Mundus Program.

References

1. Carey, M. E., Herz, M., Corner, B., et al. (2000). Ballistic helmets and aspects of their design. *Neurosurgery, 47*, 678–689.
2. Bajaj, P. (1997). Ballistic protective clothing: An overview. *Indian Journal of Fibre & Textile Research, 22*, 274–291.
3. UK Ministry of Defence. (2005). *Proof of ordnance, munitions, armour and explosives: Part 2 guidance, defence standard 05-101 part 2 issue 1*. Defense Procurement Agency. Glasgow.
4. Lewis, E., & Carr, D. J. Personal armor. In: *Lightweight ballistic composites* (pp. 217–229). Elsevier Ltd.
5. Chen, X., & Chaudhry, I. (2005). Ballistic protection. In R. Scott (Ed.), *Textiles for protection* (p. 229). Woodhead Publishing Limited.
6. Saxtorph, M. (1972). *Warriors and weapons of early times*. Macmillan Co.
7. Sun, D. *Ballistic performance evaluation of woven fabrics based on experimental and numerical approaches*. Elsevier Ltd. Epub ahead of print 2016. https://doi.org/10.1016/B978-1-78242-461-1.00014-5.
8. Carlson, R. A. (2010). *Pleated ballistic package for soft body armor*. US20100313321 A1.
9. Messiry, M. El, El-tarfawy, S. (2015). Performance of weave structure multi-layer bulletproof flexible armor. In *The 3rd conference of the National Campaign for Textile Industries, NRC Cairo, '"Recent Manufacturing Technologies and Human and Administrative Development"'* (pp. 218–225).
10. Wanger, L. Military and law enforcement applications. In A. Bhatnagar (Ed.), *Lightweight ballistic composites* (pp. 1–25). Woodhead Publishing.
11. Smith, W. C. (1999). *An overview of protective clothing-markets,materials, needs*.
12. Bhatnagar, A., Arvidson, B., & Pataki, W. (2006). *Lightweight ballistic composites, military and law-enforcement applications* (pp. 213–214). Woodhead Publishing.
13. Erlich, D. C., Shockey, D. A., & Simons, J. W. (2003). Slow penetration of ballistic fabrics. *Textile Research Journal, 73*, 179–184.
14. Zahid, B., & Chen, X. (2014). Impact performance of single-piece continuously textile reinforced riot helmet shells. *Journal of Composite Materials, 48*, 761–766.
15. Jang, B. Z., Chen, L. C., Hwang, L. R., et al. (1990). The response of fibrous composites to impact loading. *Polymer Composites, 11*, 144–157.
16. Yang, C., Tran, P., Ngo, T., et al. (2014). Effect of textile architecture on energy absorption of woven fabrics subjected to ballistic impact. *Applied Mechanics and Materials, 553*, 757–762.
17. Othman, A. R., & Hassan, M. H. (2013). Effect of different construction designs of aramid fabric on the ballistic performances. *Materials and Design, 44*, 407–417.

18. Tran, P., Ngo, T., Yang, E. C., et al. (2014). Effects of architecture on ballistic resistance of textile fabrics: Numerical study. *International Journal of Damage Mechanics, 23*, 359–376.
19. Park, J. L., Yoon, B. il., Paik, J. G., Kang, T. J. (2012). Ballistic performance of p-aramid fabrics impregnated with shear thickening fluid; Part I – Effect of laminating sequence. *Textile Research Journal, 82*, 527–541.
20. Ha-Minh, C., Boussu, F., Kanit, T., et al. (2012). Effect of frictions on the ballistic performance of a 3D warp interlock fabric: Numerical analysis. *Applied Composite Materials, 19*, 333–347.
21. Brantner, S. B. policeone.com. *Police History: The evolution of women in American law enforcement*, https://www.policeone.com/police-history/articles/8634189-Police-History-The-evolution-of-women-in-American-law-enforcement/ (2015, accessed 23 December 2016).
22. Monrique, M. (2004). *Place des femmes dans la professionnalisation des armées*, http://www.c2sd.sga.defense.gouv.fr/IMG/pdf/c2sd_synth_loriot_choix_femmes.pdf.
23. Zufle, T. T. (1986). *Body armor for women*. US 4578821 A, US Patent: US 4578821 A.
24. Mellian S. A. (1980). *Body armor for women*. US4183097 A, US.
25. Nelson, J. S, Price, A. L. (1997). *Ballistic vest*. US 5619748 A, US.
26. Hussein, M., & Parker, G. (2001). *Ballistic protective wear for female torso*. US6281149 B1, US.
27. McQueer, P. S. (2011). *Woman's bullet resistant undergarment*. US20110277202 A1, USA.
28. Vives, M. (1993). *Ballistic protection armor*. US5221807 A, USA.
29. Carlson, R. A., & Hills, C. (2012). *Female armor system*. US20120174275 A1, US Patent: US20120174275 A1.
30. Dyer, P. A. (1991). *Ballistic protective insert for use with soft body armor by female personnel*. US5020157 A.
31. Chen, X., Lo, W.-Y., & Tayyar, A. E. (2002). Mouldability of angle-interlock woven fabrics for technical applications. *Textile Research Journal, 72*, 195–200.
32. Boussu, F., Legrand, X., Nauman, S., et al. (2008). Mouldability of angle interlock fabrics. In: *FPCM -9, the 9th international conference on flow processes in composite materials*.
33. Chen, X., Yang, D. (2010). Use of 3D angle-interlock woven fabric for seamless female body armour : Part I : Ballistic evaluation. *Textile Research Journal*, 1–8.
34. Boussu, F., Cristian, I., & Nauman, S. (2015). General definition of 3D warp interlock fabric architecture. *Composites. Part B, Engineering, 81*, 171–188.
35. Bilisik, K. (2016). Two-dimensional (2D) fabrics and three-dimensional (3D) preforms for ballistic and stabbing protection: A review. *Textile Research Journal, 87*, 2275–2304.
36. Dufour, C., Wang, P., Boussu, F., et al. (2014). Experimental investigation about stamping behaviour of 3D warp interlock composite preforms. *Applied Composite Materials, 21*, 725–738.
37. Crouch, I. G. (2019). Body armour – New materials, new systems. *Defence Technology, 15*, 241–253.
38. Nilakantan, G., & Nutt, S. (2014). State of the art in the deterministic and probabilistic ballistic impact modeling of soft body armor: Filaments to fabrics. In: *Proceedings of the American Society for Composites - 29th technical conference, ASC 2014; 16th US-Japan Conference on Composite Materials; ASTM-D30 Meeting* (pp. 1–21).
39. Yavaşa, M. O., Avcıa, A., Şimşirb, M., et al. (2015). Ballistic performance of Kevlar49/UHMW-PEHB26 hybrid layered- composite. *International Journal of Engineering Research and Development, 7*, 1–20.
40. Zee, R. H., & Hsieh, C. Y. (1998). Energy absorption processes in fibrous composites. *Materials Science and Engineering A, 246*, 161–168.
41. Zhang, Y. D., Wang, Y. L., Huang, Y., et al. (2006). Preparation and properties of three-dimensional braided UHMWPE fiber reinforced PMMA composites. *Journal of Reinforced Plastics and Composites, 25*, 1601–1609.
42. Tan, V. B. C., Tay, T. E., & Teo, W. K. (2005). Strengthening fabric armour with silica colloidal suspensions. *International Journal of Solids and Structures, 42*, 1561–1576.
43. Dong, Z., & Sun, C. T. (2009). Testing and modeling of yarn pull-out in plain woven Kevlar fabrics. *Composites. Part A, Applied Science and Manufacturing, 40*, 1863–1869.

44. Karahan, M., Ulcay, Y., Eren, R., et al. (2010). Investigation into the tensile properties of stitched and unstitched woven aramid/vinyl ester composites. *Textile Research Journal, 80,* 880–891.

45. Chen, X., & Zhou, Y. Technical textiles for ballistic protection. In: *Handbook of technical textiles* (pp. 169–192). Elsevier Ltd.

46. Bilisik, K., & Korkmaz, M. (2010). Multilayered and multidirectionally-stitched aramid woven fabric structures: Experimental characterization of ballistic performance by considering the yarn pull-out test. *Textile Research Journal, 80,* 1697–1720.

47. Nayak, R., Crouch, I., Kanesalingam, S., et al. (2017). Body armor for stab and spike protection, Part 1 : Scientific literature review. *Texttile Research Journal,* 1–21.

48. David, N. V., Gao, X.-L, Zheng, J. Q. (2009). Ballistic resistant body armor: Contemporary and prospective materials and related protection mechanisms. *Applied Mechanics Reviews, 62,* 050802.

49. Azrin Hani, A. R., Roslan, A., Mariatti, J., et al. (2012). Body armor technology: A review of materials, construction techniques and enhancement of ballistic energy absorption. *Advances in Materials Research, 488–489,* 806–812.

50. Eurostat. Police, court and prison personnel statistics. https://ec.europa.eu/eurostat/statisticsex plained/index.php?title=Police,_court_and_prison_personnel_statistics#One_in_five_police_ officers_is_a_woman.

51. Hargreaves, J., Cooper, J., Woods, E., et al. (2016). Police workforce, England and Wales. *Statistical Bulletin, 05*(16), 1–38.

52. International Women's Day, police scotland. *Police Scotland 8 march 2017,* Retrived 26 July 2016.

53. Police Service of Northern Ireland (PSNI). Workforce Composition Statistics, https://www. psni.police.uk/inside-psni/Statistics/workforce-composition-statistics/ (2017, accessed 26 July 2017).

54. Crime in the United States 1995. FBI, https://ucr.fbi.gov/crime-in-the-u.s/1995 (2016).

55. Crime in the United States 2005. *Gender,* www2.fbi.gov (2006).

56. Crime in the United States 2014. *Police Employee Data,* https://ucr.fbi.gov/crime-in-the-u.s/ 2014/crime-in-the-u.s.-2014/police-employee-data/main (2014).

57. Zufle, T. T. *Body Armpr For Women.* 4,578,821, USA patent, 1986.

58. Chen, X., & Yang, D. (2010). Use of Three-dimensional angle-interlock woven fabric for seamless female body armor: Part II: Mathematical modeling. *Textile Research Journal, 80,* 1589–1601.

59. Boussu, F., & Bruniaux, P. (2012). Customization of a lightweight ballistic vest for the female form. In *Advances in military textiles and personal equipment* (pp. 167–195). *Woodhead Publishing.*

60. Boussu, F., Ragot, A., Kulinska, M., et al. (2008). Customization of a lightweight ballistic vest. In *2nd International Scientific Conference Textiles of the Future.*

61. Moureaux, B., Pfister, F. V., & Van Zijl, N. A. *Specially shaped multilayer armor.* US5943694 A, USA: Google Patents, https://www.google.sm/patents/US5943694 (1999).

62. Hostettler, I. F, Rhum, D., Forman, M. R, et al. (12) United States Patent subiectto any disclaimer : the term of this. 1.

63. Yang, D., & Chen, X. (2017). Multi-layer pattern creation for seamless front female body armor panel using angle-interlock woven fabrics. *Textile Research Journal, 87,* 1–6.

64. Mulat, A. A., Bruniaux, P., Boussu, F., et al. (2018). Female seamless soft body armor pattern design system with innovative reverse engineering approaches. *International Journal of Advanced Manufacturing Technology, 98,* 2271–2285.

65. Teijin Aramid. (2016). *Ballistics material handbook.*

66. Dong, Z., Jiang, G., Wu, Z., et al. (2015). 3D parametric human modeling for warp-knitted seamless garment. *International Journal of Clothing Science and Technology, 27,* 532–548.

67. Mahbub, R., Wang, L., & Arnold, L. (2014). Design of knitted three-dimensional seamless female body armour vests. *International Journal of Fashion Design, Technology and Education, 7,* 198–207.

68. Cuong Ha-Minh, Boussu, F., Kanit, T., et al. (2012). Effect of frictions on the ballistic performance of a 3D warp interlock fabric : Numerical analysis. *Applied Composite Materials, 19,* 333–347.
69. Yang, D., Chen, X., Sun, D., et al. (2017). Ballistic performance of angle-interlock woven fabrics. *Journal of the Textile Institute, 108,* 586–596.
70. Das, S., Jagan, S., Shaw, A., et al. (2015). Determination of inter-yarn friction and its effect on ballistic response of para-aramid woven fabric under low velocity impact. *Composite Structures, 120,* 129–140.
71. Yang, C. C., Ngo, T., & Tran, P. (2015). Influences of weaving architectures on the impact resistance of multi-layer fabrics. *Materials and Design, 85,* 282–295.
72. Karahan, M. (2008). Comparison of ballistic performance and energy absorption capabilities of woven and unidirectional aramid fabrics. *Textile Research Journal, 78,* 718–730.
73. Zhang, D., Sun, Y., Chen, L., et al. (2014). Influence of fabric structure and thickness on the ballistic impact behavior of ultrahigh molecular weight polyethylene composite laminate. *Materials and Design, 54,* 315–322.
74. Bandaru, A. K., Vetiyatil, L., & Ahmad, S. (2015). The effect of hybridization on the ballistic impact behavior of hybrid composite armors. *Composites. Part B, Engineering, 76,* 300–319.
75. Nguyen, L. H., Ryan, S., Cimpoeru, S. J., et al. (2015). The effect of target thickness on the ballistic performance of ultra high molecular weight polyethylene composite. *International Journal of Impact Engineering, 75,* 174–183.
76. Byun, J. H., & Chou, T. W. (1990). Elastic properties of three- dimensional angle-interlock fabric preforms. *Journal of Textile, 81,* 538–548.
77. Abtew, M. A., Loghin, C., Cristian, I., et al. (2018). Two dimensional (2D) P-aramid dry multi-layered woven fabrics deformational behaviour for technical applications. *IOP Conference Series: Materials Science and Engineering, 374,* 1–12.
78. De Luycker, E., Morestin, F., Boisse, P., et al. (2009). Simulation of 3D interlock composite preforming. *Composite Structures, 88,* 615–623.
79. Khokar, N. (2001). 3D-Weaving: Theory and practice. *Journal of the Textile Institute, 92,* 193–207.
80. Gokarneshan, N., & Alagirusamy, R. (2009). Weaving of 3D fabrics: A critical appreciation of the developments. *Textile Progress, 41,* 1–58.
81. Zahid, B., & Chen, X. (2013). Development of a helmet test rig for continuously textile reinforced riot helmets. *International Journal of Textile and Science, 2,* 12–20.
82. Roedel, C., & Chen, X. (2007). Innovation and analysis of police riot helmets with continuous textile reinforcement for improved protection. *Journal of Information and Computing Science, 2,* 127–136.
83. Min, S. (2016). *Engineering design of composite military helmet shells reinforced by continuous 3D woven fabrics.*
84. Mih, S., Roedel, C., Zahid, B., et al. (2012). Moulding of single-piece woven fabrics for protective helmets – A review and future work. In: *Moulding of single-piece woven fabrics for protective helmets – A review and future work.* TexEng/RWTH Aachen.
85. Smith, B. L, Chung, L. Ting. *Molded torso-conforming body armor including method of producing same.* Patent No: US 2009/0255022 A1, US, 2009.
86. T.K. LPB and G. (2002). *Apparel sizing and fit.* Oxford.
87. Elizabeth Niemczyk, S., Arnold, L., & Wang, L. (2017). Qualitative observations on the design of sports bras for wear under body armour. *KnE Engineering, 2,* 342
88. Abtew, M. A., Bruniaux, P., Boussu, F., et al. (2018). Development of comfortable and well-fitted bra pattern for customized female soft body armor through 3D design process of adaptive bust on virtual mannequin. *Computers in Industry, 100,* 7–20.
89. Abtew, M. A., Bruniaux, P., Boussu, F., et al. (2018). A systematic pattern generation system for manufacturing customized seamless multi-layer female soft body armour through dome-formation (moulding) techniques using 3D warp interlock fabrics. *Journal of Manufacturing Systems, 49,* 61–74.

90. Zhong, T., & Hu, H. (2007). Formability of weft-knitted fabrics on a hemisphere. *Autex Research Journal, 7*, 245–251.
91. Abtew, M. A., Boussu, F., Bruniaux, P., et al. (2018). Influences of fabric density on mechanical and moulding behaviours of 3D warp interlock para-aramid fabrics for soft body armour application. *Composite Structures, 204*, 402–418.
92. Abtew, M. A., Boussu, F., Bruniaux, P., et al. (2019). Ballistic impact performance and surface failure mechanisms of two-dimensional and three-dimensional woven p-aramid multi-layer fabrics for lightweight women ballistic vest applications. *Journal of Industrial Textiles, 50*, 1351–1383.

Printed in the United States
by Baker & Taylor Publisher Services